Proceedings in Life Sciences

Primary and Secondary Metabolism of Plant Cell Cultures

Edited by
K.-H. Neumann, W. Barz, and E. Reinhard

With 140 Figures

Springer-Verlag
Berlin Heidelberg New York Tokyo

Editors:
Professor Dr. KARL-HERMANN NEUMANN, Justus-Liebig-Universität, Institut für Pflanzenernährung, Abteilung Gewebekultur, Südanlage 6, 6300 Gießen, FRG

Professor Dr. WOLFGANG BARZ, Westfälische Wilhelms-Universität, Lehrstuhl für Biochemie der Pflanzen, 4400 Münster, FRG

Professor Dr. ERNST REINHARD, Universität Tübingen, Institut für Pharmazie, Auf der Morgenstelle, 7400 Tübingen, FRG

Editorial Assistants:
Dr. L. BENDER (Gießen), Dr. W. HÜSEMANN (Münster), and Dr. A. W. ALFERMANN (Tübingen)

ISBN-13: 978-3-642-70719-3 e-ISBN-13: 978-3-642-70717-9
DOI: 10.1007/978-3-642-70717-9

Library of Congress Cataloging in Publication Data. Main entry under title: Primary and secondary metabolism of plant cell cultures. (Proceedings in life sciences). Papers from a symposium held in Rauischholzhausen, Germany in 1981. Bibliography: p. . Includes index. 1. Plant cell culture – Congresses. 2. Plants – Metabolism – Congresses. 3. Cell metabolism – Congresses. I. Neumann, K.-H. (Karl-Hermann), 1936– . II. Barz, W. (Wolfgang), 1935– . III. Reinhard, E. (Ernst), 1926– . IV. Series. QK725.P77 1985 581'.0724 85-17257

Softcover reprint of the hardcover 1st edition 1985

Preface

It must have been some feeling of frustration invoked by the genus loci of big conferences which brought Don Dougall, Mike Fowler, and me over a cup of coffee to think of organizing a small meeting on cell culture metabolism. The basic aim was to bring people working on "primary metabolism" and those who work on "secondary metabolism" to interact with each other, hopefully to induce some new approaches to utilize the cell culture technique more efficiently in basic research and in its practical application. The outcome of this was the small symposium at Schloß Rauischholzhausen near Gießen (Germany), and the reader of this small volume of articles on topics discussed will judge to what extent this aim was realized.

Although F.C. Steward was unable to attend the meeting, we were happy that he contributed a review on some important turning points of the cell culture method in the past, and at the end of the book some participants of the meeting attempted some prognosis on possible future developments, summarized by Mike Fowler.

It was the wish of our publisher to organize the contributions as a summary of the major topics of research of the individual groups represented at the meeting. Due to this, some readers will certainly feel that important areas of research in the cell culture field are missing. This is regrettable, but space had to be limited to keep this volume at a reasonable price.

Finally, I should like to thank my coorganizers of the meeting and my "crew" in Gießen as well as all the participants for their contribution to the organization of those interesting 3 days in Rauischholzhausen in fall 1984 as well as the sponsors of financial assistance, so successfully organized by E. Reinhard; and last but not least, my thanks also go to Dr. B. Pauler and my secretary Frau H. Müller for arranging the subject index by means of a computer program and to our publisher, Springer-Verlag, Heidelberg, for producing this small book so speedily.

Gießen, Summer 1985 *K.-H. Neumann*

Financial assistance from the following sponsors was greatly appreciated:

Deutsche Forschungsgemeinschaft, Bonn-Bad Godesberg; Land Hessen, Wiesbaden; Gießener Hochschulgesellschaft, Gießen; BASF Aktiengesellschaft, Limburgerhof; BAT Cigarettenfabriken GmbH, Hamburg; Boehringer Mannheim GmbH, Mannheim; Boehringer Ingelheim GmbH, Ingelheim; Ciba Geigy AG, Basel, Schweiz; Consortium für Elektrochemische Industrie GmbH, München; "elf" Bio Recherches, Castanet Tolosan, Frankreich; Ergo Forschungsgesellschaft mbH, Hamburg; Flow Laboratories GmbH, Meckenheim bei Bonn; Gist Brocades NM, Delft, Niederlande; Höchst AG, Frankfurt; Homburg Degussa Pharma Gruppe, Frankfurt; E. Merck, Darmstadt; A. Nattermann and cie. GmbH, Köln, Schering AG, Berlin

Contents

V Herbicides

List of Contributors

You will find the addresses at the beginning of the respective contribution

I From Metabolism and Osmotic Work to Totipotency and Morphogenesis: A Study of Limitations Versus Multiple Interactions

F.C. STEWARD[1]

Preamble

This brief paper is the outcome of an invitation to speak at the end of your symposium. My inability to accept was dictated by events over which nature, not I, had control. Had I been able to speak I would have traced the principle events of my long research career (since 1924). Inevitably one moved from the simplistic single variable experiments inherited from chemistry in which answers could be unequivocal and precise, albeit usually over a very restricted range of problems. In the first studies into the mechanism by which plant cells acquire and retain their solutes and their water one soon learned to respect living cells as integrated working machines. The events of nutrition and metabolism, synthesis and breakdown, active growth and quiescence are so interlocked that the single variable experiment or approach can rarely tell the whole truth. Nevertheless, the organismal, holistic, approach places greater demands on experimental planning and interpretation. In these circumstances it is easy to be more obsessed with the question "how" events occur than with the more deep-seated question "why" they occur. Preoccupation with the problems of growth and development in general and with the behavior of cells and tissues in culture fostered an enlightened "holism" to supplement the prevalent reductionist trend by which we tend to know "more and more about less and less".

1 Leaching, Diffusion, Salt Accumulation in Cells

This account, in a narrative style, begins essentially in 1946 when research work was resumed in the University of Rochester N.Y., after a long gap during the war years. Earlier research in Leeds in 1924 and in the U.K. had aroused an interest in the ability of cells in life to retain their solutes against the outward diffusion gradients that were all too evident when viability was impaired. Prolonged leaching of slices of storage organs caused virtually no removal of solutes, especially if aseptic conditions with a minimal access to oxygen were ensured. The converse problem of how the solutes

1 1612 Inglewood Drive, Charlottesville, VA 22901, USA
Professor em., Cornell University, Ithaca, NY, USA

Primary and Secondary Metabolism of Plant Cell Cultures, ed. by Neumann et al.

entered the cells in the first place and how they were mobilized later to satisfy needs elsewhere, prompted the investigation at Cornell University of diffusion of solutes (organic and inorganic) through membranes composed of parenchyma cells. This resulted in the realization that the apparent diffusion coefficients through such tissue membranes were orders of magnitude lower than for free diffusion in water – a result attributed to the resistance of the limiting surface membranes of the cells. But the transition from chemistry to botany established a persistent trend to devise and construct novel apparatus adapted to the task in hand, as in the diffusion studies. This characteristic was maintained throughout the later work on aseptic cultures of cells.

The diffusion experiments showed that solutes, whether nutrient ions from the ambient media, or non-electrolytes needed in metabolism, need to traverse regions of unspecialized parenchyma very much more rapidly than diffusion alone could maintain. For example, they must traverse the regions of unspecialized cells between the tips of organized vascular tissue and the apical growing regions proper and also cross the cortical parenchyma of roots to the stele, or in leaves, from mesophyl cells to the tips of vascular bundles. This implied that in situ such unspecialized cells greatly expedite movements of solutes over substantial distances.

In Berkeley, Cal., therefore one set out to ascertain whether bromide ions, which freely accumulated in *Nitella* cells in the light, could traverse membranes of potato tuber cells and also accumulate therein. In the outcome, under the conditions of the passive diffusion experiments the bromide ions passed through the membranes at the expected very slow rate without entering the cells, even to equality of concentration with the ambient medium. This posed the question under what conditions would bromide, as in *Nitella* in the light and like chloride already in the tuber cells, accumulate in the cells of potato tuber? This dilemma prompted the use of very thin (< 1.0 mm) discs of potato tuber, cut as for the diffusion experiments, in very dilute solutions of KBr (order of 1.0 mgEq. per litre). It was found that, whereas the *Nitella* cells accumulated the bromide in the light, the potato cells accumulated the bromide rapidly, even in the dark, when the solutions were aerated and maintained at least in equilibrium with air. Moreover this process could continue for long periods of time (over 100 h) at linear rates with time to reach high degrees of accumulation over the ambient medium. Clearly this was not a physicochemical equilibrium. Moreover in this uptake bromide was accompanied by potassium. Unlike the *Nitella* cells, however, the bromide accumulated in the discs, at rates that were a function of oxygen concentration in the medium and oxygen tension in the flowing gas stream and also of the temperature. Whereas *Nitella* cells obtained this energy and oxygen via photosynthesis in the light, the potato cells could do so via respiration and with the use of their stored carbohydrates.

Profiting from prior knowledge of the "healing" at a cut potato surface, it was immediately evident that the cells near the surface, having access to dissolved oxygen, were undergoing renewed metabolic activity that would normally result in cell division, and to do this they not only mobilized their respiratory metabolism but they also could use endogenous soluble nitrogen compounds in the renewed protein synthesis and nitrogen metabolism that the conditions for bromide accumulation required. So several pre-war years of work ensued in the U.K. (at Leeds University and later at Birbeck College in London) to investigate the biochemical background of salt accumulation in the cells of storage organs.

Also, in Berkeley, Cal., the bromide accumulation in a system comprised of excised roots of barley was investigated. The barley roots were so grown that they were long and unbranched over a sufficient length from the apex so that the bromide accumulation could be examined at different distances from the tip. These diverse systems yielded results that were compatible in that the act of accumulating salts was not a simple physical phenomenon, for they each invoked the energy of their metabolism to drive a metabolic machine in which the cells, capable of active growth and metabolism, did osmotic work. This was very far from simplistic ideas based upon passive diffusion and physicochemical equilibria.

But during this period work on the large algal cells (strictly multi-nucleate coenocytes) was not neglected and especially because W.J.V. Osterhout at the Rockefeller Institute and at Bermuda had erected a hypothesis which left no room with respect to *Valonia* for the general ideas that emerged from the work on root systems or discs of potato tuber etc. In this dilemma work was done at the Dry Tortugas Fla. to study Valonias as they grow in their habitats and, by contrast as they perform when isolated under aquarium conditions. Briefly this work emphasized that the large so-called-"cells", especially of *Valonia ventricosa*, are actually multi-nucleate coenocytes which, when isolated from the substratum on which they occur in nature, do not really grow as such, nor do they multiply directly. They expand only sluggishly and respond to superimposed changes in the osmotic value of the medium. In other words, the valonias were not in fact as simple as at first supposed, a result emphasized when it was found how to culture them in very large numbers by first causing their peripheral cytoplasms to form aplanospores, which can be re-attached to grow independently on small marble blocks in still sea water, in diffuse light, and in large glass vessels. Therefore, to make use of the *Valonia* systems effectively one had first to understand its morphology and how it grows in nature.

2 Growth Induction, Aseptic Cultures, Totipotency

Therefore, just prior to the outbreak of World War II, all was ready at Birkbeck College in London to continue the work on two fronts. First, the laboratory conditions were established for studying metabolism in the cells of discs of storage organs and relating their ability to accumulate solutes from their ambient media recognizing that this event, simple as it may seem, invokes interactions with the many features of the cells that are compatible with their renewed growth. Secondly, a "Valonia Farm" had been established in large glass tanks of seawater in the adjacent greenhouses of the Chelsea Physic Garden, to pursue the uptake of inorganic ions throughout the growth and development of this organism. But war and the air-raids on London intervened. Birkbeck College was "blitzed" and the Valonia Farm disappeared with damage to the greenhouses of the Chelsea Physic Garden. Damage to the laboratory established at Birkbeck College, effectively put a stop to the work there. Nevertheless, a turning point had been reached.

Results from all the systems studied thus far, cut discs of storage organs, excised roots of crop plants (e.g. barley), the large algal cells whether in laboratory conditions or in natural habitats, had emphasized the futility of interpreting nature in terms of preconceived physical ideas based on equilibria not on biological realities and which are not compatible with the living systems as they grow and develop. It was left to the postwar period to put these ideas into practice.

The immediate postwar aim was to devise a system well adapted to reveal the relations of soluble (non-protein-N) compounds to protein and the involvement of protein synthesis, and of turnover, with respiration and other aspects of cell physiology, including ion absorption. But a primary consideration was that the system should be capable of being used under strictly aseptic conditions so that it could operate with single non-nutrient salt solutions, or a range of organic solutes. Also it should be capable of comparing and contrasting relatively slow or non-growing cells with those growing as fast as they are able. These requirements indicated the use of a tissue culture system, although the ones then in use employed relatively large tissue explants with slow proliferating on semi-solid media. Also, at that time, there was still much pre-occupation with "potentially unlimited growth" as stimulated by the accepted natural auxin (IAA) or even naphtalene acetic acid. A search was therefore made by testing a range of potentially useful source materials and also a range of potentially effective growth-promoting agents. This survey revealed its most dramatic results when small explants of carrot root tissue were tested in a complete basal medium and also when stimulated by the fluid (liquid endosperm) from within well-developed coconuts (coconut milk or as it is termed in the tropics coconut water). The only other system that attracted attention in this survey was one in which potato tuber tissue in a basal nutrient medium responded to contact with 2.4 D. Some years later this system was adapted to study the synergism between phenoxy-acetic acids and the component parts of the coconut milk (CM) growth-promoting complex.

The carrot system selected was standardized for experiments in several ways. First, reproducible, small cambium-free explants from secondary phloem were cut at a standard distance from the cambium. This was done aseptically, yielding large numbers that constitute a clone within which many randomly sampled batches could receive specified treatments. Experiments were carried out in a liquid medium which comprised a typical range of inorganic nutrients to which were added vitamins and sugar or any other stimulant (e.g., coconut-milk) required to foster rapid growth. Profiting by earlier experience of aerobic and anaerobic leaching and of diffusion through tissue membranes and of thin discs of potato tissue as used in the salt accumulation work, special tubes (T-tubes) were devised that alternately submerged the explants under liquid and then, as the vessels slowly rotated, end over end, exposed them to air. Each tube was vented aseptically through a cotton plug via a short central side-neck. These slowly rotated T-tubes mounted radially on wheels (about 24 tubes to a wheel and many wheels on a single shaft and several shafts geared to one motor) has provided the basis over the years for the aseptic work on carrot root explants. Again, once the desired conditions for controlled growth experiments were established, a system for its exploitation on a large scale was devised. Many "wheels" on horizontal shafts, slowly rotated (order of 1 rpm), in a temperature-controlled room provided the mechanism for many simultaneous parallel cultures. Explants (2–3 mg) which, in the absence of the coconut milk stimulus showed only minimal "growth" by cell enlargement, in its presence increased in fresh weight (order of $\times$ 80 in 21 days with a selected strain of carrot roots) and also multiplied their cells. But also, by controlled maceration of explants the growth could be expressed in terms of cell number and, by calculation, in terms of their average size and the course of growth traced out in sigmoidal curves with time. A feature not to be underestimated was the need for well-trained technicians capable of maintaining the fastidious asepsis that work on this scale requires. Thus, the special techniques devised for the aseptic culture of carrot tissue departed drastically from the time-honoured methods of tissue culture which had in fact become a "tissue cult"!

3 Soluble Nitrogen Compounds of Growing and Quiescent Cells: Chromatography

By a fortunate circumstance during the wartime period, the technique and principals of chromatography on paper had been devised. Dr. C.E. Dent had pioneered its application to the analysis of animal urines. Again by a fortunate coincidence, we were able, in collaboration with Dr. Dent, to apply it first to the detection of the ninhydrin-reactive compounds in the non-protein alcohol-soluble extracts of potato tuber tissue. Later the technique was adapted to the routine and quantitative analysis of the carrot and other plants before and after they had been subjected to experimental treatments. Also this technique permitted carrot tissue explants to be analyzed in this way, when they became a favored subject for the assay of the growth-promoting fractions of coconut milk or for any other active, even inhibitory, compounds or sources.

Even in retrospect it is difficult to over-estimate the sense of excitement and anticipation aroused by the combined use of cell division active compounds and extracts, such as those in coconut milk, together with the first use of the paper chromatography procedures to trace their effect on the soluble nitrogen compounds (with and without radioactive labelling) upon the study of metabolism of growing and non-growing cells in relation to their salt and water uptake.

In parallel with this work, the whole knowledge of the non-protein-N compounds in plants was being greatly extended by the isolation and identification of many substances, some of which were new to science, others newly recognized as plant constituents. The details of all this can be found elsewhere, but the point to be stressed here is that these great gains were the results, not of adding more in detail to already well-trodden paths, but of radical departures in both technique and experimental design from what had become traditional. But there was more to come.

To make the biochemistry of the carrot assay system more feasible, while retaining the advantages of the rotated T-tubes, use was made of special "nipple" flasks. Made from 1-liter glass flasks with short necks, these flasks accommodated 250 ml or more of liquid medium and, instead of 2–3 small explants, in 10 ml of medium per tube, they could accommodate up to 100 or more per flask. To this end short "nipples" (about 3/4 inch in diameter) were blown radially around the flask and these represented the ends of the typical T-tubes. When the flasks were rotated (1 rpm), their liquid content flowed into, and out of the nipples, simulating its flow from end to end of the T-tubes. Thus the explants like those in T-tubes, spent part of the time submerged and part in air as they adhered to the walls of the flask. This device worked well and provided a much larger amount of uniform cultured explants (green tissue, in the light) under the influence of the complete coconut milk medium. However, an unexpected and dramatic dividend occurred.

Under the conditions described in the light the cultured green explants gently rolled against each other and, as if in a very gentle "ball-mill", some surface cells were "sloughed off" to remain freely suspended in the otherwise clear medium. When first observed it was feared that these freely suspended particles might be infections (since to cut and inoculate 100 explants per flask without infection demands a very fastidious aseptic technique). On examination, however, the suspended particles were the most beautiful free cells imaginable, with well-defined cytoplasmic strands along which vigorous streaming occurred. At first such cells were regarded as curios but, on inoculation into similar flasks of media, they produced growing subcultures of free cells and small aggregates. But then on an occasion when experiments were carried out, not in continuous light, but with a light-dark diurnal regime, the surprise was that, although literally hundreds of carrot explants had been cultured without forming a single root, roots from free cells appeared. From this point on the challenge was to develop shoots and so, without elaborating upon all the details, the technique of somatic free cell carrot embryogenesis emerged. Thus it became feasible to produce hundreds, even thousands of "embryoids" from cells in a single flask and then to develop these into plantlets.

It should be stressed, however, that this dramatic result could not have been anticipated. There was no prior thought of demonstrating the "totipotency" of carrot cells. It was a fortunate circumstance that the means to expose small explants of carrot to the conditions conducive to salt accumulation also led them to form free cells and were also conducive to their development into embryoids and plantlets. Had one been overawed by auxin as *the* plant growth hormone and followed contemporary practices of culturing relatively large excised pieces of carrot root on semi-solid media (despite their capacity for potentially unlimited, but slow, growth), none of these exciting developments would have occurred. Too often we try to force nature to follow our prescribed rules instead of "listening to" or observing the "messages" which, though they may conflict with popular theories, are nevertheless offered to us by nature.

4 Cells, Solutes, and Growth of Aseptic Cultures: Trace Elements, Growth Factors

But one should return to the salt accumulation problems for the investigation of which the carrot system was developed. Carrot explants in dilute salt solutions, free from growth-promoting substances can increase their cell size without division and in so doing may establish high internal concentrations, not only of salts, but first of *organic* solutes (sugars). Provided with appropriate exogenous media, such cells will reversibly store either non-electrolytes or salts. When provided with the stimuli that cause cells and explants to trace out a sigmoidal curve of growth there is a first phase in which cells divide more than they expand and average cell sizes remain small. The first solutes that are then absorbed are not inorganic (e.g., KCl or NaCl) but are preferentially organic. Potassium ion (K^+), however, is the most distinctively required, but it does not conspicuously accumulate in the small cells, nor is it accompanied by chloride. Later, however, when the divided cells also enlarge, inorganic solutes (with K^+ in excess of Na^+) may prevail over organic solutes. Meanwhile endogenous non-protein-N compounds may nourish the protein synthesis which also occurs.

The philosophy that emerges is as follows. This growing system is not one that develops its solutes by acting selectively upon one ion after another, but it exercises its control over *solutes*, organic and inorganic, and cells develop and maintain the internal activity of their water by the colligative properties of their solutes. Although one could arrange in the tissue slice system that specific ions (K^+, Br^-) were the only ones being accumulated, and that they even achieve for Br^- accumulation ratios approaching infinity, this should not encourage the view that cells utilize their energy selectively to accumulate *individual* ions. The real drive, however, is to produce *solute* concentrations and in so doing, maintain the reduced activity of water which turgor demands. In the time course of growth, organic solutes may have the first preference with K^+ unaccompanied by Cl^- playing a key and specific role. Later K^+ and Cl^- or Br^- may substitute for the organic solutes which are first accumulated.

The essential point is that work with the growing cultured cells impells one to take note of the entire cell system; one can hardly be satisfied by analysis of the presumptive electrogenic effects at a membrane, or with special ion pumps specifically operating upon a given ion. At best these mechanisms bear upon one aspect of the whole process, i.e., the actual passage across a limiting membrane. A further development of the carrot culture system is, however, noteworthy. In the nutrition of the growth-induction process by the collective factors in coconut milk, it was recognized quite early that without iron neither the whole growth-promoting system nor its now known component parts could function. After Fe, Mo and other trace elements (e.g., Mn) may exert their specific effects, observable in an otherwise trace-element-freed basal and coconut milk medium. But in the study of all such systems it is necessary to set up symmetrically balanced treatments which, on a per cell or per unit weight basis, may reveal the effects of different interacting variables which can be evaluated in terms of metabolic parameters that can be measured. When these considerations involve not only the trace elements singly and the growth factor systems collectively, the number of parallel treatments that are necessary is large. But if the analysis is extended to include all the interactions of trace elements (singly and in combination) with the component parts of the growth induction systems (singly and in combination) the number of treatments taxes the resources available. Nevertheless sufficient has been done to indicate how far-reaching upon the growth and metabolism are the consequences of trace element growth factor interactions.

In fact one may now say that mere demonstrations of the consequences of trace elements singly in limiting amounts evade most of the problem. It is the multiple interactions of nutrient elements with each other and their interactions with other variables (e.g., light, temperature etc.) that can be most illuminating.

It would take this discussion too far to deal fully with the question of the nature of the growth factors in the total complex which causes carrot explants to resume maximum activity and grow. Suffice it to say that it has component parts that have been designated I and II. Each part has active fractions (AF_1, AF_2) which work in conjunction with co-factors. A co-factor for AF_1 is myo-inositol, for system II it is the familiar auxin, IAA. The activity of system II can be simulated by the combination with zeatin, that is AF_2 + IAA as its co-factor. The active fraction of system I is found in an ethyl acetate soluble part of which the only identified portions are glucosides, and its co-factor is myo-inositol. In experiments to survey the effect of all these separate components (singly and in combination) with some attention to their interactions with trace elements, measuring not only the fresh weight of the explants but cell number and cell size, the soluble nitrogen and its main component parts and protein nitrogen, the numbers of replicated cultures becomes large. The conclusion that can be briefly stated is this. The range of behavior of the carrot cells that is encompassed by the control of the growth-induction medium is unexpectedly wide, and all this is within the range of genetically identical clones.

5 From Cultured Cells to Whole Plants: Somatic Embryogenesis

However, before closing the book on the factors that govern the behavior of carrot cells as they grow, reference should be made to the feature which is biologically most significant and most difficult to comprehend. It relates to the effects superimposed upon genetically uniform totipotent cells as they develop in the different organs of the plant body and, since the development of the organs is subject to responses to environmental factors during morphogenesis, these also enter prominently as determing factors, not only in the control of form but also of composition.

Therefore, reference should now be made to the soluble content of parenchyma cells as they develop in different organs of the plant body. This has been especially studied with respect to potato and carrot plants, carrying over from the cell culture studies the analytical methods on the one hand and the demonstrable totipotency of their living somatic cells on the other. In other words differences in composition observed between parenchyma in different parts of the plant body or induced by development under different environmental conditions are modified (a) during development and (b) by interaction with environmental factors during growth. Earlier experiments with mint plants had traced the effects of long and short days and day and night temperatures on the growth and branching habit of mint plants and had correlated these responses with drastic changes in the relative composition of the soluble nitrogen fractions and the relations of soluble non-protein nitrogen to protein. At that time (1946 to 1950), the alternating environments had to be maintained in different rooms by moving the plants in their nutrients on carts from one room to another. But later, with access to automatically controlled growth chambers, this has been done more effectively and similar operations have been carried out using selected clones of potato and carrot. The work on potato focussed attention upon the form and composition of leaves and tubers. As in the case of mint so also in potato, the composition in terms of the constituents of the soluble-N and its relations to total protein-N responded quite sharply to the photo period during day light and to the temperature during the hours of darkness. But the interesting feature is that the morphogenetic responses of potato could be discerned quite early in the study of their growing points.

6 Growth and Composition as Affected by Environments

These results pose a difficult question. Leaves and tuber are all part of the shoot and, presumptively, if their cells were separately cultured all could give rise again to plants with the same genetic constitution and form. What is it then that predetermines cells in the environments conducive to leaves or tubers, even on the same axis, to differ so greatly in their soluble contents? And why should the environmental factors that affect the form of their growing points also modify so decisively the content of their cells in leaves or tuber with respect to their soluble-N constituents? Similar considerations apply to carrot leaves and storage roots with special reference to their salt content. In the 1930's, it may be recalled, variations in K^+/Na^+ content of *Valonia* sap were attributable to subtle environmental factors affecting it at the Dry Tortugas, Fla.

What is then the final message? One should be aware that simplistic, single variable, unit responses, especially where the impact of growth and development is ignored or eliminated, may tell only part of the truth. To side-step the "complications of growth and development" may indeed simplify experiments but it may also evade the rich harvest of truth revealed by the *interactions* of all the factors and responses which can make their impact upon living plants and their cells as they grow. The totipotency of the living cells of angiosperms throughout the development of their organs presents problems not easily disposed of even by the knowledge that flows from modern genetics. Parenthetically it is far easier to make those genetic applications where they relate solely to microorganisms and especially if their consequences can be described only in chemical terms.

Viewed as metabolic machines, autotrophic green plants in their environments represent highly improbable and remarkable physical systems; they are truly feats of biological engineering. Their cells need usable energy in the form of the universal biological energy currency (which is phosphate bond energy) and they produce and maintain their highly complex and improbable physical and chemical structures without violence to the physical principles that apply throughout the inanimate world. To do this they surely need to be able to balance the increased free energy that they build into their content of water and solutes against the negative entropy of their structure and organization in ways which we do not yet fully comprehend. Is this not a challenge for the plant cell physiology of the future?

References to Sources by Sections 1 to 6 of the Text

1 Leaching, Diffusion, Salt Accumulation in Cells

Hoagland DR, Hibbard PL, Davis AR (1926) The influence of light, temperature and other conditions on the ability of Nitella cells to concentrate halogens in the cell sap. J Gen Physiol 10: 121–146

Steward FC (1928a) The maintenance of semi-premeability in the plant cell during leaching experiments. Proc Leeds Philos Soc 1:258–270

Steward FC (1928b) A preliminary study of the diffusion of sugars through membranes of living plant tissue. Am J Bot 15:629–630

Steward FC (1930) Diffusion of certain solutes through membranes of living plant cells and its bearing upon problems of solute movement in the plant. Protoplasma 11:521–557

Stewart FC (1931–1933) The absorption and accumulation of solutes by living plant cells I–V. Papers on potato discs. In: Protoplasma, vols 15, 16, 17, 18

Steward FC (1940–1941) 4 papers with C. Preston and PR Stout. In: Plant physiology, vols 15, 16. On metabolic processes of potato discs under conditions conducive to salt accumulation.

Steward FC, Martin JB (1937) Valonia at the Dry Tortugas, with special reference to the problem of salt accumulation in plants. Pap. Tortugas Lab 11:89–170

Steward FC, Sutcliffe JF (1959) Inter-relations of salt accumulation with growth and development of the plant body. In: Plant Physiology: A treatise, vol 11, pp 362–367, 450–464

2 Growth Induction, Aseptic Cultures, Totipotency

Caplin SM, Steward FC (1948) Effect of coconut milk on the growth of explants from carrot root. Science 108:655–657

Steward FC, Caplin SM, Millar FK (1952) Investigations on growth and metabolism of plant cells. I. New techniques for the investigation of metabolism, nutrition and growth in undifferentiated cells. Ann Bot (London) 16:57–77

Steward FC, Shantz EM (1955) The chemical induction of growth in plant tissue cultures. I. Methods of culture and analysis of growth; II. The chemical nature of the growth promoting substances in coconut milk and similar fluids. In: Wain RL, Wightman F (eds) The chemistry and mode of action of plant growth substances. Butterworth, London, and Academic Press, London New York, pp 165–186

Steward FC, Mohan Ram HY (1959) Determining factors in cell growth, some implications for morphogenesis in plants. In: Abercrombie, Bracket JD (eds) Advances in morphogenesis, vol I. Academic Press, London New York, pp 189–265

Steward FC (1968) Totipotency of angiosperm cells: Its significance for morophology and embryology. Phytomorphology 17:499–507

Steward FC, Krikorian AB (1971) Plants, chemicals and growth. Academic Press, London New York, p 232 (see chapters 4, 5, 6, pp 44–84)

3 Soluble Nitrogen Compounds of Growing and Quiescent Cells: Chromatography

Steward FC, Pollard JK (1957) Nitrogen metabolism in plants: Ten years in retrospect. Annu Rev Plant Physiol 8:65–114

Steward FC et al. (1959) Nutritional and environmental effects on the nitrogen metabolism of plants. Soc Exp Biol 13:148–176

Steward FC, Durzan JD (1965) Metabolism of nitrogen compounds. In: Steward FC (ed) Plant Physiology: A treatise, vol 4A. Academic Press, London New York, pp 379–656

Steward FC, Bidwell RGS (1966) Nitrogen metabolism, respiration and growth of plant tissue. IV. The impact of growth on protein metabolism and respiration of carrot tissue explants: General discussion of results. J Exp Bot 9:285–305

4 Cells, Solutes, and Growth of Aseptic Cultures: Trace Elements, Growth Factors

Mott RL, Steward FC (1972a) Solute accumulation in plant cells. Ann Bot (London) 36:621–670, 897–914

Mott RL, Steward FC (1972b) Solute accumulation in cultured plant cells. V. An aspect of nutrition and development. Ann Bot (London) 36:915–937

Steward FC, Mott RL (1970) Cells, solutes and growth: Salt accumulation in plants re-examined. Int Rev Cytol 28:275–370

Steward FC, Neumann KH, Rao KVN, Craven GH, Mott RL (1968–1973) Investigations on the growth and metabolism of cultured explants of Daucus carota. Planta (Berlin) 81:333–335; 91:129–145; 99:240–264; 111:219–243

5 From Cultured Cells to Whole Plants: Somatic Embryogenesis

Steward FC, Mapes MO, Smith J (1958) Growth and organized development of cultured cells. I. Growth and division of freely suspended cells. Ann J Bot 45:693–703

Steward FC (1958) Interpretations of the growth from free cells to carrot plants. Am J Bot 45: 709–713

Steward FC (1969) From cultured cells to whole plants: The induction and control of their growth and morphogenesis. The Croonian Lecture. Proc Soc London, Ser B 175:1–30

Steward FC (1970) Totipotency, variation and clonal development of cultured cells. Endeavour 29:117–124

6 Growth and Composition as Affected by Environments

Krikorian AD, Steward FC (1978) Morphogenetic responses of carrot (Daucus carota var. carota) at zero gravity. Science 200:67–68

Steward FC (1962) Growth, nutrition and metabolism of *Mentha piperita*. Mem (379) Cornell Univ, Agric Exp Stn Parts I–VII, pp 144

Steward FC, Moreno U, Roca R (1981) Growth and composition of potato plants as affected by environments. Ann Bot (London) Suppl 2 48:1–45

II Primary Metabolism

Photosynthetic Carbon Metabolism in Photoautotrophic Cell Suspension Cultures of *Chenopodium rubrum* L.

H. HERZBECK[1] and W. HÜSEMANN[1]

1 Introduction

Photoautotrophic cell cultures "reduce" a complex photoautotrophic plant organism to a single cell level as concerns its growth requirements and photosynthetic capacities. This facilitates studies on the regulation of the photosynthetic carbon metabolism, measurements of chloroplast associated metabolic pathways, correlations between photosynthetic and dark respiratory metabolic pathways as well as relationships between cell growth and chloroplast development.

Photoautotrophic growth of in vitro cultured plant cells has been achieved for an increasing number of cell cultures derived from different plant species. Recent progress on photoautotrophic growth of cultured plant cells has been reviewed by Horn and Dalton (1984), Yamada and Sato (1984), Horn and Widholm (1984), Hüsemann (1985, in press).

The present report on photosynthetic carbon metabolism in photoautotrophic cell suspensions of *Chenopodium rubrum* summarizes the following aspects:

- Photosynthetic product formation and carbon flow from photosynthetic intermediates into different metabolites during different phases of batch growth,
- changes in the pattern of ^{14}C-labeled photosynthetic products in cells which have been exposed to different pH values of the incubation medium during $^{14}CO_2$-photosynthesis,
- metabolism of photosynthetically formed organic acids and exogenously supplied malate with respect to the operation of the tricarboxylic acid cycle in green cells in the light,
- in vitro activities of enzymes related to different metabolic pathways.

2 Capacity and Pathway of Photosynthetic CO_2 Assimilation

Photoautotrophic cell suspension cultures of *Chenopodium rubrum* were grown in a simple mineral salt medium under continuous white light (19 W m^{-2}) at 2% (v/v) CO_2 in the gaseous atmosphere using the two-tier culture vessel method (Hüsemann and Barz 1977, Hüsemann 1984). The cells assimilated about 85 μmol CO_2 (mg chlorophyll)$^{-1}$

1 Institute of Plant Biochemistry, University of Münster, Hindenburgplatz 55, 4400 Münster, FRG

Primary and Secondary Metabolism of Plant
Cell Cultures, ed. by Neumann et al.

Table 1. Sugar and starch content in photoautotrophic cell suspensions of *Chenopodium rubrum* during different phases of batch growth

	Culture age in days Day 5	Day 14	Day 21
Glucose μg $(10^6$ cells$)^{-1}$	57 ± 16	85 ± 8	157 ± 17
Fructose	44 ± 10	68 ± 7	151 ± 19
Sucrose	12 ± 1	6 ± 1	27 ± 5
Starch	35 ± 2	82 ± 5	156 ± 11

h^{-1} during exponential growth, and about 60 μmol CO_2 (mg chlorophyll)$^{-1}$ h^{-1} during stationary growth phase (Hüsemann et al. 1984).

Enhanced accumulation of sugar and starch in stationary growth phase cells was accompanied by a reduction in the rate of photosynthetic CO_2 assimilation (Table 1). Similar effects of the carbohydrate status of the cells on the rate of photosynthetic CO_2 assimilation are known for mature plant leaves (Herold 1980, Azcon-Bieto 1983).

Short-term (10 s) $^{14}CO_2$ fixation experiments demonstrated a light-dependent CO_2 assimilation that confirmed the C_3-pathway by a predominant C_1–C_5 carboxylation. But ^{14}C-labeling of malate during photosynthesis was rather high and varied between 10% and 5% of total radioactivity incorporated during both the exponential and stationary phases of growth (Hüsemann 1981).

High ^{14}C-labeling of malate during $^{14}CO_2$ photosynthesis has also been reported for photosynthetic cell cultures of tobacco (Yamada et al. 1982, Nato et al. 1984) and for carrot (Neumann et al. 1982, Bender 1984).

In vitro activities of ribulosebisphosphate carboxylase and phosphoenolpyruvate carboxylase were strictly dependent on specific growth phases in both photoautotrophic and in photoheterotrophic cell cultures. PEP-carboxylase reached maximum activities during exponential growth, whereas the highest activities of the RuBP-carboxylase were found in non-dividing stationary growth phase cells (Hüsemann 1981, Nato et al. 1984).

The physiological state of photoautotrophic cell suspensions of *Chenopodium rubrum* at different phases of batch growth has further been characterized by in vitro activities of enzymes that are involved in CO_2 assimilation, glycolysis, citric acid cycle and in nitrate reduction (Table 2).

Rapidly dividing cells possessed higher activities of cytosolic enzymes involved in non-photoautotrophic CO_2 assimilation (PEP-carboxylase), nitrate reduction (nitrate reductase), in glycolytic reactions (NAD-dependent glyceraldehyde-3-phosphate dehydrogenase, enolase, pyruvate kinase) as well as in citric acid cycle reactions (NAD-linked malate dehydrogenase and isocitrate dehydrogenase) and in cytochrome C oxidase.

In contrast, photoautotrophic cells from stationary growth phase contained higher activities of chloroplastic enzymes like RuBP-carboxylase, NADP-dependent glyceraldehyde-3-phosphate dehydrogenase and nitrite reductase.

In addition, ultrastructural investigations confirmed growth-phase-dependent changes in the fine structure of cells and chloroplasts in photoautotrophic cell cultures of *Chenopodium rubrum*. These changes paralleled alterations in the metabolic state of

Table 2. In vitro activities of enzymes related to different metabolic pathways in photoautotrophic cell suspension cultures of *Chenopodium rubrum* at different phases of batch growth[a]

	nmol Substrate (mg protein)$^{-1}$ min^{-1}		
	Day 5	Day 14	Day 21
RuBP-Carboxylase	107 ± 9.5	193 ± 12.7	225 ± 12.6
PEP-Carboxylase	86 ± 2.5	64 ± 5.4	55 ± 4.4
NADP-Glyceraldehyde-3-phosphate dehydrogenase	221 ± 46	398 ± 58	363 ± 100
NAD-Glyceraldehyde-3-phosphate dehydrogenase	975 ± 146	685 ± 58	678 ± 160
Nitrate reductase	6 ± 1.0	5 ± 0.6	3 ± 0.1
Nitrite reductase	19 ± 4.5	30 ± 4.5	31 ± 3.5
Enolase	290 ± 10.0	230 ± 8.0	228 ± 11.0
3-Phosphoglycerate kinase	3,150 ± 150	3,270 ± 400	3,140 ± 380
Pyruvate kinase	123 ± 5.5	108 ± 6.8	42 ± 6.8
NADP-Malate dehydrogenase	100 ± 9.8	194 ± 22.0	180 ± 11.0
NAD-Malate dehydrogenase	5,850 ± 633	5,280 ± 770	3,260 ± 130
NADP-Isocitrate dehydrogenase	109 ± 9.0	89 ± 11.3	47 ± 3.6
NAD-Isocitrate dehydrogenase	14 ± 2.1	14 ± 1.8	6 ± 0.9
Cytochrome C oxidase	34 ± 4.2	10 ± 2.5	5 ± 0.8

[a] Based on Hüsemann et al. (1984)

the cells during the growth cycle (Hüsemann et al. 1984). In rapidly dividing cells, the chloroplasts were always surrounded by numerous large mitochondria. The tight grouping of these organelles might indicate an intensive metabolite transfer between chloroplasts and mitochondria. On the other hand, in stationary growth phase cells, fewer mitochondria surrounded the chloroplasts which were loaded with numerous starch grains.

In accordance with Nato et al. (1981) and Brangeon and Nato (1981), we also propose that during the growth cycle of photosynthetic cell cultures an increase in cytosolic and mitochondrial enzyme activities will always precede chloroplast maturation and chloroplast-associated metabolic activities. Only recently, Nato et al. (1984) found by immunological studies that growth phase-dependent changes of carboxylating activities were correlated with changes in the amount of enzyme protein.

3 Pattern of Photosynthetic Carbon Metabolism Under Photoautotrophic Growth Conditions

Growth phase-correlated changes in the metabolic activities of the chloroplastic and extrachloroplastic compartments in photoautotrophic cell suspensions of *Chenopodium rubrum* were determined by the pattern of ^{14}C-labeled photosynthetic products and by following the metabolic flow of ^{14}C from photosynthetic intermediates into different metabolites by $^{14}CO_2$-pulse/$^{12}CO_2$-chase labeling (Hüsemann et al. 1984).

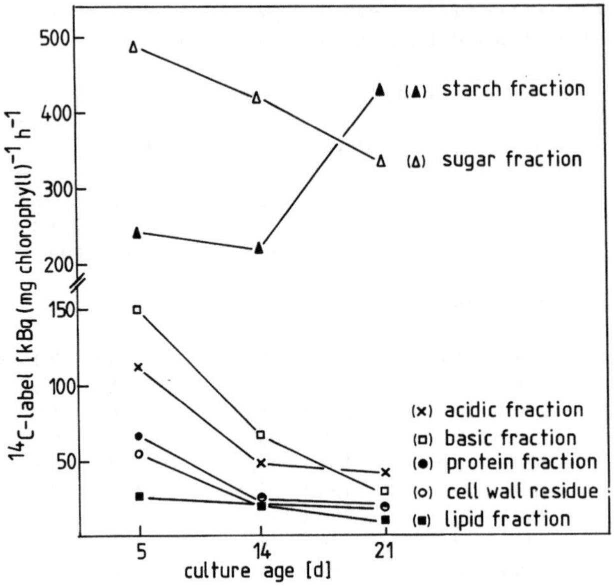

Fig. 1. ^{14}C-label of photosynthetic products after 1 h $^{14}CO_2$ photosynthesis by photoautotrophic cell suspension cultures of *Chenopodium rubrum* during different phases of batch growth. Each growth cycle was started with cells from early stationary growth phase (day 14). Therefore, the pattern of ^{14}C-labeled products in cells at day 0 and day 14 are identical. The data are the mean of fived independent experiments. (After Hüsemann et al. 1984)

To avoid unspecific changes in the photosynthetic carbon assimilation by changes in the pH of the incubation medium, light-dependent CO_2 assimilation was always measured under the actual growth conditions, using the same culture medium in which the cells had been previously grown. The pH of the culture medium was 4.5 and 5.5 during exponential and stationary growth phase, respectively.

Figure 1 shows the pattern of ^{14}C-labeled products after 1 h $^{14}CO_2$ photosynthesis. Free sugars (glucose, fructose, sucrose) and starch were the most heavily labeled compounds, thus representing "primary sinks" for photosynthetically fixed carbon. ^{14}C-labeling of amino acids, organic acids, sugar phosphates and protein was highest during exponential growth and was reduced by 50%–60% in non-dividing cells from the stationary phase of batch growth (day 14 and 21).Concomitant with a decrease in the content of ^{14}C-labeled free sugars was a drastic increase in ^{14}C incorporation into starch, as is known for mature plant leaves (Giaquinta 1978).

Carbon flow from photosynthetic intermediates into different cell compounds was measured during a 5-h chase with unlabeled CO_2 following a 1-h $^{14}CO_2$ photosynthesis (Hüsemann et al. 1984).

Carbon was preferentially shunted into insoluble, lipid and protein fractions, as documented by the increase in ^{14}C-label during the chase period. The basic, acidic and neutral fractions functioned as carbon sources as documented by the loss of ^{14}C during the chase (Table 3). The metabolism of sugars (glucose, fructose, sucrose), measured by ^{14}C-loss from the neutral fraction during the chase period, was approximately three to four times as high as compared to the acidic fraction (organic acids, sugar-phosphates). During the chase, ^{14}C loss from the basic fraction was only half of the corresponding increase of ^{14}C-label in protein. This supports our assumption, that amino acid pools were supplied with newly formed ^{14}C-labeled compounds, probably by keto-acids from the acidic fraction.

Table 3. Changes in ^{14}C-label distribution in photosynthetic metabolites in photoautotrophic cell suspensions from *Chenopodium rubrum* after a 5 h $^{12}CO_2$ chase period following a 1-h $^{14}CO_2$ pulse. (After Hüsemann et al. 1984)

Chemical fractions	Changes in ^{14}C-label after a 5-h $^{12}CO_2$ chase 10^3 Bq mg^{-1} chlorophyll		
	Culture age Day 5	Day 14	Day 21
Chloroform fraction	(+) 30.5 ± 2.1	(+) 39.4 ± 0.4	(+) 5.5 ± 0.4
Ion-exchange separation			
Basic fraction	(–) 52.8 ± 2.3	(–) 19.0 ± 4.3	(–) 5.1 ± 0.9
Acidic fraction	(–) 68.0 ± 6.3	(–) 20.6 ± 3.9	(–) 23.8 ± 3.0
Neutral fraction	(–) 251.0 ± 37.0	(–) 117.0 ± 23.6	(–) 111.0 ± 22.8
Starch fraction	(+) 49.7 ± 7.8	(+) 61.0 ± 16.0	(+) 67.1 ± 13.5
Protein fraction	(+) 115.5 ± 6.6	(+) 57.6 ± 8.1	(+) 36.0 ± 5.5
Insoluble residue	(+) 19.5 ± 1.0	(+) 68.0 ± 12.8	(+) 25.0 ± 4.9

(+) increase in ^{14}C-label after chase; (–) decrease in ^{14}C-label after chase
The values are the average of five independent experiments. Variations of the average are expressed as the standard error of the mean

Finally, carbon flow from photosynthetic intermediates into proteins, lipids and structural components was drastically reduced by about 50% after the cells had entered the stationary phase of batch growth. This corroborates with the low activities of glycolytic and citric acid cycle enzymes in photoautotrophic cell cultures during stationary growth as listed in Table 2, confirming low metabolic activities of the cells.

4 Preliminary Evidence for pH-Dependent Changes in the Formation of ^{14}C-Labeled Photosynthetic Products

Photosynthetic $^{14}CO_2$ assimilation in rapidly dividing photoautotrophic cell suspensions of *Chenopodium rubrum* was measured at different pH values of the incubation medium. For experimental purposes, the pH of the culture medium during early exponential growth phase (4.5) was adjusted to 6.0 (50 mM MES–NaOH buffer) and 7.0 (50 mM HEPES–NaOH buffer).

We found that the rate of light-dependent $^{14}CO_2$ assimilation, as well as the pattern of ^{14}C-labeled products, changed upon different pH values of the incubation medium (Table 4). An increase in the pH of the incubation medium from 4.5 to 6.0 and 7.0, drastically increased the ^{14}C-content in organic acids, especially malate, and in amino acids. Whether the increase of ^{14}C-label in malate mainly resulted from an enhanced synthesis and/or reduced turnover rates of the acid can only be answered by $^{14}CO_2$-pulse/$^{12}CO_2$-chase measurements.

Though the buffer substances we used (MES and HEPES) are considered to be metabolically inert, we cannot exclude that they may exert some unspecific effects on the photosynthetic capacities of the cells to fix CO_2.

Table 4. pH Dependency of changes in the pattern of ^{14}C incorporation into photosynthetic products in photoautotrophic cell suspensions of *Chenopodium rubrum* during exponential phase of batch growth

	pH of the incubation medium 4.5	6.0	7.0
Total $^{14}CO_2$ incorporation: 10^4 Bq mg^{-1} chlorophyll h^{-1}	24.2	25.7	29.4
^{14}C-label distribution (%):			
Chloroform fraction	5.0	4.8	3.2
Basic fraction	12.0	17.2	24.1
Acidic fraction	7.8	31.3	36.3
Neutral fraction	48.4	31.4	20.0
Starch	17.7	7.3	7.1
Protein	5.2	5.4	6.5
Insoluble residue	3.9	2.6	3.1
Malate content: μg $(10^6$ cells$)^{-1}$	3.3	70.0	71.0

$^{14}CO_2$ photosynthesis and ^{14}C-labeling of photosynthetic products under photoautotrophic growth conditions was measured as described by Hüsemann et al. (1984). Under standard culture conditions, the pH of the culture medium of exponentially growing cells (day 5 of the growth cycle) is 4.5. The pH of the incubation medium was adjusted to 6.0 and 7.0 in the presence of 50 mM MES-NaOH and 50 mM HEPES–NaOH, respectively. The data are the mean of three independent experiments

Similar pH dependencies of ^{14}C-labeling of photosynthetic products and a preferential ^{14}C-labeling of C_4-acids and its derivatives by increasing the pH of the incubation medium was found with isolated leaf slices of *Spinacea oleracea* (Böcher and Kluge 1977). The underlying reasons for the observed pH effects on the pathway of photosynthetically fixed carbon is still obscure.

Raising the pH of the incubation medium from 4.5 to 7.0 may increase total availability of dissolved inorganic carbon (CO_2) with preference to HCO_3^- (Buch 1960), thus enhancing photosynthetic CO_2 uptake (Espie and Colman 1981; Volokita et al. 1983).

Until now we do not know to what extent the internal pH of photoautotrophic cells from *Chenopodium rubrum* is increased when the external pH of the incubation medium is raised from 4.5 to 7.0. An increase in the intracellular pH might result in quantitative as well as qualitative changes in CO_2 fixation reactions. In isolated photosynthetic mesophyll cells of *Asparagus officinalis* (Espie and Colman 1981), the intracellular pH remained relatively constant at 6.9 to 7.2 over the external pH range of 5.5 to 7.2.

5 Metabolism of Photosynthetically Formed Organic Acids and Exogeneously Fed Uniformly ^{14}C-Labeled Malate

As mentioned, the formation of proteins in batch-propagated photoautotrophic cell cultures of *Chenopodium rubrum* was highest in rapidly dividing cells. This was documented by maximum values for ^{14}C-label in the protein fraction as well as by maximum rates of turnover of ^{14}C-labeled amino acids during the chase (Fig. 1, Table 3). Sustained amino acid synthesis depends on the supply of keto acids, which are mostly intermediates of the citric acid cycle. The metabolism of malate, a potential intermediate of the citric acid cycle, was studied to document its function in the metabolism of photoautotrophic cells.

After 1 h $^{14}CO_2$ photosynthesis under photoautotrophic growth conditions (Hüsemann et al. 1984), ^{14}C-label in malate varied between 3% and 1.5% of total radioactivity incorporated by the cells during exponential and stationary growth phase respectively. About 90% to 75% of photosynthetically formed ^{14}C-malate was metabolized as measured by ^{14}C-loss from the acid during a 5-h chase with unlabeled CO_2 in exponential and stationary growth phase cells. This corroborated well with maximum values of ^{14}C-labeling of amino acids and proteins in the cells of the respective growth phases.

The turnover of photosynthetically formed organic acids in photoautotrophic cell cultures of *Chenopodium rubrum* thus mirrors the functioning of the citric acid cycle in photosynthesizing cells in the light.

In order to determine the metabolic fate of malate in photosynthesizing cells in the light without any interference with the turnover of other radioactively labeled photosynthetic intermediates, the metabolism of exogenously supplied uniformly ^{14}C-labeled malate was measured. *Chenopodium rubrum* cell suspensions were incubated under photoautotrophic growth conditions (Hüsemann et al. 1984) in the presence of 10 mM malate containing 3.8×10^4 Bq uniformly ^{14}C-labeled malate.

Malate was rapidly taken up and metabolized by photoautotrophic cell suspensions of *Chenopodium rubrum* (Table 5). In exponentially dividing cells, maximum propor-

Table 5. Metabolism of exogenously supplied (U-^{14}C)-malate in photoautotrophic cell suspension cultures of *Chenopodium rubrum*

	Exponential growth	Stationary growth	
	Day 5	Day 14	Day 21
^{14}C-Malate uptake:			
(% of ^{14}C-malate supplied)	70	80	45
^{14}C-Malate metabolized:			
(% of ^{14}C-malate uptake)	64	52	50
^{14}C-Distribution between metabolites:			
(% of ^{14}C-malate metabolized)			
Amino acids	58	34	33
Citrate, succinate, fumarate	11	29	22
Protein	10	8	8
Lipids	6	5	11
Starch	5	7	10
Sugars	4	14	13
Not identified	6	3	3

The cells were exposed for 30 min to 10 mM malate containing 3.77×10^4 Bq (U-^{14}C)-malate. Incubation of the cells under photoautotrophic growth conditions, product isolation and identification were performed as described by Hüsemann et al. (1984)
The date are the mean of three independent feeding experiments

tions of exogenously supplied malate were transferred into other organic acids (citrate, succinate, fumarate), amino acids and protein. Approximately 10% of ^{14}C derived from ^{14}C-malate was recovered in carbohydrates.

In non-dividing stationary growth phase cells, the turnover of malate into amino acids was drastically reduced, while at the same time the metabolism of malate into other organic acids and into carbohydrates including starch increased markedly (Table 5).

Since both photosynthetically produced as well as exogenously supplied malate were rapidly metabolized, malate can be regarded as an important source of carbon skeletons for synthetic reactions in the photoautotrophic cell. Furthermore, the metabolism of malate clearly indicated the operation of tricarboxylic acid cycle reactions in the photoautotrophic cell in the light. Similar observations on malate metabolism are reported for photoautotrophic tobacco cell cultures (Yamada et al. 1982).

As listed in Table 5, exogenously fed (U-^{14}C)-malate is rapidly metabolized into amino acids. For example, the formation of ^{14}C-alanine from ^{14}C-malate requires decarboxylation as well as transamination reactions mediated by malic enzyme and alanine aminotransferase. Substantial in vitro activities of both enzymes were found in photoautotrophic cell suspensions of *Chenopodium rubrum* (Herzbeck and Hüsemann, unpublished).

The formation of ^{14}C-labeled sugars and starch from (U-^{14}C)-malate can be realized by a refixation of $^{14}CO_2$ evolved from decarboxylation of ^{14}C-malate and/or by recycling of respiratory $^{14}CO_2$ by recapturing by the RuBP-carboxylase. Studies on this topic are in progress.

Acknowledgements. The investigations were supported by a grant from the Deutsche Forschungsgemeinschaft to W. Hüsemann. Some of the experimental data presented here are part of a doctoral thesis by H. Herzbeck.

References

Azcon-Bieto J (1983) Inhibition of photosynthesis by carbohydrates in wheat leaves. Plant Physiol 73:681–686

Bender L (1984) The photosynthetic system of chlorophyllous *Daucus* cell cultures. Symposium on primary and secondary metabolism of plant cell culture, 6–8 Sept 1984, Gießen

Böcher M, Kluge M (1977) Der Weg der C-Fixierung bei *Spinacea oleracea* I. ^{14}C-Markierungsmuster suspendierter Blattstreifen unter dem Einfluß des Suspensionsmediums. Z Pflanzenphysiol 83: 347–361

Brangeon J, Nato A (1981) Heterotrophic tobacco cell cultures during greening. I. Chloroplast and cell development. Physiol Plant 53:327–334

Buch K (1960) Dissoziation der Kohlensäure. Gleichgewichte und Puffersysteme. In: Ruhland W (ed) Handbuch der Pflanzenphysiologie. Springer, Berlin Heidelberg New York, pp 1–11

Espie GS, Colman B (1981) The intracellular pH of isolated photosynthetically active *Asparagus* mesophyll cells. Planta 153:210–216

Giaquinta R (1978) Source and sink leaf metabolism in relation to phloem translocation. Plant Physiol 61:380–385

Herold A (1980) Regulation of photosynthesis by sink activity – the missing link. New Phytol 86: 131–144

Horn ME, Dalton CC (1984) Photosynthetic cell cultures and their biotechnological applications. In: Newsletters, vol 43. IAPTC, Univ Illinois, Dep Agron, Urbana, USA, pp 2–6

Horn ME, Widholm J (1984) Aspects of photosynthetic tissue cultures. In: Collins GB, Petolino JF (eds) Applications of genetic engineering to crop improvement. Nijhoff/Junk, The Hague

Hüsemann W (1981) Growth characteristics of hormone and vitamin independent photoautotrophic cell suspension cultures from *Chenopodium rubrum*. Protoplasma 109:415–431

Hüsemann W (1984) Photoautotrophic cell cultures. In: Vasil (ed) Cell culture and somatic cell genetics of plants, vol I. Laboratory procedures and their applications. Academic Press, London New York, pp 182–191

Hüsemann W (1985) Photoautotrophic growth of cells in culture. In: Vasil (ed) Cell culture and somatic cell genetics of plants, vol II. Cell growth, nutrition, differentiation and preservation. Academic Press, London New York, in press

Hüsemann W, Barz W (1977) Photoautotrophic growth and photosynthesis in cell suspension cultures of *Chenopodium rubrum*. Physiol Plant 40:77–81

Hüsemann W, Herzbeck H, Robenek H (1984) Photosynthesis and carbon metabolism in photoautotrophic cell suspensions of *Chenopodium rubrum* from different phases of batch growth. Physiol Plant 62:349–355

Nato A, Mathieu Y, Brangeon J (1981) Heterotrophic tobacco cell cultures during greening. II. Physiological and biochemical aspects. Physiol Plant 53:335–341

Nato A, Hoarau J, Brangeon J, Hirel B, Suzuki A (1984) Regulation of carbon and nitrogen assimilation pathways in tobacco cell suspension cultures in relation with ultrastructural and biochemical development of the photosynthetic apparatus. Symposium on primary and secondary metabolism of plant cell cultures, 6–8 Sept. 1984, Gießen

Neumann KH, Bender L, Kumar A, Szegoe M (1982) Photosynthesis and pathway of carbon in tissue cultures of *Daucus* and *Arachis*. In: Fujiwara A (ed) Plant tissue culture 1982. Mazuren, Tokyo, pp 251–252

Volokita M, Kaplan A, Reinhold L (1983) Nature of the rate-limiting step in the supply of inorganic carbon for photosynthesis in isolated *Asparagus* mesophyll cells. Plant Physiol 72:886–890

Yamada Y, Sato F (1984) Selection of photoautotrophic cells. In: Evans DA, Sharp WR, Ammirato PV, Yamada Y (eds) Handbook of plant cell culture, vol I. Techniques for propagation and breeding. McMillan, New York, pp 489–499

Yamada Y, Sato F, Watanabe K (1982) Photosynthetic carbon metabolism in cultured photoautotrophic cells. In: Fujiwara A (ed) Plant tissue culture 1982. Mazuren, Tokyo, pp 249–250

On the Photosynthetic System and Assimilate Metabolism of *Daucus* and *Arachis* Cell Cultures

L. BENDER[1], A. KUMAR[2], and K.-H. NEUMANN[1]

1 Introduction

Most cell and tissue culture systems are set up to be nourished by heterotrophic nutrition drawing energy and carbon from sugar and reduced nitrogen from amino acids, both supplemented to the nutrient medium. In the light many cultured cells turn green and mixotrophic conditions may be established, as reported by some authors in recent years (Hanson and Edelman 1972; Neumann and Raafat 1973; Nato et al. 1977; Neumann et al. 1978; Kumar et al. 1983a,b, 1984). Eventually even some reports appeared describing photoautotrophic cultures of cells of some plant species (Hüsemann and Barz 1977, Yamada et al. 1978; Hüsemann 1981; Bender et al. 1981). In mixotrophic cultures, illumination contributes considerably to growth of cultured cells. Possibly, such tissue cultures could serve as model systems to study the nutrition of young differentiating leaf cells before their photosynthetic system is sufficiently developed. Here organic nutrition depends on organic solutes derived either from the seed during germination or from older leaves during ontogenesis. As will be shown, such model systems are probably better suited to follow the development of the photosynthetic apparatus than the hitherto employed etiolated seedlings upon illumination. As pointed out earlier (Kumar et al. 1983a,b), in cultured cells as well as in the cells of the growing shoot apex etioplasts characteristic of shoot cells of dark-grown seedlings are absent. Finally it remains to be seen to what extent fully autotrophic cell cultures may some day find application in biotechnology. However, profound knowledge of the photosynthetic apparatus of cultured cells, of its development and of its interaction with the metabolic machinery of the cells is in many ways still lacking. In this paper a short summary of data and ideas derived from experiments on primary metabolism, notably photosynthesis of cultured *Daucus* and *Arachis* cells will be given.

1 Institut für Pflanzenernährung, Abteilung Gewebekultur, Justus-Liebig-Universität, Südanlage 6, 6300 Gießen, FRG

2 Department of Botany, University of Rajasthan, Jaipur, India

Primary and Secondary Metabolism of Plant Cell Cultures, ed. by Neumann et al.

2 Material and Methods

2.1 Cultivation Technique

Secondary phloem explants from *Daucus carota* tap roots (3 mg fr.wt.) were grown for 21 days in light (21 $W m^{-2}$) in a liquid medium containing IAA, inositol and kinetin (2 ppm, 50 ppm, 0.1 ppm resp.) and 2% sucrose as an initial supplement, as was described earlier (Kumar et al. 1983a, b). Established autotrophic *Arachis hypogaea* callus, maintained on a solidified medium as described before (Bender et al. 1981) was transferred to the above-mentioned nutrient solution and cultured for about 6 weeks in the light (21 $W m^{-2}$) as is described in detail elsewhere (Bender 1984).

2.2 Electron Microscopy

Daucus and *Arachis* callus material was fixed in glutaraldehyde/osmium tetroxide, dehydrated in acetone/water series and embedded in Epon 812 resin. The sections were stained with lead citrate and examined with a Philips Mod. 300 electron microscope at 60 kV. Detailed descriptions of the technique is given by Kumar et al. (1983a, b) and by Bender (1984).

2.3 PEP Carboxylase Assay

Carrot tissue cultures were grown for 21 days as described and darkened for 1.5 h by wrapping the culture vessel with aluminum foil, whereas control cultures were not darkened. Enzyme extraction and determination of the in vitro activity was carried out as described elsewhere (Kumar et al. 1983a, b; Bender 1984).

2.4 CO_2 Fixation

Pulse $^{14}CO_2$ fixation experiments where carried out as recently described (Bender 1984, 1985) in $H^{14}CO_3^-$ containing nutrient solutions (pH 7) for various length of time at steady-state conditions in light of 21 $W m^{-2}$ and also in darkness. Total CO_2 concentration (about 82% as HCO_3^-, see Buch 1960) per l medium was 187 μmol with a spec. radioactivity of 53.5 Ci mol^{-1}. Four-minute fixation experiments using gaseous $^{14}CO_2$ (Bender 1984, 1985) were carried out in a gas cuvette with 34 μmol CO_2 l^{-1} air (50 Ci mol^{-1}). All fixations were terminated by freezing and grinding the tissues in liquid N_2. Analysis of labelled compounds was performed by means of ethanol (80%) extraction, ion exchange column chromatography, liquid scintillation spectrometry, TLC/HV-electrophoresis or two-dimensional TLC, both techniques in combination with autoradiography as described recently (Bender 1984, 1985) and by radio gas liquid chromatography (Pertzsch et al. 1981). Isolation and digestion of ^{14}C malate with malic enzyme was also done as described elsewhere (Bender 1984, 1985).

3 Results and Discussion

Within 4 to 5 days after explantation and transfer to the liquid nutrient medium carrot root explants (sec. phloem, cambial area, pith or outer cortex) start to turn green and at the end of a subsequent culture period of 2 or 3 weeks a considerable contribution of illumination (21 W m^{-2}) to the growth performance of the cultures (Table 1) can be noticed.

Table 1. Influence of light and sucrose on the growth of explants of carrot tissue cultures in relation to the hormonal content of the nutrient medium. (Neumann and Raafat 1973)

Treatment	Dark + 58 mM sucrose	Light + 58 mM sucrose	Light – sucrose
A.			
Fresh wt in mg/explant original tissue	2	2	2
Nutrient solution	12	22	16
Nutrient solution + inositol + IAA	23	48	38
Nutrient solution + inositol + IAA + kinetin	84	197	109
B.			
Number of cells 10^3			
– Original tissue	13.0	13.0	13.0
– Nutrient solution	12.5	35.2	29.6
– Nutrient solution + inositol + IAA	25.4	146.1	148.5
– Nutrient solution + inositol + IAA + kinetin	444.0	752.4	802.6

Tissues were precultured for 10 days in a sucrose-containing medium and then transferred to culture vessels containing a fresh nutrient medium of the same composition or free of sucrose [at transfer: (+)kinetin cultures = 28 mg/explant, (–)kinetin cultures = 12 mg/explant]. After another 19 days of culture the tissues were harvested. The hormonal content per litre of medium was: inositol 50 mg, IAA 2 mg, and kinetin 0.1 mg

In the original root explants chromoplasts containing carotene cristals could be detected as the only existing typ of plastids and during the first 10 days of culture these chromoplasts are transformed via "amylo-chromoplasts" (with starch besides carotene), into amyloplasts and finally chloroplasts. The influence of various factors on the development of plastids can be derived from Fig. 1 and a detailed description of these ultrastructural developments is given by Kumar et al. (1983a,b). The photosynthetic apparatus established after 10 days of cultivation in a sucrose containing medium is sufficient to support further growth of the explants in a sugar free medium for at least another 6 weeks (Bender et al. 1981).

The initiation of the transformation of chromoplasts into chloroplasts requires rather inspecifically the supplement of either IAA, kinetin or inositol singly or in

Fig. 1. The influence of sucrose and some phytohormones (IAA, inositol, kinetin) on plastid developmental pattern in cultured carrot root explants (average number of plastids/cell in *brackets*)

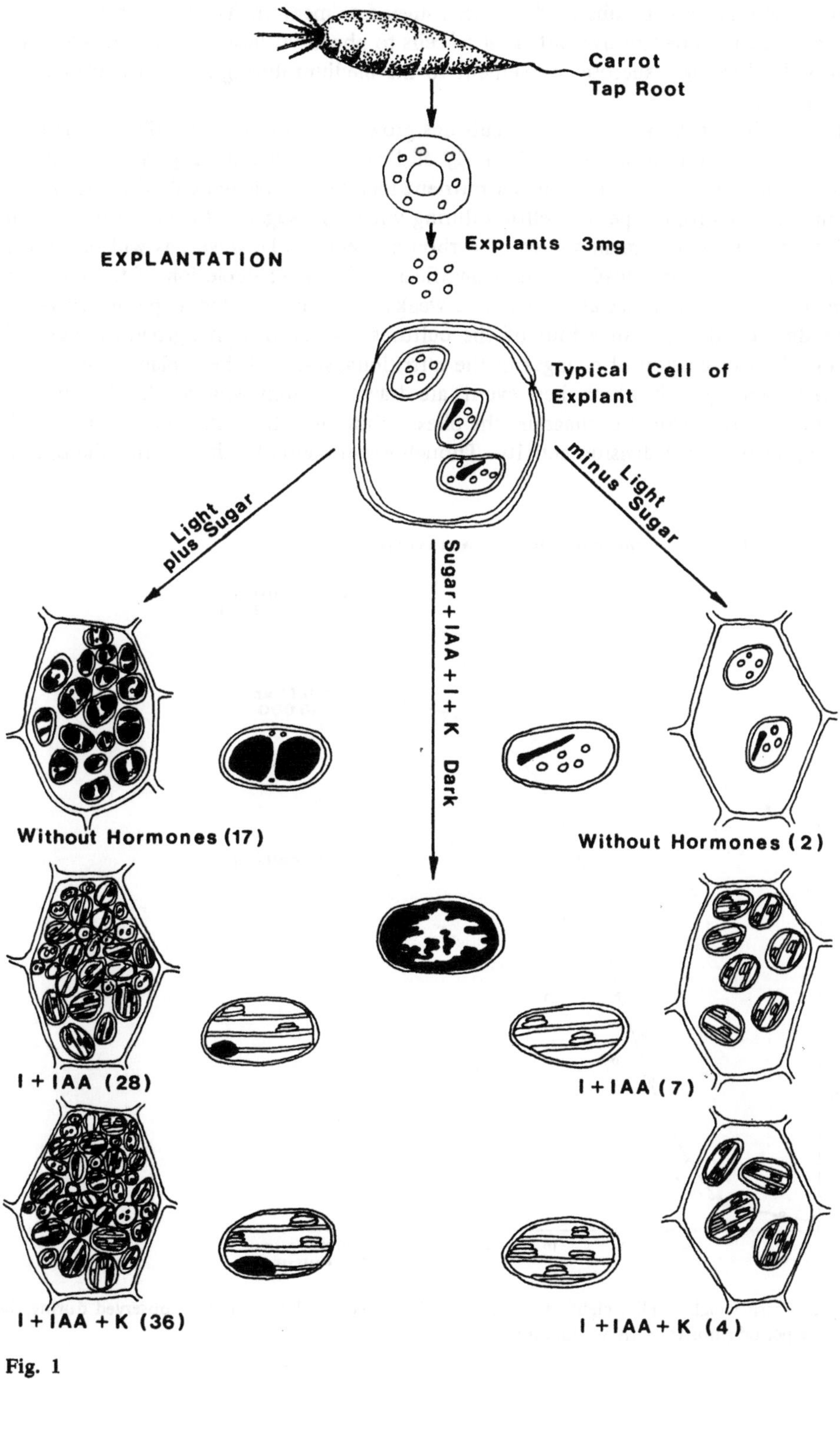
Carrot
Tap Root
Explants 3mg
EXPLANTATION
Typical Cell of
Explant
Light
plus Sugar
Sugar + IAA + I + K Dark
Light
minus Sugar
Without Hormones (17)
Without Hormones (2)
I + IAA (28)
I + IAA (7)
I + IAA + K (36)
I + IAA + K (4)

Fig. 1

combination and can be observed also in a sugar free medium. The number of plastids per cell, i.e., the quantitative potential of cells to photosynthesize, however, is greatly increased following a sucrose supplement to the medium during the initial culture period of 10 days.

During these first 8–10 days the cultures grow heterotrophically utilizing the sucrose of the medium as a source of energy and carbon, while chloroplasts develop. If these cultures remain in the original nutrient medium supplemented with sucrose at t_0, then a mixotrophic period follows during which the sugar of the nutrient medium still dominates as the main source of carbon and energy. However, as will be shown later, photosynthesis already plays a significant role. After depletion of the medium from sucrose – sometimes already after 3 weeks of culture, in other experiments later, depending on the growth vigour of the cultures – an autotrophic growth phase will follow. This succession of changes in the nutritional status of the explants can be also derived from Fig. 2, in which these events are drawn in a somewhat idealized form.

As can be seen from this diagram, the phase of mixotrophic nutrition coincides with the log phase of cell division activity. Though experiments to characterize the photo-

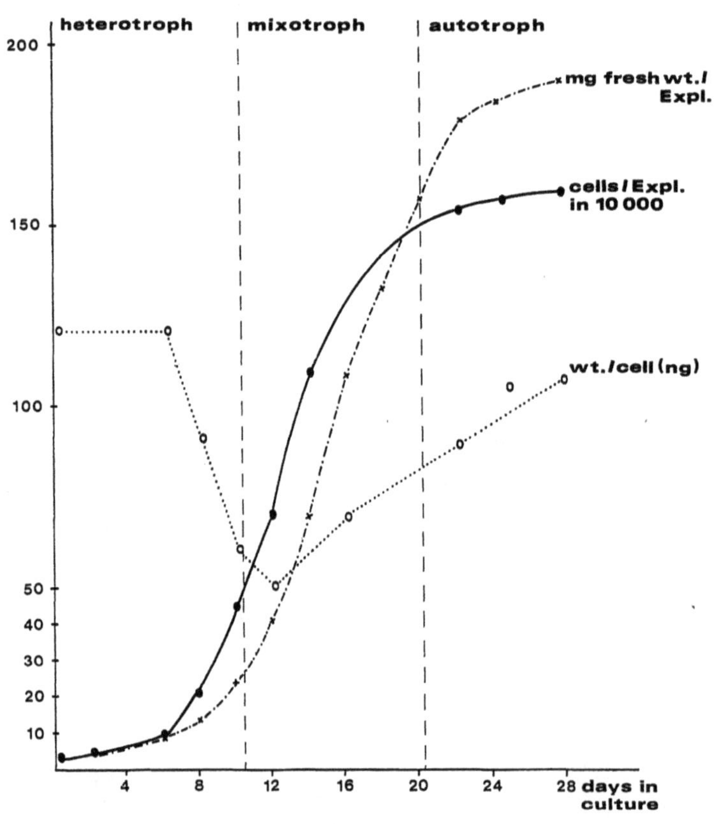

Fig. 2. Fresh weight, cell weight, and number of cells per explant as can be observed during the culture period (inositol + IAA + kinetin)

synthetic system were carried out at various stages during the culture period, most of the data reported in this paper were obtained after 3 weeks of culture, i.e., at the late mixotrophic stage, which corresponds more or less to the transition from the log phase of cell division into the stationary phase. In many tissue culture systems this period coincides with the onset of the production or the accumulation of secondary products. However, carbon metabolism in this cultural stage differs considerably from that during other phases. In Table 2 an estimation of the contribution of sugar in the medium and photosynthesis to dry matter production of the cells is attempted for the period from day 18 to day 22. More than 40% of carbon taken up during this period is derived from inorganic carbon and about 4/5 of this is due to light-dependent CO_2 fixation (see also Neumann et al. 1982; Kumar et al. 1983a,b; Bender 1984). These cultures are also able to grow photoautotrophically (Table 3) at carbon dioxide concentrations in the nutrient solution as it establishes in equilibrium with *normal air* (based on Bunsen's absorption coefficient for CO_2 at 20 °C, the equilibrium comes to about 11.8 μmol CO_2 l^{-1} nutrient solution versus 13.4 μmol CO_2 l^{-1} air at 0.03%).

Table 2. Estimated carbon balance during the late mixotrophic phase of carrot tissue cultures (per culture vessel 15 ml nutrient medium, 75 mg dry matter increment over 144 h). (Neumann et al. 1982)

C Uptake total	49.7 mg C	C	Uptake from exog. sugars	28.0 mg C
C Respired total	12.0 mg C	C	Uptake form CO_2	21.7 mg C
C Retained	37.7 mg C	CO_2	Uptake due to light	18.3 mg C
		CO_2	Uptake in darkness	3.4 mg C

Table 3. Autotrophic and mixotrophic growth of *Arachis hypogaea* and *Daucus carota* cultures (mg fresh weight increment per day) grown in a 5-l laboratory fermenter at 30 W m^{-2} continuous light and a CO_2 conc. as it develops in equilibrium with normal air at 21 °C (12 μmol l^{-1} nutrient solution). The fermenter was inoculated with 3 l nutrient solution (inositol + IAA + kinetin) and 7–8 g fresh weight of green callus material and run for 6 weeks. (Bender et al. 1981)

	– Sucrose		+ Sucrose	
	mg fr.wt.	% Dry matter	mg fr.wt.	% Dry matter
Arachis hypogaea	250	3.88	1,409	7.9
Daucus carota	932	6.04	3,264	6.28

The data in Table 4 show the total CO_2 fixation in light during 15 s and 4 min of exposure to $H^{14}CO_3^-$ in nutrient solution at pH 7 and the labelling of some metabolites is given in Table 5. After 15 s, besides PGA as the first stable product of the C_3 carbon fixation pathway also malate and aspartate as the primary products of the C_4 pathway (and also citrate) were labelled. Since after 15 s labelled PEP could not be detected, but radioactive PGA on one hand and aspartate and malate on the other, these results were

Table 4. Carbon incorporation into various organic fractions by photosynthesizing 21-day-old carrot tissue cultures during different fixation periods in $H^{14}CO_3^-$ containing nutrient solution (pH 7). Data given as μmol carbon atoms × 10^{-4} mg^{-1} Chl

	ETOH-soluble			ETOH-insoluble		Total
	CH	OA	AA	Protein	Others	
15 s	3.3	28	4.3	0.06	2.2	38
4 min	344	1,080	355	3.6	57	1,840

(CH = carbohydrates, OA = organic acids, AA = amino acids)

Table 5. Radioactivity of various metabolites in 21-day-old photosynthetically active carrot tissue cultures after different incubation periods in $H^{14}CO_3^-$ containing nutrient solution (pH 7). Data given as μmol carbon atoms × 10^{-4} mg^{-1} Chl

	15 s (light)	4 min (light)	4 min (dark)
PGA	10.4	88	0
Glucose-P	0	214	0
Fructose-P	0	72	0
Glucose	0	108	0
Fructose	0	74	0
Sucrose	0	0	0
PEP	0	22	0
Glycerate	trace	47	0
Glycine + serine	0	20	0
Alanine	0	19	0
Malate	7.5	188	17
Aspartate	3.5	198	21
Citrate	2.4	81	13
Glutamate	0	17	2.3

taken as an indication that in the cultured carrot cells a C_4 fixation mechanism operates at an elevated level, besides the C_3 fixation pathway (see also Neumann et al. 1982; Bender 1984). After 4 min ^{14}C was then detected also in PEP, hexose phosphates, mostly glucose-6-P, in glucose and fructose but not in sucrose, and also in glycerate, alanine, serine, glycine, and glutamic acid. No labelling, however, was detected in C_3 pathway products after 4 min of $^{14}CO_2$ fixation in darkness (Table 5).

The acceptor of inorganic carbon in the C_4 fixation mechanism PEP could be derived from triose phosphates being exported from the chloroplast in the light via the so-called phosphate translocator (Heber and Heldt 1981), thus linking the Calvin cycle and C_4 fixation in these cultures as it is depicted in Fig. 5, and also from metabolism of exogenous sugars. Since the HCO_3^-/CO_2 ratio was about 4.5 in the nutrient solution at pH 7 (Bender 1985), one could argue that the observed high CO_2 fixation by PEPC would be due to this ratio because of the low K_m of the enzyme for HCO_3^- (Coombs et al. 1975). However, fixation experiments using gaseous $^{14}CO_2$ (Fig. 3) revealed, that even here the ^{14}C distribution pattern is quite similar to that observed after $H^{14}CO_3^-$ supply. For either substrate about half of the label appears in the "C_3" and in the "C_4"

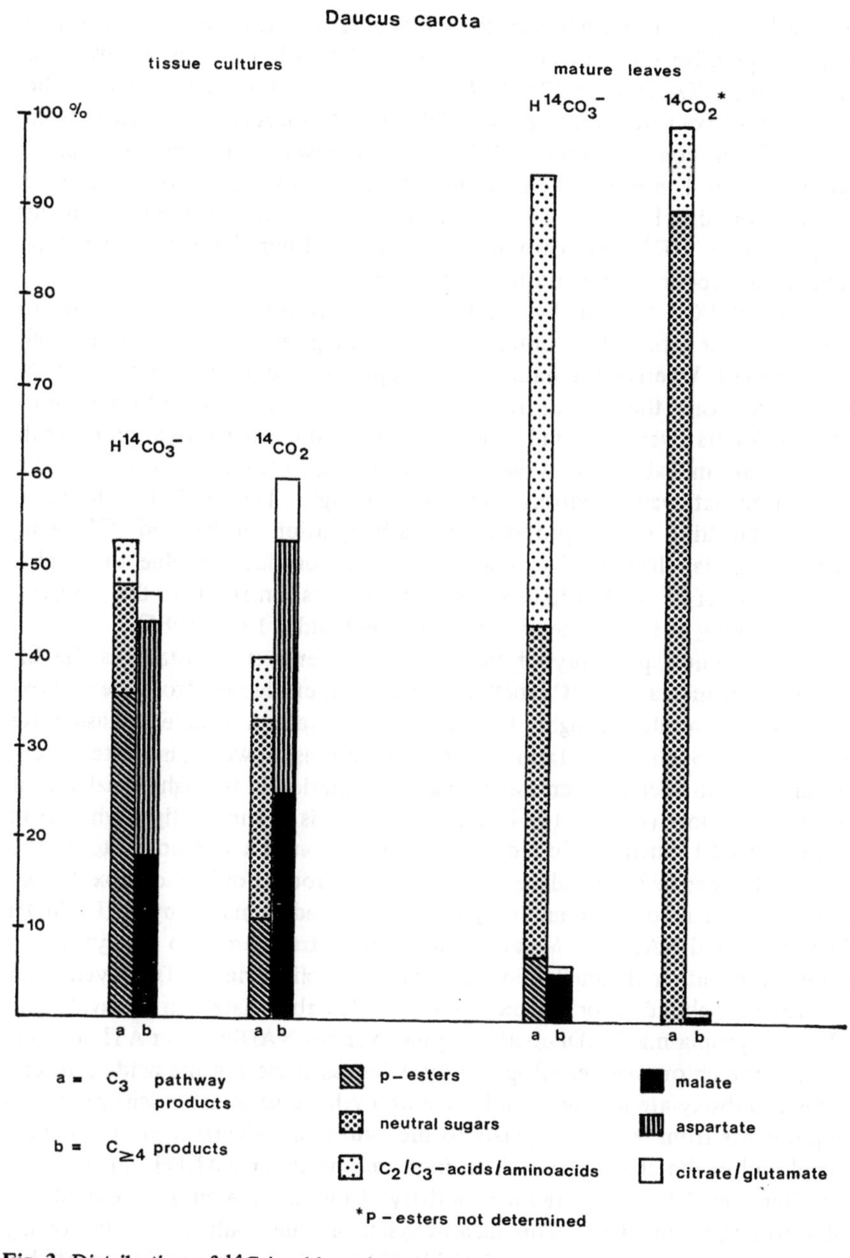

Fig. 3. Distribution of ^{14}C in chlorophyllous *Daucus carota* tissue cultures and mature leaves after $H^{14}CO_3^-$ or $^{14}CO_2$ supply for 4 min in the light

products. Contrary to this mature carrot leaves incorporated more than 90% of ^{14}C fixed into C_3 pathway products irrespective whether gaseous $^{14}CO_2$ or $H^{14}CO_3^-$ in a nutrient solution of pH 7 was provided. This indicates that cell differentiation may be more responsible for the mode of CO_2 fixation than the CO_2 species supplied.

Based on the ^{14}C distribution between the metabolites of the C_3 and the suggested at least partially operative C_4 pathway after 4 min of fixation as they are listed in Table 5, and the total CO_2 fixation (Table 4), an attempt was made to calculate the proportion of $^{14}CO_2$ fixed by either pathway (Bender 1984). As can be derived from the data in Table 8, the contribution of ^{14}C by the C_3 pathway to the partial C_4 pathway is about 15% after 4 min of CO_2 fixation. Taking this into account, the original CO_2 fixation was found to be 295×10^{-4} μmol carbon atoms mg^{-1} Chl min^{-1} for the C_3 pathway and 165×10^{-4} μmol carbon atoms mg^{-1} Chl min^{-1} for the "partial C_4 pathway" under the experimental conditions employed.

Thus the ratio of CO_2 fixation by the C_3 and a "partial C_4 pathway" comes to about 1.8 and a similar ratio can be obtained by the data given in Table 9 for the high phosphorus treatment. Whereas the data in Fig. 3 represent the labelling pattern of C_3 and C_4 products inclusive the ^{14}C contributed by the C_3 pathway to the formation of C_4 products, the results given above indicate the extent of the original fixation by both mechanism. The data in Table 5, however, indicate that also the CO_2 fixation by the "partial C_4 fixation pathway" is strongly increased by light. This could be either due to some double labelling of C_4 products by carboxylation of labelled PEP being derived from the metabolism of C_3 fixation pathway products or due to a light-dependent activation of PEPC. As will be shown, both effects contribute to the observed light enhanced labelling of the C_4 metabolites (see also Bender 1984, 1985).

Due to what we know presently of the metabolic events summarized as the C_4 fixation pathway, an increase in ^{14}C labelling of C_4 products derived from the carboxylation of PEP should result in roughly the same labelling ratio in malate and aspartate in light and darkness. Such a calculation for carrot cultures, however, indicates a considerable variation with specific increases in malate formation in the light usually after 30 min exposure or more (Bender 1984). Here, this ratio is > 1 in the light whereas in darkness it comes to < 1. Such a ratio indicates the operation of a dicarboxylate shuttle in these cultures (Heber 1974). Oxalacetate, as the first product of PEP carboxylation, can be taken up by the chloroplast and may then be reduced to malate by NADP-MDH utilizing light-generated $NADPH_2$. Malate then could be transferred to the cytoplasm from where it can be either channelled into general metabolism via the TCA cycle after its uptake into mitochondria or be oxydized and decarboxylated to pyruvate and $NADPH_2$ by the cytoplasmic NADP-malic enzyme. Neither $NADPH_2$ nor ATP are able to pass through the chloroplast envelope easily, whereas these organic acids can, and therefore this dicarboxylate shuttle could serve as a vehicle to transfer light-generated reducing equivalents from the chloroplasts to the cytoplasm whenever production exceeds demand within the chloroplasts or an increased need for $NADPH_2$ in the cytoplasm exists (fatty acid synthesis, reduced activity of the pentose phosphate oxidative pathway due to permanent light). This dicarboxylate shuttle could work additionally to the phosphate translocator (Heber and Heldt 1981), which is generally accepted to be the most important photosynthetate export shuttle bound to the inner chloroplast envelope. The dicarboxylate shuttle may serve as a supplementary mechanism, as will be discussed later. If the malate/aspartate labelling ratio is considered as an indication for the operation of this mechanism, then an application of kinetin to the medium might be a prerequisite to its induction (see Table 6). In Table 7 the activity of a number of enzymes, mostly related to the dicarboxylate shuttle and to carbohydrate metabolism

Table 6. The labelling ratios of malate/aspartate in chlorophyllous carrot tissue cultures after exposure to $H^{14}CO_3^-$ and ^{14}C-fructose as influenced by light and kinetin

Substrates	(–) Kinetin	(+) Kinetin	
	Light	Light	Dark
$H^{14}CO_3^-$	0.4	1.5	0.5
^{14}C-fructose	0.3	1.5	n.d.

Table 7. Activity pattern of some enzymes (U μg^{-1} protein) in mixotrophic carrot tissue cultures (21-days of culture, NL + I + IAA + kin., acetone powder) (Scharrer and Lein, unpubl. data of our laboratory)

	^{a}U μg^{-1} Protein	K_m ($M\ l^{-1}$)
RuBPC RuBP + CO_2 → 2 PGA	202×10^{-3}	–
PEPC PEP + CO_2 → OAA	154.9	–
Invertase		
Sucrose → Gluc.-Fruct.	$1{,}075.7 \times 10^{-3}$	2.5×10^{-4}
Hexoseisomerase		
D-Gluc-6-P ⇋ -Fruct-6-P	38.4	1.6×10^{-4} 1.9×10^{-4}
Gluc-6-P-dehydrogenase		
Gluc-6-P + $NADP^+$ → Gluconat-6-P + NADPH + H^+	5.0	2.5×10^{-4}
Aldolase		
Fruct-1.6-BisP ⇋ PGA + DHAP	7.9	1.4×10^{-5}
Glutamate-oxalacetate-transaminase		
OAA + Glut.A ⇋ Asp.A + α Keto Glut.A	65.1	8.8×10^{-5} 8.9×10^{-3}
Glutamate-pyruvate-transaminase		
Pyruvate + Glut.A ⇋ Alanine + α-Keto Glut.A	4.1	3.0×10^{-4} 2.5×10^{-2}
Fumarase		
Fumarate + H_2O ⇋ Malate	Low	1.6×10^{-6} 3.8×10^{-6}
Malate-dehydrogenase		
OAA + NADH + H^+ ⇋ Malate + NAD^+ NADPH + H^+ $NADP^+$	10,160.9	1.4×10^{-3}
Malic enzyme		
Malate + $NADP^+$ ⇋ Pyruvate + CO_2 + NADPH + H^+	844.1	2.4×10^{-6}

^{a} U = μmol min^{-1}

are listed and as can be seen from the data, the enzymes required for the operation of the proposed scheme (Fig. 4) are active in carrot tissue cultures at this growth phase.

Malate produced via $H^{14}CO_3^-$ fixation by PEPC is labelled in the C_4 position of the molecule and the C_4 of malate is specifically cleaved off by malic enzyme. To investigate the origin of ^{14}C in malate in detail, malate extracted from tissue exposed to $H^{14}CO_3^-$ for 15 s and 4 min was purified and then subjected to malic enzyme digestion (Bender 1984, 1985). As can be seen from the data in Table 8, after 15 s about 88% of ^{14}C could

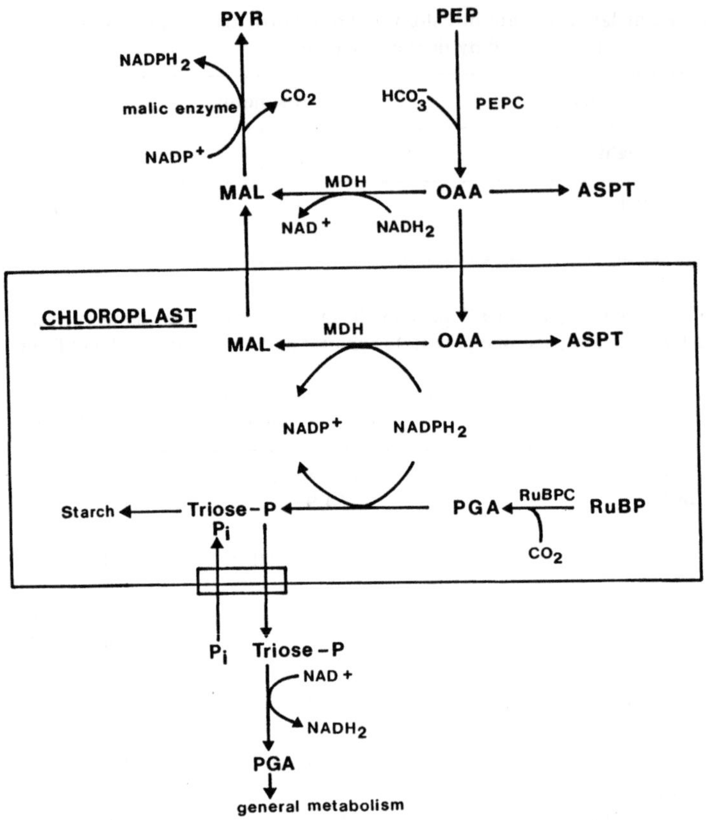

Fig. 4. Suggested alternative mechanisms for the export of reducing equivalents from the chloroplasts of photosynthetically active carrot tissue cultures

be found in the C_4 position. This is in close agreement with the percentage of ^{14}C in this molecular position after $H^{14}CO_3^-$ fixation for 4 min in darkness. After 4 min of fixation in the light, however, only about 72% of ^{14}C incorporated could be detected in the C_4 position. These data indicate that the vast majority of ^{14}C found in malate is derived from PEP carboxylation, as would also be required if an operating dicarboxylate shuttle exists. Only about 15% of the ^{14}C should then be derived from labelled PEP as a result of the simultaneous operation of the C_3 fixation pathway, which is certainly too low to account for the strong increase in ^{14}C labelling of malate due to illumination as it was found after 4 min of fixation in light (Table 5).

However, the PEP carboxylase activity in light-grown, but light- or dark-adapted cultures indicates also some rather strong though not yet understood direct influences of light, i.e., a five times higher activity of the enzyme in illuminated tissue. This could be demonstrated by in vitro measurements, and with the same results by calculating the relative in vivo activity of PEPC on the basis of the ^{14}C incorporation into the C_4 position of malate/aspartate, as outlined in Table 8. Nevertheless such a direct light influence on PEP carboxylase still does not explain the disparity of the ^{14}C ratio in malate/aspartate in light and darkness as discussed above.

Table 8. Distribution of label within the malate molecule after light and dark $^{14}CO_2$ fixation and the dependence of PEPC activity in light and darkness in photosynthetically active carrot tissue cultures. (Bender 1985)

	dpm		^{14}C in	
	Control	Malic enzyme treatment	C_4	C_{1-3}
15 s light fixation	661	81 ± 1.8	87.8%	12.2%
4 min light fixation	4,066	1,149 ± 98	71.7%	28.3%
4 min dark fixation	1,682	220 ± 15	86.9%	13.1%

	dpm × 10^4 g^{-1} dry matter		Relative PEPC activity
	Malate total	Malate C_4 position	(in vivo)
4 min light fixation	134	71.7% = 96.1	100%
4 min dark fixation	20	86.9% = 17.4	18.1%

PEPC Activity	μmol $^{14}CO_2$ g^{-1} fresh wt min^{-1}		Relative PEPC activity (in vitro)	
Experiment	a	b	a	b
Light	0.141 ± 0.014	0.086 ± 0.004	100%	100%
Darkness	0.033 ± 0.004	0.016 ± 0.003	23.4%	18.6%

Fixation was done with cultures which were darkened 1.5 h before ^{14}C-bicarbonate was added. ^{14}C malate was degraded by the malic enzyme procedure. PEPC activity was determined in light-grown cultures and in light-grown cultures which had been darkened for 1.5 h prior to the extraction.
Rel. "in vivo" PEPC activity was determined as follows: The incorporation of ^{14}C into the C_4 position of malate during light fixation was taken as 100% and the respective ^{14}C incorporation during dark fixation was calculated as the percentage of the light fixation value

In summary from the data available so far it is concluded that in cultured carrot explants during the late mixotrophic phase, which coincides with the transition from the log phase of cell division activity to the stationary growth phase, an at least partially active C_4 fixation pathway operates besides the C_3 fixation pathway. The primary carbon metabolism including the metabolic interactions of the two pathways in photosynthetically active and growing carrot tissue cultures is depicted in Fig. 5. In this scheme it is shown that primary photosynthetates (triose-P) are transferred to the cytoplasm by the phosphate translocator, where they may be used for glycolytic oxidation as well as for sugar formation. Triose-P derived directly from photosynthesis or alternatively from sugar metabolism in turn serves as precursor for the generation of PEP. By the carboxylation of this substrate and further metabolism of the products of this "partial C_4 pathway", biosynthesis and consequently growth is anaplerotically supported. To some extent malate may be utilized by a dicarboxylate shuttle to export light generated reducing equivalents from the chloroplast to the cytoplasm if $NADPH_2$ would be produced in excess of demand, e.g., in conditions of a disturbed phosphate translocator due to P deficiency in the cytoplasm.

The following experiments could serve to get some first ideas on parameters and factors which may influence carbon and reducing power export from the chloroplasts in cultured cells. The passage of DHAP or also PGA is mediated by the so-called phosphate translocator, which for its operation requires free phosphate ions in the cytoplasm to be taken up by the chloroplast in exchange for the excreted phosphor-

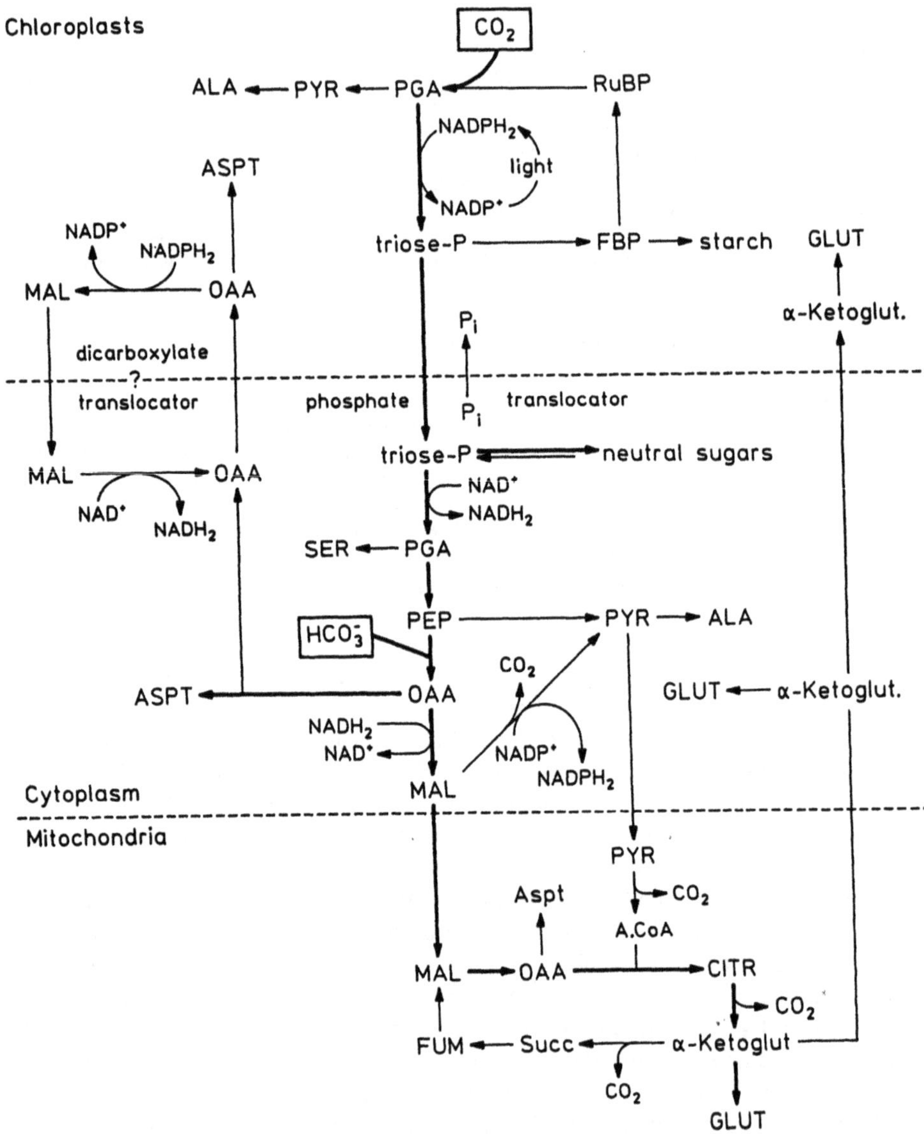

Fig. 5. Proposed scheme of the primary carbon metabolism in photosynthetically active *Daucus carota* tissue cultures. (Bender 1984, 1985)

ylated C_3 compounds (see also Heber and Heldt 1981). Phosphorus deficiency conditions and consequently a low phosphate level in the cells should slow down the operation of this shuttle. Experiments were carried out comparing growth as well as CO_2 fixation at different phosphate levels in the nutrient medium. Whereas the higher phosphate supply corresponds to that used in our standard nutrient medium, in the low phosphorus treatment the concentration was reduced to 25%. However, this is still about three times as high as that in Whites Basal medium. The data on dry weight, P uptake, and chlorophyll concentration as well as ^{14}C fixation data of these experiments are summarized in Table 9. Obviously even at the lower P level, quite substantial growth could be observed, which was only increased by about 40% at the higher P level while the P concentration was increased by about 150%.

The cultures were exposed to $^{14}CO_3^-$ for 4 min and, as the data show, the total carbon fixed during this period is about the same at both phosphorus levels. Though quite low in percentage of total ^{14}C fixed, ^{14}C incorporated into the insoluble fraction is almost exclusively found in starch and is at the lower P level about twice as high as at the higher P supply. Though no difference in ^{14}C in the total soluble fraction due to an enhanced P supply was detected, however, significant changes could be observed in the distribution of ^{14}C between soluble compounds (see Table 10). While ^{14}C in the products of the C_3 fixation pathway, like P esters, neutral amino acids and sugars, is drastically reduced at phosphorus deficiency, a strong increase in label of the C_4-products (aspartate, malate) could be observed. Apparently at P deficiency conditions the PEP-carboxylation dominates.

To interpret these data, some speculation is required for which many data are as yet lacking but which, however, should help to design the experiments still required. Due to a low concentration of P_i which results in a low $[P_i]/[\text{triose-P}]$ ratio in the cytoplasm,

Table 9. Growth, P-uptake and chlorophyll accumulation of carrot root sec. phloem explants (3 mg fr.wt. at t_0) during a 21-day culture period as well as CO_2 fixation (μmol carbon atoms $\times 10^{-4}$ mg^{-1} Chl) of 21-day-old cultures as influenced by phosphorus nutrition

	mg P l^{-1} Nutrient solution 11.6	46.5
Dry matter (mg/vessel)	56.6 ± 6.8	79.1 ± 6.4
P Uptake (mg/vessel)	0.17 ± 0.02	0.62 ± 0.03
P Concentration (mg/g dm)	3.14 ± 0.14	7.27 ± 0.25
Chlorophyll (mg/g dm)	0.75 ± 0.04	0.91 ± 0.03
CO_2 Fixation in light	1,428	1,338
CO_2 Fixation in darkness	206	90
Carbon incorporation into the soluble fraction (light)	1,301	1,278
Carbon incorporation into the insoluble fraction (light)	127	61
[a]CO_2 fixation Calvin cycle (light)	557	880
[a]CO_2 fixation "partial C_4-pathway"	871	458

[a] The calculation of CO_2 fixation by the C_3 and the "partial C_4 pathway" was based on the ^{14}C distribution ratio between the C_3 and the "partial C_4 pathway" metabolites as given in Table 10 and a 15% contribution of ^{14}C by the C_3 pathway to the labelling of C_4 products was taken into account

Table 10. Incorporation of carbon (μmole carbon atoms × 10^{-4} mg^{-1} Chl^{-1}) into various metabolites of the C_3 and the "partial C_4 pathway" by carrot tissue cultures after 4 min of $^{14}CO_2$ fixation in light and darkness

	Light			Darkness		
mg P l^{-1} nutrient solution	11.6	23.3	46.5	11.6	23.3	46.5
C_3 Pathway products:						
PGA	8.6	11	25			
Glucose P	104	135	187			
Fructose P	45	74	73			
Glucose	31	87	54			
Fructose	11	30	21			
Sucrose	Trace	Trace	Trace	Not radioactive		
PEP	Trace	Trace	11			
Pyruvate	0	0	3.4			
Glyzerate	0	6.0	7.5			
Glyzine + serine	10	30	25			
Alanine	10	13	8.0			
Total	220	386	415			
Products of the "partial C_4 pathway":						
Malate	228	127	106	41	14	17
Aspartate	281	201	150	31	19	21
Aspargine	6.8	3.8	3.9	0	0	0
Citrate	29	12	17	27	12	13
Glutamate	14	6.0	3.5	3.6	2.7	2.3
Total	559	350	280	103	48	53

the operation of the phosphate translocator is slowed down, i.e., its capacity is less then the full operation of the C_3 fixation pathway requires. Therefore photosynthetates are first used to produce starch, especially since ADP glucose pyrophosphorylase is increased at low stromal $[P_i]/[PGA]$ ratios (Heldt et al. 1977). After the storage capacity of the chloroplasts is reached, the excess of photosynthetates inside the chloroplasts slows down the activity of the C_3 fixation pathway in the same way as can be observed in experiments with a high sugar supplement (Neumann and Raafat 1973; Bender 1984). If the C_3 fixation pathway operates at a reduced rate, some excess of reducing equivalents might pile up in excess of demand. These, however, could then eventually be exported via the dicarboxylate shuttle serving as some kind of an anaplerotic reaction. Nevertheless, during this short pulse-labelling period (Table 10), the excess of malate over aspartate, as would be required by a light-operated dicarboxylate translocator, is not yet apparent (see Bender 1984). However, these ideas are based on data derived from experiments designed for purposes other than to check the influence of phosphate deficiency on the operation of the dicarboxylate translocator. This problem would require specifically designed experiments planned for the future.

To study the significance of the C_4 fixation pathway in cell cultures of C_3 plants further, an *Arachis* cell culture system was employed. Whereas in carrot cultures in all experiments freshly cut explants from root tissue were used, the *Arachis* explants are an established system derived from cotelydonary explants isolated about 15 years ago and subcultured since (Kumar 1974). These cultures are able to grow autotrophically at normal air CO_2 concentrations over long periods, as was shown for some samples now growing on a sugar-free agar-medium during the last 7 years. Autotrophic growth, however, is much less than mixotrophic growth on a nutrient medium supplemented with 1% or 2% sucrose (Table 3).

As in cultured carrot cells, in the chloroplasts of *Arachis* cultures heavy starch accumulation can be observed following a sucrose supplement to the nutrient medium (Fig. 6). The chloroplasts indicate stacking of thylakoids and except for some deviations in the blue region, chlorophyll absorption spectra of these cultures are quite comparable to that of leaves. Also fluorescence induction profiles of leaves and cultured cells are allmost identical (Bender et al. 1981). The ^{14}C distribution pattern after $^{14}CO_2$ uptake of *Arachis* cultures, in particular after 15 s of exposure, is quite similar to that observed in carrot cultures, indicating also the operation of a partial C_4 fixation pathway besides the Calvin cycle. This is also supported by the ^{14}C isotope labelling pattern of the malate molecule with about 80% of ^{14}C in the C_4 position (Table 11).

Table 11. CO_2 incorporation into various metabolites of *Arachis hypogaea* cultures after short-term $^{14}CO_2$ fixation in $H^{14}CO_3^-$ supplemented nutrient solution in light. (Data given as μmol carbon atoms × 10^{-4} mg^{-1} Chl)

	15 s		4 min
PGA	7.7		236
PEP	0		41
Glucose-P	2.0		1,320
Fructose-P	0.8		346
Glucose	0		336
Fructose	0		81
Sucrose	0		39
Pyruvate	0		Trace
Glycerate	Trace		46
Glycine	4.8		100
Serine	0		51
Alanine	0		22
Total	15.3		2,618
Malate	8.4		348
Aspartate	11.2		295
Citrate	Trace		75
Glutamate	0		22
Total	19.6		740
Malate	C_{1-3}	20.3%	
	C_4	79.7%	

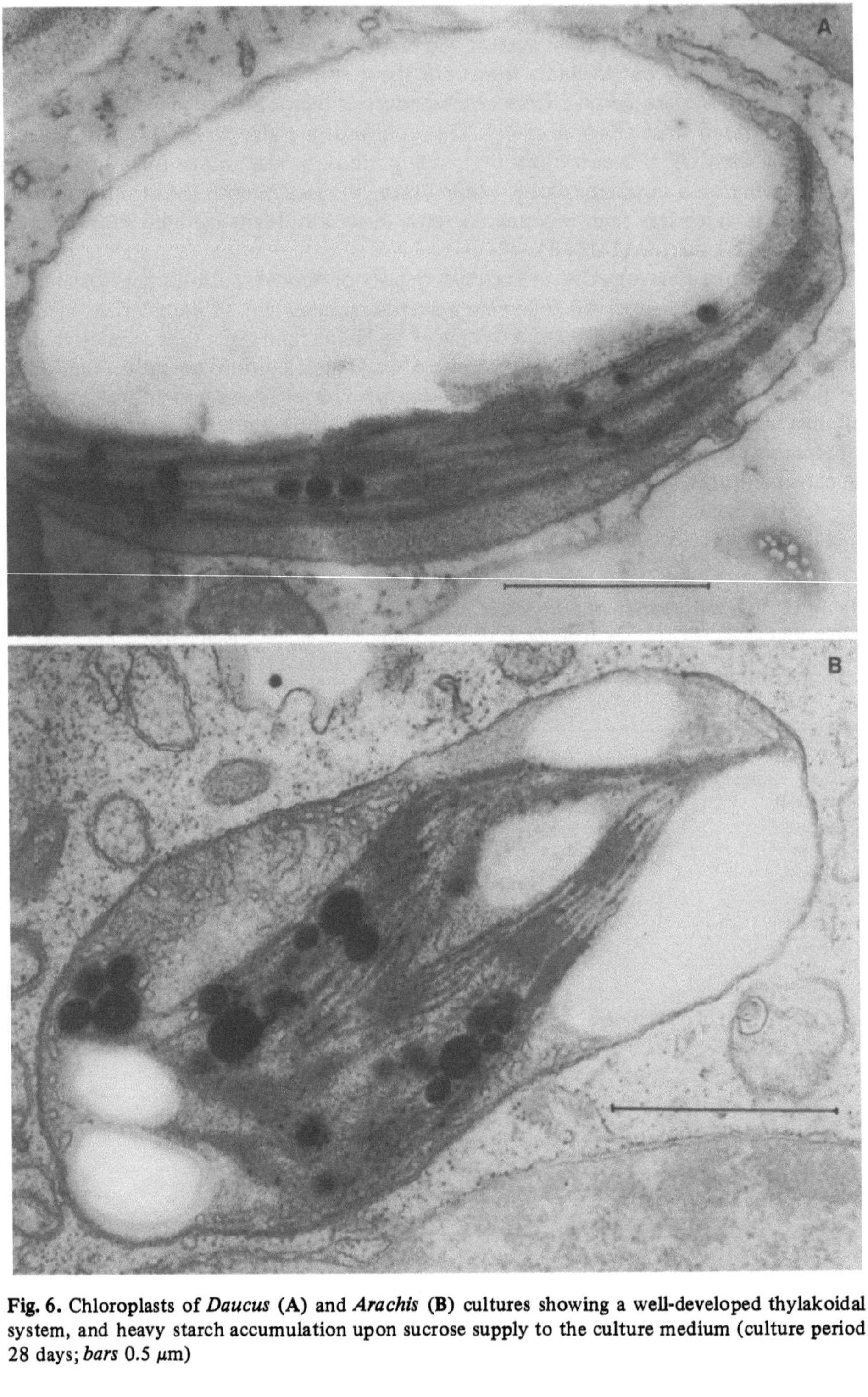

Fig. 6. Chloroplasts of *Daucus* (**A**) and *Arachis* (**B**) cultures showing a well-developed thylakoidal system, and heavy starch accumulation upon sucrose supply to the culture medium (culture period 28 days; *bars* 0.5 μm)

Some differences, however, seem to exist between *Daucus* and *Arachis* cultures in respect to the fate of C_4 fixation products. Whereas in carrot cultures some disparity in the labelling ratio of malate/aspartate in light and darkness was observed, pointing to the operation of a light-dependent dicarboxylate shuttle between chloroplasts and cytoplasm, this is apparently not the case in *Arachis* cultures (Bender 1984). Here this ratio is the same in light and in darkness. These results indicate variations in primary carbon metabolism in cultured cells of different species, and it remains to be seen to what extent these variations account for the differences in growth performance as can be derived from the data in Table 3.

4 Conclusions

The investigations on carbon fixation of *Daucus* and *Arachis* cell cultures indicate, that in cultured cells of these two "C_3 plant species" at least a partially operating C_4 fixation pathway functions besides the Calvin cycle accounting for about 30%–40% of the CO_2 fixed. Similar data were also reported for tobacco cell cultures by Nato et al. (1985). However, we do not wish to imply that this PEP carboxylation pathway serves as a CO_2 trap for the Calvin cycle as in the classical C_4 and CAM type plants.

Both pathways are metabolically linked by PEP as is depicted in Fig. 5. The advantages of such a metabolic pattern for actively growing systems may be:

- Additional CO_2 fixation including refixation of respiratory CO_2.
- Increased utilization of photosynthetates for biosynthesis because of the higher demand for PEP due to PEPC activity.
- Decarboxylation of malate by cytoplasmic "malic enzyme" provides additional $NADPH_2$ in the cytosol.
- Anaplerotic support of the TCA cycle.

In *Daucus* cultures, besides the phosphate translocator, a dicarboxylate shuttle seems to be in operation which is fueled by products of the C_4 fixation pathway to channel light-generated reducing equivalents from the chloroplasts if these are produced in excess of demand. However, the formation of excess $NADPH_2$ by the light reaction system would require equimolar generation of ATP. Nothing can as yet be said about a mechanism that would allow the utilization or the export from the chloroplasts of such excess adenine nucleotide energy in case of a reduced chloroplastic CO_2 fixation.

Daucus carota and *Arachis hypogaea* cell cultures are able to grow **photoautotrophically at ambient CO_2 concentrations.** Considerable growth on sugar-free media in light and ambient CO_2 concentrations was also observed with chlorophyllous *Nicotiana* and *Datura callus* cultures (unpublished), however, no detailed investigations on photosynthesis of these cultures have been made in our laboratory yet.

Acknowledgement. The investigations were supported by grants of the Deutsche Forschungsgemeinschaft (K.H.N.) and the Alexander von Humboldt-Stiftung (A.K.).

References

Bender L (1984) Photosynthese und Assimilatverwertung bei Gewebekulturen und intakten Pflanzen von Daucus carota L. Habilitation Thesis, Justus-Liebig-Univ Gießen

Bender L (1985) Fixation and metabolism of carbon dioxide by photosynthetically active carrot callus cultures. Plant Cell Tissue Org Cult (in press)

Bender L, Kumar A, Neumann KH (1981) Photoautotrophe pflanzliche Gewebekulturen in Laborfermentern. In: Lafferty RM (ed) Fermentation. Springer, Berlin Heidelberg New York, pp 193–202

Buch K (1960) Dissoziation der Kohlensäure, Gleichgewichte und Puffersysteme. In: Ruhland et al. (eds) Encyclopedia of plant physiology, vol V/1. Springer, Berlin Heidelberg New York, pp 1–11

Coombs J, Maw S, Baldry CN (1975) Metabolic regulation in C4 photosynthesis: The inorganic carbon substrate for PEP carboxylase. Plant Sci Lett 4:97–102

Hanson AP, Edelman J (1972) Photosynthesis by carrot tissue cultures. Planta 102:11–25

Heber U (1974) Metabolic exchange between chloroplasts and cytoplasm. Ann Rev Plant Physiol 25:393–421

Heber U, Heldt HW (1981) The chloroplast envelope: Structure, function and role in leaf metabolism. Ann Rev Plant Physiol 32:133–138

Heldt HW, Chon CJ, Maronde D, Herold A, Stankovic ZS, Walker DA, Kraminer A, Kirk MR, Heber U (1977) Role of orthophosphate and other factors in the regulation of starch formation in leaves and isolated chloroplasts. Plant Physiol 59:1146–1155

Hüsemann W (1981) Growth characteristics of hormone and vitamin independent photoautotrophic cell suspension cultures from Chenopodium rubrum. Protoplasma 109:415–431

Hüsemann W, Barz W (1977) Photoautotrophic growth and photosynthesis in cell suspension cultures of Chenopodium rubrum. Physiol Plant 40:77–81

Kumar A (1974) In vitro growth and chlorophyll formation in mesophyll callus tissues on sugar-free medium. Phytomorphology 24:96–101

Kumar A, Bender L, Pauler B, Neumann KH, Senger H, Jeske C (1983a) Ultrastructural and biochemical development of the photosynthetic apparatus during callus induction in carrot root explants. Plant Cell Tissu Org Cult 2:161–177

Kumar A, Bender L, Neumann KH (1983b) Photosynthesis in cultured plant cells and their autotrophic growth. Plant Physiol Biochem 10:130–140

Kumar A, Bender L, Neumann KH (1984) Growth regulation, plastid differentiation and the development of a photosynthetic system in cultured carrot root explants as influenced by exogenous sucrose and various phytohormones. Plant Cell Tiss Org Cult 3:11–28

Nato A, Batzetoux S, Mathieu Y (1977) Photosynthetic capacities and growth characteristics of Nicotiana tabaccum (cv. Xanthi) cell suspension cultures. Physiol Plant 41:116–123

Nato A, Hoaroun J, Brangeon J, Hirel B, Suzuki A (1985) Regulation of carbon and nitrogen assimilation pathways in tobacco cell suspension cultures in relation with ultrastructural and biochemical development of the photosynthetic apparatus. In: Neumann et al. (eds) Primary and secondary metabolism of plant cell cultures. Springer, Berlin Heidelberg New York

Neumann KH, Raafat A (1973) Further studies on the photosynthesis of carrot tissue cultures. Plant Physiol 51:685–690

Neumann KH, Pertzsch C, Moos M, Krömmelbein G (1978) Untersuchungen zur Charakterisierung des Ernährungssystems von Karottengewebekulturen: Photosynthese, Kohlenstoff- und Stickstoffernährung, Mineralstoffverbrauch. Z Pflanzenernähr Bodenkd 141:299–311

Neumann KH, Bender L, Kumar A, Szegoe M (1982) Photosynthesis and pathways of carbon in tissue cultures of Daucus and Arachis. Proc 5th Int Congr Plant Tissue Cell Cult "Plant Tissue Culture 1982", Tokyo

Pertzsch C, Bender L, Neumann KH (1981) Radiogaschromatographie von Kohlenhydraten, organischen Säuren und Aminosäuren aus Pflanzen. Z Pflanzenernaehr Bodenkd 144:231–240

Yamada Y, Sato F, Hagimori M (1978) Photoautotrophism in green cultured cells. In: Thorpe TA (ed) Frontiers of plant tissue culture, Calgary, pp 453–461

Regulation of Carbon and Nitrogen Assimilation Pathways in Tobacco Cell Suspension Cultures in Relation with Ultrastructural and Biochemical Development of the Photosynthetic Apparatus

A. NATO[1], J. HOARAU[1], J. BRANGEON[1], B. HIREL[2], and A. SUZUKI[2]

1 Introduction

Over the last few years, a great number of reports have been published dealing with in vitro chlorophyllous culture cells [3, 12, 13, 21, 30, 33, 37]. Green cell suspension cultures appear to be a suitable system for analyzing cytoplasm-plastid relationships during differentiation [1, 4, 18]. This experimental model is also useful for investigations of plant cell physiology in relation with different cell growth phases [21, 23]. Many studies have been reported using unicellular algae systems, but aspects of physiological events in actively dividing higher plant cells are not well documented [11, 26, 27]. The difficulties reside in the establishment and maintenance of stable green cells during the successive subcultures. Moreover, sucrose, the main carbon source present in cell culture media, is known to be implicated in the repression of chloroplast development and photosynthetic capacity [5, 14, 31]. Nowadays, selection procedure and improved culture conditions make it possible to obtain cell lines with functional chloroplasts [16, 35, 36] and high photosynthetic potentials [28, 32, 34, 40].

The metabolic changes in green cell cultures could illustrate the physiological steps which occur during leaf development [6, 9, 15]. The dependence of cells on the organic culture medium during the early growth phase may be a situation comparable to the heterotrophic state of a young leaf cell prior to its autotrophic gain of carbon.

The carbon flow regulation pathways in relation to cell growth phases have been emphasized by the measurements of chloroplastic RuBPCase and cytosolic PEPCase enzyme capacities in varied cell line cultures [2, 23–29].

Besides, recent investigations [2, 19] suggest that the enhancement of secondary metabolites synthetized by cell suspension cultures requires the presence of differentiated chloroplasts. This fact opens new interesting perspectives for understanding the regulation of alkaloid production. Moreover, there is good evidence for a relationship between key enzyme activities in the nitrogen assimilation pathway and chloroplast photochemical activity [17, 39]. With the exception of nitrate reductase, enzymes such as nitrite reductase, glutamine synthetase and glutamate synthase are all located in chloroplasts.

1 Structure et Metabolisme des Plantes, CNRS LA 40, Université Paris-Sud, Bat. 430, 91405 Orsay Cedex, France
2 Physiologie Végétable Metabolique, CNRS ERA 799, Université Paris-Sud, Bat. 430, 91405 Orsay Cedex, France

Primary and Secondary Metabolism of Plant Cell Cultures, ed. by Neumann et al.

All these considerations imply the need for further investigations on the plastid ultrastructural, physiological, and biochemical features in plant cell cultures.

The present report summarizes the following aspects studied with tobacco cell suspension cultures:

– The ultrastructural and physiological changes of plastids during the greening process of etiolated cells. These modifications have been analyzed from green cells in batch culture conditions [4, 21, 25].

– The photosynthetic capacity changes in relation to cell growth phases and the possible role of chloroplast activity [20, 23].

– The developmental patterns of PEPCase and RuBPCase of cells in different culture conditions [24, 26].

– The characteristics of nitrogen assimilation pathways linked to nitrate reductase, glutamine synthetase and glutamate synthase in wild type cells and mutant strains deficient in RuBPCase activity [7, 10, 22, 38].

2 Results

2.1 The Greening Process of Cultured Tobacco Cells

Generally, plastid biogenesis studies have made use of the greening of etiolated plant leaves. However, the transformation from etioplasts to chloroplasts does not necessarily copy the sequence of events of plastid differentiation in developing leaf cells.

The developmental processes during the greening of cultured tobacco cells resemble rather closely the differentiation of meristematic cells into mature cells. A rapid transformation of amyloplast structure into proplastid-like organelles, characterized by rudimentary single thylakoids is noted after 2–3 days of culture in the light (Fig. 1). The typical cell is rich in cytoplasm and contain an abundance of proplastids, mitochondria, extensive rough endoplasmic reticulum and spiral ribosomes (Fig. 2). This density of cellular organelles, as well as the close associations between the outer plastid envelope and the endoplasmic reticulum, suggest that cytoplasmic events are called upon to supply energy and material for plastid biogenesis [4].

An increased cellular respiratory activity, a high accumulation of soluble proteins, a low photosynthetic O_2 evolution and the inverse behaviour of cytosolic PEPCase and chloroplastic RuBPCase activity [25] are noted at the early stages of plastid development. These physiological data are consistent with the idea that the activation of cytoplasmic events and the initiation of the chloroplast activities occur at separate stages of the light-dependent development, as illustrated in the schematic drawing (Fig. 3). This view is also supported by the increase of ferredoxin and methyl viologen-dependent glutamate synthase activity in the later greening stage [38], whereas the NADH or NADPH dependent enzyme activity remains constant during the greening process (Table 1).

Fig. 1A–C. Ultrastructure of cell and plastids at lag phase (T_0, T_2). **A** Vacuolated cell at end of dark period (T_0). **B** Cell after 2 days of light (T_2). **C** Details of proplastid at T_2. *st* starch; *T* time (in days) exposed to light; *N* nucleus; *Pp* proplastid

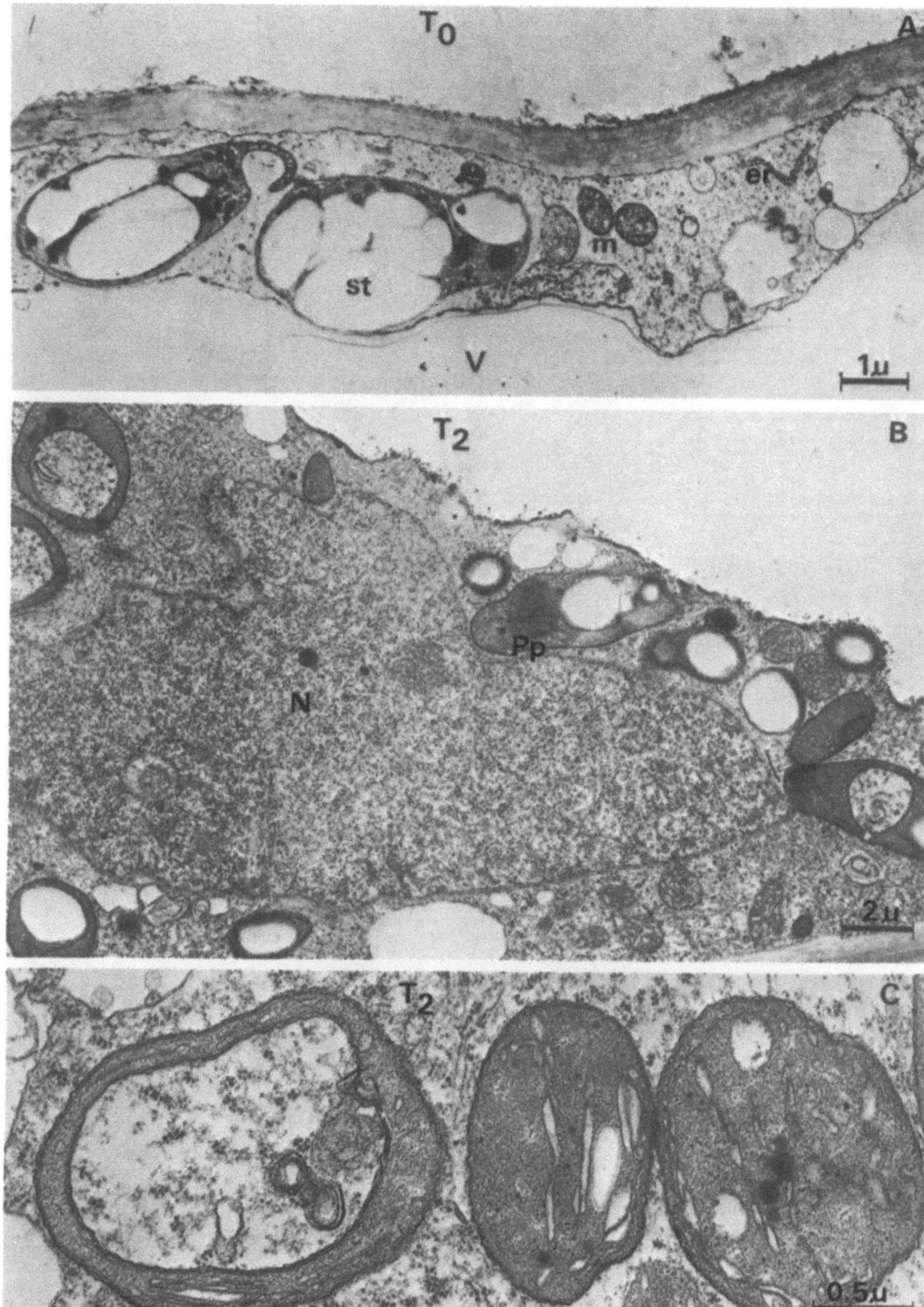

Fig. 1

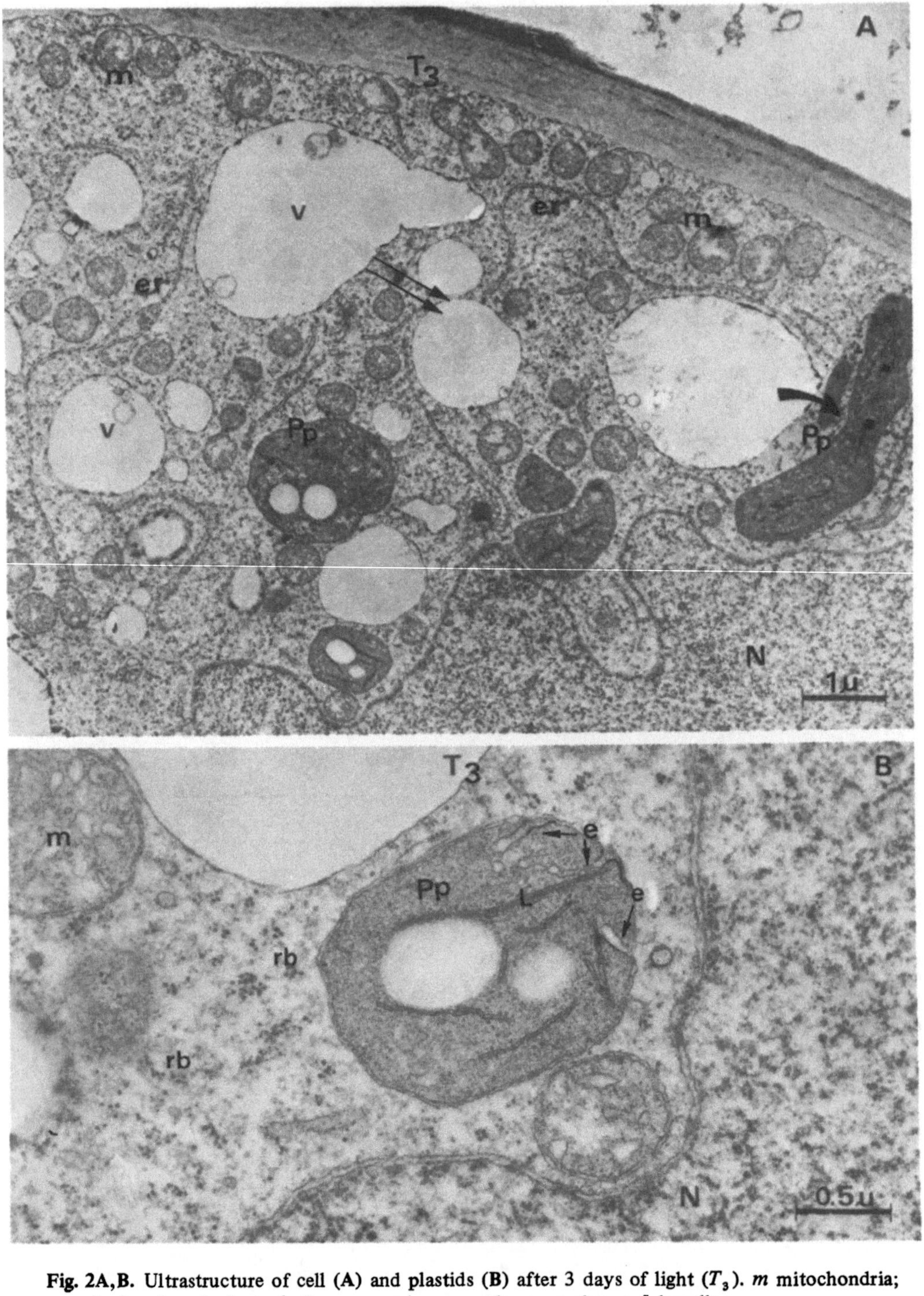

Fig. 2A,B. Ultrastructure of cell (**A**) and plastids (**B**) after 3 days of light (T_3). *m* mitochondria; *er* endoplasmic reticulum; *rb* ribosomes; *e* inner envelope membrane; *L* lamellae

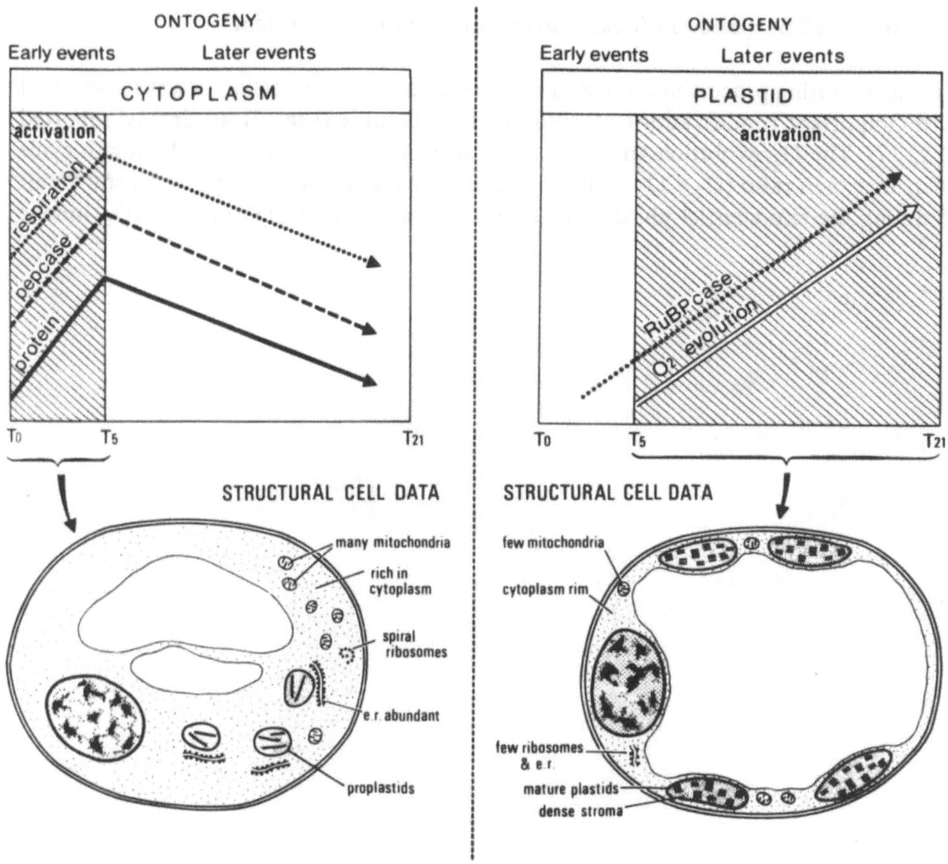

Fig. 3. Schematic representation of sequential events in cytoplasmic and plastid compartments during the greening of tobacco cells

Table 1. Glutamate synthase activities assayed with different electron donors during greening of tobacco cells

Greening days	0	7	14	21	Green
Ferredoxin	8.2	15.5	19.6	31.0	31.3
Methyl viologen	4.5	8.7	13.7	16.5	16.3
NADH	2.9	4.9	3.4	4.1	4.3
NADPH	2.7	4.4	2.9	3.5	3.7
Chlorophylls	0	1,280	1,680	2,830	2,870

Activities are given as nmoles glutamate formed/min g^{-1} fresh weight and chlorophylls as μg chl (a + b) g^{-1} dry wt

2.2 Green Cell Suspension Characteristics in Batch Culture System

The subculturing process in sucrose-enriched medium of green cells is always associated with important modifications of chloroplast structures (Fig. 4) as already reported [21, 37]. The O_2 evolution and CO_2 fixation is modified during the cell growth cycle such that their ratio (O_2/CO_2) is near 1 during the lag and stationary phases and about 2 during the exponential phase. This is concomitant with a high rate of NO_3^- uptake

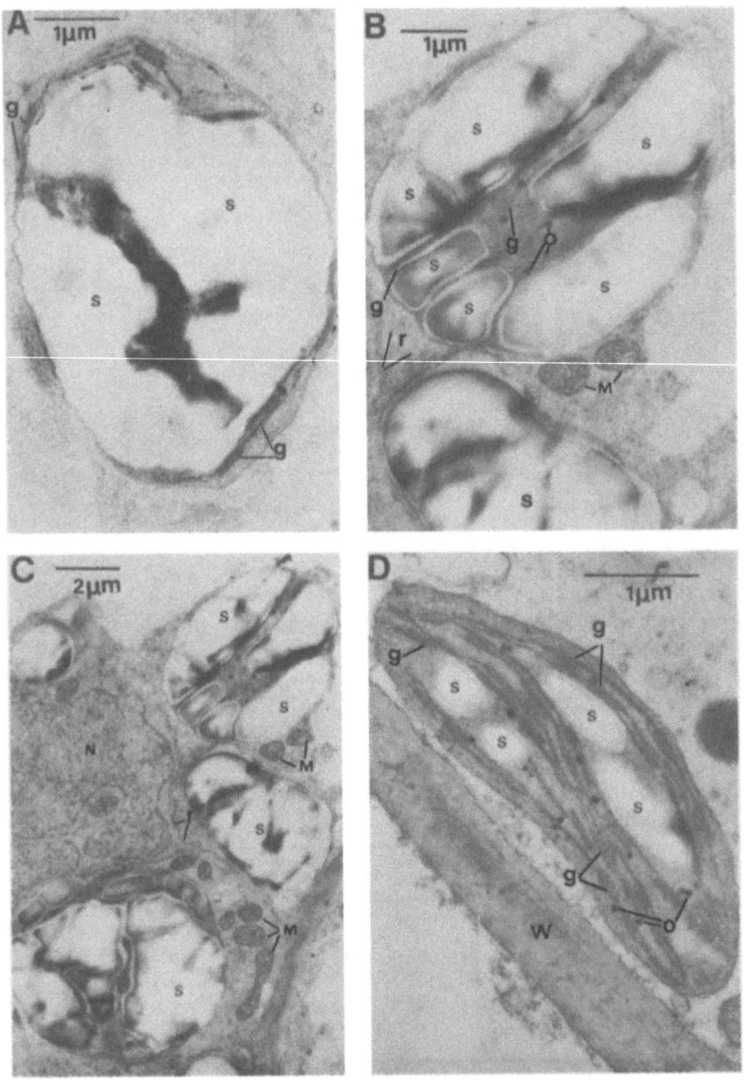

Fig. 4. Plastid morphology of *Nicotiana tabacum* cells during the growth cycle. **A** 2 days of growth; **B, C** 5 days of growth; **D** inoculum (day 0 or day 21 of culture); *g* granum; *M* mitochondria; *N* nucleus; *o* osmiophilic body; *r* ribosomes; *s* starch; *w* cell wall

from the medium [21] and suggests that photosynthetic electron flow can be used for the nitrate assimilation pathway in cultured green cells.

2.3 Effect of Light Growth Intensity in Cultured Cells

Although the chloroplasts confer high photosynthetic capacity to cell populations, the contribution of this process in growth conditions, particularly in sucrose-enriched medium has not been clearly defined.

The growth pattern of tobacco cells cultured in sucrose medium under 3.9 and 39 W m^{-2} shows that high light intensity reduced the doubling time from 3.6 to 2.4 days, stimulated the cellular biomass by twofold, and increased the efficiency of sucrose carbon recovery from 50% to 90% [23].

The CO_2 content variations in culture flasks (Fig. 5) indicate that photosynthesis can be operative in cells grown at 39 W m^{-2} during the first 1 1/2 days. Between 2 and 5 days, the increase of CO_2 content, high PEPCase (Table 2) capacity and enhanced respiratory activity [23] suggest that the intense metabolic processes preceding cell division are sucrose-energy dependent. Thereafter, a low CO_2 content in culture flasks indicates a good cellular photosynthetic efficiency, which allows the fixation of major

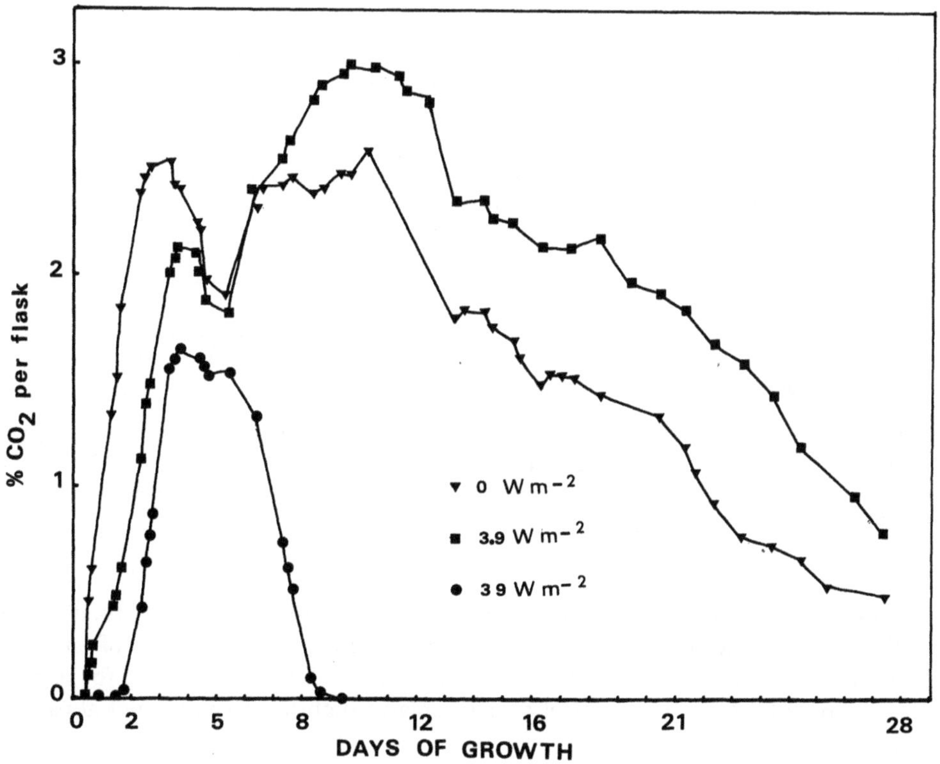

Fig. 5. CO_2 content in culture flasks during a culture cycle of tobacco cells. ▼ cells in darkness; ■ 3.9 W m^{-2}; ● 39 W m^{-2}

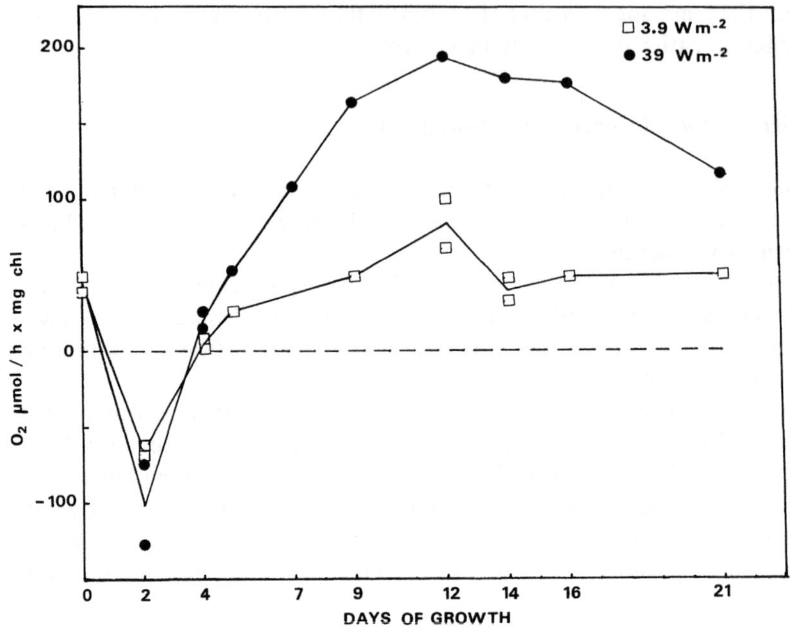

Fig. 6. Changes in endogenous net oxygen evolution under saturating white light (≅ 2.000 µE m^{-2} s^{-1}) in tobacco cells during the growth cycle in two light intensities (□ 3.9 W m^{-2}; ● 39 W m^{-2})

Table 2. Variations in PEPCase and RuBPCase specific capacities in crude extract of tobacco cells during the growth cycle under two light intensities

	µmol CO_2 h^{-1} mg^{-1} sol prot			
	PEPCase		RuBPCase	
D_0	4.63 ± 0.70		2.26 ± 0.53	
	3.9 W m^{-2}	39 W m^{-2}	3.9 W m^{-2}	39 W m^{-2}
D_2–D_8	7.15 ± 0.72	8.32 ± 1.94	1.10 ± 0.29	3.04 ± 0.72
D_9–D_{21}	5.72 ± 1.32	7.27 ± 1.80	2.29 ± 0.50	5.39 ± 0.82

Standard error was calculated at n = 4

respiratory CO_2 [23]. The higher rates of O_2 evolution (Fig. 6) and CO_2 fixation via RuBPCase in cells grown at 39 W m^{-2} (Table 2) support the hypothesis of an autotrophic growth potentiality of green tobacco cells. These data are promising for further investigations of growth light control on chloroplast structural and physiological parameters.

2.4 Carboxylase (PEPCase, RuBPCase) Developmental Pattern Studies in Cultured Cells

An analysis of label distribution among the products of in vivo $^{14}CO_2$ fixation (Table 3) shows a marked increase in PEPCase capacity during the exponential phase [21]. The different behaviour in vitro activities of PEPCase and RuBPCase (Table 2) during the growth cycle under 3.9 W m^{-2} have been investigated by immunological approaches. Using *Sorghum vulgare* leaf PEPCase antibodies [26], the immunoprecipitation curves of enzyme extracted from cells either at the exponential or stationary phase (Fig. 7) show an increased amount of PEPCase molecules in dividing cells. The partial purification of the enzyme by chromatography and the Ouchterlony double diffusion test (Fig. 8) confirm this relative enrichment of the PEPCase molecules in the total pool of soluble proteins [26]. A comparable developmental pattern of PEPCase activity is noted in dividing cells during the greening process or during the photoautotrophic growth in 4% CO_2-enriched atmosphere. These observations emphasize the primordial role of this enzyme in plant cells at the early stages of growth [26].

The immunotitration and the Ouchterlony double diffusion tests using tobacco leaf RuBCase antibodies (Fig. 9) suggest a decline in the enzyme amount in dividing cells. The carboxylase partial purification by sucrose gradient procedure (Fig. 10) confirms the drop in the enzyme amounts. Lescure [18] has reported a low capacity for protein synthesis by isolated chloroplasts from tobacco cells in the active division phase. In ad-

Table 3. CO_2 fixation capacity in the light of cell suspension of *Nicotiana tabacum* and percentage distribution of ^{14}C among the products during the growth cycle

	Percentage distribution	
	5 days	20 days
Hexose-P (G6P + F6P)	10.4	31.0
Trioses-P (PGA + DHAP + PEP)	3.7	5.8
Sucrose	1.5	3.0
Ser + Gly	17.2	31.0
α-Ala	9.6	17.2
Glycolate	1.8	0.3
Glycerate	4.0	0.6
Leu + Tyr + Val	3.4	1.7
Total "C_3 Compounds Group"	52.3	90.6
Glu + Glu (NH_2)	6.1	1.9
Asp + Asp (NH_2)	6.6	2:2
Malate	28.6	3.4
Fum + Succ + Cit	2.9	0.4
Threonine	0.8	0.3
Total "C_4 Compounds Group"	45.0	8.2
Insolubles	0.8	0.5
Non identified	1.9	0.7
CO_2 fixation capacity (μmol CO_2 h^{-1} g^{-1} dry wt.)	140–150	90–100

Ci t, citrate; DHAP, dihydroxyacetone phosphate; F6P, fructose-6-phosphate; Fum, fumarate; G6P, glucose-6-phosphate; PGA, phosphoglycerate; SUCC, succinate. Measures made after 5 and 20 days of growth

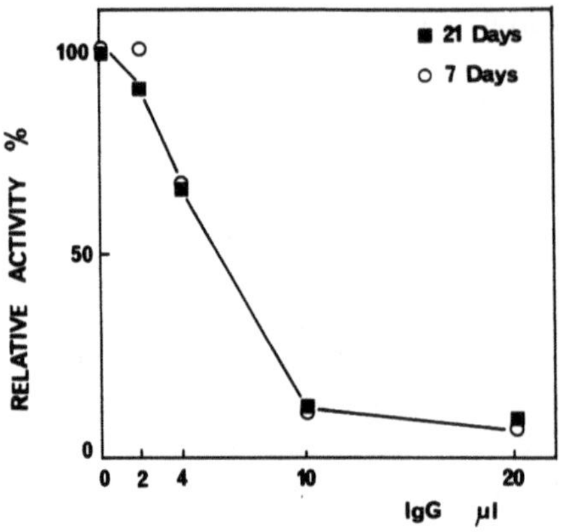

Fig. 7. Immunoprecipitation curves of PEPCase for two different cell extracts (7 and 21 days of culture)

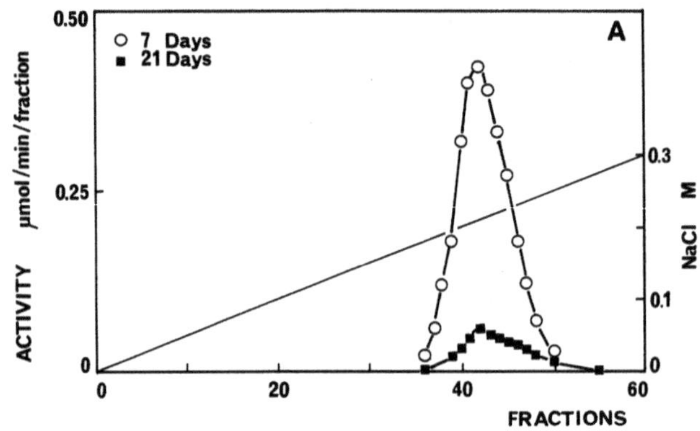

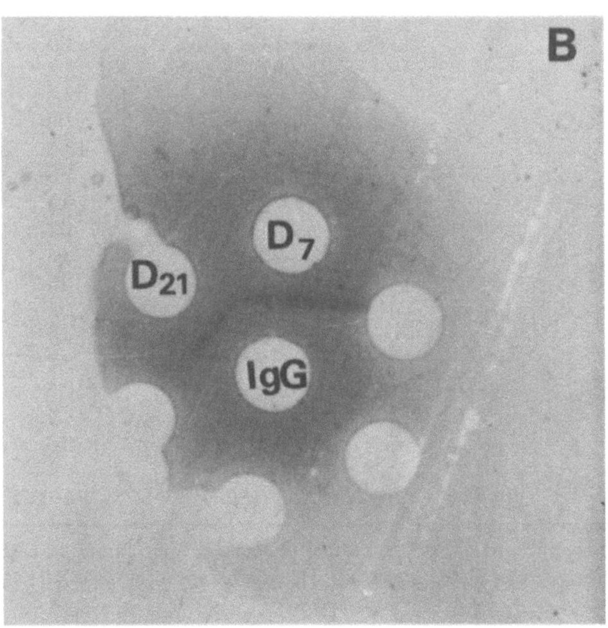

Fig. 8A,B. Partial purification and immunological characterization of PEPCase.

A Chromatography of PEPCase on DEAE-cellulose. The enzymatic extracts (5 enzyme units, 17 mg of soluble proteins for cells grown during 7 days; 1.7 enzyme units, 11 mg of soluble proteins for cells grown during 21 days) were fixed on the top of the column. Fixed proteins were eluted with 0.1 to 0.3 M NaCl.

B Ouchterlony double diffusion test for tobacco PEPCase. The central well contained 10 μl of PEPCase antibodies and the other wells contained 70 μg (day 7 of culture) or 140 μg (day 21) eluted from DEAE column

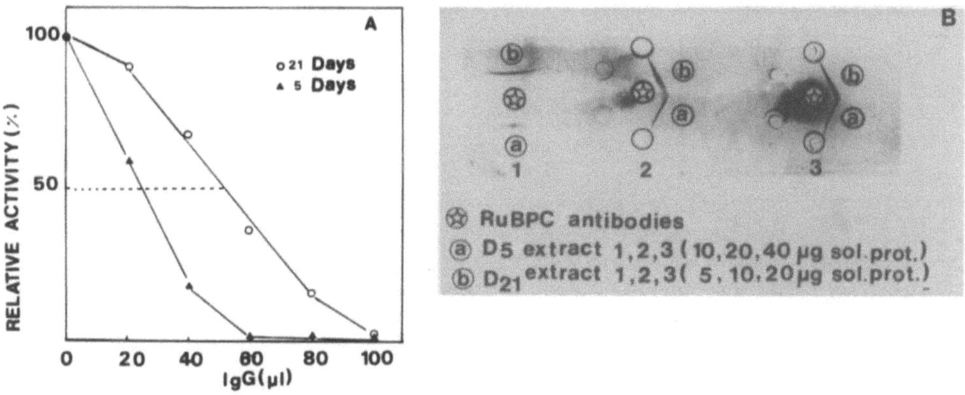

Fig. 9. **A** Immunoprecipitation curves of RuBPCase for two different cell extracts (5 and 21 days of culture). **B** Ouchterlony double diffusion tests for tobacco RuBPCase. The central well contained 10 µl of RuBPCase antibodies and the other wells (*a* day 5 of culture; *b* day 21 of culture)

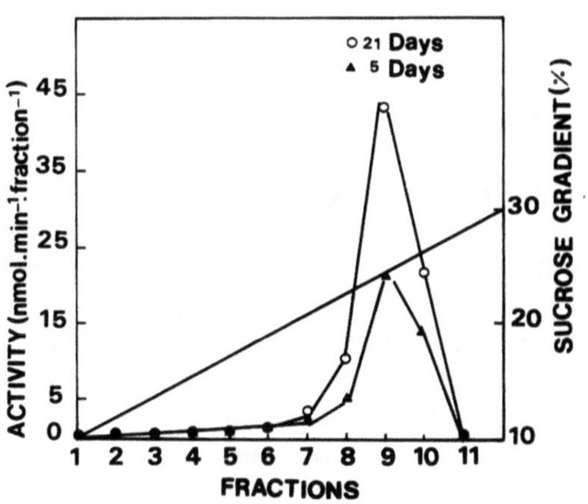

Fig. 10. Sucrose gradient procedure of RuBPCase purification: 11 mg (day 5 of culture) or 5.5 mg (day 21) soluble proteins were deposed on the top of sucrose gradient (0.3–0.1 M). After 16 h ultracentrifugation (45.000 vpm), fractions (1 ml) were collected and assayed for enzymatic tests

dition, important changes in RuBPCase synthesis by isolated tobacco mesophyll protoplasts in the first days of culture have been pointed out by Fleck et al. [8]. A high growth light intensity can prevent this drop in RuBPCase activity during the exponential phase (Table 2). These data suggest that light growth intensity and sucrose, or a metabolite of it, are involved in the regulatory mechanism of the RuBPCase in cultured cells.

In stationary growth phase, the restored initial RuBPCase activity is due to a preferential *de novo* synthesis of enzyme molecules (Fig. 10). Immunotitration tests of an different sucrose gradient fractions (Fig. 11) reveal a small accumulation of an inactive enzyme form in mature cell extracts. The degradative process of RuBPCase molecules suggests that this major protein can be remobilized as a nitrogen source in the latter growth phase of cells.

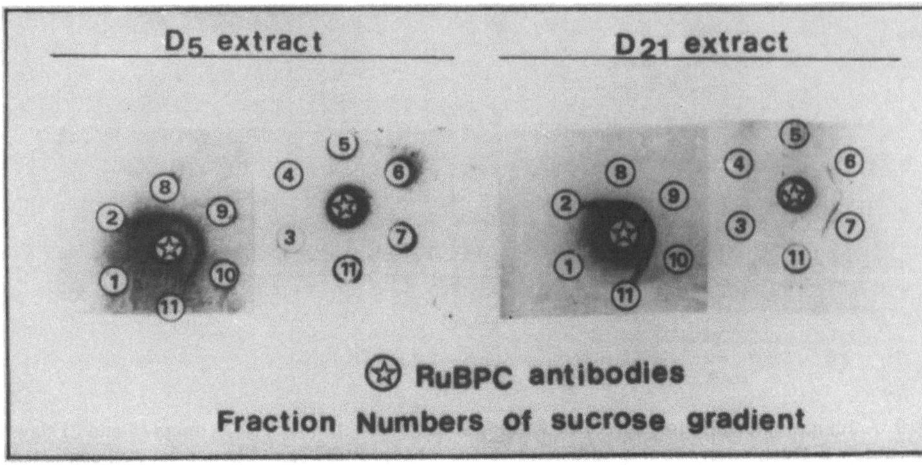

Fig. 11. Ouchterlony double diffusion tests for tobacco RuBPCase. The central well contained 10 μl of RuBPCase antibodies and the other wells the different sucrose gradient fractions of cell extracts from 5 or 21 days of culture

2.5 Nitrate Reductase and Glutamine Synthetase Characteristics in Cultured Cells

An unusual green mutant SP_{25} deficient in RuBPCase activity [7, 22] has been used to study the relationship between the nitrate reductase and the plastid photochemical activity. The low net endogenous O_2 evolution capacity of mutant cells is stimulated 10 fold by the addition of 2 mMp-benzoquinone. The activity is then comparable to that obtained with wild-type cells [22]. Mutants cells grown in continuous light during 4–7 days exhibit an in vivo nitrate reductase (NR) activity about 2 fold higher than that in dark-grown cells. During a 12-h light/dark transition period, an intermediary value between the two conditions is found. However, the in vitro measurements of NR capacity under optimal conditions reveal no statistically significant difference between the three culture conditions (Table 4). These results suggest that in green cell mutants lacking photosynthetic CO_2 fixation, the use of $NADPH_2$ is more clear-cut than in

Table 4. Nitrate reductase activity in mutant cells (4–7 days of culture)

Growth conditions	In vivo activity nmol NO_2^- min^{-1} g^{-1} DW	In vitro activity nmol NO_2^- min^{-1} g^{-1} DW
Light	410 ± 45 (N = 6)	747 ± 173 (N = 5)
Dark	257 ± 64 (N = 5)	586 ± 369 (N = 4)
Transition light dark (12 h)	300 ± 68 (N = 4)	716 ± 172 (N = 4)
Test F Snedecor	$F_{calcul.} = 16.2 >$ F^2_{12} Table = 12.97 At $\alpha = 0.01$	$F_{calcul.} = 1.13 <$ F^2_{10} Table = 4.10 At $\alpha = 0.05$

Mean and conficende interval $\alpha = 0.05$

Table 5. Glutamine synthetase activity and protein content in various types of tobacco cell suspension culture (5 days of culture)

	Activity (μmol min^{-1} g^{-1} dry weight)	GS content (μg g^{-1} dry weight)
Wild type (green cells)	6.6 ± 0.4	132 ± 12
Wild type (etiolated cells)	1.2 ± 0.2	26 ± 5
α Mutant	1.2 ± 0.2	26 ± 5
SP_{25} Mutant	0.8 ± 0.2	18 ± 4

wild-type cells. We now focus our attention on glutamine synthetase (GS) activity and quantity as determined by immunochemical techniques in green SP_{25} cells and in albino α mutant cells [10]. The absence of RuBPCase activity in these mutant cells is correlated to a six-fold lower GS activity compared to wild-type green cells (Table 5). A similar situation is found in etiolated wild-type cells. However, PEPCase and glutamate synthase activities in these different cell strains are comparable to wild-type cells. Thus, it would appear that these mutations affect specifically the mechanism controlling the light-dependent synthesis of chloroplastic GS and RuBPCase enzymes [10]. An hypothesis concerning a common control mechanism of these two proteins in tobacco cells is under investigation.

3 Conclusion

- The cell suspension culture system may be conveniently used for studies on chloroplast biogenesis and for analyzing cytoplasm-plastid relationship during the differentiation process.
- The subculturing process of green cells in sucrose-enriched medium leads to chloroplast morphological and cell photosynthetic capacity changes during the growth cycle. The different behaviour activities of the two carboxylases at various stages of cell growth suggest that tobacco cells, normally thought to have a C_3-type metabolism, exhibit a predominant synthesis of C_4 compounds during the early growth phase.
- Plant cell culture systems have been useful in studying light-dependent neosynthesis and the molecular regulatory mechanisms of carbon and nitrogen assimilation proteins.
- Lastly, the mutant cell suspension cultures have been used as experimental models for the physiological and molecular investigations of key enzymes in nitrogen assimilation pathways.

Acknowledgements. The authors are grateful to Mrs M. Balde for her skillful assistance in culturing tobacco cells and Mrs D. Gillion for typing the manuscript. We are also indebted to Profs. R. Bourdu and P. Gadal for their helpful comments. A. Nato is particularly grateful to Drs J. Fleck and J. Vidal for supplying RuBPCase and PEPCase antibodies.

References

1. Axelos M, Péaud-Lenoël C (1980) The apoprotein of the light-harvesting chlorophyll a/b complex of tobacco cells as a molecular marker of cytokinin activity. Plant Sci Lett 19:33–41
2. Barz W, Hüsemann W (1982) Aspects of photoautotrophic cell suspension cultures. In: Fujiwara A (ed) Plant tissue culture. Mazuren, Tokyo, pp 245–247
3. Berlyn MB, Zelitch I, Beaudette PD (1978) Photosynthetic characteristics of photoautotrophically grown tobacco cells. Plant Physiol 61:606–610
4. Brangeon J, Nato A (1981) Heterotrophic tobacco cell cultures during greening. I. Chloroplast and cell development. Physiol Plant 53:327–334
5. Dalton CC, Street HE (1977) The influence of applied carbohydrate on the growth and greening cultured spinach (*Spinacia oleracea* L.) cells. Plant Sci Lett 10:157–164
6. Dickmann DI (1971) Chlorophyll, ribulose-1,5-diphosphate carboxylase and Hill reaction activity in developing leaves of *Populus deltoides*. Plant Physiol 48:143–145
7. Dulieu H (1981) L'induction de variabilité génétique dans le plastome. Recherche de protéines modifiées dans les chloroplastes. In: Proc Int Atom Energy Agency 251:67–80
8. Fleck J. Durr A, Lett MC, Hirth L (1979) Changes in protein synthesis during the initial stage of life of tobacco protoplasts. Planta 145:279–285
9. Hedley CL, Rowland AD (1975) Changes in the activities of some respiratory and photosynthetic enzymes during the early leaf development of *Antirrhinum majus* L. Plant Sci Lett 5: 119–126
10. Hirel B, Nato A, Martin F (1984) Glutamine synthetase in ribulose 1,5-bisphosphate carboxylase/oxygenase deficient tobacco mutants in cell suspension culture. Plant Cell Rep 3:106–108
11. Hunt L, Fletcher JS (1976) Estimated drainage of carbon from the tricarboxylic acid cycle for protein synthesis in suspension cultures of Paul's scarlet rose cells. Plant Physiol 57:304–307
12. Hüsemann W, Barz W (1977) Photoautotrophic growth and photosynthesis in cell suspension culture of *Chenopodium rubrum*. Physiol Plant 40:77–81
13. Katoh K (1983) Photosynthesis and photoassimilation of glucose during photomixotrophic growth of *Marchantia polymorpha* cells in suspension culture. Physiol Plant 57:67–74
14. Kaul K, Sabharwal PS (1971) Effects of sucrose and kinetin on growth and chlorophyll synthesis in tobacco tissue cultures. Plant Physiol 47:691–695
15. Kisaki T, Hirabayashi S, Yano N (1973) Effect of the age of tobacco leaves on photosynthesis and photorespiration. Plant Cell Physiol 14:505–514
16. Kumar A, Bender L, Neumann KH (1984) Growth regulation, plastid differentiation and the development of a photosynthetic system in cultured carrot root explant as influenced by exogenous sucrose and various phytohormones. Plant Cell Tissue Org Cult 3:11–28
17. Lea PJ, Miflin BJ (1974) An alternative route for nitrogen assimilation in higher plants. Nature (London) 251:614–616
18. Lescure AM (1978) Chloroplast differentiation in cultured tobacco cells: in vitro protein synthesis efficiency of plastids at various stages of their evolution. Cell Differ 7:139–152
19. Lindsey K, Yeoman MM (1983) The relationship between growth rate, differentiation and alkaloid accumulation in cell cultures. J Exp Bot 34:1055–1065
20. Miginiac-Maslow M, Mathieu Y, Nato A, Hoarau A (1981) Contribution of photosynthesis to the growth and differentiation of cultured tobacco cells. The regulatory of inorganic phosphate. In: Akoyunoglou G (ed) Photosynthesis, vol V. Chloroplast development. Balaban Int Sci Ser, Philadelphia, pp 977–984
21. Nato A, Bazetoux S, Mathieu Y (1977) Photosynthetic capacities and growth characteristics of *Nicotiana tabacum* (cv Xanthi) cell suspension cultured. Physiol Plant 41:116–123
22. Nato A, Hoarau J, Bhatnagar R, Dulieu H (1984) Some characteristics of a ribulose-1,5-bisphosphate carboxylase deficient green tobacco mutant in cell suspension cultures. In: Sybesma C (ed) Advances in photosynthesis research. Nijhoff/Junk, The Hague Boston Lancaster, pp 799–802

23. Nato A, Hoarau J, Bourdu R (1983) Effect of light intensity during growth on physiological and photosynthetic characteristics of cell suspension cultures of *Nicotiana tabacum*. Biol Cell 47:213–217
24. Nato A, Mathieu Y (1978) Changes in phosphoenolpyruvate carboxylase activities during the photoheterotrophic growth of *Nicotiana tabacum* (cv Xanthi) cell suspensions. Plant Sci Lett 13:49–56
25. Nato A, Mathieu Y, Brangeon J (1981) Heterotrophic tobacco cell cultures during greening. II. Physiological and biochemical aspects. Physiol Plant 53:335–341
26. Nato A, Vidal J (1983) Phosphoenolpyruvate carboxylase activity in relation to physiological processes during the growth of cell suspension cultures from *Nicotiana tabacum*. Physiol Veg 21:1031–1039
27. Nesius KK, Fletcher JS (1975) Contribution of non-autotrophic carbon dioxide fixation to protein synthesis in suspension cultures of Paul's scarlet rose. Plant Physiol 55:643–645
28. Neumann KH, Bender L, Kumar A, Szegoe M (1982) Photosynthesis and pathway of carbon in tissue cultures of *Daucus* and *Arachis*. In: Fujiwara A (ed) Plant tissue culture. Mazuren, Tokyo, pp 251–252
29. Nishida K, Sato F, Yamada Y (1980) Photosynthetic carbon metabolism in photoautotrophically and mixotrophically cultured tobacco cells. Plant Cell Physiol 21:47–55
30. Ohta Y, Katoh K, Miyake K (1977) Establishment and growth characteristics of *Marchantia polymorpha* L with high chlorophyll content. Planta 136:229–232
31. Pamplin EJ, Chapman JM (1975) Sucrose suppression of chlorophyll synthesis on tissue cultures: changes in the activity of the enzymes of chlorophyll biosynthesis pathway. J Exp Bot 26:212–220
32. Peel E (1982) Photoautotrophic growth of suspension cultures of *Asparagus officinalis* L cells in turbidostates. Plant Sci Lett 24:147–155
33. Sato F, Asada K, Yamada Y (1979) Photoautotrophy and the photosynthetic potential of Chlorophyllous cells in mixotrophic cultures. Plant Cell Physiol 20:193–200
34. Sato F, Nakagawa N, Tanio T, Yamada Y (1981) An improved medium for the photoautotrophic culture of *Cytisus scopavius* Link cells. Agric Biol Chem 45:2463–2467
35. Seeni S, Gnanam A (1980) Photosynthesis in cell suspension cultures of the CAM plant *Chamaecereus sylvestrii* (Cactaceae). Physiol Plant 49:463–472
36. Seeni S, Gnanam A (1982) Growth of photoheterotrophic cells of peanut (*Arachis hypogaea* L) in still nutrient medium. Plant Physiol 70:816–822
37. Seyer P, Marty D, Lescure AM, Péaud-Lenoël C (1975) Effect of cytokinin on chloroplast cyclic differentiation in cultured tobacco cells. Cell Differ 4:187–197
38. Suzuki A, Nato A, Gadal P (1984) Glutamate synthase isoforms in tobacco cultured cells. Plant Sci Lett 33:93–101
39. Wallsgrove RM, Keys AJ, Led PJ, Miflin BJ (1983) Photosynthesis, photorespiration and nitrogen metabolism. Plant Cell Environ 6:301–309
40. Yamada Y, Sato F (1978) The photoautotrophic cultures of chlorophyllous cells. Plant Cell Physiol 19:691–699

Application of Gas Analysis to Continuous Culture

C.C. DALTON[1]

1 Introduction

For fermentation processes involving micro-organisms such as those producing antibiotics, baker's yeast or single cell protein, analysis of the exit gas stream has provided a means of estimating important parameters (e.g. carbon conversion efficiency, volatile substrate concentration, yield factor, productivity, biomass concentration and specific growth rate). Gas analysis has several advantages over other methods of estimating these parameters: it is non-invasive, non-destructive, accurate, continuous, and can easily be automated or interfaced to data-manipulating devises. Gas analysis has also formed an integral part of feed-back control systems in fermentations such as those used to produce alcohol, or those consuming substrates which are toxic above a critical concentration.

Until recently, gas analysis had only been applied to a limited number of fermentations involving plant cells. For instance King (1973) used an infra red gas analyzer to measure the partial pressure of CO_2 in the gas exiting from a fermenter containing *Acer pseudoplatanus* cells. The data were used in the construction of a carbon balance.

For plant cell fermentations it is of interest to measure more than one gas. CO_2 and O_2 are of primary interest, but for special studies other gases such as ethylene would also be of interest. The problem has been that for fermentation involving plant cells the partial pressure difference between the inlet gas stream and the exit gas stream is very small. This is because plant cells have rates of metabolism which are low relative to those of micro-organisms and plant cells are grown in reactors which are specially designed to give low mass transfer rates (i.e. low shear).

For the measurement of CO_2 this is less of a problem except in photosynthetic systems where additional CO_2 is supplied. Normally for heterotrophic systems the inlet gas stream (air) contains very little CO_2 (ca. 0.03%) and the partial pressure difference between the inlet and exit gas streams is sufficiently large for accurate measurement by infra red gas analysis. Measurement of oxygen is very much more difficult because the inlet gas stream (air) contains about 21% O_2. Typically the gas stream exiting from a plant cell fermenter will contain 20.96% O_2. Thus the problem is to measure 0.04% O_2 difference against a background of 21 % O_2. This was approached for the work reported here by using a differential paramagnetic oxygen analyzer which was calibrated using gas mixing pumps.

1 PA Technology, Cambridge Laboratory Melbourn, Royston Hertfordshire, SG8 6DP, United Kingdom

Primary and Secondary Metabolism of Plant Cell Cultures, ed. by Neumann et al.

At the start of this work there was some doubt as to whether the gas analyzers and the methods used were going to provide accurate data; a means of checking was required. One approach was to conduct a carbon balance, the test would be to balance carbon entering the reactor with carbon exiting. Another approach was to apply the methods of Cooney et al. (1977) and Wang et al. (1979) for estimating yield based on gas analysis and material balancing, then to test the gas analysis approach by comparing these estimates of yield with estimates of yield based on wet chemistry for the same fermentations.

For both approaches the continuous culture system was used. Steady states in continuous culture provide an opportunity to collect sufficient data for statistical comparisons. Further, in steady-state culture, variables which influence yield such as specific growth rate, biomass composition, substrate composition, and photosynthetic development can be controlled. However, it was hoped that if proved successful for steady-state cultures of plant cells, then gas analysis would be applied to batch or fed-batch cultures and to study other phenomena such as transient changes and product formation.

2 Methods

2.1 Cultures and Conditions

Ocimum basilicum cultures were used for all experiments. These cells were green and had the capacity to become photosynthetic although for the conditions reported here were grown photoheterotrophically (deriving all their carbon from the fructose supplied in the fresh medium feed). Continuous cultures were established in 1.7 l laboratory fermenters as has been described elsewhere by Dalton (1983) and Dalton (1984). Seven steady states are reported here (Table 1).

2.2 Gas Analysis

Gas analysis was conducted on the fermenter inlet and outlet gas streams both in the normal lighted conditions of culture and for brief periods (2 h) after the fermenter

Table 1. Main features of the steady states

Steady state characteristics	Steady state number 1	2	3	4	5	6	7	8
Limiting[a] nutrient	P	F	F	F	F	F	P	F
Fructose[b] (mg ml^{-1})	10	5	2.5	5	5	5	10	10
KH_2PO_4 [b] ($\mu g\ ml^{-1}$)	85	85	85	85	85	85	85	170
Dilution rate ($\times 10^{-3}\ h^{-1}$)	9.1	9.1	9.1	1.4	4.0	6.7	7.1	7.1

[a] P = phosphate, F = fructose
[b] Fresh medium concentration

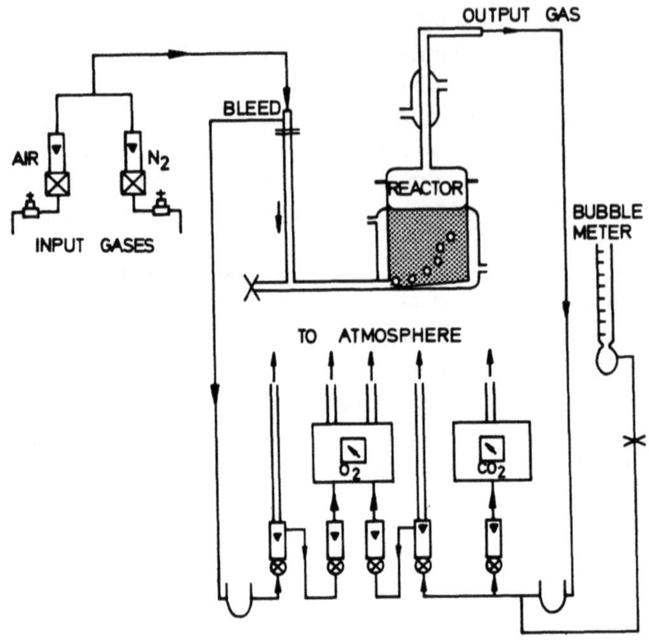

Fig. 1. Flow of gas during analysis

had been put into darkness for 2 h. The partial pressure difference between the dried inlet and outlet gas stream for carbon dioxide was measured using a GP Instruments (Longbenton, UK) infra-red gas analyzer 120 (0–2,500 ppm range). The zero was backed off to the inlet carbon dioxide partial pressure. For oxygen a Taylor Analytics (Sybron Corp., Crowborough, UK) 184 differential paramagnetic analyzer was used. Both gas analyzers were carefully calibrated using gas mixing pumps (H. Wosthoff, Bochum, FRG) supplied with nitrogen and carbon dioxide or oxygen. The flow of gas during analysis is illustrated in Fig. 1.

3 Results Evaluating the Method of Gas Analysis

3.1 Carbon Balance

For the seven different steady states a carbon balance was conducted as previously reported by Dalton (1984). The results are summarized in Table 2. For all steady states except numbers 5 and 7, 100% of carbon entering the fermenter was recovered in the spent medium, plus biomass plus gas exiting from the fermenter. However, for steady state numbers 3–6 the bioproduct yield was high, consequently the percentage of carbon in the exit gas was close to or less than the 95% confidence limits for the value of "total recovered". Thus the carbon balance was not a good test of the accuracy of gas analysis in these steady states.

Table 2. The overall carbon balance for eight steady states[a]

Source of carbon	Steady state number[c] 1	2	3	4	5	6	7	8
Spent medium	39	13	15	11	13	12	15	12
Biomass	39	64	80	83	78	74	50	59
Gas	24	25	5	4	3	6	26	26
Total recovered[b]	104 ± 4	103 ± 4	101 ± 5	98 ± 2	92 ± 2	95 ± 8	92 ± 3	97 ± 3

[a] Expressed as a percentage of carbon in the fresh medium
[b] The ± value refers to the 95% confidence limits
[c] See Table 1

3.2 Yield Measurement

Three equations were derived relating yield to:

1. The rate of CO_2 production (RCO_2),
2. the rate of O_2 consumption (RO_2), and
3. the ratio of RCO_2/RO_2 referred to here as the gaseous exchange quotient (GEQ).

This was based on the theory of Cooney et al. (1977). Each of these three questions was then used to calculate estimates of yield based on gas analysis at several times during each steady state. For similar times estimates of yield based on culture samples and wet chemistry were also determined. For each steady-state time the difference was calculated between the yield based on wet chemistry and the yield base on RCO_2, RO_2, or GEQ. The average differences are expressed as a percentage of the yield based on wet chemistry for each steady-state (Table 3).

For three out of the eight steady states (number 1-3) the yield based on RCO_2 was not different from the yield based on wet chemistry (using the 5% level of significance). For yield based on RO_2 this was the case for two out of eight steady states (numbers 2 and 3) and for GEQ only one steady state (number 2).

Table 3. Values of the difference between yield based on wet chemistry and yield based on gas analysis for three different methods[a]

Gas analysis method[b]	Steady state number[c] 1	2	3	4	5	6	7	8
RCO_2 (%)	0	0	0	− 9	−7	−18	−9	−9
RO_2 (%)	−12	0	0	−107	−7	−12	−9	−9
GEQ (%)	−24	0	−9	− 14	−7	−12	−9	−9

[a] Expressed as a percentage of the yield based on wet chemistry
[b] Based on three equations which were derived relating yield to the rate of CO_2 produced (RCO_2), the rate of O_2 consumed (RO_2) or the Gas Exchange Quotient ($GEQ = RCO_2/RO_2$)
[c] See Table 1

For steady state number 4 the smallest partial pressure differences between inlet and exit gas streams were recorded. This resulted in the greatest inaccuracy in measuring RO_2 and hence GEQ.

For all steady states where the values in Table 3 are different from zero, the values are also negative, i.e. yields based on gas analysis were higher than yields based on wet chemistry.

The most likely explanation for this is that part of the bioproduct was lost during wet chemical analysis (e.g. washed off the filtered biomass during sample preparation). Thus it could be concluded that the yield based on wet chemistry underestimated the true yield by about 10% in five out of the eight steady states. These results also suggest that the measurement of yield based on RCO_2 is more accurate than the measurement yield based on RO_2. This is to be expected from the specification of the gas analyzers used. However, for most steady states the difference in accuracy was not important as estimates of yield based on RO_2, RCO_2 or GEQ were similar. Only in steady state number 4 (operating at the lowest dilution rate), and steady state 1 (no explanation) was a major discrepancy recorded.

3.3 Can Gas Analysis Be Applied with Confidence to Plant Cell Fermentation?

It can be concluded from the aforementioned work that gas analysis involving CO_2 and O_2 can be applied to plant cell fermentations and provide results which inspire confidence. However, great care has to be taken in calibrating the gas analyzers accurately, especially for paramagnetic oxygen analysis. Attention also has to be paid to the control of gas pressure to the fermenters (by using oil damped pressure regulators), the elimination of water vapour prior to analysis (by absorption using magnesium perchlorate) and for rate measurements the accurate measurement of gas flow (using a bubble meter). These and other considerations have been discussed by Brooks et al. (1982).

4 Results of Applications of Gas Analysis to Plant Cell Cultures

There are two types of data which can be aquired by using gas analysis: kinetic and stoichiometric. Kinetic examples include the rates of O_2 uptake or CO_2 evolution, and the specific rate of oxygen uptake.

Stoichiometric examples include the measurement of bioproduct yield on a particular substrate, and the change in elemental composition of the substrates or bioproducts.

It is of interest to apply gas analysis to the estimation of the maximum specific growth rate in batch culture. This has been done for micro-organisms, but it was assumed that the stoichiometry of the reaction and the yield remained constant throughout the period of measurement.

Let us consider some other examples of how gas analysis has been applied in studies involving plant cells.

4.1 Measurement of in situ Photosynthesis Rate

In photoheterotrophic cultures photosynthesis rate is invariably limited by the photon flux density. It is of interest to estimate the contribution which photosynthesis is making towards growth, but the fermentation system must not be perturbed or else the photon flux density per unit biomass will be altered. Gas analysis offers an ideal means of estimating photosynthesis in these circumstances. In order to determine the extent to which photosynthesis rate could be improved by increasing the photon flux density, culture samples were taken from the fermenter and photosynthesis rate (based on RO_2) was measured in a Clark electrode (Table 4).

Table 4. Photosynthesis rate of *O. basilicum* cells measured by gas analysis in the fermentation conditions[a] and measured on samples in a Clark electrode where additional photons were supplied

Conditions	Steady state number[b] 1	2	3	4	5	6	7	8
Fermenter (μmol O_2 mg^{-1} dry biomass h^{-1})	0.08	0.12	0.24	0.12	0.13	0.18	0.06	0.12
Fermenter (μmol O_2 mg^{-1} Chl h^{-1})	180	120	140	30	49	92	62	61
Clark electrode (%)[a]	− 28	7	12	150	140	52	45	140

[a] Expressed as the percentage improvement of photosynthesis rate when additional photons were supplied i.e. 100. (Clark electrode rate − fermenter rate) ÷ fermenter rate
[b] See Table 1

The results show that in most steady states an improvement in photosynthesis rate could be obtained by supplying additional photons. In other words the cells were photon limited and had the capacity to use additional photon when these were supplied. The capacity to use photons tended to be greater for steady states limited by sugar supply (numbers 2, 3, 4, 5, 6, and 8) and steady states at lower dilution rates (numbers 4 and 5).

4.2 Measurements of Photosynthetic Efficiency in Photoheterotrophic Conditions

The yield of biomass in the light can be measured with relatively little difficulty using methods based on RCO_2, RO_2, and GEQ (Table 5). However, the measurement of photosynthetic efficiency is fraught with problems. First the heterotrophic contribution to yield in photoheterotrophic conditions has to be estimated. This was done here for each measurement of yield in the light by putting the reactor into the dark, allowing for equilibration and measuring the heterotrophic yield by gas analysis. As with most estimates of "dark" respiration in the light this method also has disadvantages. For instance the specific growth rate of the biomass can be expected to immediately decline

Table 5. Estimates of the yield[a] in photoheterotrophic conditions by three methods of gas analysis

Gas analysis method[b]	Steady state number[c] 1	2	3	4	5	6	7	8
RCO_2	0.66	0.75	0.94	0.93	0.93	0.91	0.65	0.68
RO_2	0.66	0.78	0.96	1.77	0.95	0.97	0.65	0.69
GEQ	0.76	0.78	1.02	0.98	0.95	0.94	0.65	0.71

[a] Yield is the gramme of bioproduct g^{-1} sugar consumed
[b] See Table 3
[c] See Table 1

in the dark causing a concomitant decline in the yield. Second, problems are encountered when the photon flux absorbed by the biomass is estimated. For the calculation used here it was assumed that the average value measured incident on the inside of an empty reactor closely approximated the value absorbed by the biomass. (It should be remembered that an over estimate here would reduce the estimated value of the photosynthetic efficiency.)

The result of this approach (Table 6) give values of photosynthetic efficiency for *O. basilicum* cells in photoheterotrophic conditions of between 0.5% and 5%. In general, values for one steady state estimated by the three different methods were very similar.

Table 6. Estimates of photosynthetic efficiency (%) based on three methods of gas analysis

Gas analysis method[a]	Steady state number[b] 1	2	3	4	5	6	7	8
RCO_2	2.4	2.1	3.1	1.7	3.3	4.2	1.9	5.0
RO_2	2.4	2.3	3.1	2.6	3.3	4.2	1.9	5.0
GEQ	2.4	2.7	3.1	0.5	3.3	4.2	0	3.4

[a] See Table 3
[b] See Table 1

5 Discussion

These results establish that gas analysis can be applied to plant cell fermentations with utility, particularly for the measurement of in situ photosynthesis and yield. This has several advantages:

1. Sampling errors can be reduced because the whole population of cells is measured and logging of data can be prolonged until satisfactory standard errors have been obtained.
2. The bioproduct is not destroyed and the reaction is not disturbed.
3. Transient phenomena can be studied.

4. Changes in the stoichiometry of growth can be followed, e.g. product formation on transition from trophase to idiophase in batch culture.

Measurement of RO_2 and GEQ would be much improved by more accurate measurement of the partial pressure of O_2 (pO_2), especially small differences in pO_2 in the 10%–21% range. It should be possible to apply mass spectrometry to plant cell fermentations in the same way as this has been applied to fermentations involving microorganisms. It would be particularly interesting to measure "dark" respiration of photoheterotrophic cultures in the light without perturbing the normal growing conditions.

References

Brooks JD, Maclennan DG, Barford JP, Hall RJ (1982) Design of laboratory continuous culture equipment for accurate gaseous metabolism measurements. Biotech Bioeng 24:847–856

Cooney CL, Wang HY, Wang DIC (1977) Computer-aided material balancing for prediction of fermentation parameters. Biotech Bioeng 19:55–67

Dalton CC (1983) Photosynthetic development of *Ocimum basilicum* cells on transition from phosphate to fructose limitation. Physiol Plant 59:623–626

Dalton CC (1984) The effect of sugar supply rate on photosynthetic development of *Ocimum basilicum* (sweet basil) cells in continuous culture. J Exp Bot 35:505–516

King PJ (1973) The continuous culture of plant cells. Thesis, Univ Leicester, UK

Wang HY, Cooney CL, Wang DIC (1979) On-line gas analysis for material balances and control. Biotech Bioeng Symp 9:13–23

Carbohydrate Source, Biomass Productivity and Natural Product Yield in Cell Suspension Cultures

M.W. FOWLER[1] and G. STEPAN-SARKISSIAN[1]

1 Introduction

A large number of carbon sources have now been tested for their ability to support the growth of plant cell cultures (Stepan-Sarkissian and Fowler 1984). The materials tested range from refined sources such as sucrose, glucose, and galactose to often complex carbohydrate containing waste materials such as milk whey and molasses (Table 1). They also cover a wide spectrum in regard to structure, properties, and molecular size. Of all the different carbon sources tested, however, none has been found to be as effective as sucrose or glucose in supporting both high growth rates and biomass yields. The use of these sugars also tends to result in high carbon conversion efficiency, often approaching and exceeding levels of 50%. In spite of this, it would appear that the two sugars, although structurally related, do not necessarily affect or influence the physiological status of all cell cultures in quite the same way. For instance in the case of *Catharanthus roseus* cell cultures maximum biomass yields were achieved in shorter periods of time with glucose than with sucrose (Fowler 1982). At the same time the efficiency of sucrose utilisation declined with increasing sugar concentration. In contrast, it has been shown that sycamore cells will utilise sucrose or glucose with equal facility (Simpkins et al. 1970). In this case the biomass yield and conversion efficiency rose [up to 6% (w/v) sucrose] as the level of sucrose supplied was increased.

In recent years an additional and recurring observation has been that sucrose appears to have a more pronounced effect on natural product synthesis than glucose at an

Table 1. Carbon sources tested for their ability to support cell growth and division

Arabinose	Inositol	Raffinose
Cellobiose	Inulin	Rhamnose
Dulcitol	Lactose	Ribose
Fructose	Maltose	Sorbitol
Galactose	Mannitol	Stachyose
Glucose	Mannose	Starch
Glycerol	Melibiose	Trehalose
High fructose syrup	Molasses	Xylose

1 Wolfson Institute of Biotechnology, The University of Sheffield, Sheffield, S10 2TN, United Kingdom

Primary and Secondary Metabolism of Plant Cell Cultures, ed. by Neumann et al.

Table 2. Examples of cell cultures where sucrose supports a "high" level of natural product synthesis

Species	Product	Sucrose conc. range (% w/v)	Ref.
Populus sp.	Anthocyanins	1–5	Matsumoto et al. 1973
Acer pseudoplatanus	Lignin	2–15	Carcellar et al. 1971
Morinda citrifolia	Anthraquinone	2	Zenk et al. 1975
Nicotiana tabacum	Ubiquinone	2–5	Ikeda et al. 1976
Coptis japonica	Berberine	3–10	Sato and Yamada 1984
Digitalis purpurea	Digitoxin	3–10[a]	Hagimori et al. 1982
Vitis	Anthocyanin	1.5–10[b]	Yamakawa et al. 1983
Rosa sp.	Polyphenols	2–4	Davies 1972

[a] Declined beyond 3% initial sucrose
[b] Increased markedly between 5%–10% initial sucrose

equivalent initial concentration (Table 2). Quite obviously such observations are important in the development of our understanding of cell culture physiology and biochemistry and may be of great use as a means of manipulating the carbon source in both quality and quantity to acheive higher natural product yields. The question then arises as to what differences exist between sucrose and glucose, which in turn mark out sucrose in some way as being different? This short article sets out to explore those differences, in so far as this is possible with the limited data available. The areas for discussion are as follows: structural aspects and properties, cellular uptake, and finally intracellular utilisation.

2 Structural and Related Properties of Sucrose and Glucose

Sucrose and glucose are of course structurally closely related and their structures well established. Sucrose is a non-reducing disaccharide of glucose and fructose, and glucose is a six-carbon reducing sugar. Some workers view sucrose as a protected form of glucose, possessing as it does a glucose unit joined to a fructose unit through an anomeric link between carbon 1 of glucose and carbon 2 of fructose. This linkage creates a non-reducing system. This is in contrast to the other fairly common disaccharide, maltose, which is a glucose disaccharide and possesses a free aldehyde group. The question arises as to whether or not the structural form of sucrose and the combination of being a disaccharide and a non-reducing sugar conveys to the molecule a uniqueness which allows it to take part in biochemical/physiological control mechanisms, perhaps in concert with other molecules, not possible with even closely related sugars? (Such a situation would also of course have to take into account the rôle of sucrose as the major translocatable carbon source of plants).

In the 1960's there were a number of suggestions that sucrose may act as a phytohormone (Carceller et al. 1971). While there would not appear to be a great deal of support today for this view, there are nonetheless indications of complex interactions between sucrose and both synthetic and natural plant growth regulators. For example while sucrose utilisation was negatively affected by increasing kinetin levels in sycamore cell cultures, glucose utilisation was not (Simpkins et al. 1970). At the same time cell yield and the degree of cell clumping were significantly affected by high kinetin and 2,4-D levels in sucrose-based cultures. A number of similar effects with the natural auxins are also reported in the literature. Little is understood of the underlying mechanism behind these observations.

There is a further aspect to this line of discussion. Sucrose is, of course, the major translocate of higher plants. A number of workers have argued that glucose is unsuitable for such a role because of its rapid and wide-ranging pattern of utilisation. Is there a possibility therefore that because of its specific and to a degree unique function in long-range carbon transport, sucrose may also act in a "signal" or "message" system in combination with other "information molecules?"

One further possible difference between glucose and sucrose which may be of importance relates to the difference in osmotic potential of equimolar concentrations of the two sugars. Unfortunately no information relating osmotic phenomena to either biomass yield or natural product synthesis is available.

3 Uptake Mechanisms for Sucrose and Glucose

Fairly extensive studies have been carried out into the mode of cellular uptake of both glucose and sucrose by cell cultures and certain differences are obvious from the data (for review see Stepan-Sarkissian and Fowler 1984).

In the case of glucose, rapid intracellular utilisation of the sugar makes it particularly difficult to determine the mechanism of uptake. Two basic models are extant, one involves "active" movement across the plasmalemma (Maretzki and Thom 1978) while the other suggests that glucose enters the cell passively (Parr and Edelman 1976; Opekarova and Kotyk 1973) (Fig. 1). Unfortunately, the experiments suggesting an "active" mechanism of uptake were carried out with suspension cultures of *Saccharum* sp., while those indicating passive uptake were carried out with callus cultures of *Daucus carota* and *Nicotiana tabacum*, making strict comparison difficult.

External medium	Cell wall	Plasmalemma	Cytosol
a) Glucose →	→	Permease →	Glucose + ATP ↓ hexokinase Glucose-6-P
b) Glucose →	→	Permease + "Kinase" →	Glucose-6-P

Fig. 1. Possible mechanisms for the "active" uptake of glucose into plant cells

From their work on active uptake, Maretzki and co-workers (see Maretzki et al. 1974) proposed two possible mechanisms for glucose entry. The first envisages uptake of the hexose sugar across the plasmalemma, mediated through an energy-driven process, followed by cytosolic phosphorylation, presumably by hexokinase, to glucose-6-phosphate. A second possibility is of a combined "permease" plus "phosphorylase" on the plasmalemma. Both mechanisms have precedents in work from bacterial systems. "Soluble" and "particulate" hexokinase activities have both been reported in cell cultures. Detailed studies by Fowler and Clifton (1974, 1975) with *Acer pseudoplatanus* cell suspension cultures, however, failed to show any hexokinase activity associated with the plasmalemma. While a not insubstantial level of hexokinase activity was found to be "particulate", it was predominantly associated with the mitochondrial fraction. It is difficult to see how this "mitochondrial" enzyme could be involved in a phosphorylative entry of glucose into the cell through the plasmalemma. A "passive" uptake of glucose envisages entry of the hexose "on demand" into the cell, the rate and phasing depending upon metabolic activity. Inoculation of cells into fresh glucose-containing medium presumably results in diffusion of glucose into the cells, which in turn fuels increased metabolic activity. Increased activity rapidly results in glucose utilisation and so further glucose enters the cell down a diffusion gradient.

The situation or situations which may pertain for sucrose are in marked contrast to glucose, and are potentially much more complex. Mechanisms of sucrose entry into the cell can broadly be divided into two categories, hydrolytic and non-hydrolytic (Fig. 2). There are numerous examples of sucrose hydrolysis occurring prior to uptake in a wide range of cell culture systems. This probably represents the most simple system, where hydrolysis takes place in the external medium, on the cell-wall surface or within the cell wall. The indications from the literature and from our own work are that total external hydrolysis of sucrose is not an essential prerequisite for cell growth and metabolism. However, the rapidity with which, and level to which hydrolysis may occur is to be seen in Fig. 3. Here *Catharanthus roseus* cells were supplied and grown under

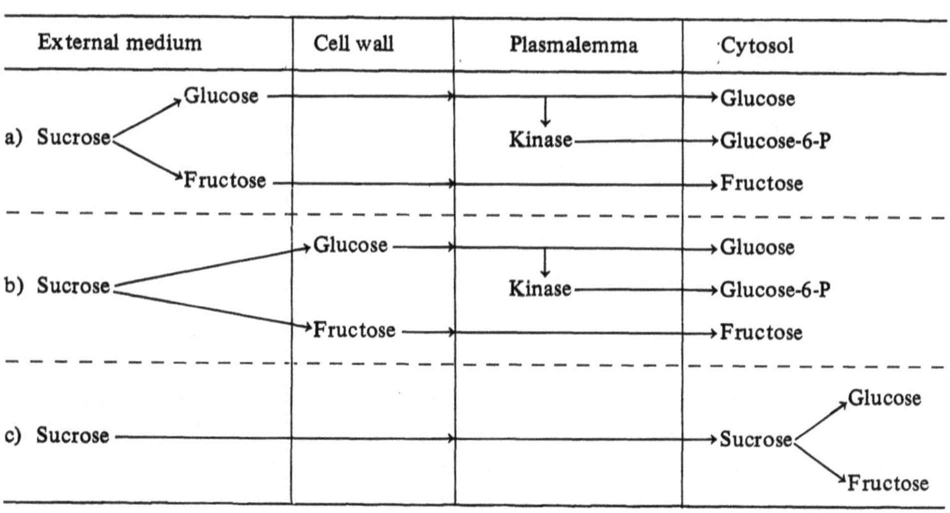

Fig. 2. Possible routes of sucrose uptake into plant cells

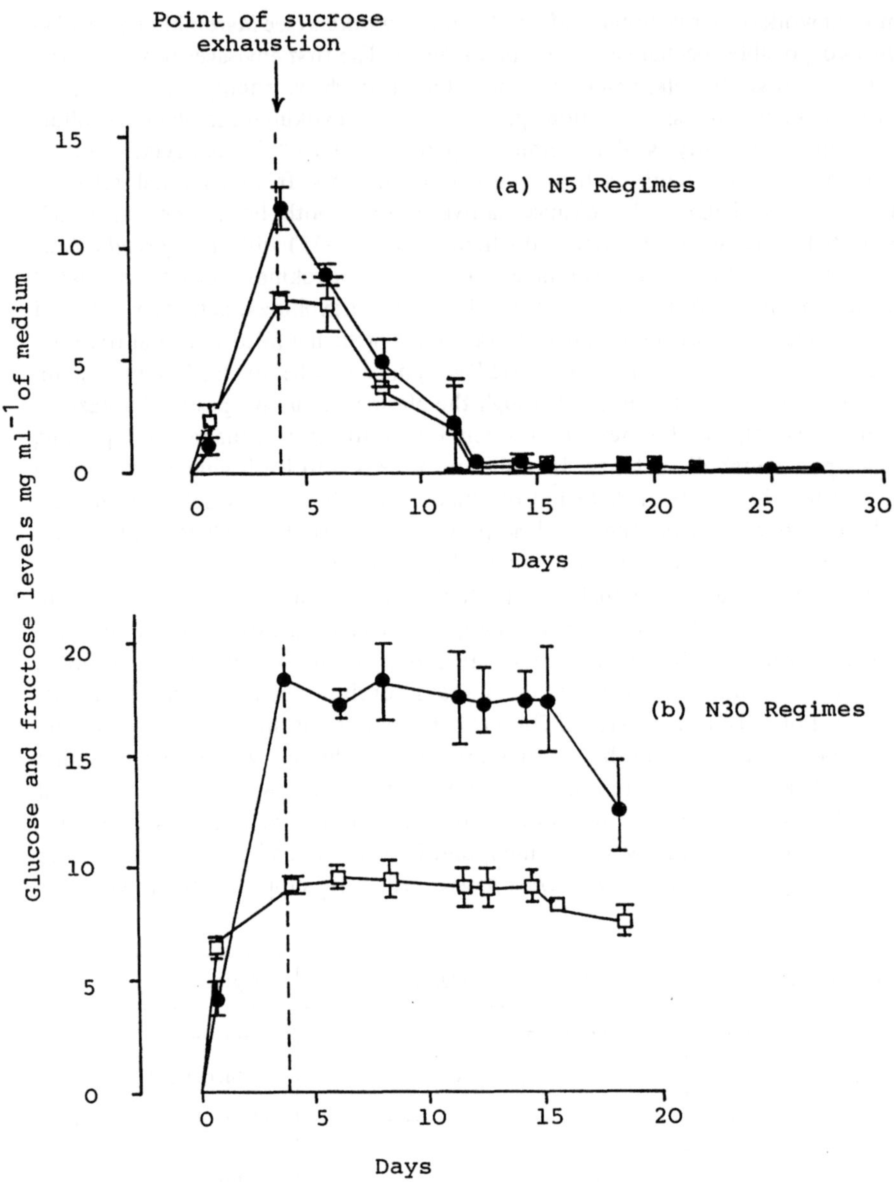

Fig. 3. Medium content of glucose and fructose in sucrose-fed cultures of *Catharanthus roseus.* Key: • fructose; □ glucose

either nitrogen limitation (N_5) or carbon limitation (N_{30}) with sucrose at either 2% (N_{30}) or 3% (N_5). Measurements were made of total sucrose in the external medium together with glucose and fructose levels. Two points are immediately obvious; first there is a rapid and consistent decline of sucrose level in the medium such that the disaccharide supply is exhausted in about 4 days, secondly almost as soon as cell are inoculated into fresh medium, hydrolysis of sucrose takes place to its constituent monosaccharides. Two further points also arise. First the rate of hydrolysis would appear to be initially at least, independent of the cellular demand for carbohydrate, secondly in excess of 90% of the medium sucrose is probably hydrolysed before cell entry. (The only cell lines which so far seem to differ from this situation are of *Saccharum* and *Daucus*, both of which store sucrose in the whole plant). Once released by hydrolysis the monosaccharides may enter the cell through a variety of mechanisms; as mentioned previously for glucose and further outlined in Fig. 2.

Non-hydrolytic uptake of sucrose involves the intact molecule passing through the cell wall and plasmalemma with inversion or other modification occurring intracellularly. There is definitive evidence for such a mechanism in carrot callus cultures, although, as mentioned above, this species stores sucrose naturally and may therefore have different carbohydrate physiology. In our own experience with species which habitually store starch, it is unlikely that cell cultures derived from these species would take up more than a very small amount of sucrose as the intact molecule. There is a possibility also that in some cases the ingoing sucrose may be phosphorylated during its passage through the bounding layers of the cell. There is no clear evidence of this occurring.

From the large amount of evidence available it is certain that external hydrolysis of sucrose is effected by a cell wall located invertase. Copping and Street (1972) demonstrated the presence of both "acid" and "neutral" invertases in the cell wall of suspension-cultured sycamore cells and also observed that the enzyme activity was tightly bound to the cell wall matrix. In work with protoplasts of carrot cells we (Watson and Fowler 1982, unpublished data) have shown that the vast majority of "external" invertase activity is lost once the cell wall has been removed. This activity subsequently returns with cell wall regeneration.

That the "external" hydrolysis of sucrose occurs to a large degree independently of cell physiology is supported by data contained in the two plots in Fig. 3. The pattern and rate of release of the two monosaccharides into the medium is very similar up to the point of sucrose exhaustion, although of course the medium levels of the sugars are lower in carbon-limited N_{30} cultures. Also there are strong indications in both nutrient regimes that fructose is taken up into the cell much less rapidly than glucose. A number of questions arise from this observation. Is glucose utilised preferentially to sucrose; what happens if cells are grown with fructose as sole carbon source or if sucrose is replaced by equimolar amounts of glucose and fructose? In experiments where we grew cells of *Catharanthus roseus* on fructose, the first passage has typically given growth rates with biomass yields equivalent to these observed with glucose and sucrose. However, transfer to a second passage often results in total lack of growth (Fowler 1982). In some exceptional cases, growth has subsequently been re-established and the cultures have become fructose adapted. Where glucose and fructose have been supplied to cultures of *Catharanthus roseus* in place of sucrose, a similar situation to that seen from the data in Fig. 3 has been indicated. No comparable information is available relating to natural product synthesis.

4 Metabolism of Sucrose, Glucose, and Fructose

Few detailed studies have been carried out on the metabolism of these sugars subsequent to uptake by cell cultures. In the case of sucrose, it is more than likely, as outlined above, that the disaccharide enters the cell as its monosaccharide constituents. Analyses of our own cell lines of *Catharanthus roseus* and *Nicotiana tabacum* suggest that once hydrolysed externally, the sucrose is not totally reformed, much of the released glucose and fructose passing directly into the central metabolic pathways of the cell or being converted into storage polymer as starch. The route into the central pathways for glucose is via hexokinase and glucose-6-phosphate. From here the glucose units can enter rapidly into a myriad of reactions. The situation for fructose is not well understood. The K_m of hexokinase for fructose is known to be higher than for glucose, and this may to a degree explain the differential between glucose and fructose uptake noted earlier. Alternatively fructose may enter metabolism via a specific fructokinase with the synthesis of fructose-1-phosphate, a reaction known from animal systems. It is doubtful if the "influence" of sucrose in natural product synthesis is operational at this level, since these primary pathways of entry are essentially ubiquitous to cellular metabolism in general and secondary metabolite synthesis in particular. It is noteworthy in this context that the effects noted for sucrose-promoted enhancement of natural product synthesis do not appear to be confined to any one group of natural products.

5 Concluding Remarks

As will be obvious from the preceding discussion, the data available do not allow formulation of a precise hypothesis as to the nature of the apparent interrelationship between sucrose availability and natural product synthesis. On the other hand, one can begin to see possible openings through which experimental explorations may be mounted. These suggest themselves as follows:

1. Does sucrose exert its influence through differential osmotic potential? (It is of interest to note that the disaccharide maltose does not appear to have the same level of effect on natural product synthesis as sucrose).
2. Is the conformation of the sucrose molecule (perhaps as "protected" glucose) of crucial importance?
3. Does the sucrose molecule interact, and if so in what way, with plant growth regulators?
4. Does glucose have any interactive role as in (3) above?
5. Is the role of fructose and its position in the sucrose molecule crucial? (Experiments where glucose and fructose replaced sucrose tend to indicate that this might be so).
6. Does sucrose exert its influence extra- or intracellularly? (It would appear that the former is more likely, as most of the sucrose is hydrolysed to glucose and fructose, prior to entry into the cell).

Quite obviously these are but a few of the questions which could be asked. It is hoped, however, that they will stimulate discussion and investigation and will help to provide insights into whether or not the "sucrose effect" is real, and if so, what is the mechanism of its operation.

Acknowledgements. The authors wish to express their thanks Dr. Stafford and Mr. Lyons for much valuable discussion and the provision of information.

References

Carcellar M, Davey MR, Fowler MW, Street HE (1971) The influence of sucrose, 2,4-D and kinetin on the growth, fine structure and lignin content of cultured sycamore cells. Protoplasma 73:367

Copping L, Street HE (1972) Properties of the invertases of cultured sycamore cells and changes in their activity during culture growth. Physiol Plant 26:346

Davies ME (1972) Polyphenol synthesis in cell suspension cultures of Paul's Scarlet rose. Planta 104:50

Fowler MW (1982) Substrate utilisation by plant cell cultures. J Chem Tech Biotechnol 32:338

Fowler MW, Clifton A (1974) Activities of enzymes of carbohydrate metabolism in cells of *Acer pseudoplatanus* L. maintained in continuous (chemostat) culture. Eur J Biochem 45:445

Fowler MW, Clifton A (1975) Hexokinase activity in cultured sycamore cells. New Phytol 75:533

Hagimori M, Matsumoto T, Obi Y (1982) Studies on the production of *Digitalis* cardenolides by plant tissue culture III. Effects of nutrients on digitoxin formation by shoot-forming cultures of *Digitalis purpurea* L. grown in liquid media. Plant Cell Physiol 23:1025

Ikeda T, Matsumoto T, Noguchi M (1976) Effects of nutritional factors on the formation of ubiquinone by tobacco plant cells in suspension culture. Agric Biol Chem 40:1765

Maretzki A, Thom M (1978) Characteristics of a galactose adapted sugar cane cell line grown in suspension culture. Plant Physiol 61:544

Maretzki A, Thom M, Nickell LG (1974) Utilisation and metabolism of carbohydrates in cell and callus cultures. In: Street HE (ed) Tissue culture and plant science 1974. Academic Press, London New York, pp 329–361

Matsumoto T, Nishida K, Noguchi M, Tamaki E (1973) Some factors affecting the anthocyanin formation by *Populus* cells in suspension cultures. Agric Biol Chem 37:561

Opekarova M, Kotyk A (1973) Uptake of sugars by tobacco callus tissue. Biol Plant 15:312

Parr D, Edelman J (1976) Passage of sugars across the plasmalemma of carrot callus cells. Phytochemistry 15:619

Sato F, Yamada Y (1984) High berberine producing cultures of *Coptis japonica* cells. Phytochemistry 23:281

Simpkins I, Collin HA, Street HE (1970) The growth of *Acer pseudoplatanus* cells in a synthetic liquid medium: response to the carbohydrate, nitrogenous and growth hormone constituents. Physiol Plant 23:385

Stafford A, Fowler MW (1983) Effect of carbon and nitrogen growth limitation upon nutrient uptake and metabolism in batch cultures of *Catharanthus roseus* L. A. Don. Plant Cell Tissue Org Cult 2:239

Stepan-Sarkissian G, Fowler MW (1984) The metabolism and utilisation of carbohydrates by suspension cultures of plant cells. In: Morgan MJ (ed) Carbohydrate metabolism in cultured cells. Plenum Press, New York, London, in press

Yamakawa T, Kato S, Ishida K, Kodama T, Minoda Y (1983) Production of anthocyanins by *Vitis* cells in suspension culture. Agric Biol Chem 47:2185

Zenk MH, El-Shagi H, Arens H, Stöckigt J, Weiler EW, Deus B (1975) Formation of the indole alkaloids serpentine and ajmalicine in cell suspension cultures of *Catharanthus roseus*. In: Barz W, Reinhard E, Zenk MH (eds) Plant tissue culture and its biotechnological application. Springer, Berlin Heidelberg New York, pp 27–43

Nitrogen Metabolism of Leaf and Microspore Callus of *Betula pendula*

LIISA KAARINA SIMOLA [1]

1 Introduction

The main problems in the in vitro propagation of trees are principally the same as in herbs, i.e., induction of callus, cultivation of subculturable callus lines and induction of differentiation. In this paper, the second problem has been studied in more detail because poor growth of callus cultures of trees usually results in a limited quantity of material for experiments on cytodifferentiation and physiological regulation of organogenesis. Nitrogen sources have a central role in callus growth but a less well-known function in differentiation.

The importance of the right nitrogen source for good growth response has been known for a long time in plant physiology, and much interest has been paid to the activities of the key enzymes of nitrogen metabolism (e.g. nitrate reductase, glutamate dehydrogenase, and glutamate synthase) during the past 10 years. However, it is rather surprising that there are only a couple of papers in which the intracellular levels of NO_3^- and NH_4^+-ions have been analyzed in tissue cultures after cultivation on media containing different nitrogen sources. More is known about intact agricultural plants (*Glycine max*, Brevedan et al. 1977; *Zea mays*, Mengel et al. 1983) grown in solution cultures. Only the total nitrogen content has been analyzed (*Daucus carota*, Veliky and Rose 1973; *Ipomoea purpurea*, Martin et al. 1977) in the case of tissue cultures. If NH_4^+/NO_3^- was 1, there was only a slight increase in the total nitrogen content in a suspension culture of *Daucus carota* (from 50 mg N g^{-1} dry wt. to 70 mg N g^{-1} dry wt.) when the nitrogen content of the medium increased tenfold (14.3 mM N and 143 mM respectively).

During induction of callus production, many nutrient media are supplemented with a rather high level of organic nitrogen in the form of casein hydrolysate or coconut milk. After induction of callus, the cultures are cultivated on variable nutrient media usually based on M & S medium. However, the authors do not describe the medium in detail and so it is difficult to trace the presence of edamin, which is in the complete medium according to Murashige and Skoog (1962). This medium contains a high inorganic nitrogen level (60 mM N) and the ratio NH_4^+/NO_3^- is 1:2.

It is known that birch seedlings have optimal growth on a medium containing a rather low level of nitrogen (55–100 mg l^{-1}) and ammonium can be used as main nitrogen source if the pH of the medium is controlled (Ingestad 1971). The nutrient media

1 Department of Botany, University of Helsinki, Unioninkatu 44, 00170 Helsinki 17, Finland

Primary and Secondary Metabolism of Plant Cell Cultures, ed. by Neumann et al.

for callus cultures usually contain NO_3^- as the sole nitrogen source or contain ammonium only 4%–33% of the total nitrogen content. Some of the nutrient solutions contain high quantities of inorganic nitrogen (490 mg l^{-1} for *Atropa*, Wood and Braun 1961; 500 mg l^{-1} for carrot, Veliky and Rose 1973; 840 mg l^{-1} for tobacco and rose, Murashige and Skoog 1962; Bergmann et al. 1976). Thus there exists a discrepancy in the nitrogen source used for callus and in what is favourable for seedlings.

In this paper, the growth of callus has been tested using different NH_4^+/NO_3^- ratio, and the macronutrient level was used in three concentrations. Comparisons were made with cultures cultivated on a modified N_6 medium containing 490 mg l^{-1} (35 mM) nitrogen.

The callus cultures of *Betula pendula* have been mainly cultivated on media having a high salt content but induction of differentiation is difficult (Huhtinen 1976; Srivastava and Steinhauer 1981). Therefore, the macronutrient medium in which the ion balance and nitrogen sources simulate those favourable for seedlings were studied in this paper and the intracellular content of NH_4^+ and NO_3^- was analyzed. If the callus cultures of trees respond to mineral nutrients, as do seedlings, they could be used as good test items for screening the best combinations of fertilizers.

2 Materials and Methods

2.1 Cultivation of Callus

The new macronutrient medium was designed so that a stock solution could be made in 10 × concentration, and so that media containing different nitrogen sources could be obtained with the addition of a few nitrogen-containing salts (NH_4Cl, $NaNO_3$ or NH_4NO_3), all without changing the ratio of macronutrient and without difficult pH adjustments. The basic solution already contains a small amount of nitrate. The concentration of the macronutrient solution which can be diluted or used at a higher concentration has been given for a solution containing 36 mM (504 mg l^{-1}) nitrogen (Table 1). The amount of total nitrogen was adjusted so that it was 3.6 times the sum

Table 1. Macronutrient media with different nitrogen sources (basal media 36 mM N, 504 mg N l^{-1})

Nutrient salts	mg l^{-1}	mM	NH_4^+/NO_3^- 5:1	NH_4^+/NO_3^- 1:1	NO_3^-
KNO_3	404	4	+	+	+
KH_2PO_4	272	2	+	+	+
K_2SO_4	174	1	+	+	+
$Ca(NO_3)_2$	164	1	+	+	+
$MgSO_4$	241	2	+	+	+
$NaNO_3$	2,550	30			+
NH_4Cl	1,605	30	+		
NH_4NO_3	960	12		+	
NH_4Cl	321	6		+	

of molarity of K and P. $NaNO_3$ was added to the medium containing only NO_3^- as a nitrogen source, NH_4NO_3 and NH_4Cl were added to obtain medium with NH_4^+ and NO_3^- in equal proportion, and NH_4Cl added to obtain a medium containing NH_4^+/NO_3^- 5:1. The total nitrogen contents of the media tested were 25, 50, 100, and 200 mg l^{-1} (1.8, 3.6, 7.2, and 14.3 mM N respectively). All the media contained macronutrients in the following ratios: N 3.6:K 0.8:P 0.2:Mg 0.2:Ca 0.1:S 0.3. The ratios are very near those of Ingestad (1971), but the medium is more simple than his.

The micronutrients and vitamins were according to Murashige and Skoog (1962) but iron was added as FeNaEDTA (50 mg l^{-1}). The medium contained: 20 g l^{-1} sucrose, 6 gl^{-1} agar, 100 mgl^{-1} myoinositol and kinetin (0.5 mg l^{-1}), and 2,4-D (2 mg l^{-1}) as auxin. The pH was adjusted to 5.6 before autoclaving.

N_6 macromineral solution containing 490 mg N l^{-1} (35 mM, NH_4^+/NO_3 1/4) (Chu et al. 1975) was used as the control.

The callus was of leaf or microspore origin. Five or seven pieces of callus were transferred to plastic Petri dishes (ϕ 9 cm) containing the autoclaved media. Each experiment included five replications and the cultures were incubated at room temperature (22°–24 °C) in rhythmic fluorescent light (18 h) for 10 weeks. After this, the material was rinsed with distilled water and lyophilized for 2 days. The lyophilized weight was used as dry weight.

2.2 Ammonia, Nitrate, and Amino Nitrogen Assays

Powdered tissue freeze dried (20 mg) was extracted twice with 0.01 N HCl. The supernatant was used for analyses. Ammonia was estimated by a modification of the Weatherburn method (1967). $(NH_4)_2SO_4$ was used as standard. Nitrate was analyzed, after reduction to nitrate using zinc dust, in an alkaline solution (Young 1973). Nitrate was estimated according to Filner (1966). $NaNO_3$ was used as a reference. Amino nitrogen was analyzed according to Yemm and Cocking (1955). Alanine was used as the reference.

3 Results

3.1 Growth of Callus Strains

The growth of microspore callus was faster than that of leaf origin, but it had a rather similar relative growth response to different salt concentrations and nitrogen sources. A high salt solution containing NO_3^- as the main nitrogen source was the most favourable (Fig. 1). The possible effects due to changes in the pH of the agar medium could not be eliminated. The low salt solution containing only 25 mg l^{-1} (1.8 mM) nitrogen mainly in the form of ammonium did not promote growth. The same ratio (5:1) of NH_4^+/NO_3^- gave nearly linearly increasing growth when higher concentrations were tested, apparently due to the increasing amount of the total content of NO_3^-, which was the best nitrogen source for the microspore callus at all concentrations tested (Table 2).

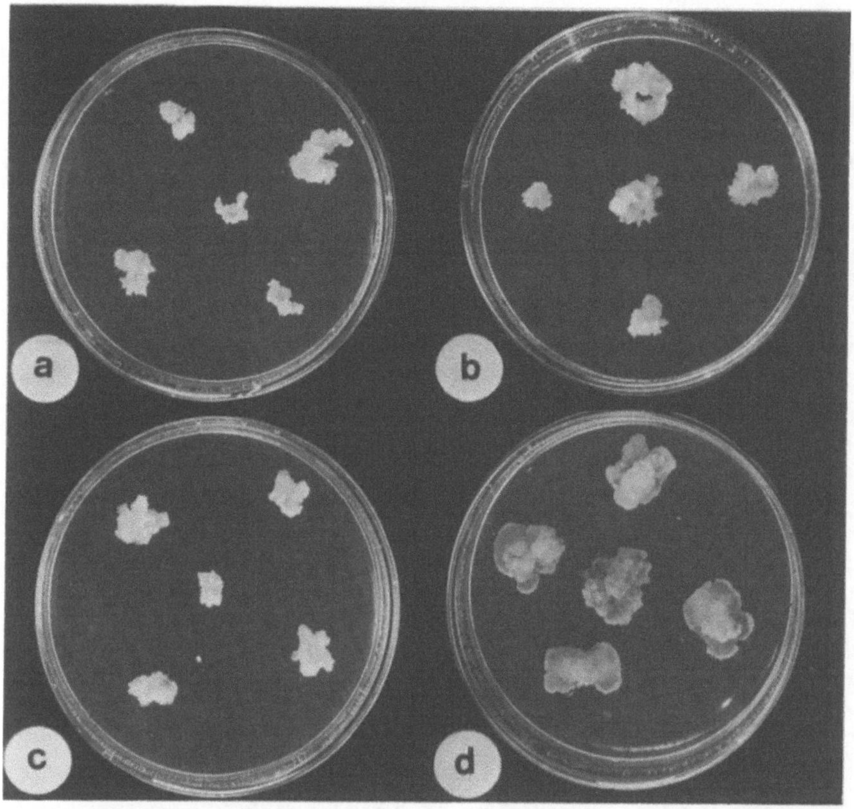

Fig. 1. Growth of microspore callus of ***Betula pendula***, on nutrient media having different salt concentrations and NO_3^- as the sole source of nitrogen. Macronutrient elements in the same proportions. The concentration of nitrogen: **a** 50 mg l^{-1}; **b** 100 mg l^{-1}; **c** 200 mg l^{-1}; **d** N_6 medium 490 mg l^{-1}

Table 2. The effect of nutrient level and different NH_4^+/NO_3^- ratios on the growth of leaf and microspore callus of *Betula pendula*

N mg l^{-1}	Strain	NH_4^+/NO_3^- 5:1	NH_4^+/NO_3^- 1:1	NO_3^-
25	L	–	–	26.7
	M	–	12.4	34.8
50	L	5.9	17.7	39.4
	M	9.5	15.9	43.5
100	L	9.2	20.7	46.7
	M	18.1	48.0	65.0
200	L	21.8	59.0	37.5
	M	33.1	66.1	84.7

L = leaf callus; M = microspore callus. Results in percentages of the controls, 490 mg l^{-1} N (NH_4^+/NO_3^-, 1:4)

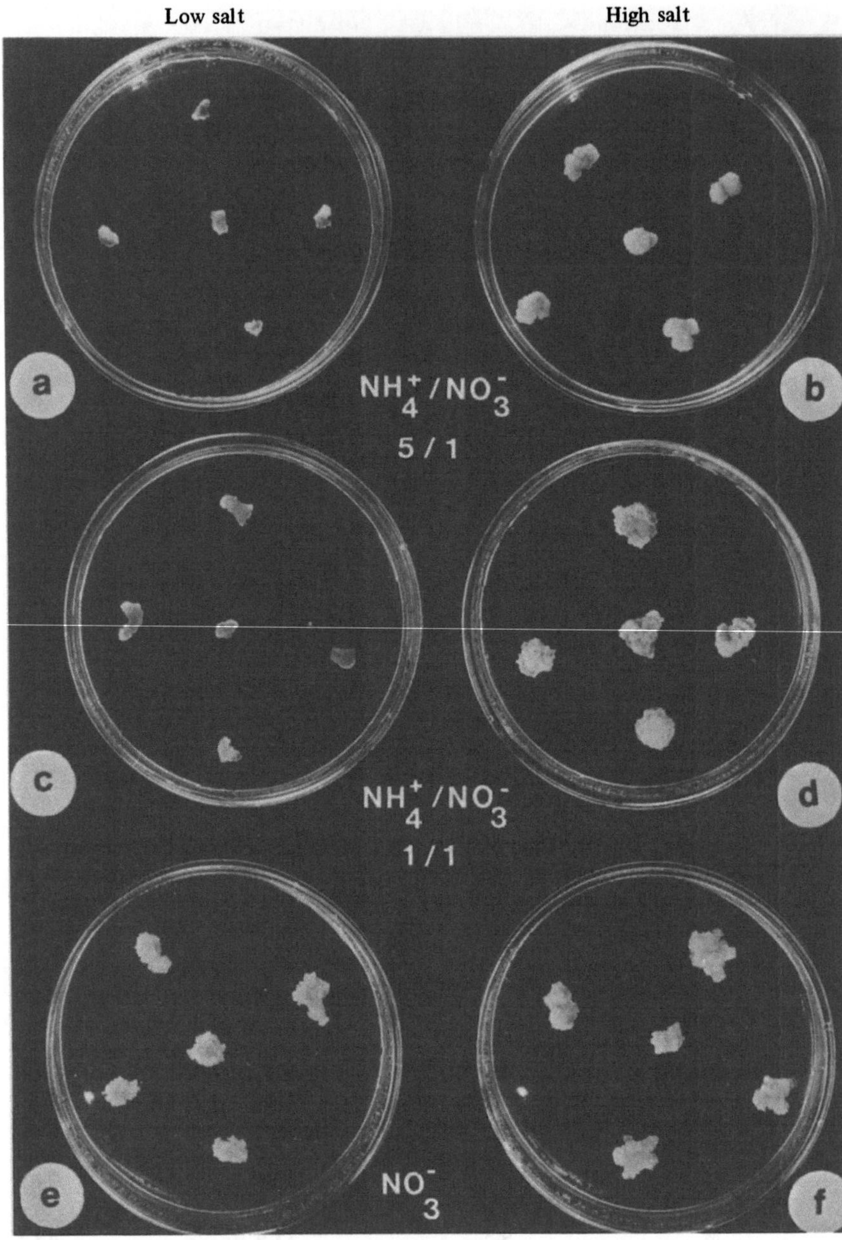

Fig. 2. Growth of microspore callus cultures of *B. pendula*, on nutrient media having different concentrations of nitrogen and NH_4^+/NO_3^- ratios. **a**, **c**, and **e** 25 mg l^{-1} N; **b**, **d**, and **f** 200 mg l^{-1} N

Low salt High salt

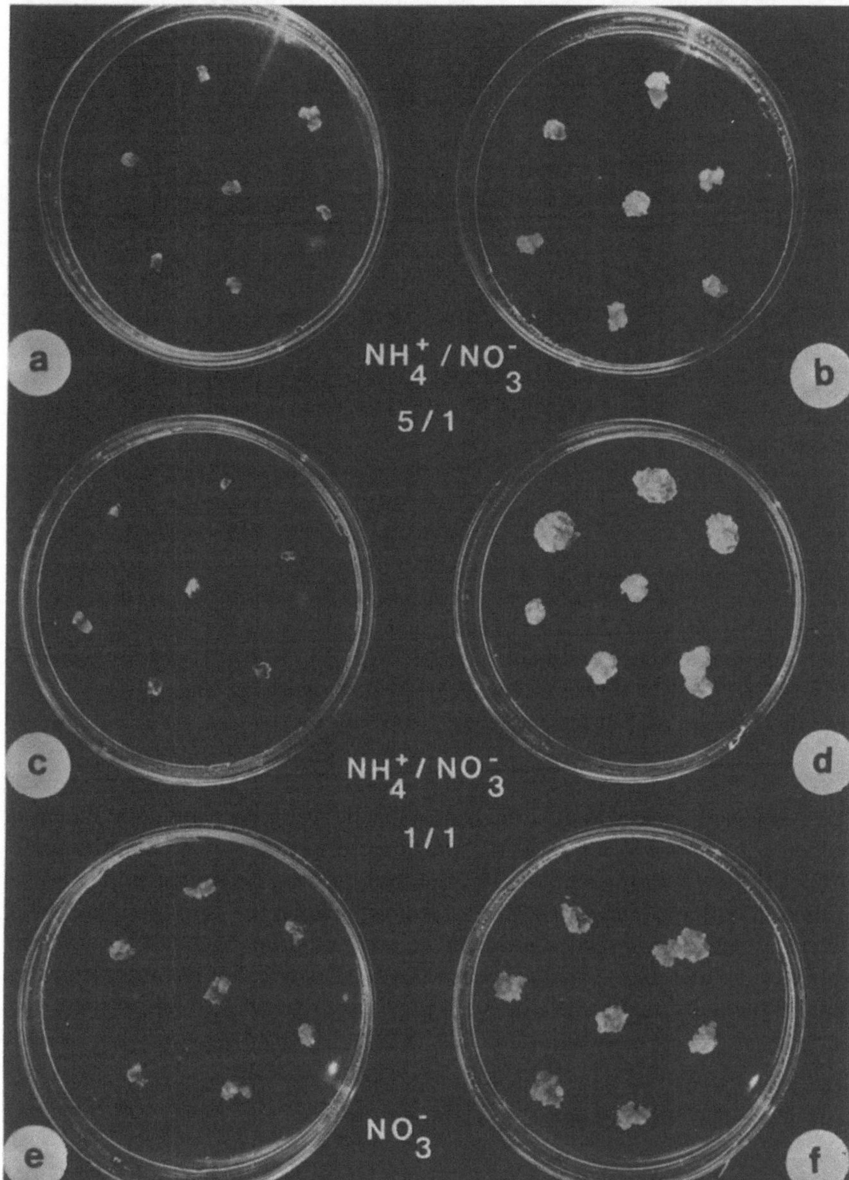

Fig. 3. Growth of leaf callus cultures of *Betula pendula* f. *purpurea* on nutrient media containing different concentrations of nitrogen and having different NH_4^+/NO_3^- ratios. **a**, **c**, and **e** 25 mg l^{-1} N; **b**, **d**, and **f** 200 mg l^{-1} N

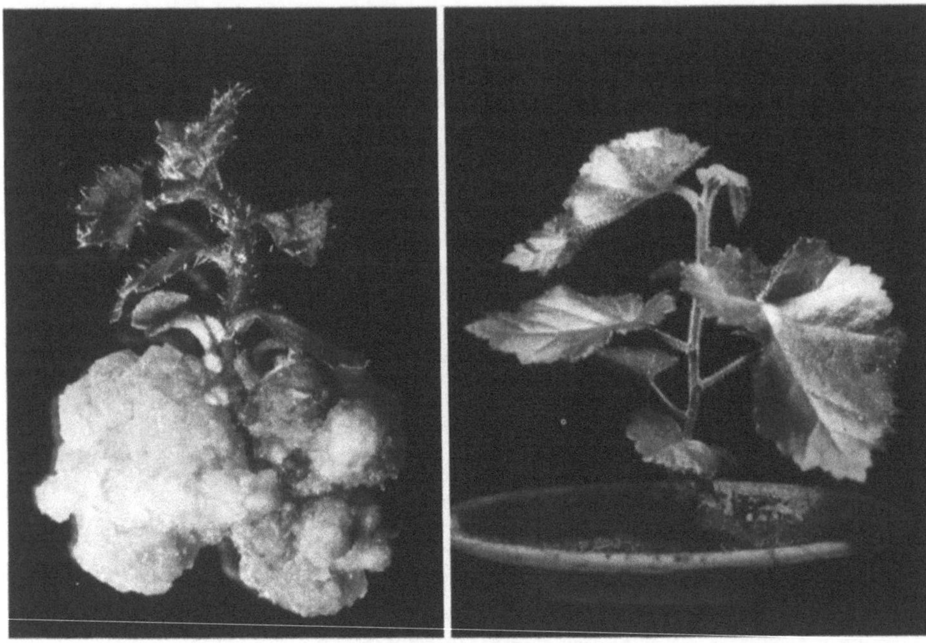

Fig. 4. Plantlet formation following subculture of leaf-derived callus of *Betula pendula* f. *purpurea* on a high salt medium containing zeatin (10 mg l^{-1}) and NOA (0.2 mg l^{-1}). × 4

Fig. 5. Tissue culture derived plant under green house conditions. 1:1

When the NH_4^+/NO_3^- ratio was 1:1, the leaf callus did not grow if the total nitrogen content was only 1.8 mM (25 mg l^{-1}), but the microspore callus could use it to some extent. Both calli grew well if a medium salt solution (200 mg N l^{-1}) containing NH_4^+/NO_3^- 1:1 was used. This medium was better for leaf callus than the corresponding NO_3^--medium. The leaf callus grew significantly better in the modified high salt N_6 medium containing nitrate as the main nitrogen source. Leaf callus was able to form shoots on this medium, if zeatin or zeatin riboside were used as cytokinin (Simola, in press and Figs. 4 and 5).

3.2 Content of Soluble Nitrogen in Callus Cultures

Both callus strains of *Betula pendula* accumulated relatively much ammonium N in the cells if the ammonium ions dominated in the nutrient medium (Table 3). Nitrate was found at even lower concentration than its relative proportion indicated. It cannot be concluded if the uptake of nitrate is repressed or if it is reduced to ammonium nitrogen. The amount of amino acid nitrogen was higher in leaf callus than in the microspore callus. The slow growth rate of callus tissue is apparently not due to repression of amino acid synthesis, because the leaf callus contains the same amount of amino nitrogen when it grows rapidly on nitrate medium, as when it grows slowly on a medium

Table 3. The effect of the nitrogen sources of the nutrient medium (200 mg N l^{-1}) on the nitrogen content (mg g^{-1} dry wt.) of *Betula*

Medium	Strain	NH_4^+	NO_3^-	Amino-N	Soluble N
NH_4^+/NO_3^- 5:1	L	1.78	0.06	0.64	2.48
	M	1.76	0.03	0.22	2.01
NH_4^+/NO_3^- 1:1	L	0.13	0.08	1.34	1.55
	M	0.65	0.07	0.72	1.44
NO_3^-	L	0.12	0.14	0.60	0.86
	M	0.09	0.16	0.51	0.76

L = leaf callus of *B. pendula* f. purpurea; M = microspore callus of *B. pendula*

containing mainly ammonium as a nitrogen source. However, there may be qualitative differences in the amino acid pools (cf. carrot explants, Craven et al. 1972).

If ammonium and nitrate ions are given in equal concentration, the microspore callus accumulates more ammonium nitrogen than the leaf callus. This is apparently due to a more effective uptake of ammonium. Amino-N concentration in microspore callus is lower than in leaf callus (Table 3). Microspore callus may be more tolerant towards high intracellular ammonium content than the leaf callus and less effective in the amino acid synthesis.

Nitrate nitrogen does not accumulate in high concentration in the cells. When ammonium and nitrate ions occur in about equal intracellular concentration, the growth rate is good in the leaf callus. The cells of *Betula* callus can tolerate nitrate stress better than a high amount of ammonium. This is the case also in conditions in which carbohydrates are available in sufficient amounts.

4 Discussion

Both callus cultures of *Betula* were able to use nitrate as the only nitrogen source. The intracellular NO_3^- levels in callus cultures of *Betula* was lower than those found in leaves of *Zea mays* (Mengel et al. 1983). The nitrate content of the tissue is not always in accordance with their nitrogenase activity (e.g. roots of *Zea mays*). This may be due to the storage of nitrate in vacuoles and only a smaller part is used as the substrate for nitrate reductase in the cytoplasm. There is general agreement that nitrate enhances nitrate reductase activity but ammonium and amino acids have the reverse effect in most cases. Interpretations of the effect of metabolites are speculative because their concentrations in the cytoplasm are not known (Mengel et al. 1983).

Tissue cultures of different plants vary widely in their tolerance of ammonium. If NH_4^+ is the dominant nitrogen source in the nutrient medium, both callus strains of *Betula* contain a relatively high level (~1.8 mg g^{-1} dry wt.) of this ion and the growth is poor (cf. Table 2). The same level of ammonium has also been found in *Spirodela oligorrhiza* if cultivated on a medium containing 8 mM N (NH_4^+/NO_3^-, 1:1) (Ferguson and Bollard 1969). The poor results with NH_4^+ as the main nitrogen source for callus of *Betula* may be due the decrease of pH in the solid medium. If pH has been frequently adjusted, the growth response of suspension cultures has been quite good (Martin et al. 1977).

The callus cultures of *Atropa belladonna* are able to grow at a rather high extracellular ammonium concentration containing the corresponding concentration of NO_3^- (Salonen 1984). In leaf callus of *Betula pendula*, a higher concentration of amino-N was found on a medium containing equal levels of NH_4^+ and NO_3^- (Table 3). Synthesis of glutamine and asparagine is associated with the tolerance of ammonium in many species (Givan 1979). Glutamic acid is one of the main components in the amino acid pool of leaves in *Betula* (Darrall and Wareing 1981).

The optimal NH_4^+/NO_3^- ratio for *Betula* seedlings (Ingestad 1971) is high compared to callus cultures. The young shoot tip explants of *Betula papyrifera* are able to use ammonium and nitrate (1–20 mM) as sole source of nitrogen (Minocha 1982). The older shoot tips of this plant were less tolerant at a high level of ammonium than the younger ones. Only a few species have so flexible nitrogen metabolism that they can effectively utilize both NH_4^+ or NO_3^- as the sole nitrogen source, and the combination of these ions gives also a good growth response (*Spirodela*, Ferguson and Bollard 1969; cell culture of *Ipomoea*, Zink and Veliky 1977).

A high salt nutrient medium containing NO_3^- as the main nitrogen source gave the best growth response in both callus strains of *Betula* (Table 2). Using this medium it was possible to induce shoot differentiation, but this depends on the good choice of growth regulators (Simola, in press). In the case of wild carrot suspension cultures, the growth and embryogenesis were stimulated if the nitrate-containing medium (12–40 mM KNO_3) was supplemented with 10 mM NH_4Cl. Urea or a suitable amino acid (e.g. glutamine and alanine) could partially replace NH_4Cl as supplement to KNO_3 (Wetherell and Dougall 1976).

5 Conclusion

Both callus lines of *Betula pendula* grew very similarly on all media tested. The growth was nearly linearly proportional to the salt concentration of the nutrient medium. The optimal salt concentration was clearly higher than that found for birch seedlings (Ingestad 1971). Nitrate as a sole source of nitrogen could better support growth of microspore callus than a combination of NH_4^+ and NO_3^- 1:1. When NH_4^+ was present as the main source of nitrogen, growth was quite poor.

The cellular level of NO_3^- was relatively low in all cultures analysed. Accumulation of NH_4^+ was very high (~130 μmol g^{-1} dry wt.), if NH_4^+ was the main nitrogen source. Leaf callus cells did not accumulate NH_4^+ ions when the growth was good, but the amino nitrogen level was rather high. The callus cultures of *Betula* respond to nitrate in a way almost similar to old shoot tips cultivated in vitro (Minocha 1982), but results are difficult to compare due to the different macronutrient media used. Nitrogen metabolism of callus cultures of *Betula* seems to be rather different from that found in intact seedlings in which photosynthesis affects the interrelationship between carbohydrate metabolism and nitrogen assimilation. The callus cells of *Betula* have usually relatively small plastids (Simola and Koskimies-Soininen 1984).

A correct understanding of the physiological and biochemical role of the components of nutrient solutions is necessary for the choice of optional conditions for effective growth of tissue cultures. It may be that regulation of organogenesis and secondary metabolism depends more on mineral nutrients than is realized at present. Tables of enzyme reactions dealing with specific aspects of nutrient assimilation and synthesis of primary and secondary metabolites will be of great value to applied plant physiology. It is surprising that many basic substrate levels in the cells are not known.

The cheapest way to obtain valuable secondary products deriving from amino acids may be to feed the cells with a good combination of inorganic nitrogen sources so that the amino acid pool in the cells is not the limiting factor of synthesis.

References

Chu CC, Wang CC, Sun CS, Hsü C, Yin KC, Chu CY, Bi FY (1975) Establishment of an efficient medium for anther culture of rice through comparative experiments on the nitrogen sources. Sci Sin 18:659–668

Craven GH, Mott RL, Steward FC (1972) Solute accumulation in plant cells. IV. Effects of ammonium ions on growth and solute content. Ann Bot (London) 36:897–914

Bergmann L, Grosse W, Koth P (1976) Einfluß von Ammonium und Nitrat auf Stickstoff-Metabolismus, Malatanhäufung und Malatenzym-Aktivität in Suspensionskulturen von *Nicotiana tabacum* var. *"Samsun"*. Z Pflanzenphysiol 80:60–70

Brevedan RE, Egli DB, Leggett JE (1977) Influence of N nutrition on total N, nitrate, and carbohydrate levels in soybeans. Agron J 69:965–969

Darrall NM, Wareing PF (1981) The effects of nitrogen nutrition on cytokinin activity and free amino acids in *Betula pendula* Roth. and *Acer pseudoplatanus* L. J Exp Bot 32:369–379

Ferguson AR, Bollard EG (1969) Nitrogen metabolism of *Spirodela oligorrhiza*. I. Utilization of ammonium, nitrate and nitrite. Planta 88:344–352

Filner P (1966) Regulation of nitrate reductase in cultured tobacco cells. Biochim Biophys Acta 118:299–310

Givan CV (1979) Metabolic detoxification of ammonia in tissues of higher plants. Phytochemistry 18:375–382

Huhtinen O (1976) Early flowering of birch and its maintenance in plants regenerated through tissue cultures. Acta Hortic 56:243–249

Ingestad T (1971) A definition of optimum nutrient requirements in birch seedlings. II. Physiol Plant 24:118–125

Martin SM, Rose D, Hui V (1977) Growth of plant cell suspension cultures with ammonium as the sole source of nitrogen. Can J Bot 55:2838–2843

Mengel K, Robin P, Salsac L (1983) Nitrate reductase activity in shoots and roots of maize seedlings as affected by the form of nitrogen nutrition and the pH of the nutrient solution. Plant Physiol 71:618–622

Minocha SC (1982) Role of the source of nitrogen in the growth of shoot tips and callus cultures of wooay plants in vitro. Colloq Int Cult In Vitro Essences Forest. IUFRO, Fontainebleau 1981, pp 307–311

Murashige T, Skoog F (1962) A revised medium for rapid growth and bioassays with tobacco tissue cultures. Physiol Plant 15:473–497

Salonen M-L (1984) Influence of ammonium and nitrate nutrition on levels of ammonium nitrate and total nitrogen in the callus of *Atropa belladonna* (Solanaceae). Ann Bot Fenn 21:367–381

Simola LK (1985) Propagation of plantlets from leaf callus of *Betula pendula* f. purpurea. Sci Hortic (in press)

Simola LK, Koskimies-Soininen K (1984) Comparison of glycolipids and plastids in callus cells and leaves of *Alnus* and *Betula*. Plant Cell Phys 25:1329–1340

Srivastava PS, Steinhauer A (1981) Regeneration of birch plants from catkin tissue cultures. Plant Sci Lett 22:379–386

Veliky IA, Rose D (1973) Nitrate and ammonium as nitrogen nutrients for plant cell cultures. Can J Bot 51:1837–1844

Weatherburn MW (1967) Phenol hypochlorite reaction for determination of ammonia. Anal Chem 39:971–974

Wetherell DF, Dougall DK (1976) Sources of nitrogen supporting growth and embryogenesis in cultured wild carrot tissue. Physiol Plant 37:97–103

Wood HN, Braun AC (1961) Studies on the regulation of certain essential biosynthetic systems in normal and crown-gall tumor. Proc Natl Acad Sci 47:1907–1913

Yemm EW, Cocking EC (1955) The determination of amino-acids with ninhydrin. Analyst 80: 209–213

Young M (1973) Studies on the growth in culture of plant cells. XVI. Nitrogen assimilation during nitrogen-limited growth of *Acer pseudoplatanus* L. cells in chemostat culture. J Exp Bot 24: 1171–1185

Zink MV, Veliky IA (1977) Nitrogen assimilation and regulation of nitrate and nitrite reductases in cultured *Ipomoea* cells. Can J Bot 55:1557–1568

V Secondary Metabolism

4-Coumarate: CoA Ligase in Wild Carrot Cell Culture Clones Which Accumulate Different Amounts of Anthocyanin

C.L. CHENG [1], D.F. WETHERELL [2], and D.K. DOUGALL [3]

1 Introduction

Single cell clones isolated from an anthocyanin-producing suspension culture of wild carrot *(Daucus carota)* differed widely in their anthocyanin accumulation under standard conditions (Dougall et al. 1980). Recloning of the high-accumulating and low-accumulating clones showed that the subclones from each displayed a wide range of anthocyanin accumulation. These and other data were interpreted as ruling out traditional mutations as the mechanism leading to the differences in anthocyanin accumulated by the cells. The differences could be accounted for by the ability of the cells to exist in one of two states of anthocyanin accumulation, either high or low, which are heritable but which interconvert at high frequency relative to the time required for cloning and analysis (Meins 1983).

While the high and low anthocyanin-accumulating subclones are not pure populations of the two proposed states of anthocyanin accumulation, they do have substantial differences (greater than 130-fold) in the amounts of anthocyanin accumulated. The difference between extreme subclones tends to decrease on maintenance by serial passage. The changes associated with maintenance by serial passage are avoided by storing the subclones in liquid nitrogen (Dougall and Whitten 1980). It is, therefore, possible to test explanations for the difference in anthocyanin accumulation by subclones.

One explanation of the difference between the extreme subclones arises out of the studies of the mechanism of gibberellic acid (GA) suppression of anthocyanin accumulation in a subline of domestic carrot *(D. carota)* by Heinzmann et al. (1977). They examined the 4-coumarate:CoA ligase in anthocyanin-containing cells and in the same culture devoid of anthocyanin by inclusion of GA in the culture medium. The enzyme from each tissue was partially purified using DEAE-cellulose and Sephadex G-100 columns and they behaved identically. The partially purified preparations differed in their K_m's with hydroxycinnamic acids as a substrates. Further, on disc gel electrophoresis the enzyme from the anthocyanin-containing cells was resolved into three peaks. That from the anthocyanin-free cells was resolved into two peaks which corresponded to the slower migrating peaks in the preparations from the anthocyanin-containing cells. The authors assume that the rapidly-migrating isozyme of 4-coumarate:CoA ligase was responsible for anthocyanin synthesis.

1 Department of Molecular Biology, Massachusetts General Hospital, Boston, MA 02114, USA
2 Biological Sciences Group, University of Connecticut, Storrs, CT 06268, USA
3 Department of Botany, The University of Tennessee, Knoxville, TN 37920, USA

Primary and Secondary Metabolism of Plant Cell Cultures, ed. by Neumann et al.

Grisebach (1981) has reviewed studies of 4-coumarate:CoA ligase and the possibilities that isozymes are involved in directing 4-coumarate:CoA to lignin biosynthesis, flavonoid biosynthesis and other pathways. Where isozymes have been separated, they differ in the rates of reaction catalyzed with different hydroxycinnamic acids and, except for the study of Heinzmann et al. (1977), were separated by chromatography on DEAE cellulose (Knobloch and Hahlbrock 1977; Grand et al. 1983).

The available data thus suggested that the difference between the high anthocyanin- and low anthocyanin-accumulating subclones of wild carrot could be due to differences in the 4-coumarate:CoA ligases in the subclones. We therefore examined this enzyme from two subclones which differed widely in the amounts of anthocyanin they accumulated.

2 Materials and Methods

2.1 Cultures

The subclones WC63-1-9 and WC63-2-2 described by Dougall et al. (1980) and Dougall and Whitten (1980) were used in this study. These subclones are stored in liquid nitrogen (Dougall and Whitten 1980). Stock cultures of each of these sublines were maintained by serial passage, as described by Dougall et al. (1980), for 6 months and then replaced by a new culture recovered from liquid nitrogen storage.

Tissue for enzyme isolation was grown in the medium for improved anthocyanin production by these cultures (Kinnersley and Dougall 1980). The inoculation and growth conditions of shake flask cultures in this medium were the same as those for stock cultures. Larger quantities of tissue were grown in a fermentor (Microferm MF 100, New Brunswick Scientific Co.) containing 8 l of medium at 25, stirred at 200 rpm using two standard flat blade turbine impellers and aerated at 3.2 L of humidified air min^{-1}. Medium for the fermentors was sterilized by filtration through a 0.22 micron filter (Minicapsule, Gelman Scientific Inc.). Gibberellic acid was filtered through a 0.22 micron filter and added asceptically to media as needed. Other media were sterilized by autoclaving at 121 °C (15 psi) for 15 min.

Dry weight of cultures and anthocyanin content were measured as described by Dougall and Weyrauch (1980).

2.2 Buffers and Solutions

All buffers and solutions were made with degased and argon-saturated, double distilled water. The solutions were made at 25 °C and stored at 4 °C. The buffers used were:

Buffer A: 0.1 M tris-HCl, 20 mM mercaptoethanol, 30% ethylene glycol pH 7.8
Buffer B: 40 mM tris-HCl, 0.1 mM dithiothreitol (DTT), 30% ethylene glycol, pH 7.0
Buffer C: 50 mM tris-HCl, 28 mM mercaptoethanol, 25% ethylene glycol, pH 7.0
Saturated $(NH_4)_2SO_4$ solution: 697 g $(NH_4)_2SO_4$ in 1 L solution, pH 7.5.

2.3 Standard Direct Spectrophotometric Assay

Enzyme activity was measured at all steps of purification by direct spectrophotometric assay. The assay mixture (1 ml) at pH 7.5, contained 0.125 M potassium phosphate, 0.25 mM cinnamic acid, 0.5 mM $MgCl_2$, 0.5 mM ATP, 0.1 mM DTT, up to 0.2 mg protein when $(NH_4)_2SO_4$ fraction was used and up to 0.05 mg protein when DEAE-cellulose purified enzyme was used.

The reaction was started by addition of CoA (0.1 mM). The blanks contained no CoA. With 4-coumaric acid as substrate, the reaction was followed at 333 nm for 5 min at 30 °C. The absorption maximum of feruloyl CoA is 345 nm, caffeoyl CoA 363 nm, sinapoly CoA 349 nm, and 346 nm for 3,4-dimethoxycinnamoyl CoA (Gross and Zenk 1966), and reactions with these acids were followed at the appropriate wavelength.

2.4 Protein Determination

Protein was determined by Bio-Rad protein microassay (Bio-Rad Laboratory Instruction Manual 1979).

2.5 KCl Gradient Determination

The conductivities of the fractions were measured and the concentrations of KCl were calculated from a standard curve.

2.6 Enzyme Purification

All the enzyme purification steps were carried out at 4 °C.

1. Enzyme Extraction. The cells were harvested by filtering through two layers of Miracloth. The cells were frozen in liquid nitrogen and stored at −70 °C. 100 g of frozen cells were stirred in 150 ml buffer A until thawed completely. The cell debris was removed by centrifugation at 20,000 *g* for 20 min. The supernatant was stirred for 20 min with 15 g of Dowex 1 × 2 (200–400 mesh) which had been equilibrated overnight in buffer A. The resin was removed by filtration through Miracloth.

2. $(NH_4)_2SO_4$ *Fractionation.* The filtrated was brought to 45% saturation by addition of saturated $(NH_4)_2SO_4$ solution, and the precipate was removed by centrifugation at 20,000 *g* for 20 min. The supernatant was adjusted to 75% saturation with $(NH_4)_2SO_4$ solution and centrifuged again at 20,000 *g* for 20 min. The precipitate was dissolved in 6 ml buffer A.

3. Column Chromatography. The 45%–75% $(NH_4)_2SO_4$ fraction was desalted by passage through a Sephadex G25 column (2.5 × 25 cm). Enzyme was eluted with buffer B, and the fractions containing enzyme activity were combined.

Pooled, desalted enzyme was applied to DE-52 columns (2.5 × 20 cm, flow rate 0.8 ml min^{-1}) preequilibrated with buffer B. The columns were washed with 200 ml buffer B, and then KCl gradients of 0–0.3 M in buffer B (250 ml each) were applied to elute the enzyme. Seven ml fractions were collected and assayed for 4-coumarate:CoA ligase activity with 4-coumaric acid, ferulic acid, and caffeic acid. The fractions with enzyme activity were pooled, concentrated by ultrafiltration (Amicon, PM10) to 20 ml and diluted with buffer B to 60 ml and reconcentrated to 20 ml so that the KCl concentration was approximately 0.03 M. Portions were stored at −20 °C.

Enzyme (5 ml), either before or after DE-25 chromatography, was applied to Blue Sepharose column (1.5 × 6 cm, flow rate 0.5 ml min^{-1}). The columns were washed with 20 ml buffer B and eluted with 25 ml of 0.2 M KCl in buffer B, followed by a 0.2–1.5 M KCl gradient in buffer B (25 ml each). Fractions of 2 ml were collected and assayed for 4-coumarate:CoA ligase activity. The fractions of the major peak were pooled and concentrated to 2 ml by ultrafiltration. The concentrated protein solution was then dialyzed against 1 L of buffer C overnight at 4 °C and stored at −20 °C.

4. Analytical-Disc Gel Electrophoresis. All solutions were made in degased and argon-saturated, double distilled H_2O. About 0.15 mg of protein per tube (0.5 × 9 cm) was subjected to polyacrylamide disc electrophoresis at a constant current of 2 mA per tube in gel systems number 1 of Maurer (1971) with minor modifications. 10% ethylene glycol was incorporated in the separation gel, and the gel was subjected to preelectrophoresis overnight at 5 mA per tube. A layer of electrode buffer with 10% ethylene glycol was gently laid on top of the enzyme solution. After electrophoresis, 2 mm slices of the gels were eluted at 4 °C by shaking overnight in 400 μl buffer B. Hydroxamic acid assay (Stadmann 1957) was used to determine the enzyme activity.

Protein bands were fixed in 20% TCA for 20 min, then stained in 0.1% Coomassie Brilliant Blue R250 in 20% TCA (Groth et al. 1963) for 30 min. The gels were destained with 7% acetic acid overnight.

3 Results

3.1 Growth and Anthocyanin Production of the Two Cell Lines

The growth curves (determined by dry weight) of the high and low anthocyanin-yielding cell lines are shown in Fig. 1. The exponential growth period was from day 1 to day 6 for WC63-1-9 and to day eight for WC63-2-2. The specific growth rates of the two cell lines were 0.333 and 0.338 day^{-1} respectively and were not different. The anthocyanin production curves of the two cell lines are shown in Fig. 1. With cell line WC63-2-2, the exponential part (days 1–8) had a slope of 0.37 day^{-1}. With the cell line WC63-1-9, the exponential part (days 1–6) had a slope of 0.742 day^{-1}.

3.2 Gibberellic Acid (GA_3) Inhibition of Growth and Anthocyanin Accumulation

After 6 days of growth in the presence of 10^{-6} M GA_3 or more, the anthocyanin level in the high anthocyanin-yielding cell line WC63-1-9 decreased from 1.61 A_{530} to 0.07 A_{530}, which is within the range of the low anthocyanin-yielding cell line WC63-2-2. The dry weight of the culture at 10^{-6} M GA_3 was 82% of the control culture.

3.3 Comparison of Growth and Anthocyanin Production of Fermentor-Grown and Batch Cultures

When 8-L cultures grown in a 15-L fermentor were compared with 250-ml batch cultures grown in 1-L Erlenmeyer flasks, both the growth (0.84 ± 0.02 versus 1.15 ± 0.02 mg dry wt ml^{-1}) and anthocyanin yield (A_{530} 0.331 ± 0.01 versus 0.86 ± 0.04) were significantly lower in the fermentor-grown cells. The lowered yield may be due to cell destruction by the turbine impeller used (Fowler 1982), or it be may due to inadequate gas exchange, which is known to affect grown (Yamakawa et al. 1983) and anthocyanin accumulation in these cell lines (Dougall unpublished).

3.4 The Activity of 4-Coumarate:CoA Ligase from Growing Cultures

The specific activity of the enzyme from day 3 to 8 after inoculation (measured after desalting) showed no significant fluctuation (Table 1). The highest total activity of 4-coumarate:CoA ligase was found in cultures harvested 6 days after inoculation. Therefore, cell cultures of this age were used for all enzyme preparations.

3.5 Enzyme Purification

The purification of the 4-coumarate:CoA ligase from the two sublines grown in the presence and absence of GA_3 is summarized in Table 2. The preparations from the cell line accumulating low amounts of anthocyanin had more protein, greater total enzymatic activity and higher specific activity than those from the cell line accumulating high amounts of anthocyanin. With both cell lines, growth in the presence of 10^{-6} M GA increased the total 4-coumarate:CoA ligase activity and protein extracted from the tissue (Table 2).

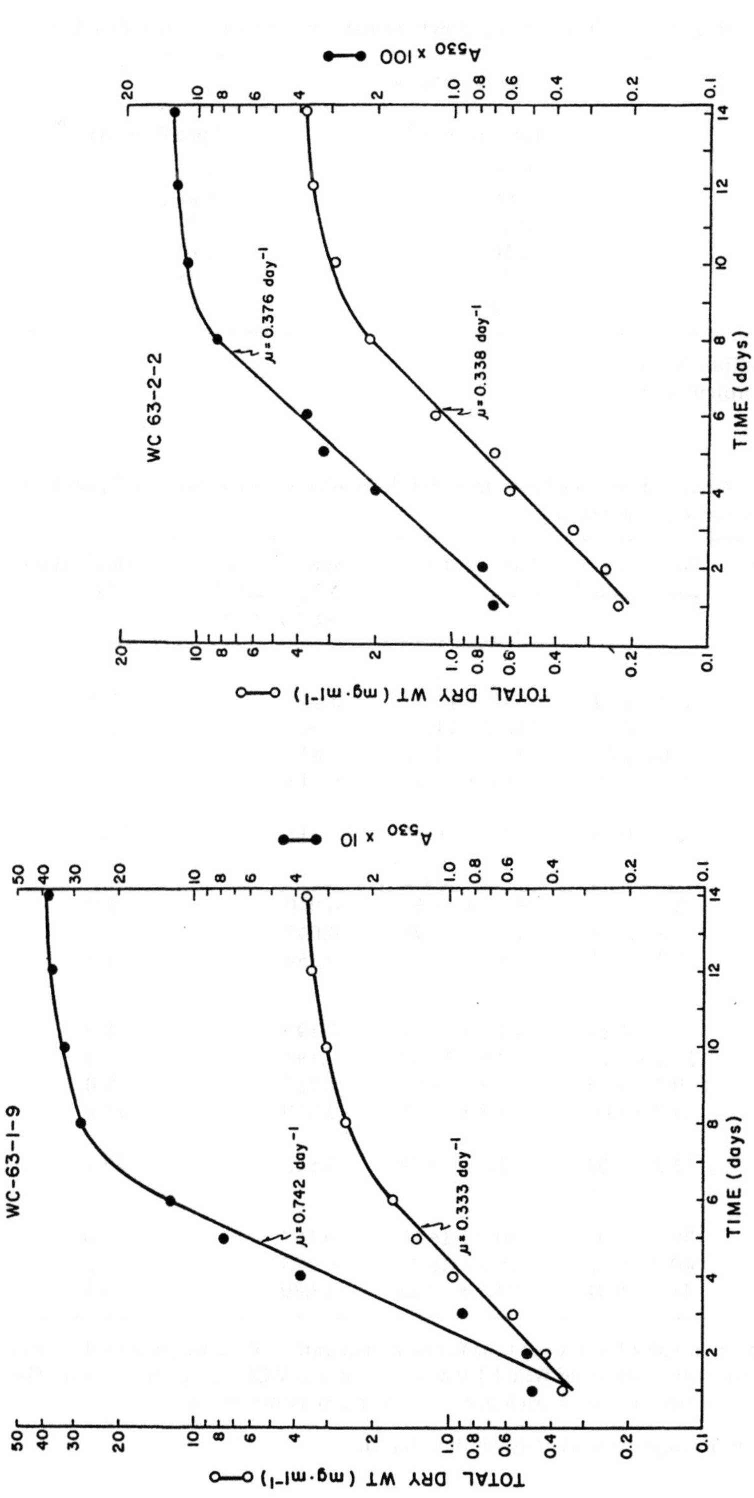

Fig. 1. The growth and anthocyanin production of high anthocyanin-yielding wild carrot cell line WC63-1-9 and low anthocyanin-yielding cell line WC63-2-2. Triplicate samples of 5 ml culture were taken from duplicate flasks for dry weight and anthocyanin determination. Each *point* represents the average of six measurements. The standard deviations are less than 15% of the means

Table 1. Activity of 4-coumarate:CoA ligase during growth of batch cultures of cell line WC63-1-9

	Enzyme activity	
Days after inoculation	Total activity[a]	Specific activity[b]
3	5.26	8.67
4	6.66	8.21
5	8.00	7.84
6	22.20	8.10
7	11.60	8.05
8	8.51	8.25

[a] A_{333} min^{-1} ml^{-1} culture × 10^4
[b] A_{333} min^{-1} mg^{-1} protein × 10^2

Table 2. Purification of 4-coumarate:CoA ligase from cell lines WC63-1-9 and WC63-2-2 grown in the presence or absence of GA_3 in a fermentor

Cell line (−) or (+) GA_3	Total activity ΔA_{333} min^{-1}	Total protein mg	Specific activity ΔA_{333} min^{-1} mg^{-1} protein	Purification fold
WC63-1-9 (−)				
$(NH_4)_2SO_4$	5.2 ± 1.02	118.7 ± 10.50	0.043	1.0
G25 Sephadex	5.1 ± 0.95	118.7 ± 11.01	0.043	1.0
DEAE-Cellulose	6.0 ± 1.60	14.1 ± 1.61	0.425	9.9
Blue Sepharose[a]	5.0 ± 0.50	6.6 ± 0.63	0.748	17.4
DEAE-Cellulose and Blue Sepharose	2.5 ± 0.48	1.9 ± 0.38	1.315	30.5
WC63-1-9 (+)				
$(NH_4)_2SO_4$	6.5 ± 1.51	162.6 ± 9.62	0.040	1.0
G25 Sephadex	5.8 ± 1.15	138.9 ± 10.99	0.042	1.0
DEAE-Cellulose	6.2 ± 1.21	17.5 ± 1.50	0.354	8.9
WC63-2-2 (−)				
$(NH_4)_2SO_4$	27.4 ± 4.14	144.0 ± 15.11	0.190	1.0
G25 Sephadex	27.1 ± 5.15	139.5 ± 13.38	0.194	1.0
DEAE-Cellulose	29.5 ± 3.39	17.2 ± 1.91	1.715	9.0
Blue Sepharose[a]	25.2 ± 4.00	7.8 ± 0.68	3.230	17.0
DEAE-Cellulose and Blue Sepharose	12.2 ± 1.07	2.2 ± 0.09	5.545	29.1
WC63-2-2 (+)				
$(NH_4)_2SO_4$	38.4 ± 5.01	230.2 ± 14.60	0.167	1.0
G25 Sephadex	40.0 ± 4.22	228.7 ± 10.77	0.175	1.0
DEAE-Cellulose	41.5 ± 5.91	24.7 ± 3.25	1.680	9.9

The values are corrected to 100 g fresh wt of cells for each preparation. Data are given as the mean ± standard deviation for three batches of WC63-1-9 or six batches of WC63-2-2 grown without GA and as the mean ± range of two batches of each line grown in the presence of GA

[a] Desalted enzyme chromatographed directly on Blue Sepharose

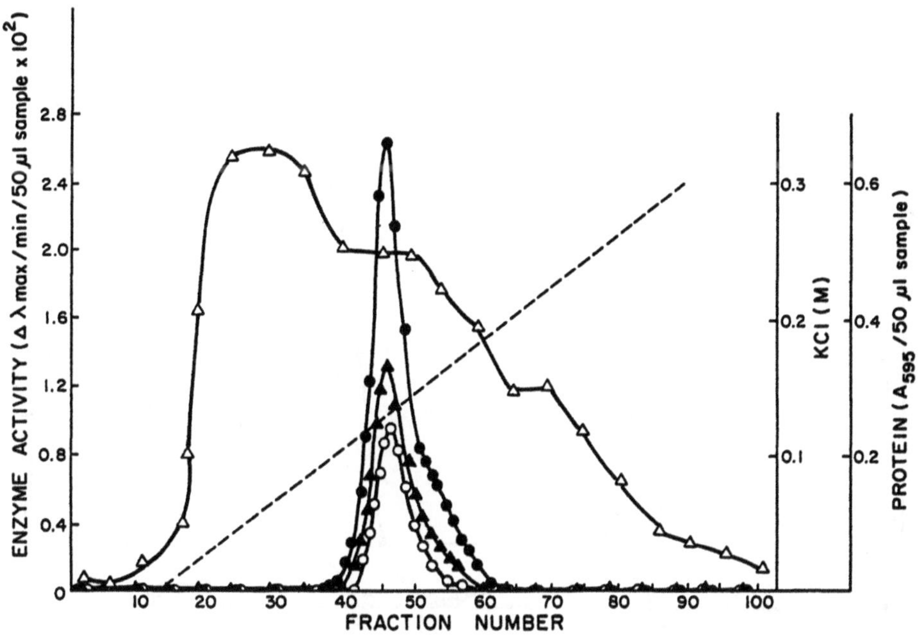

Fig. 2. Chromatography of 4-coumarate:CoA ligase of cell line WC63-1-9 on a DEAE-cellulose column. Enzyme activity with 3-coumaric acid, ●; with ferulic acid, ▲; and with caffeic acid ○ as substrate protein, △; KCl gradient, ---. See Materials and Methods for additional details

1. DEAE-Cellulose Chromatography. The enzyme from both cell lines grown in the presence and absence of GA_3 was purified approximately tenfold with no loss of activity by chromatography on DEAE-cellulose (Table 2). Each gave a single peak of 4-coumarate:CoA ligase activity eluting at the same KCl concentration when 4-coumaric acid, ferulic acid and caffeic acid were used as substrates. The elution profile for the enzyme from cell line WC63-1-9 is shown in Fig. 2.

2. Blue Sepharose Chromatography. When the DEAE-cellulose-purified material was chromatographed on Blue Sepharose, no enzymatic activity was eluted with ATP up to 0.2 M, with CoA up to 0.02 M or with both combined. When the columns were eluted with KCl, as shown in Fig. 3, two peaks of enzymatic activity were obtained. The size of the minor relative to the major peak varied from 1/3 to 1/12 with three independent enzyme preparations from cell line WC63-1-9. Similar behavior was found from the enzyme for cell line WC63-2-2. When Sephadex G25-purified enzyme from two different batches of each cell line was applied to Blue Sepharose columns, each gave two peaks and the relative sizes of the peaks varied as before.

A batch of enzyme from each cell line after chromatography on DE-52 was chromatographed on Blue Sepharose, and the fractions of each of the two peaks were separately pooled. The enzyme activity, with 4-coumaric acid, ferulic acid and caffeic acid as substrates, for each peak and the material before chromatography was 100:58:40 respectively for both cell lines.

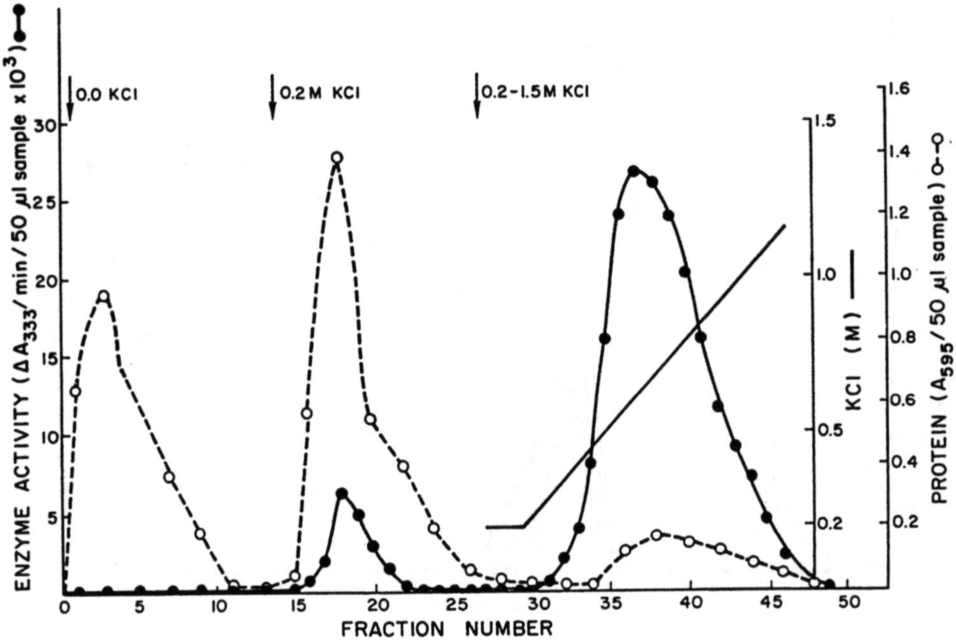

Fig. 3. Blue-Sepharose column chromatography of 4-coumarate:CoA ligase of cell line WC63-1-9 after DEAE cellulose chromatography

The specific activity of the pooled fractions of the major peak from Blue Sepharose was three times higher than that of the DEAE-cellulose purified material applied.

3.6 Gel Electrophoresis

When approximately 0.15 mg protein after DE-52 chromatography from either cell line, grown in the presence and absence of GA_3, was subjected to electrophoresis, no activity was detected in any gel slice by hydroxylamine assay. Protein staining of gels showed the presence of well resolved bands. When 0.1 mg of commercially prepared acetyl CoA synthetase (EC6.2.1.1.) was subjected to the same electrophoretic system under identical conditions, 30% of the acetyl CoA synthetase activity was recovered in three adjacent gel slices.

Various possibilities for lack of detection of the 4-coumarate:CoA ligase after electrophoresis were tested. When a known quantity of enzyme was incubated with a gel slice under the conditions used for elution, 95% of the activity could be detected, ruling out chemical inactivation by the gel and loss of activity during elution. When the tracking dye was allowed to migrate only one third of the tube length and this portion of the gel used in a single elution and assay, no activity was detected.

3.7 Stability of the Enzyme

When spectrophotometric assays were conducted with reagents that had been degassed with argon and compared with reagents that had not, the enzyme activity decreased 17% in 5 min.

An enzyme preparation after the $(NH_4)_2SO_4$ precipitation lost 20% of its activity after 2 weeks at − 5 °C. At − 20 °C no loss of enzyme activity was detected after 2 weeks.

3.8 Characteristics of 4-Coumarate: CoA Ligase From Two Cell Lines

The enzyme purified by DEAE-cellulose chromatography was used in the following experiments.

1. Substrate Specificity. The enzyme preparations from both cell lines used 4-coumaric acid most rapidly. Ferulic acid and caffeic acid were used but more slowly than 4-coumaric acid. Cinnamic acid, sinapic acid, and 3,4-dimethoxy-cinnamic acid were not used as substrates. K_m values of 4-coumaric acid and ferulic acid were determined with two independent enzyme preparations from each cell line. The enzyme from both cell lines gave the same K_m values with each substrate (Fig. 4). The K_m values of 4-coumaric acid and ferulic acid were 1.2×10^{-5} M and 2.5×10^{-5} M respectively (Fig. 4).

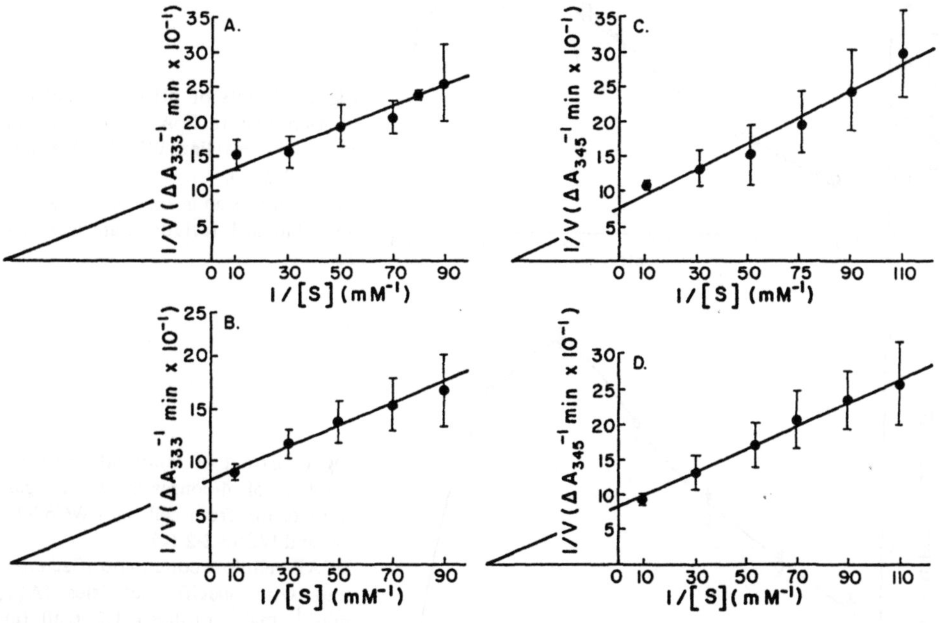

Fig. 4. Lineweaver-Burk plot of 4-coumarate:CoA ligase activity from cell lines WC63-1-9 (A&C) and WC63-2-2 (B&D) with 4-coumaric acid (A&B) and ferulic acid (C&D) as substrates. DEAE-cellulose purified enzyme was used. 2.5 μg protein per assay with enzyme from cell line WC63-1-9 and 0.7 μg protein per assay with enzyme from cell line WC63-2-2. Each *point* is an average of six readings and the *vertical bars* indicate standard deviations. The lines were determined by least-squares fit

2. Effects of Phosphate Concentration on Activity. Enzyme activities were measured with 0.1–0.7 M phosphate buffer in the spectrophotometric assay. The enzymes from both cell lines gave the same response to the phosphate buffer concentration. Highest enzyme activity was obtained at 0.3 M phosphate and half maximal activity was obtained in the absence and the presence of 0.55 M phosphate.

3. Dependence on pH. In the spectrophotometric assay the enzymes from both cell lines showed the same response to pH from 6 to 8.5 (Fig. 5). The enzyme activity was measured at 333 nm from pH 6 to pH 8. Since the absorption coefficient of 4-coumarate:CoA changes with pH, the values in Fig. 5 were corrected according to Knobloch and Hahlbrock (1975). At pH 8.5 the enzyme activity was measured at 386 nm following Knobloch and Hahlbrock (1975).

4. Dependence on Temperature. The effect of assay temperature on the activity of preparations from both cell lines was identical (Fig. 6). The temperature optimum was 40 °C.

The stability of the enzymes from both cell lines at 30 °C were compared. The half-lives of the 4-coumarate:CoA ligases from WC63-1-9 and WC63-2-2 were 28 and 31 min

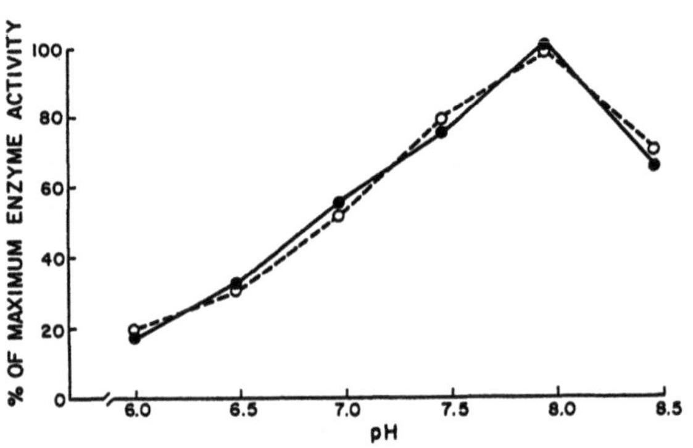

Fig. 5. Effects of pH on the activity of 4-coumarate:CoA ligase from the two wild carrot lines WC63-1-9 (•) and WC63-2-2 (○).
Each *point* is an average of 3 readings and standard deviations are less than 15%

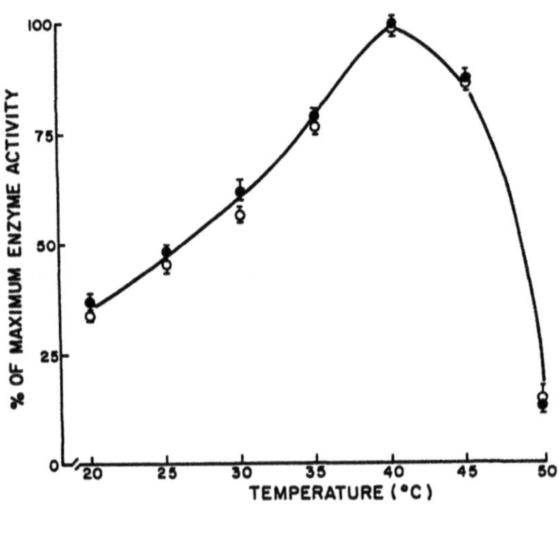

Fig. 6. Effect of temperature on the activity of 4-coumarate:CoA ligase preparation from cell lines WC63-1-9 (•) and WC63-2-2 (○).
DEAE-cellulose column purified enzymes with specific activities (A_{333} min^{-1} mg^{-1} protein) 1.2 (cell line 63-2-2) and 0.4 (cell line 63-1-9) were used. The activity of independent samples was measured over 5 min at each temperature. Bars show 1 standard deviation from triplicate measurements

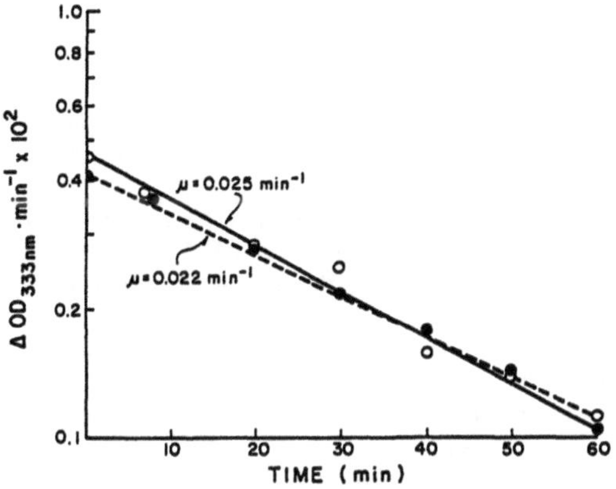

Fig. 7. Stability of 4-coumarate: CoA ligase from two wild carrot cell lines WC63-1-9 and WC63-2-2 at 30 °C. DEAE-cellulose purified enzyme from both cell lines containing 34 mg/ml protein were incubated at 30 °C. Triplicate samples were taken from each at 10-min intervals and assayed. Sample sizes from each preparation were different and chosen to given similar initial rates of reaction. The lines were fitted to the averaged rate at each time point by the method of least squares. Data for WC63-1-9, ——— for WC63-2-2 - - - - -

at 30 °C respectively (Fig. 7). Straight lines described the data when log activity was plotted against time for the enzymes from the two sublines (Fig. 7), and the slopes of these lines were not significantly different.

4 Discussion

The only difference detected in the 4-coumarate: CoA ligase from the two cell lines was that the subline which accumulated the lowest amount of anthocyanin had the highest total and specific activity of the enzyme (Table 2).

The enzymes from the two sublines were identical by the following criteria: (1) behavior on DE-52 cellulose, (2) behavior on Blue Sepharose, (3) relative activities with substrates, (4) K_m's with coumaric and ferulic acids, (5) effects of pH, temperature and phosphate buffer concentration on the reaction rates, and (6) thermal inactivation rates. When both cell lines were grown in the presence of 10^{-6} M GA_3, the total activity of the enzyme and total protein was increased but not in the same proportion so that the specific activity of the enzyme was slightly lower than that from tissue grown in the absence of GA_3.

No evidence for the presence of isozymes of 4-coumarate: CoA ligase in either subline was obtained. This evidence consists of chromatography as a single peak on DE-52 with no change of relative activities with different cinnamic acids as substrates, no change in relative activities with different substrates on chromatography on Blue Sepharose and the similar linear relationship between $\log_{10}$ (activity) and time during thermal denaturation.

We conclude that neither differences in 4-coumarate: CoA ligase nor isozyme differences account for the difference in anthocyanin accumulation between the two sublines of wild carrot examined.

Acknowledgement. Financial support for this study was provided by the W. Alton Jones Cell Science Center and by the University of Connecticut.

References

Dougall DK, Weyrauch KW (1980) Growth and anthocyanin production by carrot suspension cultures grown under chemostat conditions with phosphate as the limiting nutrient. Biotechnol Bioeng XXII:337–352

Dougall DK, Whitten GH (1980) The ability of wild carrot cell cultures to retain their capacity for anthocyanin synthesis after storage at −140 °C. Planta Med Suppl 1980:129–135

Dougall DK, Johnson JM, Whitten GH (1980) A clonal analysis of anthocyanin accumulation by cell cultures of wild carrot. Planta 149:292–297

Fowler MW (1982) The large scale cultivation of plant cells. In: Progress in industrial microbiology, vol 17. Elsevier, Amsterdam New York, pp 207–229

Grand C, Boudet A, Boudet AM (1983) Isoenzymes of hydroxycinnamate:CoA ligase from popular stems properties and tissue distribution. Planta 158:225–229

Grisebach H (1981) Lignins. In: Conn EE (ed) The biochemistry of plants: A comprehensive treatise, vol 7. Secondary plant products. Academic Press, London New York, pp 457–478

Gross GG, Zenk MH (1966) Darstellung und Eigenschaften von Coenzym A Thiolestern substituierter Zimtsäuren. Z Naturforsch 21b:683–690

Groth FS, Webster RG, Datyner A (1963) Two new staining procedures for quantitative estimation of proteins on electrophoretic strops. Biochim Biphys Acta 71:377–391

Heinzmann U, Seitz U (1977) Purification and substrate specifities of hydroxycinnamate:CoA ligase from anthocyanin-containing and anthocyanin-free carrot cells. Planta 135:313–318

Kinnersley AM, Dougall DK (1980) Increase in anthocyanin yield from wild-carrot cell cultures by a selection system based on cell-aggregate size. Planta 149:200–204

Knobloch K-H, Hahlbrock K (1975) Isoenzymes of p-coumarate:CoA ligase from cell suspension cultures of *Glycine max*. Eur J Biochem 52:311–320

Knobloch K-H, Hahlbrock K (1977) 4-coumarate:CoA ligase from cell suspension cultures of *Petroselinum hortense* Hoffm. Arch Biochem Biophys 184:237–248

Maurer HR (1971) Disc electrophoresis and related techniques of polyacrylamide gel electrophoresis. De Gruyter, Berlin New York, p 44

Meins F (1983) Heritable variation in plant cell culture. In: Briggs WR, Jones RL, Walbot V (eds) Annual review of plant physiology, vol 34. Rev Inc, Palo Alto, Calif, pp 327–346

Stadmann ER (1957) Preparation and assay of acyl coenzyme A and other thiol esters: Use of hydroxylamine. In: Colowick SP, Kaplan NO (eds) Methods in enzymology, vol III. Academic Press, London New York, pp 931–941

Yamakawa T, Kato S, Ishida K, Kodama T, Minoda Y (1983) Production of anthocyanins by *Vitis* cells in suspension culture. Agric Biol Chem 47:2185–2191

Induction of Anthocyanin Synthesis in Relation to Embryogenesis in a Carrot Suspension Culture – a Model System for the Study of Expression and Repression of Secondary Metabolism

Y. OZEKI [1] and A. KOMAMINE [2]

1 Introduction

Secondary metabolism in intact plants is expressed at specifically differentiated organs and tissues. On the other hand, in cultured cells, most secondary metabolism activities are repressed, giving rise to difficulty when production of useful secondary metabolites is attempted using cultured plant cells. The reason why secondary metabolism activities are repressed in plant cultured cells may be that the expression of secondary metabolism is closely correlated with morphological differentiation and is repressed in cultured cells under conditions of dedifferentiation such as those in actively dividing cultured cells.

Thus, the correlation between secondary metabolism and morphological differentiation must be taken into consideration in order to elucidate the mechanism of expression and repression of secondary metabolism. For this purpose, a system in which expression of secondary metabolism can be regulated under defined conditions should be established in cultured cells, and the mechanism of expression of secondary metabolism, regulated as metabolic differentiation, should be investigated in correlation to morphological differentiation. In a carrot suspension culture we established a model system in which anthocyanin synthesis was induced under the same conditions as the induction of embryogenesis (Ozeki and Komamine 1981). Using this system, the expression mechanism of anthocyanin synthesis was investigated.

2 Induction of Anthocyanin Synthesis and Embryogenesis in a Carrot Suspension Culture and the Correlation Between Anthocyanin Synthesis and Embryogenesis

Cells of a suspension culture of carrot (*Daucus carota* cv. Kurodagosun) were sieved through nylon screens and collected cells and cell clusters in the size range of 31–81 μm were transferred to a fresh medium containing 2,4-dichlorophenoxyacetic acid (2,4-D, 5×10^{-7} M) at every subculture for more than a year (sieved culture). The sieved culture was fractionated by Ficoll density gradient centrifugation and the cells in each fraction were transferred to a medium lacking 2,4-D. Figure 1 shows that somatic

1 Department of Botany, Faculty of Science, University of Tokyo, Hongo, Tokyo 113, Japan
Present address: Fermentation Research Institute, Agency of Industrial Science & Technology, Ibaraki 305, Japan

2 Biological Institute, Faculty of Science, Tohoku University, Sendai, Miyagi 980, Japan

Primary and Secondary Metabolism of Plant Cell Cultures, ed. by Neumann et al.

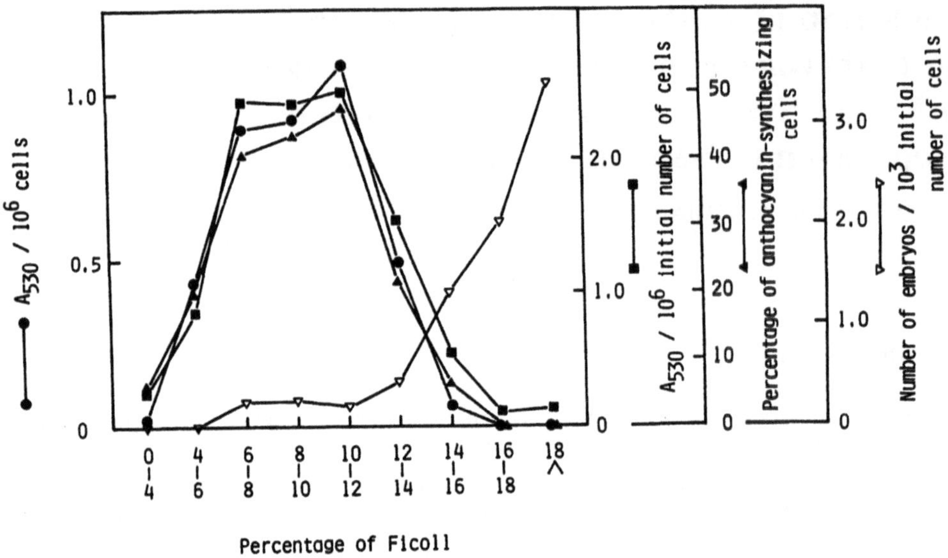

Fig. 1. Anthocyanin synthesis and embryogenesis of cells in various fractions of the sieved culture 9 days after transfer to a medium containing 10^{-7} M zeatin but no 2,4-D. $A_{530}/10^6$ cells (●), $A_{530}/10^6$ initial number of cells (■), percentage of anthocyanin-synthesizing cells (▲), number of embryos/10^3 initial number of cells (▽)

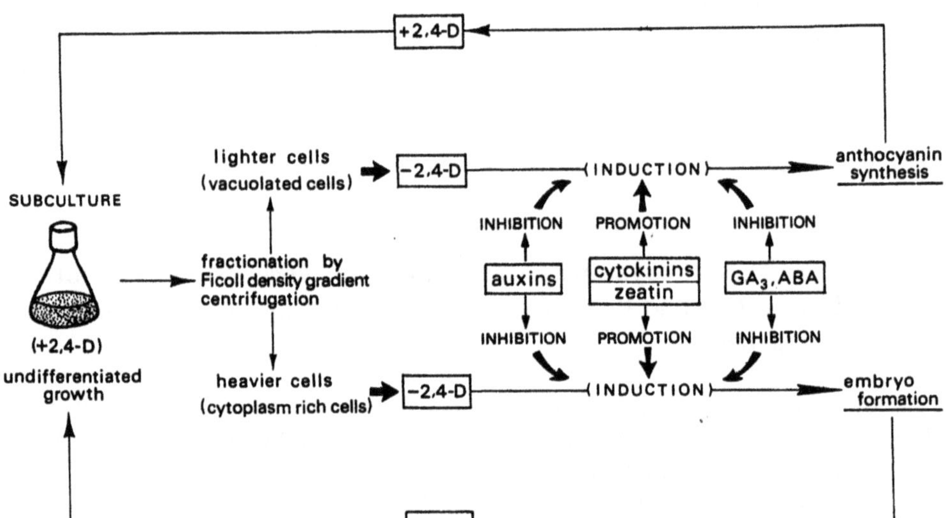

Fig. 2. Scheme of the correlation of the induction of anthocyanin synthesis with the induction of embryogenesis. The effects of growth regulators on embryogenesis are cited from Fujimura and Komamine (1975, 1979, 1980)

embryos were formed in the higher density fraction (> 14% of Ficoll). On the other hand, the cells in the lower density fractions, especially that with 6%–12% of Ficoll, synthesized anthocyanin in the medium lacking 2,4-D.

A close correlation between the induction of anthocyanin synthesis and embryogenesis was suggested from physiological observations: the time courses of both inductions were similar. All auxins inhibited the induction of anthocyanin synthesis as well as embryogenesis. Cytokinins showed a promotive effect on the synthesis of anthocyanin in the medium lacking 2,4-D. Both gibberellic acid (GA_3) and abscisic acid (ABA) were inhibitory to anthocyanin synthesis and also to embryogenesis. The effects of growth regulators on the induction of anthocyanin synthesis are summarized in Fig. 2. These observations indicated a close correlation between the induction of anthocyanin synthesis, a metabolic differentiation, and that of embryogenesis, a morphological differentiation.

3 The Relationship Between Anthocyanin Synthesis and Cell Division Regulated by 2,4-D

Serial observation revealed that, when cells were transferred from the medium containing 2,4-D to the medium lacking 2,4-D, cell division still continued for 0–4 days after the transfer. This may have been due to carried over 2,4-D. Thereafter, cell division ceased and anthocyanin synthesis occurred at 4 or 5 days after the transfer, and anthocyanin was accumulated in the vacuole. When 2,4-D was added to the anthocyanin-synthesizing cells 6 days after the transfer, the anthocyanin gradually disappeared and cell division occurred again in the same cells when anthocyanin had almost vanished. It was confirmed, therefore, that a cell in which cell division had ceased and anthocyanin synthesis was occurring in a medium lacking 2,4-D had the ability to regain cell division activity when it was transferred again to a medium containing 2,4-D.

From the results of serial observation, induction of anthocyanin synthesis and that of cell division appear to be reciprocal phenomena. However, it was shown from an investigation using an inhibitor of DNA synthesis, fluorodeoxyuridine, that when cell division was inhibited by fluorodeoxyuridine, anthocyanin synthesis was suppressed in the presence of 2,4-D and continued to occur in the absence of 2,4-D, indicating that anthocyanin synthesis was regulated by 2,4-D irrespective of cell division, and that anthocyanin synthesis and cell division may be regulated by different mechanisms.

4 Changes in the Activities of Enzymes Involved in Anthocyanin Synthesis Regulated by 2,4-D

In order to elucidate the regulatory mechanism of expression and suppression of secondary metabolism by auxins, we investigated the induction and reduction of enzyme activities involved in anthocyanin synthesis regulated by 2,4-D. Phenylalanine ammonia-lyase (PAL) and 4-coumarate:CoA ligase (4CL) activities increased 1–2 days

Table 1. Effect of 2,4-D on the activities of PAL, C4H, 4CL, CHS and CHFI, and on anthocyanin content

	Relative activity (%)		
	6 days	6 days + 24 h − 2,4-D	6 days + 24 h + 2,4-D
PAL	100	123	10
C4H	100	176	161
4CL	100	126	64
CHS	100	260	0.1
CHFI	100	185	96
	Relative anthocyanin content (%)		
	100	214	93

after transfer to media both containing and lacking 2,4-D. This was probably due to transfer effects. In the medium containing 2,4-D, the activities of both enzymes decreased and maintained a low level thereafter. Also cinnamate-4-hydroxylase (C4H), chalcone synthase (CHS) and chalcone-flavanone isomerase (CHFI) were at a low level. In particular, the activity of CHS was below a detectable level throughout the culture time. When the cells were transferred to a medium lacking 2,4-D in which anthocyanin synthesis was induced, the activities of all enzymes increased. In particular, the induction of CHS was notable. When 2,4-D was added to the culture 6 days after transfer to the medium lacking 2,4-D, the activities of PAL and CHS rapidly declined as compared to that of 4CL and CHFI. On the other hand, the activity of C4H continued to increase (Table 1). In addition, when the cells cultured in the medium lacking 2,4-D for 6 days were transferred to a fresh medium in which the density of cells was half that of the original, "super-induction" of PAL occurred irrespective of the presence or absence of 2,4-D as a result of "transfer effect" (Fig. 4). On the other hand, the activity of CHS decreased slightly in the medium lacking 2,4-D, but declined rapidly in the medium containing 2,4-D (Fig. 4). At that time, in spite of super-induction of PAL, the content of anthocyanin did not increase, due to the low level of CHS activity (Fig. 4). These results suggested that CHS may be a "key enzyme" in anthocyanin synthesis regulated by 2,4-D in this system.

5 Purification of CHS and Preparation of Anti-CHS Serum

In order to elucidate the regulatory mechanism of anthocyanin synthesis induction, it was necessary to investigate the molecular mechanism of CHS induction. Thus, we attempted the purification of CHS and the preparation of an anti-CHS serum. From 2 kg of cells cultured in a medium lacking 2,4-D for 7 days, about 8 g of crude protein was extracted. The crude extract was fractionated by $(NH_4)_2SO_4$ fractionation and 55%–80% precipitation was sedimented. The collected crude CHS was further purified by

CM-Sepharose, a hydroxyapatite column, Toyopearl HW-55F gel filtration, chromatofocusing and Butyl-Toyopearl chromatography. By means of these procedures, CHS (ca. 140 μg) was purified about 3,300-fold. The yield was about 6%, giving rise to a single band on SDS-polyacrylamide gel electrophoresis (SDS-PAGE). The molecular weight of the CHS purified from the carrot cells was about 80,000 as determined by gel filtration and native PAGE, and the molecular weight of its subunit was about 40,600 as determined by SDS-PAGE. Apparent K_m for 4-coumaroyl-CoA was 5.7 μM and that for malonyl-CoA was 18 μM.

The purified CHS was mixed with Freund's adjuvant and injected into mice (strain: C57BL ♀) under specific pathogen-free conditions (SPF). Anti-CHS serum was obtained from the mice and crossreactivity was checked by Western blotting, showing a specific reactivity to CHS.

Anti-PAL serum was obtained from a rabbit under SPF conditions using PAL that was purified by Dr. Yoshiyuki Tanaka (National Institute of Agro-Environmental Sciences) from the same carrot suspension culture used in the present work. We investigated the regulatory mechanism of CHS and PAL synthesis by 2,4-D using the anti-CHS and anti-PAL sera.

6 Regulation of Synthesis of CHS and PAL and mRNA Synthesis of CHS and PAL by 2,4-D

We investigated whether regulation of the activities of CHS and PAL by 2,4-D was caused by the activation and inactivation of enzymes or by an increase and decrease in the amounts of enzymes. Crude protein fractions were prepared and fractionated by SDS-PAGE and then Western blotting was performed using ^{125}I-Protein A and the amounts of enzymes were quantified by measuring their radioactivities. Figure 3 shows the time course of the quantitative changes of PAL and CHS. The amount of PAL increased and attained a peak 1 day after transfer to the fresh medium, but declined thereafter in the medium containing 2,4-D. In the same medium, CHS could not be detected below the background level. On the other hand, CHS was detected 4–5 days after transfer to the medium lacking 2,4-D and the amounts of CHS and PAL gradually increased. These changes correlated closely with the activity changes (Fig. 3).

Furthermore, Fig. 4 showed that the CHS synthesized decreased with addition of 2,4-D and super-induction of an amount of PAL occurred with the transfer to fresh media both containing and lacking 2,4-D. These results showed that the changes in CHS and PAL activities were probably due to the changes of amounts of the enzyme proteins, not to activation or inactivation of enzyme proteins, thus suggesting that 2,4-D regulated the synthesis of CHS and PAL protein. We therefore investigated whether the regulation of protein synthesis by 2,4-D was at the level of transcription or translation.

A total RNA fraction was prepared by the guanidine thiocyanate method and translation in vitro was performed using wheat germ extract with ^{35}S-methionine. CHS and PAL translated from mRNA were separated by antisera and Protein A-Sepharose and fractionated by SDS-PAGE. The radioactivities incorporated into CHS and PAL were measured, indicating the level of mRNAs for CHS and PAL.

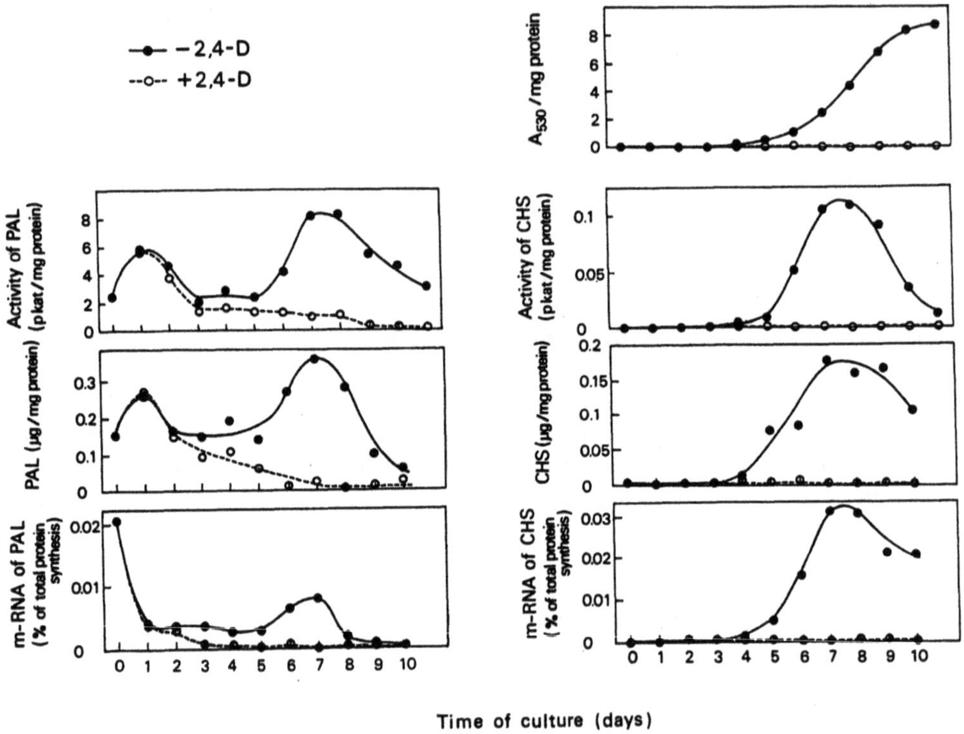

Fig. 3. Changes in anthocyanin contents, activities of PAL and CHS, enzyme proteins of PAL and CHS and amounts of mRNA for PAL and CHS. Cells were cultured in both the medium containing 2,4-D (○) and that lacking 2,4-D (●)

Figure 3 shows changes in the level of mRNAs for PAL and CHS. The amount of mRNA for PAL was very high in 0 day cultures which had been subjected to the fractionation procedure. This may have been due to the induction of mRNA of PAL by the fractionation procedure of sieving with nylon screens and Ficoll density gradient centrifugation for 2 h. The increase in the mRNA level of PAL at 0 days after transfer is most likely due to the transfer effect, resulting in the induction of protein synthesis. The level of mRNA of PAL decreased thereafter in the cells in the medium containing 2,4-D. On the other hand, the amount of mRNA for PAL increased 5–7 days after transfer to the medium lacking 2,4-D. The mRNA of CHS was below the detectable level in the cells cultured in the medium containing 2,4-D. In the medium lacking 2,4-D, however, mRNA for CHS began to be detected 4–5 days after transfer and increased thereafter. No increase caused by the transfer effect was observed for CHS.

Furthermore, when 2,4-D was added to the cells cultured for 6 days in the medium lacking 2,4-D in which anthocyanin was synthesized, the level of mRNA of both PAL and CHS decreased only 3 h after the addition. However, the mRNA for PAL rapidly increased within 3 h of transfer to either the fresh medium containing 2,4-D or that lacking 2,4-D (Fig. 4). On the other hand, no increase in mRNA for CHS was observed after transfer to the fresh medium (Fig. 4).

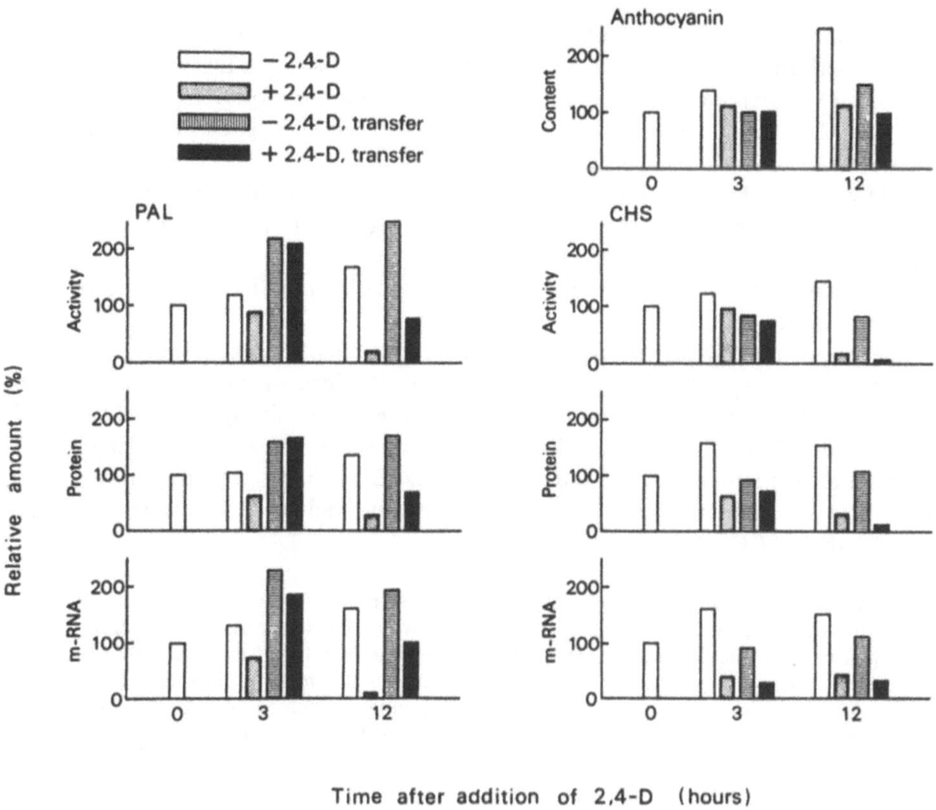

Fig. 4. Changes in anthocyanin contents, activities of PAL and CHS, enzyme proteins of PAL and CHS and amounts of mRNA for PAL and CHS. ▭: cells continued to be cultured without transfer and without addition of 2,4-D after 6 days; ▩: 2,4-D was added to the cells which were cultured in the medium lacking 2,4-D for 6 days; ▥: the cells were transferred to a fresh medium lacking 2,4-D in which cell density was half as much as that of the 6 day culture; ■: cells were transferred to a fresh medium containing 2,4-D in which cell density was half as much as that of the 6 day culture. Times in figures indicate hours after 6 days of culturing.

All the results mentioned above showed that the amounts of mRNA for PAL and CHS reflected those in the amounts of the enzymes and consequently those in their activities. It was suggested that the metabolism of anthocyanin synthesis by 2,4-D is most likely regulated by the amount of mRNA, that is, on the transcriptional level.

7 Conclusions

We established a model system in which anthocyanin synthesis was induced by the transfer of carrot cells from a medium containing 2,4-D to one lacking 2,4-D, under the same conditions as for the induction of embryogenesis. Close correlation between the induction of anthocyanin synthesis, a metabolic differentiation, and embryogenesis, a morphological differentiation, was suggested from the physiological view point.

The results obtained from experiments using an inhibitor of DNA synthesis suggested that 2,4-D regulates anthocyanin synthesis and cell division by different mechanisms.

2,4-D regulates the activities of enzymes involved in the metabolism of anthocyanin synthesis. CHS was found to be a key enzyme in the regulation of anthocyanin synthesis by 2,4-D.

CHS was purified and an antiserum against it was obtained from mice, and an anti-PAL serum from a rabbit.

Experiments using Western blotting indicated that changes in activities of CHS and PAL were caused by changes in enzyme amounts. An investigation using an in vitro translation system showed that induction and suppression of the synthesis of CHS and PAL were due to an increase and decrease respectively in the amounts of mRNA. It was concluded, therefore, that in this system 2,4-D probably regulates anthocyanin synthesis at the transcriptional level of enzymes involved in anthocyanin synthesis, especially that of CHS.

Acknowledgements. We wish to express our sincere thanks to Dr. Y. Tanaka (National Institute of Agro-Environmental Sciences) for purification of PAL. We acknowledge gratefully to Prof. U. Sankawa and Dr. H. Noguchi (Department of Phytochemistry, Faculty of Pharmaceutical Sciences, University of Tokyo) for synthesis of 4-coumaroyl-CoA and also to Dr. T. Suzuki (Kitasato Institute) for kind help in preparation of antisera. Many thanks are due to Prof. Dr. M. Tazawa and Dr. K. Sakano (Department of Botany, Faculty of Science, University of Tokyo) for helpful discussion.

References

Fujimura T, Komamine A (1975) Effects of various growth regulators on the embryogenesis in a carrot cell suspension culture. Plant Sci Lett 5:359–364

Fujimura T, Komamine A (1979) Involvement of endogenous auxin in somatic embryogenesis in a carrot cell suspension culture. Z Pflanzenphysiol 95:13–19

Fujimura T, Komamine A (1980) Mode of action of 2,4-D and zeatin on somatic embryogenesis in a carrot cell suspension culture. Z Pflanzenphysiol 99:1–8

Ozeki Y, Komamine A (1981) Induction of anthocyanin synthesis in relation to embryogenesis in a carrot suspension culture: Correlation of metabolic differentiation with morphological differentiation. Physiol Plant 53:570–577

Metabolism of Quinolizidine Alkaloids in Plants and Cell Suspension Cultures: Induction and Degradation

M. WINK [1]

1 Introduction

We have studied the physiology and biochemistry of quinolizidine alkaloids (QA) in lupins and other Leguminosae and their cell suspension cultures extensively during the last 7 years. Our goal was to analyze the various processes and their regulation which lead to the phenotype "bitter" lupin, i.e. plants which do not only synthesize but which also accumulate substantial amounts of alkaloids. A thorough understanding of the biochemistry and the control mechanisms which underlie alkaloid metabolism is a prerequisite for any planned manipulation of the biosynthetic potential of plant cells. This knowledge is especially relevant to two lines of biotechnological research: (1) Establishing cell cultures which produce and store natural products in similar or better ways than intact plants. (2) Breeding plants with a better resistance against pathogens, which is assumed to be largely mediated by plant secondary products.

2 Studies with Intact Plants

In this chapter the metabolism of QA in intact plants is summarized because this information is necessary to understand the biochemistry of QA metabolism in cell suspension cultures.

2.1 Biosynthesis

The decarboxylation of lysine to cadaverine is catalyzed by lysine decarboxylase [4, 13]. Three cadaverine units are condensed to the tetracyclic 17-oxosparteine by oxosparteine synthase, a process in which no free intermediates were found [25, 36]. Four of the six amino groups of the three cadaverine units are eliminated by transamination with pyruvate as a specific NH_2-acceptor [25, 36]. 17-Oxosparteine is further converted to lupanine [37], which serves as a precursor for most other QA [2, 6, 10, 15, 23, 26, 33]: 13-hydroxylupanine and respective esteralkaloids, tricyclic QA of the angustifoline/tetrahydrorhombifoline-type and α-pyridone alkaloids such as cytisine/N-methylcy-

1 Institut für Pharmazeutische Biologie, Technische Universität Braunschweig, 3300 Braunschweig, FRG

Primary and Secondary Metabolism of Plant Cell Cultures, ed. by Neumann et al.

tisine. Enzymes of the corresponding pathways which could be discovered include tigloyl-CoA: 13-hydroxylupanine O-tigloyltransferase [30] and S-adenosyl-L-methionine: cytisine N-methyltransferase [4, 22].

2.2 Site of QA Biosynthesis

The enzyme of QA formation could be demonstrated in leaves of *Lupinus polyphyllus* [28] and could be subsequently localized in the chloroplast (stroma) [29, 37].

2.3 Transport of QA

The alkaloids formed in lupin leaves are transported via the phloem [35, 39] to all other plant organs, including the fruits, which are especially rich in alkaloids. Since QA contribute about 8% to the nitrogen, which is translocated in the phloem, a minor function of QA might be nitrogen transport [35].

2.4 Accumulation

QA have been found in all tissues of lupin plants [23, 28, 41]. Alkaloids are generally abundant in plant parts which are important for survival and reproduction [23]. QA are not randomly distributed in lupin tissues, but seem to be concentrated in peripheral cell layers, i.e. epidermis and subepidermal cells [42]. This result implies that specific alkaloid storage cells are also present in lupins as in other plant species [17]. It is a challenging question whether these storage cells possess specific carrier proteins for QA or other mechanisms, which would enable these cells to concentrate the alkaloids, or if storage cells cells do not degrade QA whereas the other cells do. The storage of QA in lupin tissues changes during the vegetation period: From germination to flowering, all plant parts store alkaloids. During seed maturation lupin plants senesce and QA disappear from all plant parts except the fruits, which instead become even more "bitter" [23]. It is unclear if this "disappearance" is due to degradation or exportation. This result shows that alkaloid storage is also characterized by temporal and spatial changes.

2.5 Degradation

During germination and seedling development, OA are metabolized [23] to a high degree. Since QA contribute about 12% to the nitrogen stored in the seeds, they might be considered as nitrogen storage compounds, similar to the role of nonprotein amino acids in other legumes, which do not accumulate QA [12]. Degradation of QA also takes place during later stages of lupin development and seems to play an important part in QA metabolism. This is indicated by the diurnal fluctuations of QA contents observed in lupins [31, 35].

2.6 Occurrence of QA

The processes of synthesis, transport, accumulation, and degradation have to be tightly regulated before a plant is actually able to produce alkaloids. Plants which store QA are most abundant in the family of Leguminosae but have also been recorded from a few species of other plant families [5], e.g. Chenopodiaceae [14], Berberidaceae [10], Papaveraceae [14], Scrophulariaceae [9], Santalaceae [7], Solanaceae, Ranunculaceae [10, 14]. Since it could be shown recently that cell cultures of species not known for QA formation, e.g. *Conium, Daucus, Atropa, Chenopodium, Spinacia*, and *Symphytum* may accumulate QA under special conditions [34], we have postulated that the genes for the enzymes of QA biosynthesis are widely distributed in the plant kingdom [34] although the phenotype "bitter" is restricted to relatively few species.

2.7 Function of QA

As mentioned above, primary functions of QA include nitrogen transport and N-storage. On account of circumstantial and experimental evidence the main function of QA is chemical defense: QA deter the feeding of herbivores (mammals, insects, molluscs), inhibit the growth of microorganisms (bacteria, fungi) and the germination of other plants (allelopathy) [16, 18–21, 23]. Moreover, QA concentrations are up to 100-fold higher in lupins than the inhibitory concentrations against the organisms mentioned above. "Sweet" lupins are much more susceptible to predation and disease than the "bitter", alkaloid-rich lupins [23], and QA are preferentially accumulated in peripheral cell layers [42], i.e. those parts which have to ward off a herbivoral or microbial attack.

2.8 Regulation of QA Production

As a consequence of QA formation in the chloroplast, light is a major factor which regulates QA biosynthesis in lupins [27, 33] (for review see [33]). Alkaloids are only produced in the light [31] and both transport and alkaloid accumulation follow a diurnal rhythm [35].

As a result of the biological role of QA as defense compounds, we tested if alkaloid accumulation responded to environmental pressure such as a herbivoral attack. This was simulated by wounding lupin leaves: Within 2–4 h after wounding, alkaloid content of *L. polyphyllus* leaves increased by a factor of 2–4. This response would amplify the defense potential of QA: they can therefore be classified as qualitatively constitutive, but quantitatively inducible compounds [19].

3 Studies with Cell Suspension Cultures

Like many other cell culture systems, cell suspension cultures of *Lupinus, Genista, Cytisus, Sophora, Baptisia, Sarothamnus, Laburnum* fail to produce substantial amounts of QA [26, 38, 41]. In contrast to lupin leaves, which accumulate up to 5 mg alkaloids g^{-1} fr.wt. lupin cell cultures have QA levels between 0.01–1 $\mu g\ g^{-1}$ fr.wt. The failure to produce QA does not seem to be due to a suppression of the genes of QA biosynthesis, since the enzymes of lupanine formation are expressed in lupin cell cultures. Their activity is similar to or one order of magnitude lower than in the intact plant, as compared to the 3–5 orders of magnitude for alkaloid accumulation [32]. Therefore, lupin cell cultures offer a chance to evaluate the other processes which in addition to biosynthesis are important for alkaloid production.

3.1 Induction of QA Accumulation

Application of the precursor cadaverine generally leads to rapid alkaloid formation [32]. In order to check if the biosynthetic potential given can be modulated experimentally without adding precursors, lupin cell cultures were treated with compounds which are known to elicit phytoalexine response or trigger gene regulation in other systems. As can be seen from Table 1 a variety of alkaloids, which are biogenetically different from lupin alkaloids, some polyamines and metabolic inhibitors induce a substantial increase of QA concentrations in cell cultures [34]. This inductive effect could be further amplified by lowering the hydrogen ion concentration of the culture medium to pH 3.5

Table 1. Induction of quinolizidine alkaloid accumulation in cell suspension cultures of *Lupinus polyphyllus*

Condition[a]	Alkaloid content/cell culture[b] $\bar{x}$ ± s.e. μg/100 g fr.wt.	Number of experiments
Non-induced controls	1	
Control after 24 h	3.7 ± 1	13
+ 3 mM Coniine	46.8 ± 11	14
+ 1 mM Papaverine	75.5 ± 28	7
+ 3 mM Quinine	12.5	2
+ 1 mM Berberine	2	1
+ 3 mM Theobromine	10	1
+ 3 mM Ajmaline	1	1
+ 3 mM Sanguinarine	1	1
+ 3 mM Cadaverine	47.2 ± 35	6
+ 0.5 mM Spermidine	153.7 ± 105	4
+ 0.5 mM Spermine	13.3 ± 11	2
+ 0.5 mM Diaminoheptane	1	1
+ 0.3 mM Cyclic AMP	178.2 ± 104	3
+ 5 mM Diethyldithiocarbamate	174.8 ± 166	3
+ 1 mM Propargylamine	181.5 ± 22	2
+ 0.1 mM Cycloheximide	48.1 ± 19	5
+ 0.5 mM Allopurinol	17	1
+ 0.1 mM Demecolcine	12	1
+ 0.3 mM Chloramphenicol	0.7	3
+ 1 mM Arachidonic acid	14.3	1
+ 0.1% Chitosane[c]	5.5 ± 2	3
+ 1 mM Salicylic acid	1	1

[a] Filter-sterilized effectors, which were adjusted to pH 5, were added to the cell cultures directly after their transfer into fresh medium. After incubation for 24 h under standard culture conditions [38], the cell cultures were harvested and cells and medium processed separately. Following standard alkaloid extraction [41] the alkaloid extracts were analyzed by capillary gas-liquid chromatography and GLC/mass spectrometry according to Wink et al. [38–41]

[b] Includes alkaloid content of the respective culture medium

[c] Chitosane was added after alcohol sterilization

Table 2. Factors influencing the induction of quinolizidine alkaloid accumulation in cell suspension cultures of ***Lupinus polyphyllus***: A. Hydrogen ion concentration, B. Type of alkaloids, C. Coniine concentrations

Conditions[a]		Alkaloid content[d] ($\mu g\ g^{-1}$ fr.wt.)
A.[b]		
+ 3 mM Coniine	pH 3.5	9.6
	pH 4.5	0.23
	pH 5.5	0.15
	pH 6.5	0.05
	pH 7.5	0.02
B.[c]		
Control		0.04
+ Coniine		2.2
+ Papaverine		4.7
+ Reserpine		3.8
+ Quinine		7.7
C.		
Control		0.1
+ 0.1 mM Coniine		0.1
+ 1.0 mM Coniine		1.5
+ 10.0 mM Coniine		2.6
+ 20.0 mM Coniine		0.1

[a] C. 3-g cells each were transferred to 50 ml fresh medium. After 3 h the cell cultures were harvested and analyzed as outlined in Table 1
[b] The hydrogen ion concentration of the cell culture media were adjusted to the pH value given before the experiments
[c] One hour after the addition of inducers (3 mM each), the pH value of the medium was adjusted to pH 3.5
[d] Alkaloids secreted into the culture medium are incorporated in this value

(Table 2). Figure 1 shows that this induction was only short-term and was followed by alkaloid degradation. As a measure for turnover, the amount of $^{14}CO_2$ liberated from labelled QA was determined (Figs. 1, 2). From these experiments it can be concluded that the application of inducers such as papaverine stimulates QA biosynthesis. This accelerated synthesis is not due to increased enzyme concentration [34], as is the case during induction of flavonoid-derived phytoalexins [3] but to other, so far undiscovered factors [34]. It has to be taken into consideration that in addition to an accelerated biosynthesis either transport or storage phenomena may also be influenced. Although the mechanism of the induction of QA accumulation is not understood, these experiments show that alkaloid metabolism in cell suspension cultures is flexible and reactive and not static.

3.2 Degradation of QA in Cell Suspension Cultures

Since the above-mentioned experiments implied that QA degradation is important in lupin cell cultures, this process was studied in more detail. Cell suspension cultures

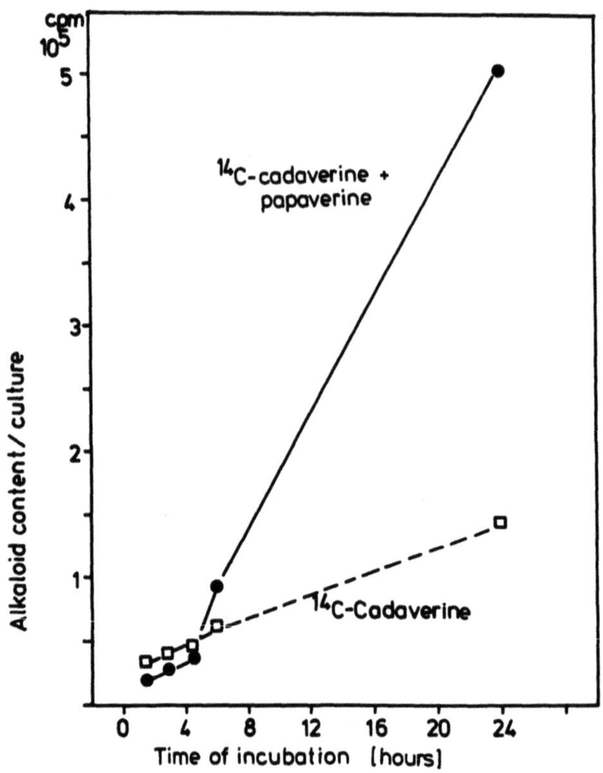
cpm
10^5
5
4
3
2
1
^{14}C-cadaverine + papaverine
^{14}C-Cadaverine
0
4
8
12
16
20
24
Time of incubation (hours)
Alkaloid content / culture

Fig. 1A

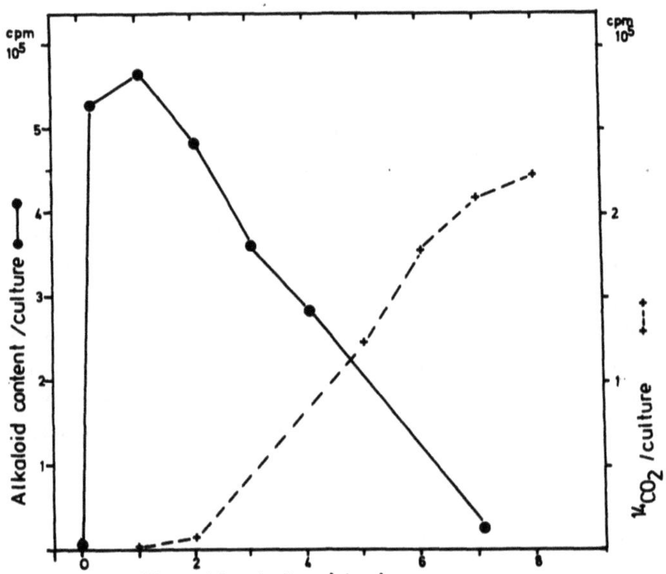
cpm
10^5
5
4
3
2
1
0
2
4
6
8
Time of incubation (days)
Alkaloid content /culture
$^{14}CO_2$ /culture

Fig. 1B

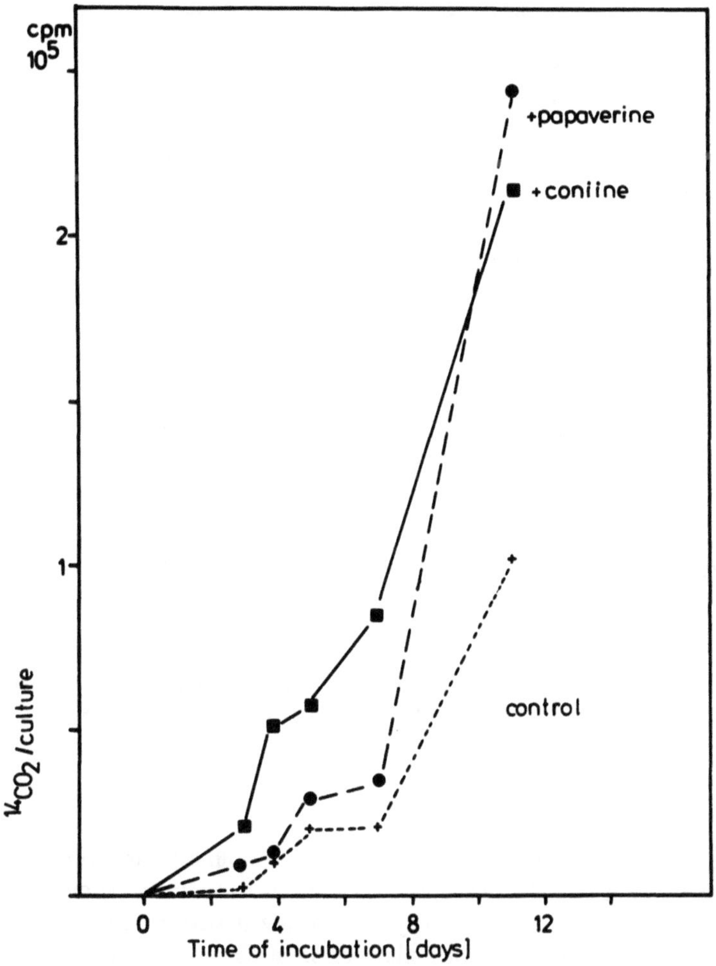

Fig. 2. Release of $^{14}CO_2$ in alkaloid-treated cell suspension cultures of *L. polyphyllus*. Cell cultures were fed with 1 μCi [1,5 ^{14}C]-cadaverine after transfer of the cells into fresh medium (control). In addition to radioactive cadaverine other cultures were simultaneously induced with 0.5 mM papaverine or 1 mM coniine. Experiments were performed in double-staged culture flasks as described in Fig. 1

◄

Fig. 1A,B. Time course of quinolizidine alkaloid accumulation in cell suspension cultures of *Lupinus polyphyllus* after induction with papaverine. About 2 g cells were transferred into 50 ml fresh medium and fed with 1 μCi [1,5 ^{14}C]-cadaverine (control) or ^{14}C-cadaverine and 0.5 mM papaverine (induction experiment). Cell cultures were harvested at the times indicated and processed separately for cells and cell culture medium as outlined in Table 1. Alkaloid extracts were separated by thin layer chromatography [ethanol/25% ammonia (7:3)] and the radioactive quinolizidine alkaloids were analyzed by liquid scintillation counting. Cpm values given account for total quinolizidine alkaloids/culture, including the alkaloids which were secreted into the culture medium and which generally accounted for 70%–90% of total alkaloids. Experiment **B** was performed as outlined for **A** but in addition radioactive $^{14}CO_2$ was also collected. For this purpose we employed double-staged culture flasks; the upper compartment contained the cell culture, the lower one 10 ml 5 N NaOH to trap $^{14}CO_2$. Culture flasks were sealed with aluminum foil and parafilm

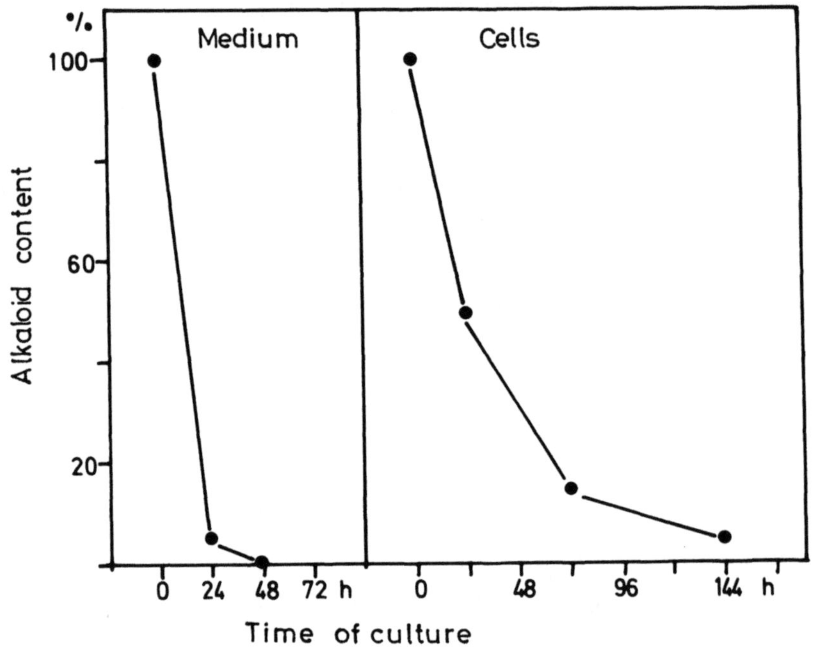

Fig. 3. Degradation of exogenous sparteine by cell suspension cultures of *Cytisus scoparius*. After transfer of the cells into fresh medium (50 ml), filter-sterilized sparteine (2 mg/flask) was added. At the intervals given, cell cultures were harvested and their alkaloids determined according to Wink et al. [40, 41] by capillary GLC

readily degrade endogenous or exogenous QA ([31], Fig. 3). Lupin cells are even able to utilize the nitrogen stored in the alkaloid molecules: Recent experiments show that lupin cells are able to survive and to grow slowly on a nutrient agar which contains sparteine as sole nitrogen source, whereas the cells very soon die on an agar without nitrogen (Wink, unpublished). Sparteine degradation could also be shown in cell-free systems ([24], Wink, unpublished). Surprisingly, QA-degrading enzymes are present both in the cells and in the culture medium [24]. Additionally, the culture medium contains substantial activities of hydrolytic and oxydizing enzymes, including peroxidase, phosphatase, esterase, glucosidase, galactosidase, amylase, lipase, and nuclease which are able to degrade a variety of primary and secondary products [24]. Since many cell cultures also secrete primary and secondary products into their medium, the culture fluid may thus function as an extracellular lytic compartment besides the vacuoles of the cells [24], an observation which might be of importance for other cell cultures which fail to produce secondary products. QA content of cell cultures, kept under light/dark regime, follows a diurnal rhythm [31]. Thereby QA are first accumulated in the cells but then released into the medium from which the alkaloids disappeared. In view of the enzymic results it is likely that part of the alkaloids were actually degraded in the medium, others were taken up by the cells and degraded later on. In contrast to the suspension-cultured cells, the alkaloids formed in the intact plants are secreted into the phloem and are translocated to peripheral tissues such as epidermal and subepidermal cell layers, which are able to store QA. In lupin cell cultures these specialized cells are obviously missing, so that instead of alkaloid storage we observe alkaloid degradation. It is often postulated that cell cultures fail to produce secondary products because the genes of the respective biosynthetic pathway are repressed [1, 8]. This may be the situation in some cell culture systems, but our data show that many more processes, such as transport, accumulation and degradation are of crucial importance as to whether a cell culture actually produces secondary products or not.

Acknowledgements. Our investigations were supported by grants of the Deutsche Forschungsgemeinschaft and the Land Niedersachsen. I would like to thank Prof. Dr. T. Hartmann for generous support and encouragement, C. Theuring and S. Schmidt for technical assistance, and Dr. Coralie Wink for reading the manuscript.

References

1. Berlin J (1984) Plant cell cultures – a future source of natural products. Endeavour (in press)
2. Cho YD, Martin RO (1971) Biosynthesis of *Thermopsis* alkaloids from Carbon-14 dioxide. Evidence for the formation of the pyridone bases from lupanine via 5,6-dehydrolupanine. Can J Biochem 49:971–977
3. Hahlbrock K, Lamb CL, Purwin C, Ebel J, Fautz E, Schäfer E (1981) Rapid response of suspension-cultured parsley cells to elicitors from *Phytophthora megasperma* var. sojae. Plant Physiol 67:768–773
4. Hartmann T, Schoofs G, Wink M (1980) A chloroplast-localized lysine decarboxylase of *Lupinus polyphyllus*, the first enzyme in the biosynthetic pathway of quinolizidine alkaloid biosynthesis. FEBS Lett 115:35–38
5. Hegnauer R (1958) Chemotaxonomische Betrachtungen. Die systematische Bedeutung des Alkaloidmerkmals. Planta Med 6:8–32
6. Leete E (1983) The biosynthesis of alkaloids. Biosynthesis 7:102–223
7. LeScao-Bogaert F, Faugeras G, Paris PR (1978) Sur la presence d'alcaloides quinolizidique chez l'*Osyris alba* L. *(Santalacees)*. Plant Med Phytother 12:315–318
8. Luckner M (1984) Secondary metabolism in microorganisms, plants and animals. Fischer, Jena
9. McCoy JW, Stermitz FR (1983) Alkaloids from *Castilleja miniata* and *Penstemon whippleanus*, two host species for the plume moth, *Amblyptilia pica*. J Nat Prod 46:902–907
10. Nowacki E, Waller GR (1977) Quinolizidine alkaloids from *Leguminosae*. Rev Latinoam Quim 8:49–56
11. Murakoshi I, Sanda A, Haginiwa J, Suzuki N, Ohmiya S, Otomasu H (1977) An S-adenosyl-L-methionine; cytisine methyltransferase in *Thermopsis* seedlings. Chem Pharm Bull 25:1970–1973
12. Rosenthal GA (1982) Plant nonprotein amino acids and imino acids. Academic Press, London New York
13. Schoofs G, Teichmann S, Hartmann T, Wink M (1983) Lysine decarboxylase in higher plants and its integration in quinolizidine alkaloid biosynthesis. Phytochemistry 22:65–69
14. Schütte HR (1969) Chinolizidinalkaloide. In: Mothes K, Schütte HR (eds) Biosynthese der Alkaloide. VEB-Verlag, Berlin, pp 324–343
15. Schütte HR (1982) Simple amines, pyrrolizidine and quinolizidine alkaloids. Prog Bot 44: 166–181
16. Waller GR, Nowacki EK (1978) Alkaloid biology and metabolism in plants. Plenum Press, New York London
17. Wiermann R (1981) Secondary products and cell and tissue differentiation. In: Conn EE (ed) The biochemistry of plants. Academic Press, London New York, pp 85–116
18. Wink M (1983) Inhibition of seed germination by quinolizidine alkaloids. Aspects of allelopathy in *Lupinus albus* L. Planta 158:365–368

19. Wink M (1984) Wounding-induced increase of quinolizidine alkaloid accumulation in lupin leaves. Z Naturforsch 38c:905–909
20. Wink M (1984) Chemical defense of lupins. Mollusc-repellent properties of quinolizidine alkaloids. Z Naturforsch 39c:553–558
21. Wink M (1984) Chemical defense of *Leguminosae*. Are quinolizidine alkaloids part of the antimicrobial defense system of lupins? Z Naturforsch 39c:548–552
22. Wink M (1984) N-methylation of quinolizidine alkaloids: an S-adenosyl-l-methionine: cytisine N-methyltransferase from *Laburnum anagyroides* plants and cell cultures of *L. alpinum* and *Cytisus canariensis*. Planta 161:339–344
23. Wink M (1984) Biochemistry and chemical ecology of lupin alkaloids. Proc 3rd Int Lupin Conf, pp 325–343
24. Wink M (1985) Evidence for an extracellular lytic compartment of plant cell suspension cultures: the culture medium. Naturwissenschaften 71:635–636
25. Wink M, Hartmann T (1979) Cadaverine-pyruvate transamination: the principal step of quinolizidine alkaloid biosynthesis in *Lupinus polyphyllus* cell suspension cultures. FEBS Lett 101: 343–346
26. Wink M, Hartmann T (1980) Production of quinolizidine alkaloids by photomixotrophic cell suspension cultures: Biochemical and biogenetic aspects. Planta Med 40:149–155
27. Wink M, Hartmann T (1981) Activation of chloroplast-localized enzymes of quinolizidine alkaloid biosynthesis by reduced thioredoxin. Plant Cell Rep 1:6–9
28. Wink M, Hartmann T (1981) Sites of enzymatic synthesis of quinolizidine alkaloids and their accumulation in *Lupinus polyphyllus*. Z Pflanzenphysiol 102:337–344
29. Wink M, Hartmann T (1982) Localization of the enzymes of quinolizidine alkaloid biosynthesis in leaf chloroplasts of *Lupinus polyphyllus*. Plant Physiol 70:74–77
30. Wink M, Hartmann T (1982) Enzymatic synthesis of quinolizidine alkaloid esters: a tigloyl-CoA: 13-hydroxylupanine 0-tigloyltransferase from *Lupinus albus* L. Planta 156:560–565
31. Wink M, Hartmann T (1982) Diurnal fluctuation of quinolizidine alkaloid accumulation in legume plants and photomixotrophic cell suspension cultures. Z Naturforsch 37c:369–357
32. Wink M, Hartmann T (1982) Physiological and biochemical aspects of quinolizidine alkaloid formation in cell suspension cultures. Proc 5th Int Congr Plant Tissue Cell Cult. IAPIC, Tokyo, pp 333–334
33. Wink M, Hartmann T (1985) Enzymology of quinolizidine alkaloid biosynthesis. In: Zalewski RI, Skolik JJ (eds) Natural products chemistry 1984. Elsevier, Amsterdam, pp 511–520
34. Wink M, Witte L (1983) Evidence for a wide-spread occurrence of the genes of quinolizidine alkaloid biosynthesis. Induction of alkaloid accumulation in cell suspension cultures of alkaloid-"free" species. FEBS Lett 159:196–200
35. Wink M, Witte L (1984) Turnover and transport of quinolizidine alkaloids. Diurnal fluctuations of lupanine in the phloem sap, leaves and fruits of *Lupinus albus* L. Planta 161:519–524
36. Wink M, Hartmann T, Schiebel HM (1979) A model mechanism for the enzymatic synthesis of lupin alkaloids. Z Naturforsch 35c:93–97
37. Wink M, Hartmann T, Witte L (1980) Enzymatic synthesis of quinolizidine alkaloids in lupin chloroplasts. Z Naturforsch 35c:93–97
38. Wink M, Witte L, Schiebel HM, Hartmann T (1980) Alkaloid pattern of cell suspension cultures and differentiated plants of *Lupinus polyphyllus*. Planta Med 38:238–245
39. Wink M, Hartmann T, Witte L, Rheinheimer J (1982) Interrelationship between quinolizidine alkaloid producing legumes and infesting insects: Exploitation of the alkaloid containing phloem sap of *Cytisus scoparius* by the Broom aphid *Aphis cytisorum*. Z Naturforsch 37c: 1081–1086
40. Wink M, Schiebel HM, Witte L, Hartmann T (1982) Quinolizidine alkaloids from plants and their cell suspension cultures. Ester alkaloids of *Lupinus polyphyllus*. Planta Med 44:15–20
41. Wink M, Witte L, Hartmann T, Theuring C, Volz V (1983) Accumulation of quinolizidine alkaloids in plants and cell suspension cultures: Genera *Lupinus, Cytisus, Baptisia, Genista, Laburnum*, and *Sophora*. Planta Med 48:253–257
42. Wink M, Heinen HJ, Vogt H, Schiebel HM (1984) Cellular localization of quinolizidine alkaloids by laser desorption mass spectrometry (LAMMA). Plant Cell Rep 3:230–233
43. Wippich C, Wink M (1985) Biological properties of alkaloids. Influence of quinolizidine alkaloids and gramine on the germination and development of powdery mildew, *Erysiphe graminis* f.sp. *hordei*. (in press)

Production of Alkaloids by Ergot (*Claviceps fusiformis* Lov.) on *Pennisetum typhoides* (Burm.) Stapf and Hubb. in Vitro

S. ROY and A. KUMAR[1]

1 Introduction

Ergot disease of pearl millet [*Pennisetum typhoides* (Burm.) Stapf and Hubb.] caused by *Claviceps fusiformis* Lov. results in the formation of sclerotia in place of normal grains. Most *Claviceps* sp. and strains, whether living as parasites or grown in axenic culture show growth phase, transition period, and finally alkaloid producing phase (idiophase). Sclerotia are reported to contain clavine group of alkaloids. Like other indole alkaloids, ergot alkaloids are derived in part from tryptophan and methyl group of methionine. The ergot poisoning which may result from consumption of ergotted grains, is characterized by symptoms such as nausea, vomiting, diarrhoea, giddiness, and even death of human beings and livestock. Several investigations have been undertaken on the estimation and identification of alkaloids produced by *Claviceps fusiformis* in vivo and in axenic cultures [1-5].

Plant tissue culture technique provides completely controlled conditions to elucidate the growth and physiology of cells and to study host-parasite interaction at cellular level [6]. Cutter [7] obtained good dual growth but only limited sporulation with rust fungus *Uromyces ari-triphylli* and callus of monocotyledon *Arisaema triphyllum*. Milholland [8] infected the healthy callus derived from scuteller node of wheat with *Puccinia graminis* f.sp. *tritici*. The auxin and cytokinin component of culture medium, together with physical and nutritional factors, influences the growth and sporulation of biotrophs grown on host tissue cultures. With the possible exception of work with *Plasmodiophora brassicae* on *Brassica* spp. [9], and *Melampsora lini* (Pers.) Lev. and flax callus [10], the effects of these factors on dual growth have not yet been widely investigated.

The investigations on the production of alkaloids by the ergot fungus in dual cultures of the host tissue are lacking. Present investigations were undertaken with the object of studying the influence of physical, nutritional, and hormonal factors on the growth of the pearl millet callus and dual cultures of *C. fusiformis* and pearl millet callus, and the production of the alkaloids by the latter.

1 Department of Botany, University of Rajasthan, Jaipur, 302004, India

Primary and Secondary Metabolism of Plant Cell Cultures, ed. by Neumann et al.

2 Materials and Methods

Bajra callus raised from hypocotyl region of cultivar P10 was maintained on Murashige and Skoog's [11] medium through periodic subculturing under dark at a temperature of 26° ± 2 °C in a culture room. The monoconidial *C. fusiformis* cultures were maintained on modified Kirchhoff's medium (KH_2PO_4 1.0 g l^{-1}; $MgSO_4 \cdot 7H_2O$ 0.2 g l^{-1}; KNO_3, 1.5 g l^{-1}; asparagine 1.0 g l^{-1}; sucrose 150 g l^{-1}, and yeast extract 10 g l^{-1}). Around 10 days old freshly subcultured callus was inoculated with a drop of uniform conidial suspension of *C. fusiformis* containing 10^6 conidia ml^{-1}. After further incubation of 3 weeks, the dual cultures were harvested and growth was recorded in terms of fresh weight. Alkaloid estimation was done following the method of Mukerji and De [12]. Optical density was recorded at 590 nm using Carl Zeiss PM6 spectrophotometer. The growth of the dual cultures is shown in Fig. 1.

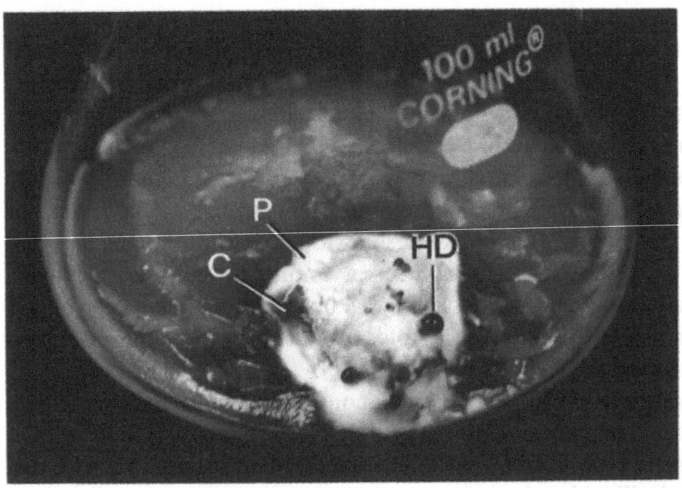

Fig. 1. Dual culture showing growth of *Claviceps fusiformis* on *Pennisetum typhoides* callus cultures. The formation of honey dew can also be observed. *C* callus; *P* pathogen; *HD* honey dew

3 Results

3.1 Effect of Temperature

In a temperature range of 10° to 40 °C, maximum growth of pearl millet callus was recorded at 25 °C, while 30 °C favoured the best growth of dual cultures. However, the maximum alkaloid production by dual cultures was recorded at 25 °C (Fig. 2).

3.2 Effect of Light

Alternate light and dark conditions (8:16 h) favoured maximum growth of callus cultures, while maximum growth of dual cultures and alkaloid production was observed in total darkness. Light including red (650 nm) and blue (450 nm) light, inhibited the growth and alkaloid production in dual cultures (Fig. 3).

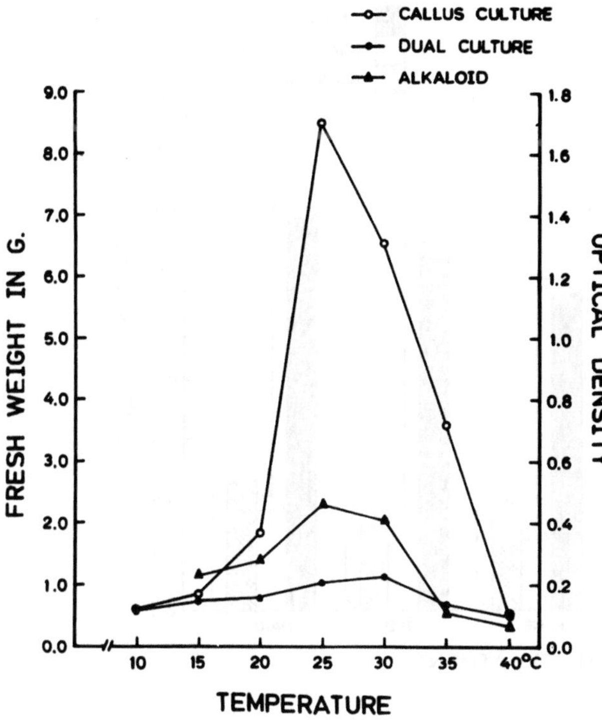

Fig. 2. Effect of temperature on the growth of pearl millet callus, dual cultures and production of alkaloids by the latter

3.3 Effect of Carbohydrates

Absence of carbohydrate source inhibited growth of normal callus as well as dual cultures. Alkaloid production was also adversely influenced in the absence of carbohydrate supply. However, the inert sugar mannitol (30 g l^{-1}) gave maximum growth and alkaloid production by dual cultures. In contrast to this, sucrose promoted maximum growth of pearl millet callus cultures. In comparison to sucrose, glucose and maltose favoured better alkaloid production (Fig. 4). Autoclaved fructose strongly inhibited growth and alkaloid production by dual cultures.

3.4 Effect of Vitamins

Different vitamins, thiamine · HCl (0.2 ppm), nicotinic acid (0.5 ppm), biotin (1.0 ppm), meso-inositol (100 ppm), and pyridoxine · HCl (0.5 ppm) when incorporated into the MS medium variously influenced the growth and alkaloid production in dual cultures. Although absence of vitamins resulted in poor growth of pearl millet callus, as well as dual cultures, alkaloid production was not significantly influenced. However, addition of pyridoxine slightly promoted the alkaloid production, which was otherwise adversely affected by supplementation of thiamine · HCl, nicotinic acid, biotin, meso-inositol (Fig. 5). The biotin promoted maximum growth of callus cultures, while thiamine · HCl favoured maximum growth of dual cultures. Apparently no direct correlation was observed in growth and alkaloid production.

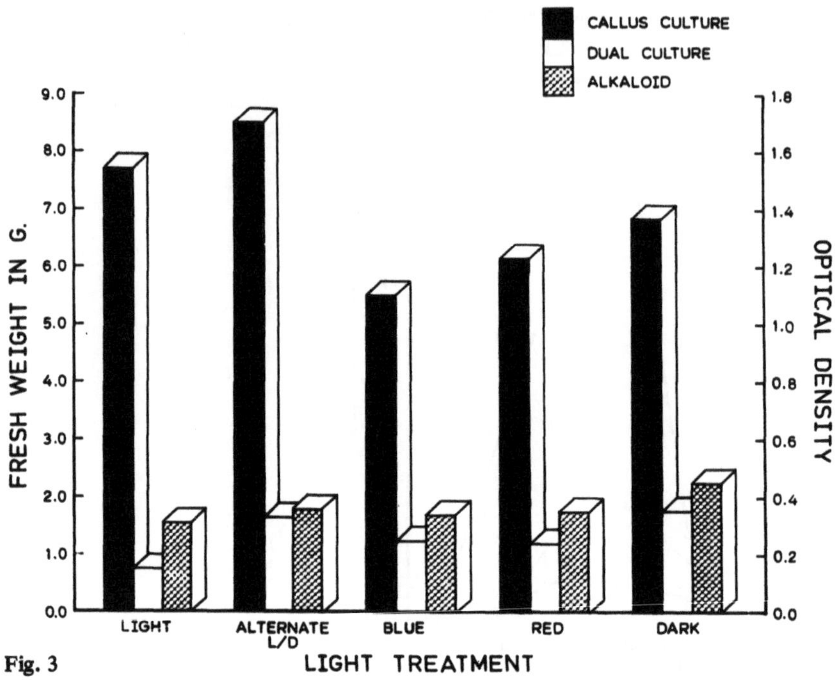

Fig. 3

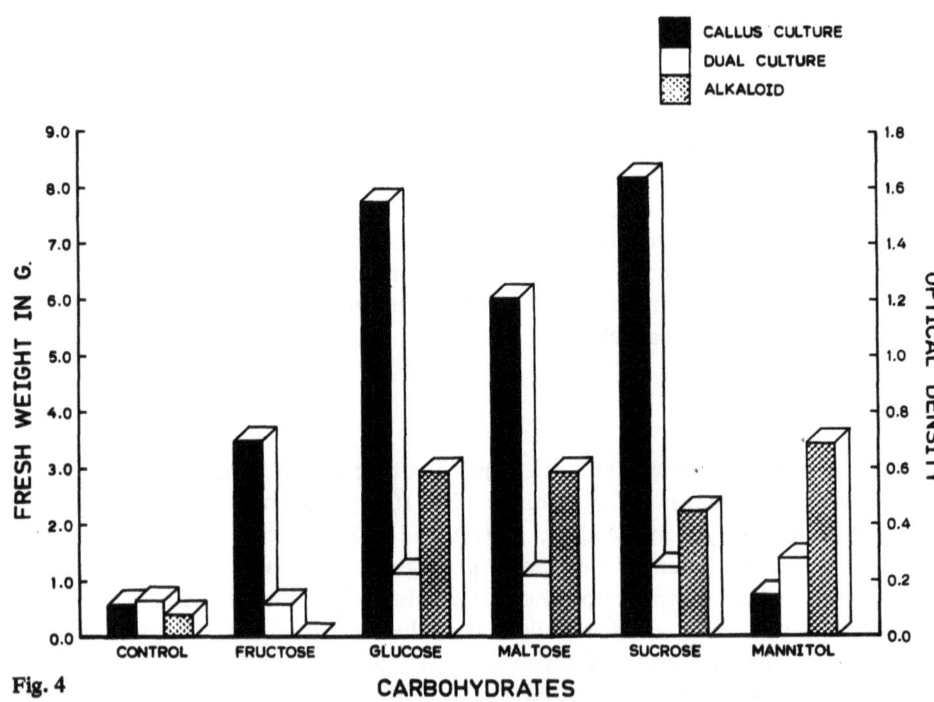

Fig. 4

Fig. 3. Effect of light on growth of callus cultures and dual cultures. The alkaloid contents in dual cultures are also recorded

Fig. 4. Effect of carbohydrates on growth of callus cultures and dual cultures. Alkaloid contents in dual cultures are also recorded

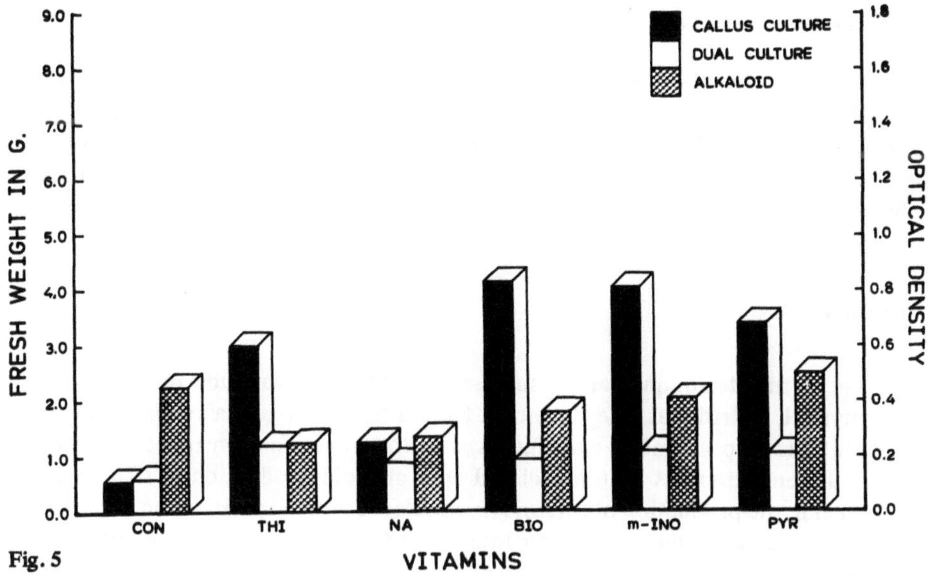

Fig. 5

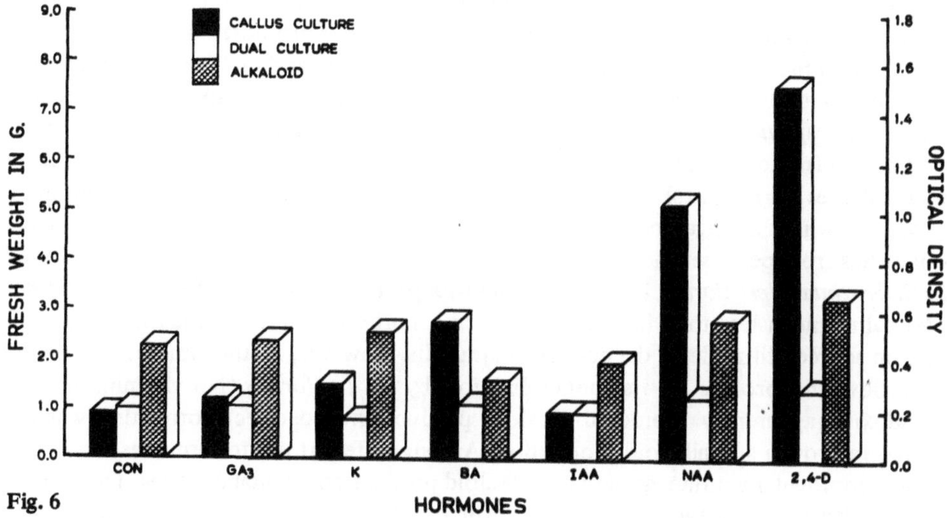

Fig. 6

Fig. 5. Effect of vitamins on growth of callus and dual cultures and the alkaloid contents in the latter

Fig. 6. Effect of phytohormones on growth of callus cultures and dual cultures. Alkaloid contents in dual cultures are also shown

3.5 Effect of Phytohormones

Among different phytohormones α-naphthaleneacetic acid (NAA) (10 ppm); benzyladenine (BA) (3 ppm); kinetin (K) (0.04 ppm); gibberellic acid (GA_3) (1 ppm); indole-3yl-acetic acid (IAA) (10 ppm), and 2,4-Dichlorophenoxy acetic acid (2,4-D) (0.5 ppm) influenced maximum growth of callus, as well as dual cultures and the alkaloid production by the latter (Fig. 6).

4 Conclusions

Plant tissue culture technique can be successfully employed to study the influence of various physical, nutritional, and hormonal factors on growth and alkaloid production by dual cultures of pearl millet and *Claviceps fusiformis* under controlled conditions. The cytodifferentiation has been associated with alkaloid production in *Claviceps* sp. [13]. In general, the production of alkaloids is correlated with a specific kind of mycelial growth known as sclerotial growth on intact plants. Histopathological investigations revealed the formation of a similar kind of tissue in dual cultures. Although alkaloids are also produced in axenic cultures, detailed qualitative studies are needed to characterize the composition of alkaloids produced on host tissue in sclerotia, in dual cultures and axenic cultures.

Physical factors such as temperature and light influenced the alkaloid production by dual cultures. Total darkness, which promoted maximum sporulation in axenic cultures, also favoured maximum alkaloid production in dual cultures. Among the different carbon sources employed, the inert sugar mannitol promoted maximum growth and alkaloid production in comparison to sucrose, glucose, and maltose. Although cane sugar is reported to be the best carbon source for the growth and sporulation of *Claviceps microcephala* [14] and it also supported maximum growth and sporulation of *C. fusiformis* axenic cultures, but the maximum growth and alkaloid production by dual cultures on mannitol needs clarification. Mannitol, a derivative of the sugar mannose is known as a fungal metabolite. It is formed in the culture medium in appreciable quantities from pentoses, hexoses, and glycerol [15]. Its biosynthesis has been recorded in *Sclerotinia sclerotiorum* [16]. Since mannitol promotes growth of *Pennisetum typhoides* callus much less than in dual cultures, apparently the fungus somehow makes this carbon source utilizable to the cultured bajra cells. However, further studies are needed to follow the process of its metabolization by *C. fusiformis*. No vitamins besides pyridoxine apparently promoted alkaloid production. No direct correlation was observed in growth and alkaloid production. Among different phytohormones used, only 2,4-D promoted maximum growth and alkaloid production by dual cultures. The normal callus cultures also grew best on 2,4-D. Apparently dual cultures can be employed to study the alkaloid production as influenced by various factors and also to develop systems for biotransformations of specific alkaloids.

Acknowledgement. We are grateful to Professor Dr. K.-H. Neumann, Institut für Pflanzenernährung der Justus Liebig-Universität, Abt. Gewebekultur, Gießen, FRG, for valuable suggestions during the course of investigation. Financial assistance from Council of Scientific and Industrial Research, New Delhi is gratefully acknowledged.

References

1. Shinde PA, Bhide IP (1958) Ergot of bajra *(Pennisetum typhoides)* in Bombay state. Curr Sci 27:499–500
2. Loveless AR (1967) *Claviceps fusiformis* sp Nov., the casual agent of an agalactia of sows. Trans Br Mycol Soc 50:15–18
3. Sivaprakasam K, Chinnadurai G, Krishnamurthy CS (1971) Alkaloid production by *Claviceps microcephala* on some varieties of pearl millet. Madras Agric J 58:431–432
4. Singh HN, Husain A (1977) Saprophytic production of ergot alkaloids by bajra ergot (*Claviceps fusiformis* Loveless). Indian J Exp Biol 15:585–586
5. Kumar A, Arya HC (1978) Estimation and identification of alkaloids produced by *Claviceps fusiformis* Loveless on some varieties of pearl millet. Curr Sci 47:633–635
6. Ingram DS, Helgeson JP (eds) (1980) Tissue culture methods for plant pathologists. Blackwell, Oxford London, pp 272
7. Cutter VM (1960) Studies on the isolation and growth of plant rusts in host tissue cultures and upon synthetic media II. *Uromyces ari-triphylli*. Mycologia 52:726–742
8. Milholland RD (1962) Callus tissue from wheat for infection studies of *Puccinia graminis* var. *tritici*. Phytopathology (Abstr) 52:21
9. Reddy MN, Williams PH (1970) Cytokinin activity in *Plamodiophora brassicae* infected cabbage tissue cultures. Phytopathology 60:1463–1465
10. Turel FLM (1969) Saprophytic development of the flax rust *Melampsora lini* race no. 3. Can J Bot 47:821–823
11. Murashige T, Skoog F (1962) A revised medium for rapid growth and bioassays with tobacco tissue cultures. Physiol Plant 15:473–497
12. Mukerji B, De NK (1944) A method for the assay of individual ergot sclerotium. Curr Sci 13: 128
13. Vaidya HC, Desai JD (1981) Cell differentiation and alkaloid production in *Claviceps* sp. strain SD58. Indian J Exp Biol 19:829–831
14. Singh SB, Bais BS, Singh DR (1972) Effect of different carbon and nitrogen sources on the growth and sporulation of *Claviceps microcephala* (Wallr.) Tull. Mycopathol Mycol Appl 46: 373–378
15. Bilgrami KS, Verma RN (1981) Physiology of fungi. Vikas, New Delhi, pp 507
16. Wang SYC, Le Tourneau D (1972) Mannitol biosynthesis in *Sclerotinia sclerotiorum*. Arch Mikrobiol 81:91–99

Compartmentation of Alkaloids in a Cell Suspension of *Catharanthus roseus*: A Reappraisal of the Role of pH Gradients

J.P. RENAUDIN, S.C. BROWN, and J. GUERN[1]

1 Introduction

Many authors have proposed that alkaloids are accumulated in acidic vacuoles, through ion trapping [4, 11, 13, 16, 17, 21] or binding to phenolic compounds [9, 13]. Furthermore, it has been claimed that vacuolar accumulation could result from a specific active transport through the tonoplast [6]. Evidence also exists for binding of alkaloids to the cell walls of whole plants [15].

We have studied alkaloid compartmentation in a suspension culture of *Catharanthus roseus* cells which synthesize mainly the indole alkaloids ajmalicine and serpentine. Earlier results with exogenous ^{14}C-alkaloids have confirmed the ion-trapping model [4, 19, 21]. Here we present new results both from tracer studies and from the titration of endogenous alkaloids in the cells, in the medium and in protoplasts derived from these cells. These studies demonstrate that, while unspecific ion trapping in the vacuole occurs for alkaloids generally, in addition other accumulation mechanisms also exist in cells.

2 Methods

Cell culture and protoplast preparation. The cell line C20 of *Catharanthus roseus* was cultivated in 0.5-l flasks containing 0.2-l Gamborg B5 medium with 1 μM 2,4-dichlorophenoxyacetic acid and 0.06 μM kinetin. Its growth characteristics are described in [3]. Protoplasts were prepared from 5- or 6-day-old suspensions according to [2].
Uptake of radioactive alkaloids. The procedure of uptake measurement with ^{14}C-indole alkaloids has already been described [4, 19, 21]. Experiments longer than 7 h were conducted aseptically. Concentrations are expressed relative to fresh weight and medium volumes.
Extraction and quantification of alkaloids. Cells were separated from their medium by filtration on a sintered glass filter (no. 2). Alkaloids were extracted from the cells by slight modification of [20], involving an extraction by 5% sulphuric acid at room temperature. The medium was made to

1 Laboratoire de Physiologie Cellulaire Végétale, CNRS-INRA, BP no 1, 91190 Gif-sur-Yvette, France

List of abbreviations and symbols. C_e extracellular concentration; C_i intracellular concentration relative to the fresh weight volume; C_v vacuolar concentration; pH_e extracellular pH; pH_s cell sap pH; pH_v vacuolar pH; pKa acidity constant of an alkaloid; CAS ceric ammonium sulphate; CCCP carbonyl cyanide m-chlorophenylhydrazone.

Primary and Secondary Metabolism of Plant Cell Cultures, ed. by Neumann et al.

pH 10 and extracted three times with chloroform (v/v). Ajmalicine and serpentine were assayed by reversed-phase high performance liquid chromatography after fractionation of the extracts on a silica column [20].

3 Evidence in Favour of the Ion-Trapping Model

3.1 Description of the Ion-Trapping Model

Alkaloids are low molecular weight amines whose neutral form is lipophilic and freely diffuses through membranes. Hence it has been suggested that alkaloids accumulate as non-diffusible cations in the acidic vacuoles of plant cells [16, 17, 21]. This model implies that, when several compartments exist, (1) there is an equilibrium distribution of the alkaloids between all the compartments, (2) the neutral base is present at the same concentration in each compartment, and (3) the cation which cannot diffuse remains trapped in the more acidic compartments, and is accumulated in these compartments according to their pH and to the acidity constant of the alkaloid.

For cell cultures, the simplest approach is to consider only two compartments: the extracellular medium, with the larger relative volume, and the vacuoles, large and acidic. The cellular concentration C_i is thus approximately that of the vacuole, C_v. At the diffusion equilibrium, the following formula gives the ratio of cell to medium concentrations:

$$C_i/C_e \cong C_v/C_e = (1 + 10^{pKa-pH_v})/(1 + 10^{pKa-pH_e})\,. \qquad (1)$$

with C_i, C_v, C_e: intracellular, vacuolar, and extracellular concentrations of the alkaloid; pH_v, pH_e: vacuolar and extracellular pH; pKa: acidity constant of the alkaloid.

The concentration ratio of an alkaloid depends on the pH gradient between the medium and the vacuole. The higher the value $pH_e - pH_v$, the more the alkaloid will be accumulated in cells. Equation (1) assumes (1) that no binding of the alkaloid occurs in any compartment and that Donnan trapping in the cell wall remains negligible, (2) that the cation fluxes through plasmalemma and tonoplast are weak compared to those of the neutral form, and (3) that active transport of the alkaloid does not exist. Any of these phenomena would lead to a different equilibrium between the cells and the medium, as previously discussed [4, 21].

3.2 Influence of the pH Gradient on the Uptake of Radioactive Alkaloids

The ion-trapping model has been validified in studies of the uptake by cells of several radioactive indole alkaloids, namely tryptamine (pKa = 10.2) [4], tabernanthine (pKa = 8.1) [19] and ajmalicine (pKa = 6.3) [21]. Tabernanthine, which is not a *Catharanthus* alkaloid, was chosen for its intermediate pKa value.

Each alkaloid was taken up by the cells and an equilibrium distribution was observed between the cells and the medium within a few hours. For a given population the concentration ratios were in the same order as the pKa values [19] which is in accordance with Eq. (1). Figure 3A presents the kinetics of influx and of efflux of ^{14}C-ajmalicine

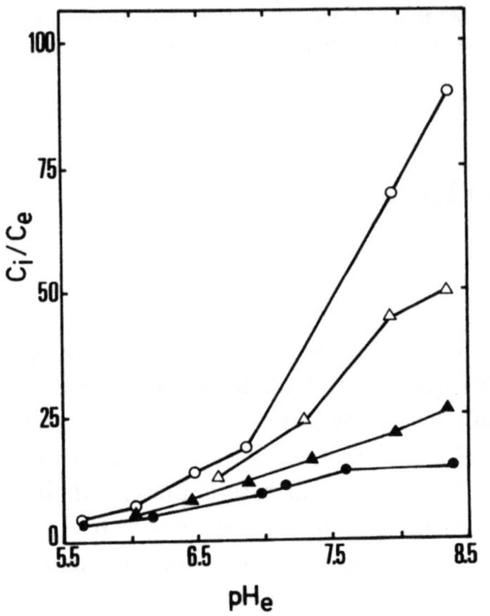

Fig. 1. Effect of the pH value of the incubation medium, pH_e, on the concentration ratio C_i/C_e of ^{14}C-ajmalicine (*solid symbols*) and of ^{14}C-tabernanthine (*open symbols*) at the diffusion equilibrium. Different symbols are used for different experiments. Adapted from [21]

in a typical experiment. Ajmalicine and tabernanthine enter the cells by a diffusion process involving mainly their neutral molecular form, as shown by the initial rates of uptake as a function of the extracellular pH [21]. Diffusion is also evident from the identical pattern of influx and of efflux curves (Fig. 3A).

The extent of the accumulation of the alkaloids within the cells depended on the pH gradient between the medium and the cells: the concentration ratios were highest when the incubation medium had a high pH value (Fig. 1). In contrast, the accumulation ratios declined when the vacuolar pH was raised by pretreatment of cells by methylamine [4, 19, 21]. This supports the ion-trapping model. The diffusional nature of the equilibrium was furthermore shown by the effect of quick pH_e modifications at the equilibrium: acidification led to efflux of a part of the cellular alkaloids, whereas alkalinization provoked influx [19, 21].

3.3 Distribution of Endogenous Ajmalicine and Serpentine Between Cells and Their Culture Medium

The distribution of alkaloids between cells and medium has been studied with 5- and 6-day-old suspensions, during the exponential growth phase. At this time, it was predicted from Eq. (1) that significant amounts of alkaloids would be found in the medium, since the vacuolar pH reached a maximum estimated at pH 6.7 with the cell sap pH value [12] whereas the extracellular pH was only 6.2. Table 1 shows that the medium contained 25%–50% of the total ajmalicine and 2%–6% of the total serpentine. The concentration ratio of ajmalicine was rather low, about 10.4, whereas that of serpentine was much higher, about 150.

The presence of these two alkaloids in the culture medium was not due to cell death, since the ratio of ajmalicine to serpentine molecules was much higher in the medium

Table 1. Ajmalicine and serpentine in cell suspensions of *Catharanthus roseus* [a]

Alkaloid	pKa	Concentration ng ml^{-1} [b]			% in the medium
		C_i	C_e	C_i/C_e	
Ajmalicine	6.3	1,480 ± 940	135 ± 65	10.4 ± 2.8	25% to 50%
Serpentine	10.8	525 ± 210	3.7 ± 0.9	150 ± 45	2% to 6%

[a] The identity of the alkaloids in the extracts was checked by their chromatographic behaviour both on silicagel and on reversed-phase, and by their specific fluorescence (serpentine) or coloration after a CAS spray [7] (ajmalicine)

[b] Mean + standard deviation of nine experiments with 5- or 6-day-old suspension cultures. C_i is the intracellular concentration relative to the fresh weight volume; C_e is the extracellular concentration. The dispersion is partly due to variation of the vacuolar and extracellular pH values

than in the cells, and in agreement with the ion-trapping model the alkaloid with the higher pKa value, serpentine (pKa = 10.8), was the more accumulated.

3.4 Flow Cytometry on Serpentine Content of Protoplasts

The level of serpentine in individual cells was assayed through the blue fluorescence of this alkaloid, and has been found to vary markedly within a population [2, 6, 17]. Fluorescence has been analyzed by flow cytometry with protoplast preparations [2]. Figure 2 is an histogram of serpentine content per individual protoplast. Once corrected for volume dispersion, the range of individual fluorescence was about 1 to 25.

A correlation between serpentine content and vacuolar pH was demonstrated by cell sorting. Protoplasts of extreme fluorescence values, accounting for 10%–25% of the total population, have been sorted and their vacuolar pH was then assessed with

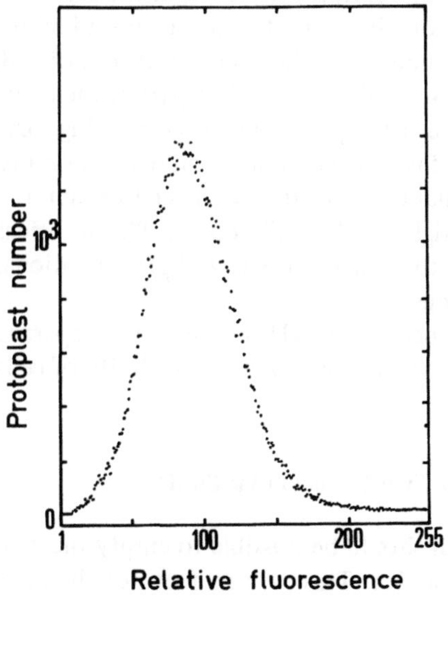

Fig. 2. Histogram of serpentine content in individual protoplasts for 10^4 protoplasts from a *C. roseus* suspension. The blue fluorescence was measured using a flow cytometer (FACS 440, details in [2]). Total cell fluorescence was divided by low-angle light scatter (proportional to diameter squared) to at least partially correct for cell volume differences: relative fluorescence was thus an approximation of intracellular concentration, distributed over 255 classes. Channel 1 approaches zero fluorescence. The mode intensity was at channel 88; channel 255, with 0.3% of the population, includes all extreme values

the fluorescent probe 9-aminoacridine and microfluorometry. The protoplasts with a high serpentine content always had a vacuolar pH value lower by 0.1–0.3 units than that of protoplasts with a low serpentine content [2].

In spite of its limitations due to quenching of serpentine fluorescence [2] and to serpentine loss from cells during protoplast isolation (Sect. 4.4), this study demonstrates the positive correlation within a population between vacuolar pH, a parameter of the ion-trapping model, and serpentine content, as already indicated with cells [17].

4 Evidence in Favour of Other Trapping Mechanisms

4.1 Poor Quantitative Fit Between the Experimental Results and the Theoretical Model

The experimental values of the accumulation ratios were in poor agreement with the theoretical ones given by Eq. (1). With the cell sap pH as one estimation of the vacuolar pH, the theoretical accumulation ratios were always one order or more below the experimental ones. Furthermore, the vacuolar pH estimated from the ajmalicine concentration ratio [Table 1 and Eq. (1)] was 5.05 and from the serpentine ratio it was 4.12, while the mean cell sap pH was 6.52. The large difference between the values for ajmalicine and serpentine indicates that the ion-trapping model cannot account fully, nor to the same extent, for the accumulation of these two alkaloids. This is irrespective of the validity of the cell sap pH as an estimate of the vacuolar pH. A more detailed analysis of such model fitting has already been published [21].

4.2 Kinetics of Isotopic Equilibration of Ajmalicine Between Cells and Their Medium Indicate the Presence of Two Pools

Figure 3B demonstrates that ^{14}C-ajmalicine reached the same distribution between the cells and the medium as endogenous ajmalicine. The isotopic dilutions in the cells and the medium became equal after about 6 h, and notably long after the apparent kinetic equilibrium reached within 1 h (Fig. 3A). After one hour, the concentration ratio of ^{14}C-ajmalicine was only 75% that of endogenous ajmalicine. The last part of the equilibration took about 5 h and was not manifest in absorption curves such as Fig. 3A. This is attributed to the fact that the kinetics of isotopic equilibration depend on the relative variations of intracellular and extracellular concentrations. This latter factor is specially important for cell suspensions with high relative cell volume. Physiological changes such as increase in total cell volume due to growth, or pH_e and pH_v variations, must also be considered during long-term experiments.

The fact that the isotopic equilibration always occurs several hours after the apparent equilibrium indicates that at least two compartments are involved in the ability of cells to store ajmalicine.

4.3 Efflux Experiments Indicate the Presence of Two Intracellular Pools

As a consequence of the diffusion of alkaloids it should be possible to empty the cells by successive dilutions in medium without alkaloids. This was in fact not the case.

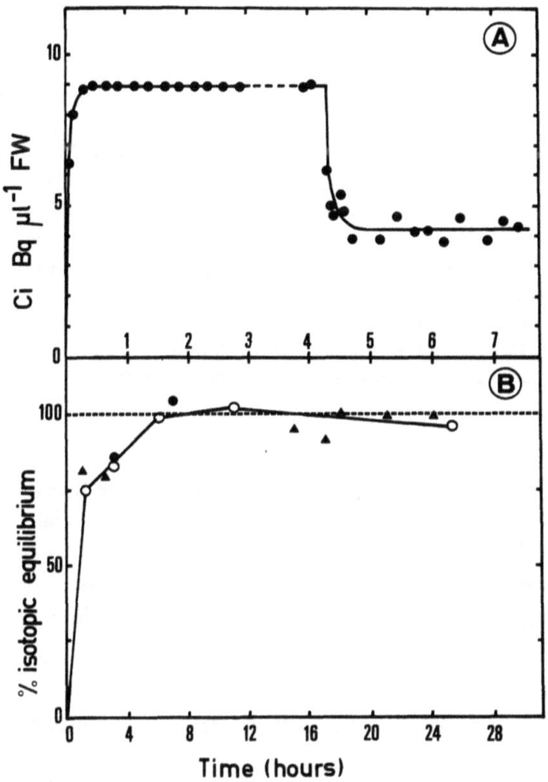

Fig. 3. Kinetics of ^{14}C-ajmalicine accumulation in *C. roseus* cells grown in suspension. ^{14}C-Ajmalicine (1.96 Bq μmol^{-1}) was added at time zero. **A** Short-term measurement of intracellular radioactivity during influx (*left*), and during efflux in the same medium without ^{14}C-ajmalicine (*right*). C_i is the intracellular concentration of the radioactive alkaloid; pH_e = 7.9; 3 ml fresh weight volume in 100 ml. The metabolism of ^{14}C-ajmalicine was negligible. At the end of the influx, the concentration ratio C_i/C_e of ^{14}C-ajmalicine was 25.1, and 70% of the radioactivity remained in the medium. These data were respectively 39.9 and 51% at the end of the efflux. **B** Longterm measurement of isotopic equilibration of ^{14}C-ajmalicine in a cell suspension. At various times the cells and the medium were analyzed and both ^{14}C-ajmalicine and endogenous ajmalicine were quantified. Percent isotopic equilibrium was calculated from the concentration ratio of ^{14}C-ajmalicine relative to that of endogenous ajmalicine. ○ points from one experiment; ● points corresponding to Fig. 3A; ▲ points from other pooled experiments

After two or three effluxes of 1.5 h, each one being in new culture medium at low cell volume density (1–3 ml fresh weight in 100 ml), there still remained 28.5% ± 5.2% ajmalicine and 114% ± 36% serpentine in the cells (mean + standard deviation of four experiments). Ajmalicine lost from cells was recovered in the medium. From curves such as Fig. 3A, the successive effluxes would have been expected to empty the cells.

Table 2 demonstrates that ajmalicine molecules which remained in the cells after successive effluxes were more strongly retained than the bulk of initial alkaloids. After either a short or a long incubation with ^{14}C-ajmalicine two successive effluxes were performed. At the end of the first most of the ajmalicine had been lost from the cells

Table 2. Efflux experiments with ^{14}C-ajmalicine[a]

Loading with ^{14}C-ajmalicine h	Influx C_i (%)	C_i/C_e	First efflux C_i (%)	C_i/C_e	Second efflux C_i (%)	C_i/C_e
2.5	100	14.2	5.2	6.2	2.4	78
17.5	100	10.9	13.8	11.9	11.8	263

[a] Cells were preloaded with ^{14}C-ajmalicine during different times; two effluxes were then successively performed, each of one hour in new culture medium at low cell volume density (1.5 ml fresh weight in 100 ml). ^{14}C-Ajmalicine was assayed in the cells and in the medium at the end of the loading period and after each efflux. C_i is the intracellular concentration of ^{14}C-ajmalicine, expressed as a percentage of the value before efflux, and C_i/C_e is the concentration ratio of ^{14}C-ajmalicine

and the concentration ratio remained in the same range as initially. During the second efflux however, a minor part of residual ajmalicine diffused into the medium, especially after a long initial preloading with ^{14}C-ajmalicine, and the concentration ratio of ajmalicine increased markedly. Furthermore, the strongly retained fraction was more abundant in cells which had been preloaded with ^{14}C-ajmalicine for a long time, in agreement with the results presented in Section 4.2.

4.4 Loss of Freely Diffusible Ajmalicine and of Serpentine During Protoplast Isolation

The alkaloid content of *C. roseus* protoplasts represented only a fraction of the initial level in cells (Table 3): only 32% of the initial ajmalicine and 49% of the serpentine were recovered in the protoplasts. Some loss of ajmalicine occurred during the prewash of cells with 0.55 M sorbitol, but the complement of effluxed ajmalicine and serpentine was recovered in the digestion medium of the protoplasts.

By rewashing the protoplasts 2.5 and 5 h after their preparation, it was shown that their serpentine content remained constant with time, whereas a slow loss of ajmalicine occurred. After 5 h, 40% of the ajmalicine initially in the protoplasts had disappeared. When protoplasts were prepared with cells incubated with ^{14}C-ajmalicine, the concentration ratio of the radioactive alkaloid in the protoplast suspension was found to be one

Table 3. Ajmalicine and serpentine in protoplasts[a]

Alkaloid	Concentration in cells		Concentration in protoplasts	
	ng ml^{-1} [b]	ng/10^6 cells[c]	ng/10^6 protoplasts	% cells
Ajmalicine	2,460 ± 2,150	226 ± 198	63 ± 68	32.4 ± 16.6
Serpentine	1,140 ± 600	104 ± 55	51 ± 28	49.2 ± 10.1

[a] Mean + standard deviation of five experiments; the time necessary for protoplast preparation was approximately 2.5 h

[b] Expressed in the fresh weight volume

[c] From the count of 10.9 ± 1.0 10^6 cells in 1 ml fresh weight, by the method described in [3]

order higher than that of the initial cells. These results recall those obtained with efflux experiments upon cells and indicate that ajmalicine molecules remaining in the protoplasts or in the cells after efflux are more strongly retained than those of the main intracellular pool.

5 Discussion

Evidence has been obtained that ajmalicine and serpentine are naturally distributed between cells and the culture medium, and that this equilibrium is a consequence of the alkaloids' ability to diffuse through membranes. The presence of alkaloids in the culture medium of plant cells has been reported many times, either for *Catharanthus roseus* cells [14, 18, 23] or for other plant species [1, 5, 8, 10], but so far no explanation has been presented for this phenomenon.

Endogenous ajmalicine can be equilibrated with ^{14}C-ajmalicine added in the culture medium. From the equilibration curve, from efflux experiments and from titration of protoplasts, it is concluded that at least two pools of ajmalicine are present inside cells. The first pools accounts for 70%–80% of the total cellular content in 5- to 6-day-old suspensions. It responds quickly to pH changes and to dilution, and thus is considered as a freely diffusible pool accumulating in the cells according to the ion-trapping model described in Section 3.1. It was suggested [21] that this pool could partly consist of bound molecules of ajmalicine, in equilibrium with the free ones. In any case, for these molecules the limiting step for the excretion into the medium is diffusion through membranes, and not dissociation of the binding.

The second pool of ajmalicine represents 20%–30% of the initial cellular content but becomes the main fraction in cells after efflux or in protoplasts. It equilibrates within 6 h with exogenous ^{14}C-ajmalicine and it is rather slowly lost from protoplasts. The identification of this pool, which we believe to be firmly bound molecules in the vacuole, is in course.

Serpentine is more strongly retained in cells than ajmalicine. Efflux experiments could not induce significant serpentine excretion within hours. This could suggest that the binding of alkaloids within cells is specific and more important for serpentine than for ajmalicine. However, the situation is more complicated as major losses of serpentine occur during isolation or CCCP treatment [2] of protoplasts. This could be due to changes in membrane permeability after plasmolysis and removal of the cell wall [22]. (A cell wall localization of serpentine cannot be concluded from these protoplast experiments since malic and citric acid are also lost from cells during protoplast isolation – data not shown). The strong retention of serpentine in cells would be mainly the consequence of a small but significant permeability through membranes of the serpentine cation, amplified by the potential difference through the plasmalemma, inside negative.

To conclude, it appears that (1) the ion-trapping model is basically valid for indole alkaloids, pH gradients being one of the forces driving alkaloid compartmentation and accumulation, (2) other forces, probably related to the binding of alkaloids to cellular components, accentuate their accumulation in cells and contribute to their compartmentation, and (3) the distribution of alkaloids between cells and their culture medium in a given system depends on the two types of force, the relative importance of which probably varying with the cell strain and the culture conditions.

Acknowledgements. The authors thank Dr. Husson and Besselièvre from the Institut de Chimie des Substances Naturelles (CNRS, Gif-sur-Yvette) for the synthesis of ^{14}C-ajmalicine.

References

1. Böhm H (1978) Regulation of alkaloid production in plant cell cultures. In: Thorpe TA (ed) Frontiers of plant tissue culture 1978, Calgary, pp 201–212
2. Brown SC, Renaudin JP, Prévot C, Guern J (1984) Flow cytometry and sorting of plant protoplasts: technical problems and physiological results from a study of pH and alkaloids in *Catharanthus roseus*. Physiol Veg 22:541–554
3. Courtois D, Guern J (1980) Temperature response of *Catharanthus roseus* cells cultivated in liquid medium. Plant Sci Lett 17:473–482
4. Courtois D, Kurkdjian A, Guern J (1980) Tryptamine uptake and accumulation by *Catharanthus roseus* cells cultivated in liquid medium. Plant Sci Lett 18:85–96
5. Delfel NE, Rothfus JA (1977) Antitumour alkaloids in callus cultures of *Cephalotaxus harringtonia*. Phytochemistry 16:1595–1598
6. Deus B, Zenk MH (1982) Exploitation of plant cells for the production of natural compounds. J Biotechnol Bioeng 24:1965–1974
7. Farnsworth NR, Blomster RN, Damratoski D, Meer WA, Cammarato LV (1964) Studies on *Catharanthus* alkaloids. VI. Evaluation by means of thin-layer chromatography and ceric ammonium sulfate spray reagent. Lloydia 27:302–314
8. Frischknecht PM, Baumann TW (1980) The pattern of purine alkaloid formation in suspension cultures of *Coffea arabica*. Planta Med 40:245–249
9. Jans BP (1973) Untersuchungen am Milchsaft des Schöllkrautes (*Chelidonium majus L.)*. Ber Schweiz Bot Ges 83:306–344
10. Kibler R, Neumann KH (1979) Alkaloidgehalte in haploiden und diploiden Blättern und Zellsuspensionen von *Datura innoxia*. Planta Med 35:354–359
11. Kurkdjian A (1982) Absorption and accumulation of nicotine in *Acer pseudoplatanus* and *Nicotiana tabacum* cells. Physiol Veg 20:73–83
12. Kurkdjian A, Guern J (1981) Vacuolar pH measurements in higher plant cells. I. Evaluation of the methylamine method. Plant Physiol 67:953–957
13. Matile P (1984) Das toxische Kompartiment der Pflanzenzelle. Naturwissenschaften 71:18–24
14. Mérillon JM, Chénieux JC, Rideau M (1983) Cinétique de croissance, évolution du métabolisme glucido-azoté et accumulation alcaloidique dans une suspension cellulaire de *Catharanthus roseus*. Physiol Veg 47:169–176
15. Mothes K (1955) Physiology of alkaloids. Annu Rev Plant Physiol 6:393–432
16. Müller (1976) Principles in transport and accumulation of secondary products. Nova Acta Leopoldina Suppl 7:123–128
17. Neumann D, Krauss G, Hieke M, Gröger (1983) Indole alkaloid formation and storage in cell suspension cultures of *Catharanthus roseus*. Planta Med 48:20–23
18. Petiard V (1980) Mise en évidence d'alcaloides dans le milieu nutritif de cultures de tissus de *Catharanthus roseus* G. Don. Physiol Veg 18:331–337
19. Renaudin JP (1981) Uptake and accumulation of an indole alkaloid, (^{14}C)-tabernanthine, by cell suspension cultures of *Catharanthus roseus* (L.) G. Don and *Acer pseudoplatanus* L. Plant Sci Lett 22:59–69
20. Renaudin JP (1984) Reversed-phase high-performance liquid chromatographic characteristics of indole alkaloids from cell suspension cultures of *Catharanthus roseus*. J Chromatogr 291: 165–174
21. Renaudin JP, Guern J (1982) Compartmentation mechanisms of indole alkaloids in cell suspension cultures of *Catharanthus roseus*. Physiol Veg 20:533–547
22. Taylor ARD, Hall JL (1976) Some physiological properties of protoplasts isolated from maize and tobacco tissues. J Exp Bot 27:383–391
23. Vinas R, Pareilleux A (1982) Production d'alcaloides par des suspensions cellulaires de *Catharanthus roseus* cultivées in vitro. Physiol Veg 20:219–225

Studies on Variability of Plant Tissue Cultures for Alkaloid Production in *Catharanthus roseus* and *Papaver somniferum* Callus Cultures

V. PETIARD, C. BAUBAULT, A. BARIAUD, M. HUTIN, and D. COURTOIS[1]

1 Introduction

Our topics in plant tissue culture concern the two types of application of these techniques: production of metabolites (biologically active substances or food additives) by mass cell culture and plant breeding of essential crops or aromatic species. In these fields, somaclonal variation can be either an advantage or a disadvantage, depending on the aim of the study (variability increase, stability of cell strains, homogeneity of micropropagated clones etc.). So the knowledge about variability in a wide extent is a fundamental bottle-neck for the success of our projects.

Concerning the production of metabolites and the biochemical phenotype of cell cultures, variability is included in all research strategies for the selection of high producing strains (Tabata et al. 1978; Constabel et al. 1982; Yamamoto et al. 1982; Zenk and Deus 1982; Sato and Yamada 1984; Schulte et al. 1984). This variability is observed either for primary metabolism such as the use of lactose as sole carbon source (Petiard 1974; Chaubet and Pareilleux 1982) or for secondary metabolism such as alkaloid production (Zenk et al. 1977; Constabel et al. 1982; Constabel 1983).

In our laboratory several strain collections were established from various species to study their ability to produce some secondary metabolites (*Papaver somniferum, Catharanthus roseus, Vinca minor, Amsonia tabernaemontana, Voacanga thouarsii, Coffea arabica, Digitalis purpurea, Medicago sativa* etc.). These strains were set up by the classical methods. Primary calli were obtained from different explants on solid media, then some strains were selected upon biochemical and morphological criteria along regular subcultures during more than 2 years. Among the different species studied a particular attention is brought in this report to *Catharanthus roseus* and *Papaver somniferum*. For the former, the variability is studied as a source of a possible cell heterogeneity within a callus from a well-established and stable strain. For the latter, the biosynthesis of new forms of alkaloids might be the expression of variability.

2 Does Somaclonal Variation Occur in a Stable Strain?

Strains, even issued from a single genotype, are phenotypically different, as was shown by numerous workers (Zenk et al. 1977; Constabel et al. 1982; Zenk and Deus 1982;

1 L.E.R.S., BP 0166, 37001 Tours Cedex, France

Primary and Secondary Metabolism of Plant Cell Cultures, ed. by Neumann et al.

Larkin et al. 1983). Furthermore it is currently claimed that variability may also be a source of instability due to cell heterogeneity within a callus or a cell suspension of a well-established strain (Bayliss 1980; Ellis 1982; Constabel 1983; Meins 1983). This heterogeneity is frequently attributed to the fact that the resulting strains do not originate from single cells. As an example, Ogino et al. (1978) showed the necessity of several successive subclonings to select tobacco cell clones producing large amounts of nicotine. As another example, Dougall et al. (1980), using *Daucus carota* cells producing anthocyanin, observed a cell heterogeneity in a suspension culture even if it is issued from a single cell by a cloning procedure. They concluded that these variations did not seem to be due to a stable modification of genetic information. On the other hand it must be noted that Yamamoto et al. (1982) obtained a stable, high anthocyanin-producing strain of *Euphorbia millii* cells by "cell agregate cloning", whereas the attempt to obtain the same result by single cell cloning failed. So concerning the nature and the origin of phenotypical variability for metabolite production, very few data are available. According to Demarly (1985) the somaclonal variation can be due to:

– a genetic phenomenon defined as a stable modification (showing or not a mendelian inheritance) and reversible with a very low frequency.

– an epigenetic phenomenon defined as a modification in structure, organization or regulation of the genotype being able to have a large inheritance in stable culture conditions but more or less easy to reverse by any modification of these conditions.

– Moreover, a more simple explanation for the cell heterogeneity within a given strain is the existence of differences in kinetics defined as an asynchrony in the differentiation process leading to an heterogenous situation at any culture time.

To evaluate the relative importance of these explanations and determine the practical interest of cloning for a preselected strain, we carried on a study of the variability existing in one of our *Catharanthus roseus* tissue strain.

3 Material and Methods

This strain (named CR2) was established in 1975. It contains a wide diversity of alkaloids (more than 50). Thirteen of them were purified and determined (Petiard et al. 1982b; Gueritte et al. 1983; Petiard and Courtois 1983). About 30% of the total alkaloids produced are extracted from the culture medium (Petiard 1980a). Moreover, the alkaloid extracts exhibit an antimitotic activity on animal and human cell cultures (Petiard 1980b). Since 1978, this strain, checked monthly, appeared to be a stable material for chlorophyll content, growth rate, alkaloid content (for the yield of crude extract and for the presence of predetected and known alkaloids).

The methodology used in this project is described in Fig. 1. We tried to obtain a clone population issued from calli of CR2 strain. Six months and 18 months after their obtention in normal culture conditions (solid media in Petri dishes), a study of these clones is carried on with regard to the three markers used to control the stability of the mother strain. As the variability expressed in the cell clone population may be due to initial heterogeneity within the calli or due to induction of variation by the cloning process itself, a subcloning must be undergone to evaluate the role of each possible cause for the observed variation.

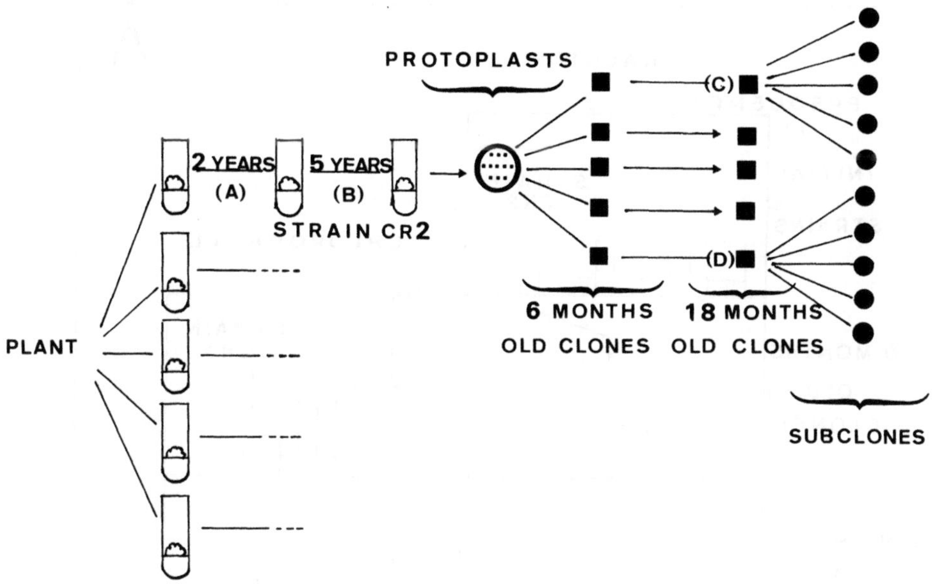

Fig. 1. Diagramatic outline showing procedure for the study of variability in *Catharanthus roseus* strains. *A* Stabilization of strains; *B* regular subculture with controls on alkaloid content and diversity, growth rate and chlorophyll content; *C* clone with a high alkaloid diversity and a high chlorophyll content; *D* clone with a low alkaloid diversity and a low chlorophyll content

The cloning procedure consist in protoplast isolation from calli and subculture at very low density (80 cells ml^{-1}) in a liquid medium conditionated by an high density (25,000 cells ml^{-1}) protoplast preculture of the same strain. To avoid variation induced by the medium there is no change in its nature except the presence of mannitol (0.25 M) during isolation, cell wall regeneration and earlier divisions.

4 Results

As examples, Fig. 2A summarizes the results of clone analysis for two markers (alkaloid diversity and chlorophyll content). It allows three essential conclusions: 6-month-old clones are quite different and a large variability is observed in the population. These 6-month-old clones are not stable and the variability decrease with time. The 18-month-old clones are distributed in two populations. One is chlorophyllous and has a high diversity of alkaloids, the other is non-chlorophyllous and has a small diversity of alkaloids. This instability of clones is also observed for the presence of some of the known alkaloids normally produced in the mother strain (Table 1). Taking into account growth rate, yield of alkaloid crude extract (results not shown) and alkaloid diversity no clone is more interesting than the mother strain for one possible practical use.

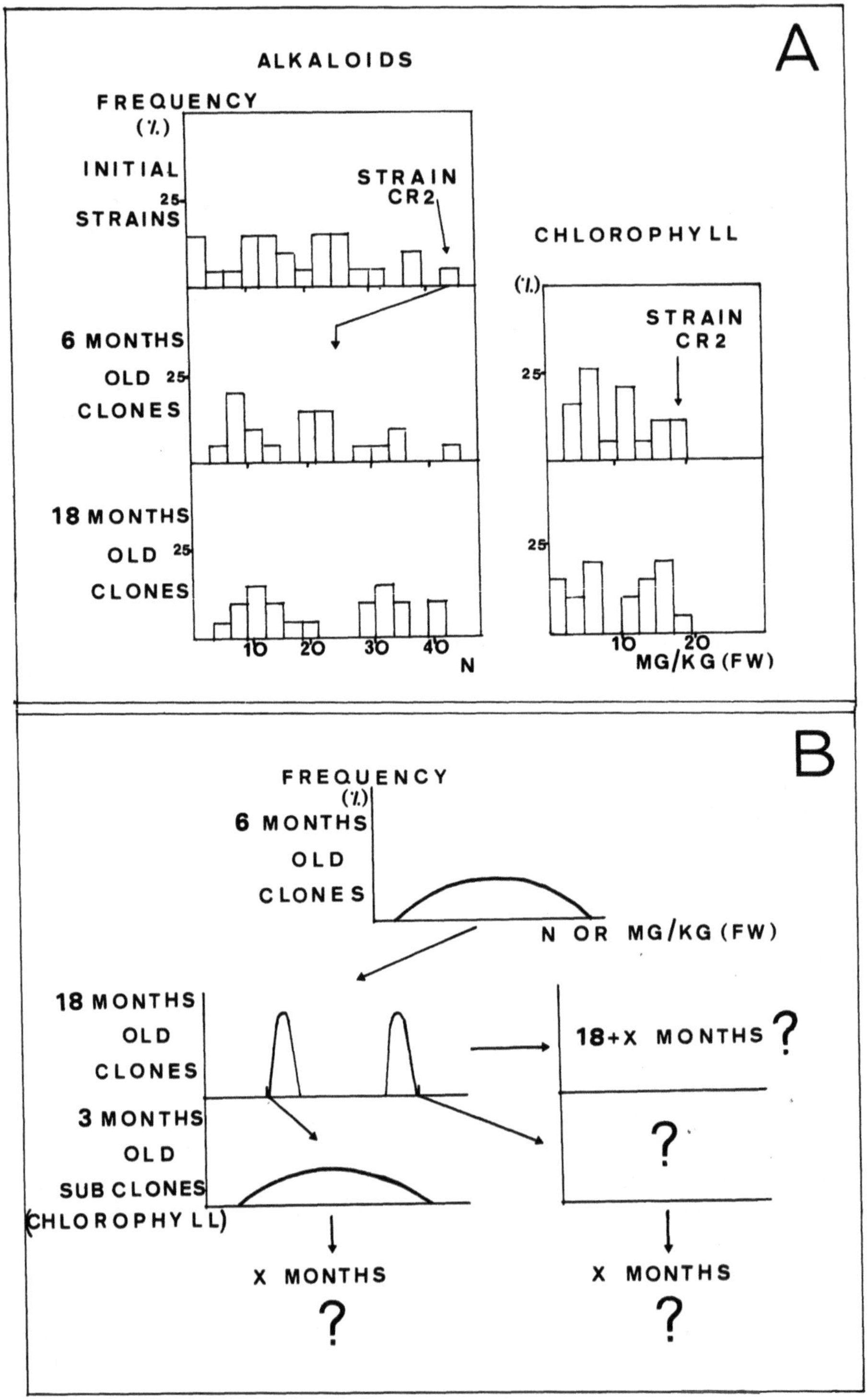
A
ALKALOIDS
FREQUENCY (%)
INITIAL STRAINS
25
STRAIN CR2
6 MONTHS OLD CLONES
25
18 MONTHS OLD CLONES
25
10 20 30 40
N
CHLOROPHYLL
(%)
STRAIN CR2
25
25
10 20
MG/KG (FW)
B
FREQUENCY (%)
6 MONTHS OLD CLONES
N OR MG/KG (FW)
18 MONTHS OLD CLONES
18+X MONTHS ?
3 MONTHS OLD SUB CLONES (CHLOROPHYLL)
?
X MONTHS
?
X MONTHS
?

Fig. 2A,B

Table 1. Qualitative analysis of clones issued from strain CR2 after 6 and 18 months of subculture

Clone / Alkaloid (A)	2		4		6		7		9		10		11		15		17	
(B)	6	18	6	18	6	18	6	18	6	18	6	18	6	18	6	18	6	18
Ajmalicine	–	+	–	+	+	+	–	–	+	+	+	+	–	–	+	+	+	–
Desacetyl-Akuamiline	–	–	–	–	–	–	–	–	–	–	–	–	–	–	–	–	+	–
Pseudo-Indoxyl-Ajmalicine	–	–	–	–	+	+	–	–	–	+	–	–	–	–	–	–	+	–
Tabersonine	–	+	–	+	–	+	–	+	–	–	–	+	–	+	–	+	–	–
Tetrahydroalstonine	–	–	–	–	–	–	–	–	+	+	–	–	–	–	–	+	+	+
Serpentine	–	+	–	–	–	–	–	–	–	–	–	+	–	+	+	–	+	–

Clone / Alkaloid (A)	1		3		5		8		12		13		14		16		18		19	
(B)	6	18	6	18	6	18	6	18	6	18	6	18	6	18	6	18	6	18	6	18
Ajmalicine	+	+	+	+	+	+	+	+	+	+	+	+	+	+	+	+	+	+	+	+
Desacetyl-Acuamiline	+	+	–	+	–	+	–	+	–	+	–	+	+	+	–	+	+	+	+	+
Pseudo-Indoxyl-Ajmalicine	+	+	–	+	+	+	+	+	+	+	+	+	–	+	–	+	+	+	+	+
Tabersonine	–	+	–	+	–	+	–	+	+	+	+	+	–	+	–	+	+	+	+	+
Tetrahydroalstonine	+	+	–	+	+	+	–	+	–	+	+	+	–	+	–	+	+	+	+	+
Serpentine	+	+	–	+	+	+	–	+	–	+	–	+	–	+	+	+	–	+	+	+

(A): The only six major alkaloids isolated from strain CR2 are represented
(B) 6: 6 Months
18: 18 Months
+: Detected on thin layer chromatograpy with C.A.S. reagent
(–: Non detected)

5 Discussion

It must first be underlined that the cloning procedure can be counter-selective for some cell types present in CR2 strain (not each cell gives a clone). However, at least the present study must give a picture of the possible initial cell diversity in the calli.

Each of the 6-month-old clones is different from the other, and one could suppose that (at least) these 19 different cell types exist in the mother strain.

However, this hypothesis does not seem very realistic for two reasons. On the one hand, it is difficult to conceive that such an important number of different cell types may be maintained in a stable strain for 5 years. On the other, considering that no clone produces more alkaloids than the initial strain, all the less productive cell clones should

Fig. 2A,B. Frequencies analysis of 6- and 18-month-old clones upon alkaloid diversity and chlorophyll content. **A** Experimental result. N = number of alkaloids detected on TLC with CAS reagent. Chlorophyll content in milligram per Kilog of fresh weight. **B** Diagramatic outline of results **A** and studies now under investigation be present at a very low density in the initial calli, unless the more productive were counter-selected.

In the 18-month-old clones, the variability is reduced to two subpopulations in which clones are either green and alkaloid-producing or white and non-alkaloid-producing. This observation shows that the cloning procedure has probably induced a destabilization of the strain leading at first to various and instable cell clones which were restabilized progressively by subculture in normal conditions. The decrease of the variability during 1 year of clone subcultures indicates that the variations observed on 6-month-old clones are not mainly due to genetic modifications (as defined previously). Furthermore this conclusion is supported by the preliminary results of a subcloning cycle from a non-chlorophyllous clone leading to some chlorophyllous subclones, thus showing a frequent reversibility for this character.

It is thus clear that variations are at least largely induced by the cloning procedure itself. One must be very careful when evaluating by this method a preexisting variability in an explant, a callus, or a cell suspension. Moreover, according to Yamamoto and Mizuguchi (1982), cell aggregate cloning may be more efficient than single cell cloning. In another way, the origin of variation seems mainly epigenetic: it is easily reversible by a subcloning that eliminates the genetic hypothesis, and it needs numerous subcultures in given culture conditions to be stabilized, which eliminates a simple physiological effect.

Moreover, it remains possible that the two cell types observed in 18-month-old clone population preexist in the original CR2 strain, and are epigenetically different cell types selected during the cloning. The proper effect of the cloning procedure prohibits any definitive conclusion. However, due to the need of strictly identical selective value for the growth of heterogeneous population in a stable strain, our present hypothesis is that the main source of phenotypical cell-to-cell variation is the asynchronous biochemical variation.

The following results of the subcloning of high-producing clones and low-producing clones (Fig. 2B) could confirm these conclusions. Moreover, cloning of a non-producing strain issued from the same *Catharanthus roseus* genotype should be of practical interest to evaluate if obtaining a strain, whatever its level of production, allows obtaining by cloning a high-producing strain.

6 Does Papaver Somniferum Callus Accumulate Unusual Forms of Morphinan Alkaloids?

For secondary metabolism, an important part of variability or plant tissue culture is revealed by yield variation and by the biosynthesis of unknown substances (Zenk 1982; Petiard et al. 1982b). With regard to this qualitative aspect one can suppose that the variability may also occur for the storage forms of various substances. Especially considering some compounds accumulated in specific differentiated structures of the plant, it is of interest to study if they are present in tissue culture under an abnormal chemical form needing no specific structures similar to those of the whole plant.

Even for Apocynaceae tissue culture, for the alkaloid content of crude extracts (Petiard et al. 1982a) or the uptake of labelled alkaloids (Renaudin and Guern 1982) the existence of bound forms for some alkaloids may be suggested. Sugar-bound monoterpenoid and phenylpropanoid substances were found in undifferentiated callus and cell suspension cultures of *Ocimum basilicum* (Lang and Horster 1977). The possible occurrence of bound forms for some secondary metabolites is also supported by their existence in the whole plant in some species (Fairbain and Steele 1980; Barz and Koster 1981).

We undertook the study a non-morphogenetic *Papaver somniferum* callus culture well known as a non-alkaloid-producing strain. For instance, the following two usual extraction procedures do not permit detection of any alkaloid. The first method consists in the extraction of tissues by $CHCl_3$ at pH 9.5 followed by a purification with successive extraction by aqueous acidic solution; acidic fractions are collected, alkalinized (pH 9.5) and then extracted by $CHCl_3$. Following another method (Paris and Cosson 1965) tissues are boiled in $EtOH/H_2O$ (7:3). Then after obtaining a dry residue of the extract, a purification, similar to the one described above is performed. Such methods do not allow detection by TLC of any alkaloid in the tissue strain, whereas they are efficient on the plant. Then unusual extraction procedures have been tried. For instance, grinding freeze-dried tissues in boiling IN HCl aqueous solution (1 h), the extract obtained shows positive reactions with Mayer's and Dragendorff's reagents. However, the further extractions with organic solvents at pH 9.5 fail to detect any alkaloid, whereas some alkaloids are extracted from the Mayer's precipitate. These surprising results show that the callus strain seems to biosynthesize some alkaloids showing an unusual behaviour. They should be under a specific form, for instance "bound form", which prevents their extraction by usual methods. These bound forms should be hydrolyzed by the acidic treatment to give free forms. However, as it was not possible to extract the alkaloids from the acidic solution by organic solvents after alkalinization, the alkaloids are possibly reconjugated.

These preliminary results led us to confirm the existence of alkaloid-bound forms by extraction, purification and structural determination of free alkaloids. After the same acidic treatment, a chromatography on a cation exchange resin (Amberlite IR 120) is performed with the acidic aqueous extract. Some alkaloids are present in MeOH/IN NaOH (1:1) mixture used to elute the column. In this crude extract codein and morphine are detected by TLC and HPLC (Hutin1983; Hutin et al. 1983a). In this way the amounts of alkaloids are evaluated as 19 mg of morphine and 15 mg of codeine per 100 g of dry weight tissues.

After numerous assays, the extraction procedure described in Fig. 3 is finally adopted. The bound forms are precipited (among other compounds) by cold MeOH from a low acidic aqueous solution, and then hydrolyzed and extracted as described previously. In this way 6.5 mg of morphine and 1.5 mg of codeine are purified. Their structural determination is performed (UV, MS, NMR, IR spectra) (Hutin 1983; Hutin et al. 1983b). The occurrence of alkaloids in the precipitate and their absence in the corresponding supernatant confirm their unusual forms.

As a control, at each stage of this extraction procedure, the feeding of authentic samples of morphine excludes the hypothesis of a chemical artefact formation. It is thus clear that this strain of *Papaver somniferum* tissue culture contains some alkaloids which are not free. The question arises as to the forms of these alkaloids.

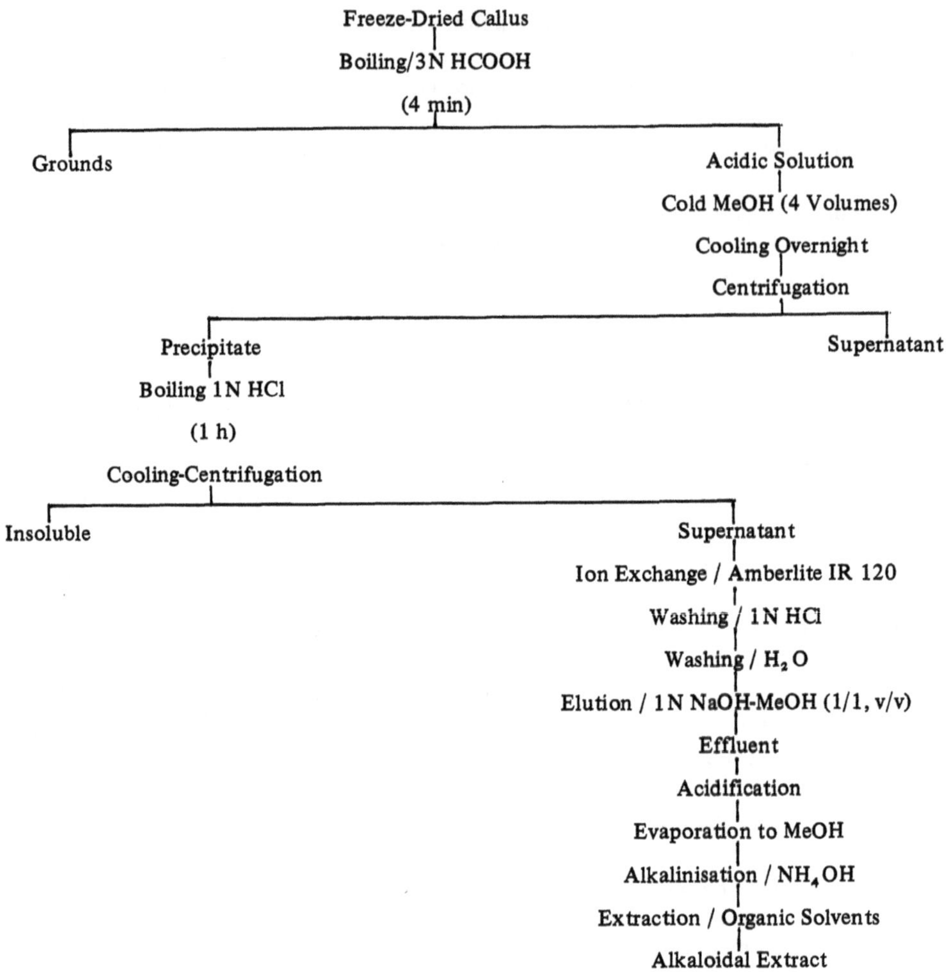

Fig. 3. Extraction procedure of the alkaloids of the *Papaver somniferum* tissue cultures

As preliminary results it seems that bound alkaloids have a relatively high molecular weight because they are excluded in the void volume of Sephadex G50 gel permeation chromatography. By an another method numerous enzymes are tested to release the alkaloids from their bound forms, avoiding a possible degradation by acidic hydrolysis. Four types of enzymatic treatment allow recovery of some alkaloids: pectinase, pectolyase, macerozyme and driselase. It must be noted that alkaloids liberated in this way are usually extracted without ion exchange chromatography. This observation led us to suppose that the enzymatic treatment damages some compounds implied in the reconjugated forms. These results suggest that alkaloids could be bound, at least partly, to pectic derivatives. Further studies are now under investigation.

7 Conclusions

From the fundamental point of view the present studies concerning biochemical variability expressed in plant tissue culture lead us to three major conclusions:

– On our plant material the cell-to-cell variation observed in a callus seems to be essentially due to the asynchrony of cell growth and differenciation.

– Due to the cloning process by protoplast isolation some non-definitive variation occurs, allowing the supposition that these variations are essentially an epigenetic phenomon.

– Some alkaloids of morphinan type may be accumulated in undifferentiated tissue culture, in unusual forms needing specific treatment for their extraction.

From the practical point of view, results imply the possibility that, screening tissue strains, some compounds may be not available to usual extraction procedures. Furthermore these new forms of alkaloid can be of interest for the screening of new biologically active substances. Moreover, to isolate high-producing strains it seems that an additional selection phase by single cell cloning is of no interest when it is undertaken on a selected strain obtained in the classical way.

References

Barz W, Koster J (1981) Turnover and degradation of secondary natural products. In: Conn EE (ed) Secondary plant products. Academic Press, London New York, pp 44–74

Bayliss MW (1980) Chromosomal variation in plant tissue in culture. Int Rev Cytol Suppl 11A: 113–143

Chaubet N, Pareilleux A (1982) Characterization of galactosidase of *Medicago sativa* suspension cultured cells growing on lactose. Z Pflanzenphysiol 106:401–407

Constabel F (1983) Protoplast technology applied to metabolite production. Int Rev Cytol Suppl 16:209–217

Constabel F, Kurz WGW, Kutney JP (1982) Variations in cell cultures of periwinkle *Catharanthus roseus*. In: Fujiwara A (ed) Plant tissue culture 1982. Proc 5th Int Congr Plant Tissue Cell Cult 11–16 July 1982. Maruzen, Tokyo, pp 395–396

Demarly Y (1985) Le determinisme épigenétique de la variabilité. Symp Euro-Arabe Cult Tissus Veg, 28–31 May 1984. Annu Meet IAPTC France. Sté Bot, France, in press

Dougall DK, Morris Johnson J, Whitten GH (1980) A clonal analysis of anthocyanin accumulation by cell cultures of wild carrot. Planta 149:292–297

Ellis B (1982) Cell to cell variability in secondary metabolite production within cultured plant cell populations. In: Fujiwara A (ed) Plant tissue culture 1982. Proc 5th Int Congr Plant Tissue Cell Cult, 11–16 July 1982. Maruzen, Tokyo, pp 395–396

Fairbain JW, Steele MJ (1980) Bound forms of alkaloids in *Papaver somniferum* and *Papaver bracteatum*. Phytochemistry 19:2317–2321

Gueritte F, Langlois N, Petiard V (1983) Metabolites secondaires isolés d'une culture de tissus de *Catharanthus roseus*. J Nat Prod 46,1:144–148

Hutin M (1983) Etude de l'extraction et de la structure des combinaisons alcaloidiques presentes dans les cultures in vitro de tissus d'une souche de Papaver somniferum. Thèse, Univ Paris Sud, Paris

Hutin M, Cavé A, Foucher JP (1983a) Utilisation de la chromatographie liquide haute performance pour la caractérisation et le dosage des principaux alcaloides du *Papaver somniferum L.* J Chromatogr 268:125–130

Hutin M, Foucher JP, Courtois D, Petiard V (1983b) Evidences for unusual forms of storage of Morphinan in a *Papaver somniferum L.* tissue cultures. CR Acad Sci 297,III:47–50

Lang E, Horster H (1977) Sugar bound regular monoterpenes part II. Production and accumulation of essential oil in *Ocimum basilicum* callus and suspension cultures. Planta Med 31:112–118

Larkin PJ, Brettel R, Ryan S, Scowcroft W (1983) Protoplasts and variation from culture. In: Potrykus I, Harms CT, Hinnen A, Hutter R, King PJ, Shillito RD (eds) 6th Int Protoplast Symp, 12–16 Aug 1983. Birkhäuser, Basel, pp 51–57

Meins F (1983) Heritable variation in plant cell culture. Annu Rev Plant Physiol 34:327–346

Ogino T, Hiraoka N, Tabata M (1978) Selection of high nicotine producing cell lives of tobacco callus by single-cell cloning. Phytochemistry 17:1907–1910

Paris RR, Cosson L (1965) Sur la biogenèse des alcaloides des Datura, études des variations ontogéniques à l'aide d'une méthode chromatographique chez les Datura metel Sims. CR Acad Sci D:3148–3159

Petiard V (1974) Culture de cellules de végétaux supérieurs sur milieux lactosés. French Patent No 7426040

Petiard V (1980a) Mise en évidence d'alcaloides dans le milieu nutritif de culture de tissus de Catharanthus roseus G. Don. Physiol Veg 18:331–337

Petiard V (1980b) Antimitotic activities of *Catharanthus roseus* tissue cultures. In: Beal JL, Reinhard E (eds) Natural products as medicinal agents. Hippocrates, Stuttgart, pp 447–469

Petiard V, Courtois D (1983) Recent advances in research for novel alkaloids in Apocynaceae tissue cultures. Physiol Veg 21, 2:217–227

Petiard V, Cosson L, Courtois D (1982a) Cultures de tissus d'Apocynaceae: problèmes posés par l'extraction du totum alcaloidique. CR Acad Sci 294,III:123–126

Petiard V, Courtois D, Gueritte F, Langlois N, Mompon B (1982b) New alkaloids in plant tissue culture. In: Fujiwara A (ed) Plant tissue culture 1982. Proc 5th Int Congr Plant Tissue Cell Cult, 11–16 July 1982. Maruzen, Tokyo, pp 309–310

Renaudin JP, Guern J (1982) Compartmentation mechanisms of indole alkaloids in cell suspension cultures of *Catharanthus roseus*. Physiol Veg 20,3:533–547

Sato F, Yamada Y (1984) High berberine producing cultures of *Coptis japonica* cells. Phytochemistry 23,2:281–285

Schulte U, El-Shagi H, Zenk MH (1984) Optimisation of 19 Rubiaceae species in cell culture for the production of anthraquinones. Plant Cell Rep 3,2:51–55

Tabata M, Ogino T, Yoshioka K, Yoshikawa N, Hiraoka N (1978) Selection of cell lines with higher yield of secondary products. In: Thorpe TA (ed) Frontiers of plant tissue culture 1978. Proc 4th Int Congr Plant Tissue Cell Cult, 20–25 Aug 1978, Calgary, pp 213–222

Yamamoto Y, Mizugushi R, Yamada Y (1982) Selection of a high and stable pigment-producing strain in cultured *Euphorbia millii* cells. Theor Appl Genet 61:113–116

Zenk MH (1982) Pflanzliche Zellkulturen in der Arzneimittelforschung. Pharm Heute 103,3:131–138

Zenk MH, Deus B (1982) Natural product synthesis by plant cell culture. In: Fujiwara A (ed) Plant tissue culture 1982. Proc 5th Int Congr Plant Tissue Cell Cult, 11–16 July 1982. Maruzen, Tokyo, pp 395–396

Zenk MH, El-Shagi H, Arens H, Stökigt J, Weiler EW, Deus B (1977) Formation of the indole alkaloids serpentine and ajmalicine in cell suspension cultures of *Catharanthus roseus*. In: Barz W, Reinhard W, Zenk MH (eds) Plant tissue culture and its biotechnological application. Springer, Berlin Heidelberg New York, pp 27–44

Biosynthesis and Accumulation of Indole Alkaloids in Cell Suspension Cultures of *Catharanthus roseus* Cultivars

W.G.W. KURZ, K.B. CHATSON, and F. CONSTABEL [1]

1 Introduction

Plant cell cultures of the Madagascar periwinkle [*Catharanthus roseus* (L.) G. Don] have received particular attention because of physiologically potent indole alkaloids they are expected to produce. The production of alkaloids in cell cultures can be regarded as a differentiation process governed both by environmental conditions and by the genotype of the material in question. While conditions which might induce cell cultures to synthesize and accumulate alkaloids have been fairly well studied, the selection of a promising genotype has not and requires continued investigation. Accordingly, the present study is focused on variations of indole alkaloid accumulation in *C. roseus* cultivars.

Initially, studies were undertaken to investigate the alkaloid pattern of serially cultured callus lines derived from uniform explants, i.e., anthers of buds identical in developmental stage. The material pertained to the cultivar *Roseus* and the pattern of indole alkaloids comprised compounds belonging to the corynanthe, strychnos, aspidosperma, and iboga families. Some of the cv. *Roseus* lines produced catharanthine, an iboga type alkaloid up to 0.005% of dry wt., i.e. in amounts ca. three times that obtainable with periwinkle plants. This compound is of particular importance as its N-oxide coupled with vindoline affords 3′,4′-dehydrovinblastine, a bisindole alkaloid and possible intermediate toward a variety of clinically important compounds, in particular vinblastine and vincristine. Bisindole alkaloids have thus far not been confirmed for periwinkle suspension cultures. Corynanthe type alkaloids have been detected most frequently. Ajmalicine, as well as its structural relative serpentine, have been isolated in amounts as high as 0.5% resp. 0.9% yield based on cell dry weight [1].

2 Results

To date over 2,000 cell lines of three *C. roseus* cultivars have been investigated for their variation in respect to physiological behavior and biosynthesis capabilities. In all cases the calli were derived from explants consisting of excised anthers of 1.5–2.0 mm buds

1 Plant Biotechnology Institute, National Research Council of Canada, Saskatoon, Sask S7N OW9, Canada

NRCC No. 23460

Primary and Secondary Metabolism of Plant Cell Cultures, ed. by Neumann et al.

of young plants grown in the greenhouse or growth room. The callus originated at the cut of the filament and in anther walls, i.e. from diploid tissue. Initially, callus pieces were cultured on Murashige and Skoog (MS) [2] nutrient agar medium with 0.1 mg l^{-1} 1-naphthaleneacetic acid (NAA). Subcultures were maintained on B5 media [3] at about 27 °C in continuous light (10 W m^{-2}). For alkaloid production callus was suspended in liquid B5 medium with 1.0 mg l^{-1} 2,4-dichlorophenoxyacetic acid (2,4-D) and 1 g l^{-1} casein hydrolyzate (1-B5C medium). These suspensions were grown in De Long flasks on gyratory shakers and subcultured weekly for a period of 2–3 months. To test for alkaloid biosynthetic capability, 2-day-old cell suspensions which had been washed in alkaloid production medium (AP-medium) [1] were used as inoculum for this medium and cultivated under identical conditions as outlined above. Samples were withdrawn at indicated intervals, the cell dry weight determined, and the cells extracted and analyzed by TLC and HPLC [4].

The alkaloid production varied with the cell line and age of the subculture and ranged from 0.1%–1.5% of cell dry weight. The relative amounts of total alkaloids produced was fairly constant under given conditions and appeared to be cell-line specific. Some cell lines did not produce any trace of alkaloids.

In order to strictly control environmental parameters and obtain cell material sufficient for the isolation and identification of alkaloids, some lines were grown in AP-medium in 7.5 l Microferm bioreactors (New Brunswick Scientific Co. Inc., Edison, N.J., USA) equipped with sintered disc spargers (5 cm O.D.). To minimize the shearing effect the bioreactors were stirred at 200 rpm and aerated with 35 ml air min^{-1} l^{-1} culture.

The growth pattern shown in Fig. 1 was essentially the same for all cultivars tested.

After incubation with an actively growing cell suspension the mitotic index (MI) dropped sharply over the first 2–3 days. Thereafter the index rose and reached its maximum within 4 days and declined again gradually over the following 10–15 days to zero. The cell dry weight over the culture period increased by a factor of 8–10.

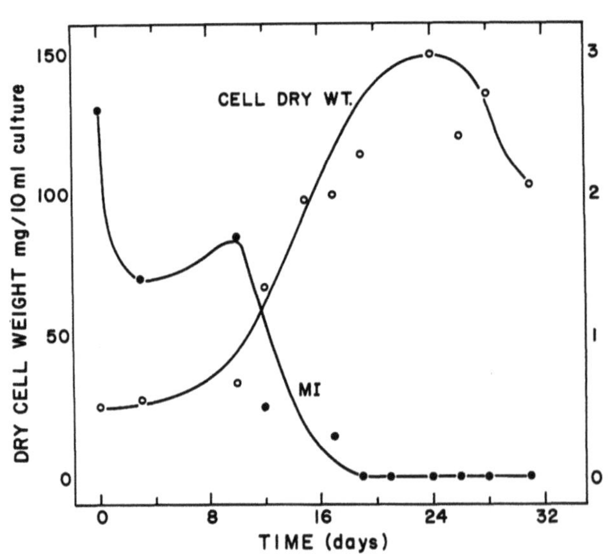

Fig. 1. *C. roseus* growth pattern in bioreactors dry cell weight (○——○); mitotic index (●——●)

The alkaloids occurred in a variety of combinations of corynanthe-, strychnos-, aspidosperma-, and iboga-type compounds. The variation affected is not necessarily genetic in nature as has been shown by the suppression of vindoline accumulation in cell cultures of *C. roseus* and its subsequent reappearance with shoot formation [5].

The cell lines of the different cultivars have been in cultivation for several years and the alkaloid pattern showed signs of stability/instability over this time period. For example, some lines retained their synthetic capability for one or two types of alkaloid and lost it for a third which initially was a part of the alkaloid spectrum, while others basically maintained the original pattern. Non-producing lines retained this character throughout. In spite of exceptions, the general observation was that alkaloid patterns change with time with the more complex compounds, for example iboga-type alkaloids fading more often than the less complex ones such as corynanthe type alkaloids. The total amount of alkaloids produced, however, is line-specific and fairly stable. The instability in alkaloid patterns reiterates the importance of developing methods to preserve and store commercially valuable cell lines in their original state of biosynthetic capability. Recent advances in cryopreservation, where periwinkle cells were frozen, stored, and thawed without losing their alkaloid-accumulating capability show one possible route of preservation [6].

2.1 Catharanthus roseus cv. Roseus

This cultivar has been the most commonly used in investigations of alkaloid biosynthesis in periwinkle.

Analysis of 458 cell lines of this cultivar revealed 312 lines to accumulate alkaloids occurring in a variety of combinations. It was of interest that the combinations were not random, but certain combinations appeared at a higher frequency than others as is shown in Table 1.

During initial screening of cell lines in De Long flasks, over 8-week culture periods alkaloids have been found as early as 2 weeks after inoculation. Most cell lines showed a maximum accumulation of alkaloids in the 3rd-5th week of culture. The alkaloid content did not seem to grow in complexity with time; rather, alkaloids appeared to be catabolized in time since after 6-8 weeks of culture the occurrence of alkaloids diminished rapidly.

Table 1. Frequency of alkaloid patterns according to structural type. *C. roseus* cv. *Roseus* (n = 458). Analysis performed ca. 1 year after callus initiation

Alkaloid	%	Alkaloid	%
None	32	STR + ASP	10
Corynanthe (COR)	7	STR + IBO	0
Strychnos (STR)	10	ASP + IBO	0
Aspidosperma (ASP)	8	COR + STR + ASP	9
Iboga (IBO)	0	COR + STR + IBO	0
COR + STR	5	COR + ASP + IBO	0.3
COR + ASP	16	STR + ASP + IBO	0
COR + IBO	0	COR + STR + ASP + IBO	0

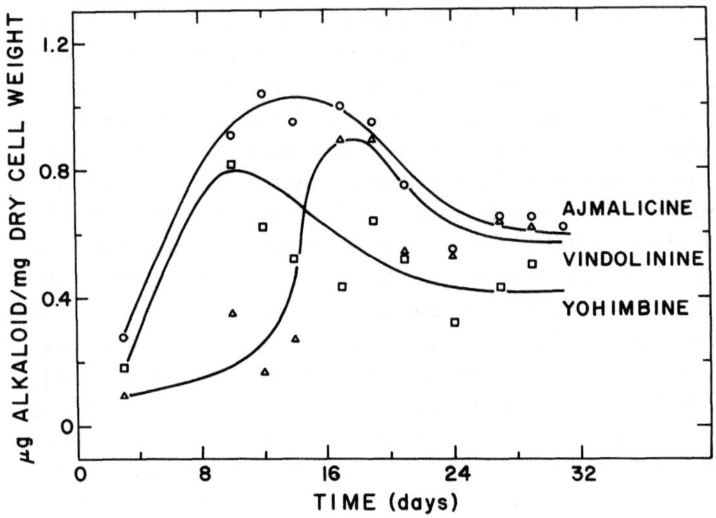

Fig. 2. Biosynthetic pattern of ajmalicine, yohimbine, and vindolinine in line # 953

Cultivations carried out in the controlled environment of bioreactors showed that the simple corynanthe alkaloids ajmalicine and yohimbine reached maximum concentration at a much earlier period in culture growth than the biosynthetically more complex vindolinine. These are presumably derived from a common key intermediate, strictosidine, reflecting the differences in complexidy of their biogenesis. After 3–4 weeks of cultivation the concentration of these alkaloids equilibrate (Fig. 2).

We present two examples of cell lines derived from cv. Roseus callus: #953 and #200.

2.2 Cell Line #953

This cell line produced an alkaloid mixture which allowed the isolation of the following alkaloids which were characterized by their physical and spectral data and by comparison with authentic materials: Ajmalicine (1), yohimbine (2), isositsirikine (3), vallesiachotamine (4), strictosidine (5), lochnericine (6), Hörhammericine (7), Hörhammerinine (8), vindolinine (9), 19-epivindolinine (10), 19-acetoxy-11-methoxytabersonine (11), 19-hydroxy-11-methoxytabersonine (12), and dimethyltryptamine (13). HPLC traces of the 3–7 week period of growth showed that each sample contained ajmalicine (1) as the major component. Furthermore, the analytical traces indicate that the other identifiable components of the mixture remained the same throughout the later period with only small changes in their relative concentrations (Fig. 3).

2.3 Cell Line #200 [7]

This cell line was found initially to produce catharanthine (14) in amounts three times that obtainable from *C. roseus* plant material (ca. 0.0017% of dry wt.). Through recent

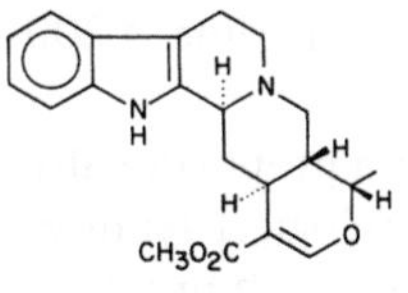
1 AJMALICINE

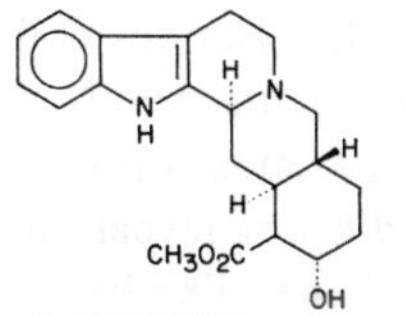
2 YOHIMBINE

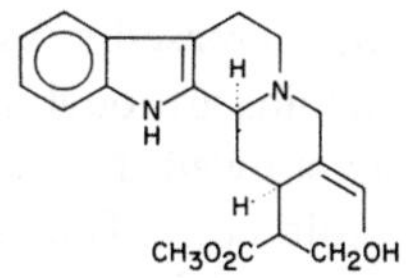
3 ISOSITSIRIKINE

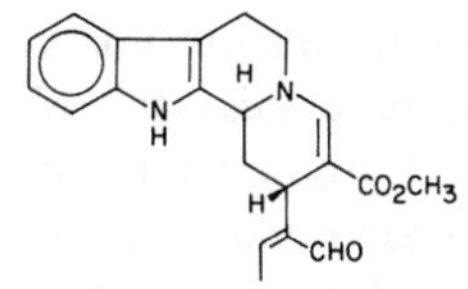
4 VALLESIACHOTAMINE

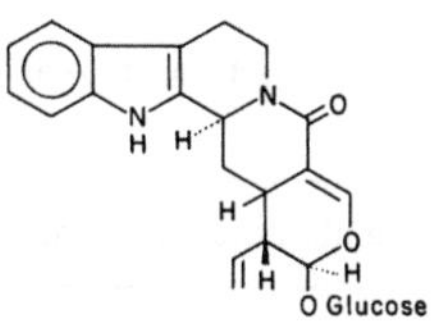
5 STRICTOSIDINE LACTAM

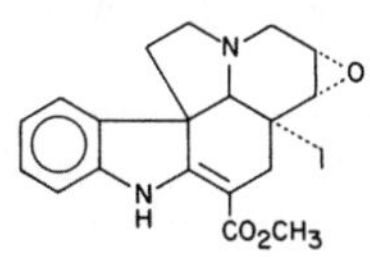
6 LOCHNERICINE

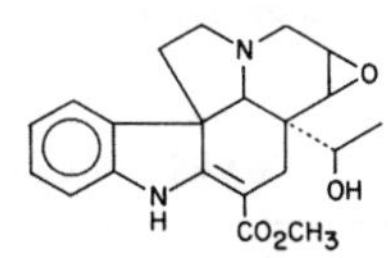
7 HÖRHAMMERICINE

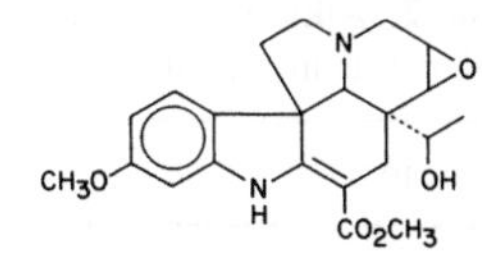
8 HÖRHAMMERININE

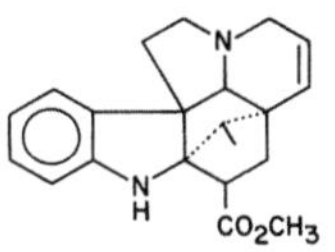
9 VINDOLININE

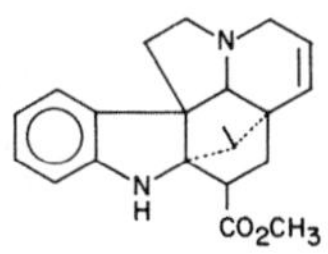
10 19-EPIVINDOLININE

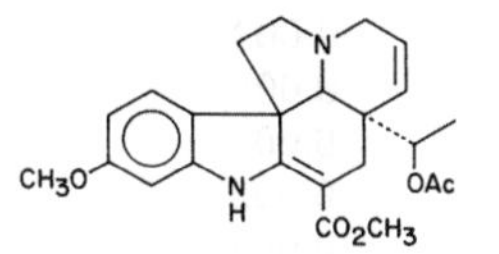
11 19-ACETOXY-11-METHOXY TABERSONINE

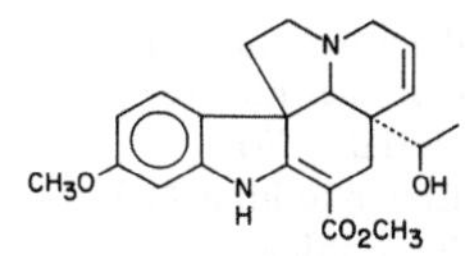
12 19-HYDROXY-11-METHOXY TABERSONINE

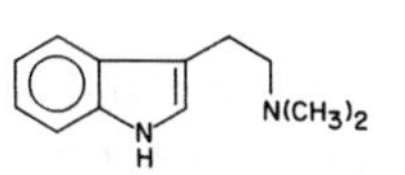
13 DIMETHYLTRYPTAMINE

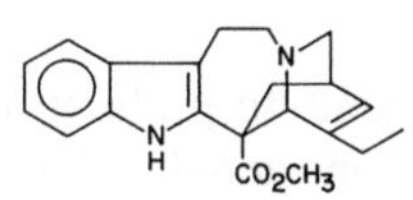
14 CATHARANTHINE

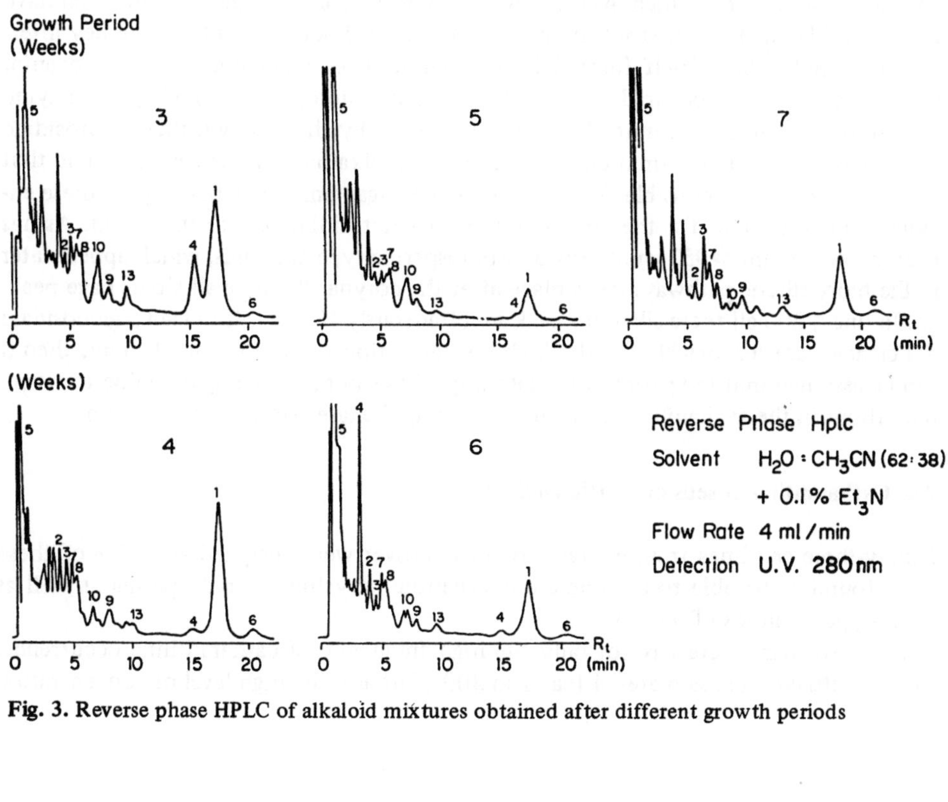

Fig. 3. Reverse phase HPLC of alkaloid mixtures obtained after different growth periods

improvements this could be increased substantially. Besides catharanthine line #200 accumulates eight other alkaloids although found in line #953.

Cytological investigations showed that this material is composed of a proportionally larger population of polyploid cells. The cells did not reveal any cytological features which would set this material apart from other cell lines, like line #953, which produces alkaloids but lack the capacity for catharanthine formation.

Table 2. Alkaloids isolated *C. roseus* cv. *Roseus* from line #200. Analysis performed ca. 1 year after callus initiation

Alkaloid	% yield of cell dry wt.	% of crude alkaloid mixture
Strictosidine lactam (5)	0.224	60.48
Vallesiachotamine (4)	0.041	11.07
Ajmalicine (1)	0.006	1.62
Hörhammericine (7)	0.002	0.54
Hörhammerinine (8)	0.005	1.44
Vindolinine (9)	0.002	0.54
19-Epivindolinine (10)	0.002	0.54
Catharanthine (14)	0.005	1.35

The line #200 appeared to preferentially synthesize and accumulate strictosidine lactam and ajmalicine. Considered in light of known biosynthetic pathways for indole alkaloids the concentration of strictosidine lactam with respect to the more complex alkaloids, was extremely high. At least two possibilities for this high concentration have to be considered. Firstly, strictosidine lactam is an artifact derived from the key intermediate strictosidine, itself formed by condensation of tryptamine and secologanin. Lactamization could occur during work up, furthermore, the possibility of enzyme catalyzed lactamization cannot be ruled out. Secondly, it is known that strictosidine undergoes further transformation to more complex alkaloids via the aglycone, and that β-glucosidases are responsible for the loss of the sugar unit. A deficiency of these enzymes might account for the large build-up of strictosidine and/or its lactam. As has been shown in line #953, synthesis of aspidosperma type alkaloids, which appear later in the biosynthetic pathway, takes place after the corynanthe type alkaloids have peaked. In line #200 all these alkaloids peak simultaneously (Fig. 4). If indeed β-glucosidase deficiencies are responsible for the high concentration of strictosidine lactam, then it can be assumed that the pattern of alkaloid synthesis in this line might be due to severe limitations in the availability of precursors during the later stages of cultivation.

2.4 Catharanthus roseus cv. Little Delicata

Ninety-three cell lines from explants of this cultivar were analyzed and 45% of these were found to be able to accumulate corynanthe, strychnos, aspidosperma as well as iboga-type alkaloids (Table 3).

Of particular interest is not only the high frequency of catharanthine occurrence, rare in cultivar Roseus material (ca. 1 in 300), but also the high level of concentration

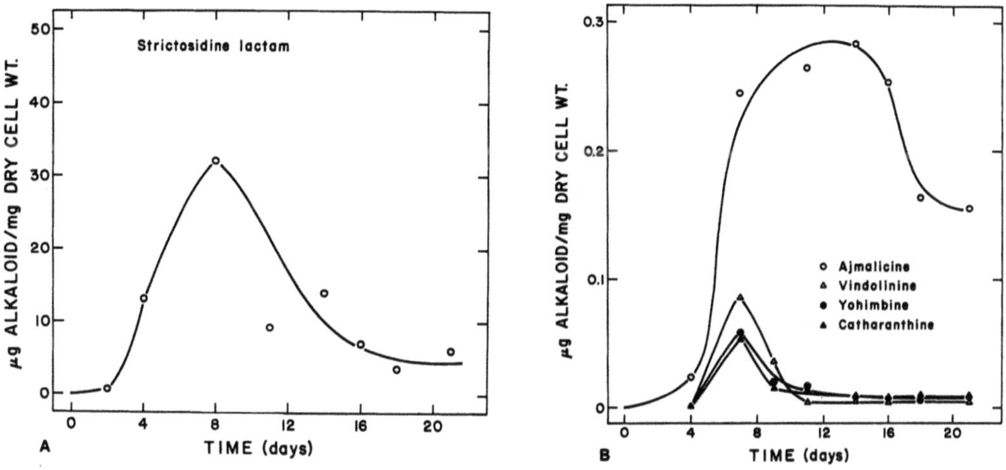

Fig. 4A,B. Biosynthetic pattern of strictosidine, ajmalicine, vindolinine, and catharanthine in line #200

Table 3. Frequency of alkaloid patterns *C. roseus* cv. Little Delicata (n = 93). Analysis performed 6–10 months after callus initiation

Alkaloids	%	Alkaloids	%
None	3	STR + ASP	1
COR	4	STR + IBO	0
STR	0	ASP + IBO	1
ASP	0	COR + STR + ASP	14
IBO	0	COR + STR + IBO	1
COR + STR	1	COR + ASP + IBO	16
COR + ASP	14	STR + ASP + IBO	0
COR + IBO	0	COR + STR + ASP + IBO	45

Table 4. Ratio of catharanthine/strictosidine (lactam) in cell lines of *C. roseus* cv. Little Delicata

Cell Line	Ratio Catharanthine/Strictosidine
cv. Roseus	
#200	1:500
#200-2	1:138
cv. Little Delicata	
#688	1:2.1
#698	1:6.1
#687	1:7.4
#647	1:4.2
#618	1:5.5

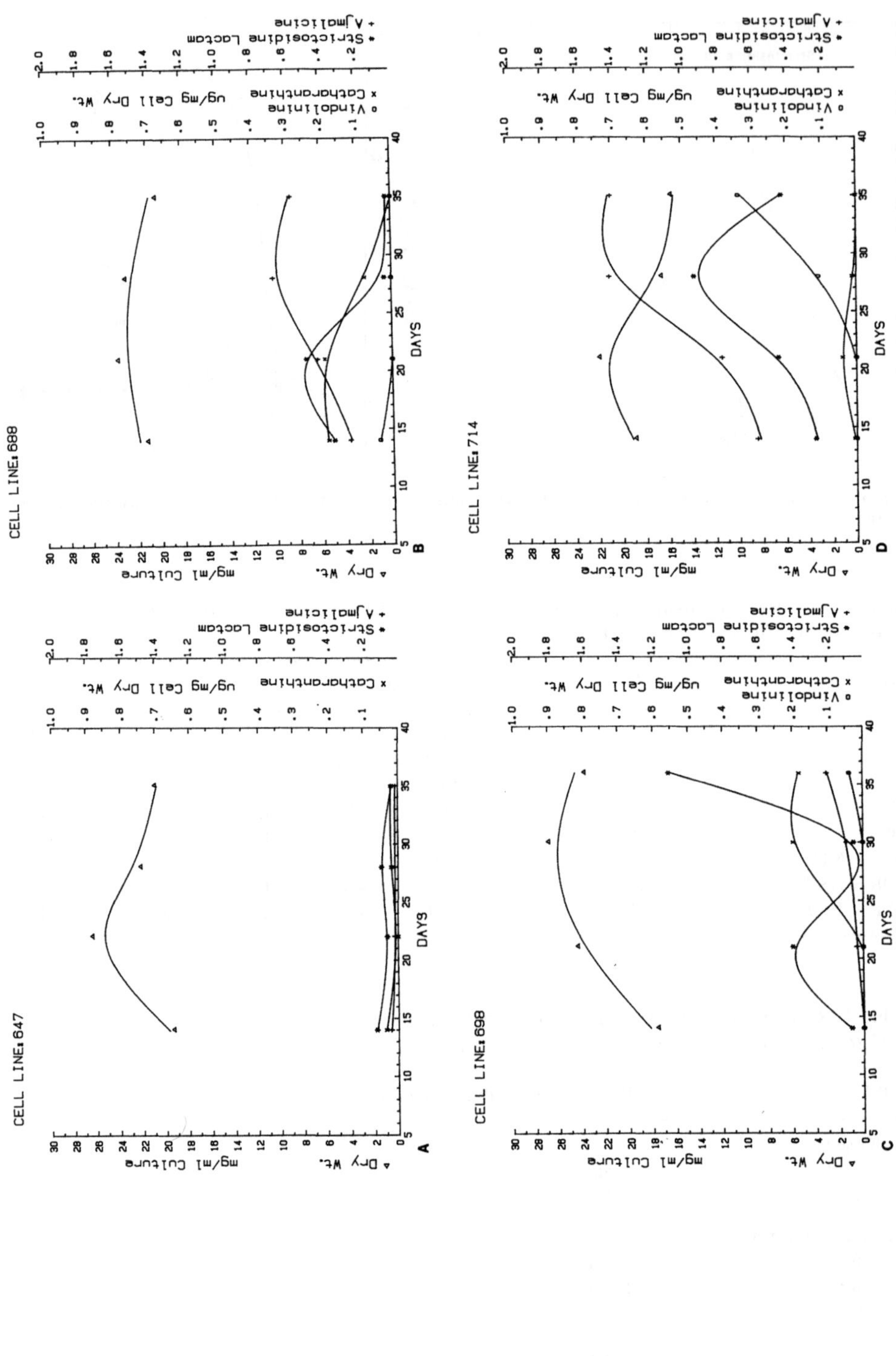

Fig. 5A–D. Biosynthetic pattern of strictosidine, ajmalicine, vindolinine, and catharanthine in lines #647, 688, 698, and 714

of this compound. The average concentration of catharanthine in 4 lines of cv. Little Delicata was 0.238 $\mu g\ mg^{-1}$ dry wt. compared to an average of 0.059 $\mu g\ mg^{-1}$ dry wt. in three lines of cv. Roseus. A significant feature of cv. Little Delicata cell lines accumulating catharanthine was a comparatively low level of strictosidine (Table 4).

In a number of these lines the level of aspidosperma type alkaloids was increased as well when compared with those of cv. Roseus. The variation regarding alkaloid biosynthesis and accumulation among lines of Little Delicata is substantial in spite of the lower strictosidine values (Fig. 5).

Cultivation in bioreactors where strict parameter control is possible could further increase alkaloid yields. For example, one cell line #618 yielded catharanthine at 0.5 $\mu g\ mg^{-1}$ cell dry wt. or 7.5 $\mu g\ ml^{-1}$ culture after a cultivation period of 17 days, outyielding other lines of the same cultivar (Fig. 6).

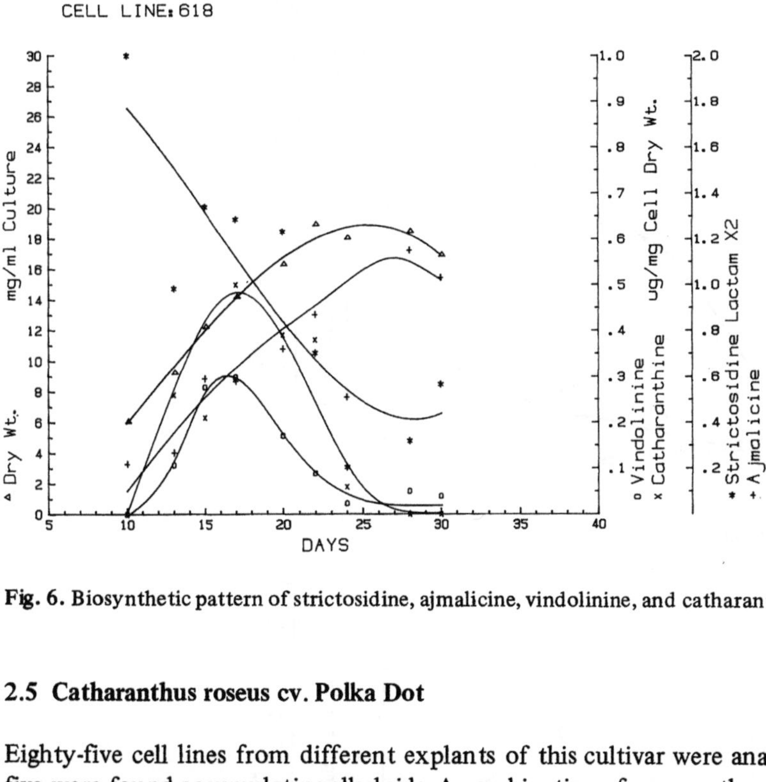

Fig. 6. Biosynthetic pattern of strictosidine, ajmalicine, vindolinine, and catharanthine in line #618

2.5 Catharanthus roseus cv. Polka Dot

Eighty-five cell lines from different explants of this cultivar were analyzed and all but five were found accumulating alkaloids. A combination of corynanthe and aspidosperma type alkaloids occurred most frequently and catharanthine was present in 20% of the lines tested (Table 5).

Strictosidine and ajmalicine were the major corynanthe-, akuammicine the strychnos-, vindolinine, 19-epivindolinine, Hörhammericine and Hörhammerinine the aspidosperma- and catharanthine the iboga-type alkaloids.

The amount of catharanthine accumulated was similar to cv. Roseus, however the ratio of catharanthine to strictosidine was of the same order of magnitude as found in cv. Little Delicata (Fig. 7).

Table 5. Frequency of alkaloid patterns *C. roseus* cv. Polka Dot (n = 85)

Alkaloids	%	Alkaloids	%
None	6	STR + ASP	2
COR	6	STR + IBO	0
STR	0	ASP + IBO	0
ASP	1	COR + STR + ASP	22
IBO	0	COR + STR + IBO	0
COR + STR	0	COR + ASP + IBO	6
COR + ASP	42	STR + ASP + IBO	0
COR + IBO	0	COR + STR + ASP + IBO	14

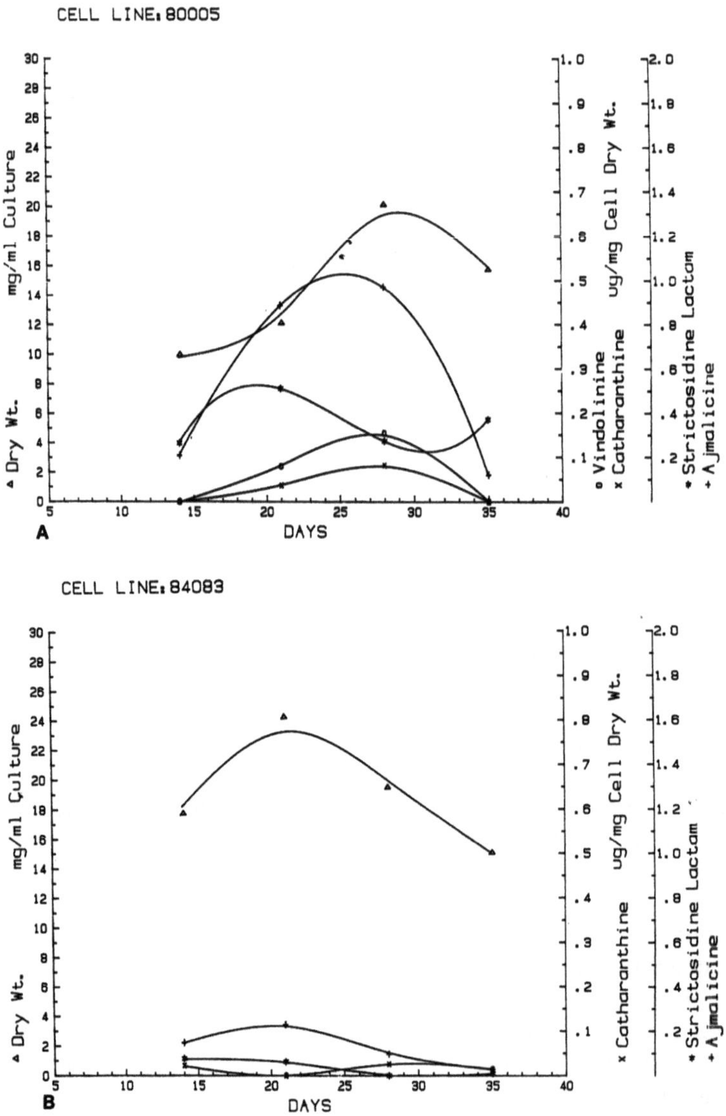

Fig. 7A,B. Biosynthetic pattern of strictosidine, ajmalicine, vindolinine, and catharanthine in lines #80005 and 84083

Cell lines of three cultivars have thus far been tested and it is evident from the results obtained, that screening for promising genotypes capable of synthesizing a desired compound should be the primal objective in any selection program involving plant cell cultures as producers of secondary metabolites.

References

1. Zenk MH, El-Shagi H, Arens H, Stöckigt J, Weiler EW, Deus B (1977) Formation of the indole alkaloids serpentine and ajmalicine in cell suspension cultures of *Catharanthus roseus*. In: Barz W, Reinhard E, Zenk MH (eds) Plant tissue culture and its bio-technological application. Springer, Berlin Heidelberg New York
2. Murashige T, Skoog F (1962) A revised medium for rapid growth and bioassays with tobacco tissue cultures. Physiol Plant 15:473–497
3. Gamborg OL, Miller RA, Ojima K (1968) Nutrient requirements of suspension cultures of soybean root cells. Exp Cell Res 50:151–158
4. Kutney JP, Choi LSL, Kolodziejczyk P, Sleigh SK, Stuart KL, Worth BR, Kurz WGW, Chatson KB, Constabel F (1980) Alkaloid production in *Catharanthus roseus* cell cultures: Isolation and characterization of alkaloids from one cell line. Phytochemistry 19:2589–2595
5. Constabel F, Gaudet-La Prairie P, Kurz WGW, Kutney JP (1982) Alkaloid production in *Catharanthus roseus* cell cultures XII. Biosynthetic capacity of callus from original explants and regenerated shoots. Plant Cell Rep 1:139–142
6. Chen THH, Kartha KK, Leung NL, Kurz WGW, Chatson KB, Constabel F (1984) Cryopreservation of alkaloid producing cell cultures of Periwinkle *(Catharanthus roseus)*. Plant Physiol 75: 726–731
7. Kurz WGW, Chatson KB, Constabel F, Kutney JP, Choi LSL, Kolodziejczyk P, Sleigh SK, Smart KL, Worth BR (1981) Alkaloid production in *Catharanthus roseus* cell cultures VIII. Characterisation of the PRL #200 cell line. Planta Med 42:22–31

Formation of Cardenolides in Cell and Organ Cultures of *Digitalis lanata*

M. LUCKNER and B. DIETTRICH[1]

1 Introduction

Cell cultures were grown from different *Digitalis species* (cf. [4, 12, 23, 31, 32, 34, 38, 44, 45, 48, 49, 51, 57]). However, in callus and suspension cultures without morphogenesis either no cardenolides were found [4, 10, 13–15, 19, 20, 33, 41, 42, 50] or with more sensitive methods, e.g., specific RIAs, only traces of cardenolides were detected ([12, 31, 34, 44] own unpublished results, but cf. [37, 39, 40]), even if potential precursors were administered [31]. Kartnig et al. reported that cardenolid production measurable with cell strains of *D. lanata* which were freshly obtained (as that of other secondary products) ceased during repeated subcultivation [21, 24, 25]. In contrast several groups of other steroids (sterols [14, 19, 20], steroidal saponins [14, 25], and progesterone [14]) were reported to be synthesized in morphologically unorganized long-term cultures of *D. lanata*.

Cultures regenerating roots also contained either no or only small amounts of cardenolides ([15, 20, 31] own unpublished results). However, triggering off the following morphogenetic programs originated the biosynthesis of cardenolides:

- the formation of compact, green, nodular tissue (non-embryoid globules, cf. Sect. 2)
- the formation of shoots (cf. Sect. 3) and
- the formation of adventive embryos (cf. Sect. 4).

2 Cardenolide Biosynthesis in Green Non-Embryoid Globules

Some cell strains of *D. lanata* formed green compact non-embryoid globules (diameter 5–10 mm) in nutrient media with low auxin activity ([42] own unpublished results).

Strain II cultivated in our laboratory grew as a fine suspension of small morphologically unspecialized colonies in media containing a high auxin-cytokinin ratio (1 mg 2,4-D and 0.02 mg Ki l^{-1}). These colonies increased in size and became morphologically heterogeneous if transferred to a medium with a reduced auxin-cytokinin ratio (1 mg NAA and 0.02 mg Ki l^{-1}). Cores of small plasma-rich meristematic cells were formed in the outer parts of the colonies, which were surrounded by parenchymatic cells. Tracheids were embedded in the matrix of parenchymatic cells, which were highly vacuolated and especially large in the center of the colonies. The meristematic areas at

1 Section of Pharmacy of the Martin Luther University Halle-Wittenberg, Weinbergweg 15, 4020 Halle, GDR

Primary and Secondary Metabolism of Plant Cell Cultures, ed. by Neumann et al.

the periphery of the nodular globules continued to grow, but did not develop into shoots or other organs, if the globules were cultivated in media with high cytokinin activity (2 mg BA and 0.1 mg NAA/l^{-1}). In the center of the colonies the cells degenerated. Upon illumination the peripheric areas formed chloroplasts and turned green.

Reinhard et al. [42] were unable to find cardenolides in the green globular structures. In contrast we detected up to 0.01 mg cardenolides g^{-1} dry wt. by means of sensitive RIAs [12] and by HPLC of the genins obtained after hydrolysis of the glycoside fraction. This value was considerably higher than the amount of cardenolides present in the colonies of unspecialized cells grown in a 2,4-D-containing medium and in unilluminated globules (< 0.001 mg g^{-1} dry wt.), demonstrating that triggering off cardenolide biosynthesis and accumulation in the simple morphological structures of strain II is possible, even if in regenerated shoots and embryoids much higher cardenolide amounts were formed (cf. Sects. 3 and 4).

3 Cardenolide Formation in Shoot Cultures

Sterile shoots suitable for in vitro cultivation were obtained from callus of *D. lanata, D. purpurea*, and other *D. species* [6, 15, 20, 51], from seedlings of *D. lanata* and *D. obscura* [31, 36], and from shoot tips (meristems with the adjacent leaf primordia) of *D. lanata* [3, 11, 30, 47]. Most convenient was shoot formation from shoot tips, a procedure which is part of the in-vitro clonation of *D. lanata* plants. Daughter shoots were formed from the axillary meristems of mother shoots in the presence of cytokinins which overcome the dominance of the apical meristeme. Cultivation of shoots was carried out either on solidified or in liquid media. In routine experiments on solidified media containing 2 mg BA l^{-1} the dry weight of the cultures increased about eight-fold [30], in submerged cultures about 20-fold within 4 weeks [31]. The in-vitro cultivated shoots grew as rosettas even if they were derived from vernalized *D. lanata* plants. Devernalization proceeded already during the establishment of the cultures [30].

The cultivated shoots synthesized cardenolides. In shoots of *D. purpurea* digitoxin and purpureaglycoside A [20], in shoots of *D. lanata* the lanatosides A, B, and C, digoxigenin and gitoxigenin were identified [31, 32]. Shoot cultures derived from *D. lanata* plants with a high content of lanatoside C in the leaves of the rosetta contained the lanatoside A and C as the main cardenolides, lanatoside B and probably the diglycosides glucodigifucoside and glucoverodoxin. After hydrolysis, in addition to digitoxigenin and digoxigenin, small amounts of gitoxigenin were detected (own unpublished results). These results demonstrated that the cardenolide pattern of the in-vitro cultivated shoots, in contrast to that of the embryoids (Sect. 4.2), resembled the pattern of the rosetta leaves of the mother plants. From the growth rate of submerged cultivated *D. lanata* shoots and their cardenolide content it was calculated that about 0.1 mg cardenolides g^{-1} dry wt. were formed within 4 weeks [31, 32]. The same value was obtained with shoots cultivated on solidified media (own unpublished results).

Addition of potential cardenolide precursors (sodium glycocholate, cholesteryl acetate, progesterone) increased the cardenolide content of *D. lanata* shoots grown submerged about three-fold [31], progesterone that of *D. purpurea* shoot cultures about two-fold [18]. Removal of BA from the medium and addition of gibberellic acid

[16, 17, 32] or abscisic acid [17] also originated higher amounts of cardenolides in the cultures. In addition, growth and cardenolide content of the in-vitro cultivated shoots of *D. lanata* depended on the carbon and nitrogen sources, the phosphate level [17], as well as on the presence of inositol, thiamine, Mn^{2+}, Zn^{2+}, H_3BO_3, MoO_4^{2-}, I^-, Fe^{2+}, and EDTA in the medium (own unpublished results).

In summary, sterile shoots may easily be obtained and propagated. They form interesting cardenolides and are suited, for instance, for biochemical experiments. However, the shoots were shown to be too complicated for cultivation in fermenters and therefore are without future biotechnological significance.

4 Cardenolide Formation in Adventive Embryos

4.1 The Selection of Embryogenic Strains

Strains able to form large numbers of adventive embryos were obtained from filaments of *D. lanata* and hypocotyls of *D. lutea* [34, 51]. It was demonstrated that callus formation in the filaments of *D. lanata* had its origin in the vascular bundle, the parenchymatic cells of which resembled the xylem parenchyma cells from which callus seems to be formed in leaf explants (own unpublished results, cf. also [43, 44, 46]).

According to their morphogenetic potential the following types of strains obtained from filament of *D. lanata* may be distinguished ([51] own unpublished results):

- strains incapable of morphogenesis under different experimental conditions, i.e., without the capacity to form roots, shoots, adventive embryos or any tissues. These strains did not turn green when illuminated (about 2%);
- strains forming nodular tissue, i.e., compact globular structures consisting of a parenchymatic matrix with tracheidal elements and meristematic areas. Upon illumination these globules were able to form chloroplasts (less than 1%, cf. Sect. 2);
- strains forming roots as main organs (about 2%);
- strains able to differentiate roots, shoots, *and* adventive embryos (about 91%), and
- strains forming *only* adventive embryos (about 4%).

4.2 Triggering off Cardenolide Formation

The embryogenic strains grew as small colonies of unspecialized cells in media with a high auxin/cytokinin ratio. Embryo development was triggered off by lowering the auxin/cytokinin ratio. It included the following stages ([12, 28] own unpublished results):

- formation of small clusters of meristematic cells in colonies of unspecialized cells. These meristematic areas developed to separate entities
- enlargement of the separated meristematic cell colonies. In later stages a core of small, polyedric, plasma-rich cells was surrounded by a layer of larger vacuolated cells
- formation of more or less round, smooth embryoid globules (ϕ 1–3 mm), unable to form chloroplasts (stage-I globules), which develop to embryoid globules forming

chloroplasts upon illumination (stage-II globules):

The embryoid globules consisted mainly of a matrix of isodiametric cells in which two strings of tracheidal elements were embedded forming a Y, the terminal parts of which were directed to the cotyledons and the root pole developing later on (own unpublished results).

– and formation of heart-shaped and torpedo-shaped embryoid structures.

Many of the embryoid structures were anomalous and during further development gave no normal plants.

The colonies of unspecialized cells and the meristematic colonies contained traces of cardenolides only (about 0.001 mg g^{-1} dry wt.). This base level increased during development of stage-I globules and reached up to 0.7 mg g^{-1} dry wt. in illuminated stage-II globules (Table 1).

Table 1. Maximum cardenolide contents of *Digitalis* in-vitro cultures

Plant/strain	Material cultivated	Cardenolide content		Reference
		(mg g^{-1} dry wt.)	(mg l^{-1} culture)	
D. lanata				
Strain II	Nonembryoid globules	0.003–0.01		Own unpublished results
Strain V	Adventive embryos	0.1	1.5	[12, 34]
Strain VII	Adventive embryos	0.1	1.0	[28]
		0.7[a]		Own unpublished results
	Plantlets	0.15	4.0	[28]
Strain S-1	Adventive embryos	0.4		[35]
Strain S-2	Adventive embryos[b]	0.05		[35]
D. lanata	Shoots[c]	0.4		[32]
	Shoots[d]	0.1		Own unpublished results
D. purpurea	Shoots	0.05	0.8	[17]
D. grandiflora	Adventive embryos	0.05		[34]

[a] In surface-grown embryoid globules or in submerged grown cultures inoculated with stage-I globules
[b] Formed in the dark
[c] Grown submerged
[d] Grown in surface cultures

The cardenolides produced differed from those present in shoot cultures (Sect. 3) and the rosetta of adult plants. With specific RIAs mainly digitoxin derivatives were detected [12]. After hydrolysis of the glycoside fraction digitoxigenin as well as smaller amounts of gitoxigenin and digoxigenin were found by means of TLC and HPLC (own unpublished results).

Although the embryogenic strains were derived from plants with a high content of lanatoside C in the leaves of the rosetta, lanatoside C could not be identified. This finding was probably due to the dependence on the developmental stage of the cardenolide pattern (and the amount of cardenolides) in *D. lanata* plants. Seedlings of *D. lanata* were shown to contain either no digoxin derivatives or only small amounts, in contrast to the adult plants [1, 2, 22, 53, 55, 56]. The same may be true relative to the normal embryos, in which only small amounts of cardenolides were found [52, 54], the composition of which was not identified as yet.

4.3 Optimum Conditions for Embryo Formation and Cardenolide Biosynthesis

4.3.1 The Influence of Hormones

Organogenesis in the cell cultures of *D. lanata* was regulated by hormones. Especially the ratio of auxins to cytokinins played an important role (cf. [26]). Triggering off adventive embryogenesis and development of embryos was optimum in the successive cultivation of embryogenic strains in the following media [28]:

1. Propagation medium (high auxin/cytokinin ratio; 1.0 mg 2,4-D and 0.02 mg Ki l^{-1}). Rapid growth of small colonies of unspecialized cells.
2. Determination medium (reduced auxin activity; 1.0 mg NAA and 0.02 mg Ki l^{-1}). Formation of meristematic cell colonies.
3. Differentiation medium (high cytokinin activity; 1.0 mg BA and 0.01 mg NAA l^{-1}). Formation of embryoid globules, heart-shaped and torpedo-shaped embryos.

The accumulation of cardenolides started in the differentiation medium (cf. Sect. 4.2).

4.3.2 The Influence of the Carbon Source

Most suitable for embryogenesis and cardenolide formation was maltose, which was slowly degraded in *D. lanata* cell cultures. Adventive embryogenesis and cardenolide accumulation was optimum when maltose was present during propagation of the cells as well as during determination and embryo formation [28].

Sucrose yielded lower amounts of embryos and cardenolides. It was rapidly degraded by saccharase, which is probably an extracellular enzyme in *D. lanata* cell cultures. In contrast to maltose-containing cultures glucose and fructose are accumulated in sucrose-containing media. These sugars yielded poor rates of embryogenesis and cardenolide formation if used as carbon sources, and may act as inhibitors also in sucrose-containing media. This inhibition was in agreement with the suppression of cell specialization, e.g., spore formation and biosynthesis of secondary products in many microbial cultures by easily degradable carbon sources, e.g., glucose [7], and with the triggering off embryogenesis in *D. lanata* cell cultures by starvation [27].

4.3.3 The Influence of the Nitrogen Source

Embryo development and cardenolide accumulation were optimal with a NO_3^-/NH_4^+-ratio of 5–10:1 [28]. Glutamine and ammonium salts of organic acids acted as inhibitors.

4.3.4 The Influence of Light

In most of the embryogenic strains of *D. lanata* accumulation of cardenolides was much larger in illuminated than in unilluminated stage-II globules and late embryoid structures [12, 28, 34, 35]. The accumulation was influenced by light intensity and light quality ([35] own unpublished results). In contrast, the cardenolide content of the colonies of unspecialized cells, meristematic cell colonies and stage-I globules was not increased by illumination indicating that the competence for the action of light develops in these strains in the stage-II globules.

However, as shown already in Sect. 4.2, illumination is no prerequisite for cardenolide accumulation. Cardenolides were also detected in non-illuminated shoot cultures of *D. purpurea* (about 30% of the amount present in illuminated cultures, [16]) and considerable quantities of cardenolides were found in dark-grown cultures of a carotene-free cell line (strain S-2) derived from one of the embryogenic strains of *D. lanata*. In this strain illumination even reduced cardenolide accumulation [35]. In addition, the chloroplast-containing nonembryoid globules described in Sect. 2 as well as "undifferentiated" green cell cultures of *D. purpurea* formed only very small amounts of cardenolides ([16] own unpublished results). Hence the morphological organization of the tissues is of higher importance for cardenolide formation and accumulation than the development of chloroplasts.

4.4 Formation of Secondary Embryoid Structures

Several attempts were made for avoiding the long lasting process of embryo de-novo formation by development of a new way for growth of the cardenolide producing embryoid structures. Lastly a method was found which based on the ability of stage-I globules for the formation of daughter globules. These globules had the same developmental state as the mother globules, separated from the latter, and grew further on as distinct entities. The development of these globules to stage-II globules was inhibited if they were grown in high density. Hence embryoid cultures of *D. lanata* may be cultivated for several weeks in nutrient solutions with a low auxin/cytokinin ratio (1.0 mg BA and 0.01 mg NAA l^{-1}) in the form of embryoid stage-I globules (own unpublished results).

Cultivation of diluted suspensions of these globules in illuminated fermenters or shaking flasks resulted in the formation of embryoid structures containing 0.7 mg cardenolides g^{-1} dry wt. within 4 weeks. The dilution of the suspension during inoculation overcame the depression of development of the globules, which after about 8 days began to grow, turned green, and after about 20 days started to accumulate cardenolides. The high purity and maturity of the embryoid globules obtained in this way resulted in cultures containing larger amounts of the cardenolide accumulating embryoid structures than cultures containing embryoid structures formed de-novo, as well as in higher cardenolide quantities (Table 1).

4.5 Optimum Cardenolide Contents in Embryoid Cultures

Table 1 demonstrates that the embryoid structures may contain the same amount of cardenolides as *Digitalis* shoot cultures. Since microscopical examination revealed that the stage-II globules in which the accumulation of cardenolides begins are rather simple in construction, they might be more suitable for investigation of the mechanisms triggering off cardenolide accumulation than plantlets or shoots, which comprise several types of cells which differ in cardenolide accumulation [52, 54] and probably biosynthesis. The adventive embryos could easily be grown in air lift fermenters and accumulated under these conditions cardenolide quantities comparable to those found in shaking-flask-cultures (own unpublished results). Therefore, embryogenic cultures may be of special value for the further investigation of in-vitro formation of cardenolides.

4.6 Attempts to Select Higher Producing Cell Lines

In some cases cell lines producing higher amounts of secondary products were selected from existing strains. The respective experiments with cell cultures of *D. lanata*, however, were without success, though examination of individual embryoid globules by specific and sensitive RIAs [12, 52] showed that their cardenolide content varies within a wide range [29]. Daughter cultures derived from high producing globules had the same mean cardenolide content as the parent culture and the same heterogeneity relative to, the amount of cardenolides accumulated in the individual globules (own unpublished results). This indicated that the differences in the cardenolide content of the globules were not determined genetically, but dependent on the developmental state.

Experiments are planned to test the cardenolide content of cell lines derived from individual cells originating from protoplasts. Colonies of unspecialized cells of *D. lanata* and *D. purpurea* gave high yields of viable protoplasts by treatment with mixtures of cellulases and pectinases [8, 9]. These protoplasts regenerated walls and formed single cells within a few days. Some of the single cells later on began to divide and after about 4 weeks individual cell colonies could be isolated [8, 9]. The cell lines obtained in this way from embryoid cell strains will be examined in further experiments with respect to their aptitude for cardenolide biosynthesis.

5 Perspectives for the Future Biotechnological Production of Cardenolides with Digitalis Cell Cultures

The results discussed in the preceding sections showed that cardenolides are synthesized in different morphological structures formed in-vitro in cell and organ cultures of several *Digitalis* species. Of importance for any future biotechnological production of cardenolides might be the embryoid globules (and probably the green non-embryoid globules) formed. In the past years methods have been developed for the isolation of strains producing large amounts of these structures and for triggering off cardenolide biosynthesis. The amounts of cardenolides synthesized and accumulated are, however, still too low. They have to be increased further by the factor $10^2 - 10^3$ before cardenolides can be produced economically. Furthermore, the embryoid structures contain cardenolides which differ in their sugar moieties from those used in medicine. It may be expected that their pharmacological actions is similar to that of the cardenolides used in therapy (cf. [5]). Until now, however, this was not demonstrated. It is an open question whether these shortcomings can be overcome in the future. Of first importance for further progress will be the selection of new cell lines, which

– have a higher productivity for cardiac glycosides;
– synthesize the cardenolides used in medicine, and probably
– form cardenolides in morphologically unorganized cell colonies.

References

1. Aldrich BJ, Frith ML, Wright SE (1956) Paper chromatographic detection of new constituents of *Digitalis lanata*. J Pharm Pharmacol 8:1042–1049
2. Balbaa SI, Hilal SH, Haggag MY (1970) A study of the yield and glycosidal content of the leaves of *Digitalis lanata* at different stages of plant growth. Planta Med 18:254–259
3. Breuel K, Poetter H, Luckner M, Diettrich B, Springer M (1984) Verfahren zur virusfreien vegetativen Vermehrung und Erhaltung von *Digitalis*-Hochleistungspflanzen. GDR-Patent No 207 731
4. Büchner SA, Staba EJ (1964) Preliminary chemical examination of *Digitalis* tissue cultures for cardenolides. J Pharm Pharmacol 16:733–737
5. Chen KK (1970) Supplementary glycosides of *Digitalis*. J Med Chem 13:1035–1036
6. Corduan G, Spix C (1975) Haploid callus and regeneration of plants from anthers of *Digitalis purpurea* L. Planta 124:1–11
7. Demain AL (1982) Catabolite regulation in industrial microbiology. In: Krumphanzl V, Sikyta B, Vanek Z (eds) Overproduction of microbial products. Academic Press, London, New York, pp 3–20
8. Diettrich B, Neumann D, Luckner M (1980) Protoplast-derived clones from cell cultures of *Digitalis purpurea*. Planta Med 38:375–382
9. Diettrich B, Neumann D, Luckner M (1982) Clonation of protoplast-derived cells of *Digitalis lanata* suspension cultures. Biochem Physiol Pflanz 177:176–183
10. Elze H, Pilgrim H, Teuscher E (1974) Die Biotransformation von Cholesterol-26-^{14}C durch Gewebekulturen von *Evonymus europaea* und *Digitalis purpurea*. Pharmazie 29:727–728
11. Erdei I, Kiss Z, Maliga P (1981) Rapid clonal multiplication of *Digitalis lanata* in tissue culture. Plant Cell Rep 1:34–35
12. Garve R, Luckner M, Vogel E, Tewes A, Nover L (1980) Growth, morphogenesis and cardenolide formation in long-term cultures of *Digitalis lanata*. Planta Med 40:92–103
13. Graves JMH, Smith WK (1967) Transformation of pregnenolone and progesterone by cultured plant cells. Nature (London) 214:1248–1249
14. Gurny L, Tabacchi R, Baud C, Kapetanidis I (1981) Cultures de cals de *Digitalis purpurea* L. 2e partie: Etude des métabolites dérivés du stérane. Pharm Acta Helv 56:49–54
15. Hagimori M, Matsumoto T, Kisaki T (1980) Studies on the production of *Digitalis* cardenolides by plant tissue culture I. Determination of digitoxin and digoxin contents in first and second passage calli and organ redifferentiating calli of several *Digitalis* species by radioimmunoassay. Plant Cell Physiol 21:1391–1404
16. Hagimori M, Matsumoto T, Obi Y (1982) Studies on the production of *Digitalis* cardenolides by plant tissue culture II. Effect of light and plant growth substances on digitoxin formation by undifferentiated cells and shoot-forming cultures of *Digitalis purpurea* L. grown in liquid media. Plant Physiol 69:653–656
17. Hagimori M, Matsumoto T, Obi Y (1982) Studies on the production of *Digitalis* cardenolides by plant tissue culture III. Effects of nutrients on digitoxin formation by shoot-forming cultures of *Digitalis purpurea* L. grown in liquid media. Plant Cell Physiol 23:1205–1211

18. Hagimori M, Matsumoto T, Obi Y (1983) Effects of mineral salts, initial pH and precursors on digitoxin formation by shoot-forming cultures of *Digitalis purpurea* L. grown in liquid media. Agric Biol Chem 47:565–571
19. Helmbold H, Voelter W, Reinhard E (1978) Sterols in cell cultures of *Digitalis lanata*. Planta Med 33:185–187
20. Hirotani M, Furuya T (1977) Restoration of cardenolide-synthesis in redifferentiated shoots from callus cultures of *Digitalis purpurea*. Phytochemistry 16:610–611
21. Kartnig T (1977) Cardiac glycosides in cell cultures of *Digitalis*. In: Barz W, Reinhard E, Zenk MH (eds) Plant tissue culture and its bio-technological application. Springer, Berlin Heidelberg New York, pp 44–51
22. Kartnig T, Hiermann A (1980) Vergleichende Untersuchungen über die Cardenolid- und Flavonoidführung der Blätter und Blüten von *Digitalis lanata* während verschiedener Entwicklungsstadien. Sci Pharm 48:193–202
23. Kartnig T, Kobosil P (1977) Beobachtungen über das Vorkommen und die Bildung von Cardenoliden in Gewebekulturen aus *Digitalis purpurea* und *Digitalis lanata*. Planta Med 31:221–227
24. Kartnig T, Kummer-Fustinioni G, Heydel B (1983) Der Einfluß des Alterns auf die Sekundärstoffbildung in Gewebekulturen aus *Digitalis purpurea*. Planta Med 47:247–248
25. Kartnig T, Russheim U, Maunz B (1976) Beobachtungen über das Vorkommen und die Bildung von Cardenoliden in Gewebekulturen aus *Digitalis purpurea* und *Digitalis lanata* 1. Cardenolide in Oberflächenkulturen aus Keim- und Laubblättern von *Digitalis purpurea*. Planta Med 29:275–282
26. Koblitz H (1972) Zell- und Gewebezüchtung bei Pflanzen. Fischer, Jena
27. Kranz E, Nover L (1981) Verfahren zur Herstellung von Cardenoliden in *Digitalis*-Zellkulturen. GDR Patent 201 460
28. Kuberski C, Scheibner H, Steup C, Diettrich B, Luckner M (1984) Embryogenesis and cardenolide formation in tissue cultures of *Digitalis lanata*. Phytochemistry 23:1407–1412
29. Luckner M, Diettrich B, Kuberski C, Schwiebode C (1981) Variation in the cardenolide content of embryogenic clumps from suspension cultures of *Digitalis lanata*. Planta Med 42:104
30. Luckner M, Diettrich B, Springer M, Breuel K, Oertel C (1984) Verklonung von *Digitalis-lanata*-Hochleistungspflanzen durch Sproßspitzenkultur. In: Internation. Vortragstagung: Methoden und Verfahren der Züchtung, des Anbaues, der Sammlung und der industriellen Verarbeitung von Arznei- und Gewürzpflanzen, Artern, 18.–22. June 1984. Vortragstexte. VEB Pharm Werk, Halle, pp 113–127
31. Lui JHC, Staba EJ (1979) Effects of precursors on serially propagated *Digitalis lanata* leaf and root cultures. Phytochemistry 18:1913–1916
32. Lui JHC, Staba EJ (1981) Effects of age and growth regulators on serially propagated *Digitalis lanata* leaf and root cultures. Planta Med 41:90–95
33. Medora R, Kosegarten DC, Tsao DPN, DeFeo JJ (1967) Cardiotonic activity in callus tissue of *Digitalis mertonensis*. J Pharm Sci 56:540–541
34. Nover L, Luckner M, Tewes A, Garve R, Vogel E (1980) Cell specialization and cardiac glycoside formation in cell cultures of *Digitalis species*. Acta Hortic 96:65–74
35. Ohlsson AB, Björk L, Gatenbeck S (1983) Effect of light on cardenolide production by *Digitalis lanata* tissue cultures. Phytochemistry 22:2447–2450
36. Perez-Bermudes P, Cornejo MJ, Segura J (1983) In vitro propagation of *Digitalis obscura* L. Plant Sci Lett 30:77–82
37. Petiard V, Demarly Y (1972) Mise en évidence de glucosides et d'alcaloides dans les cultures de tissus végétaux. Ann Amelior Plant 22:361–374
38. Petiard V, Demarly Y, Paris RR (1971) Mise en évidence d'hétérosides cardiotoniques dans les cultures de tissus de *Digitalis purpurea*. CR Acad Sci Ser D 272:1365–1367
39. Petiard V, Demarly Y, Paris RR (1972) Mise en évidence d'hétérosides et d'alcaloides dans les cultures de tissus de plantes médicinales. Plant Med Phytother 6:41–49
40. Petiard V, Demarly Y, Paris RR (1972) Mise en évidence physiologique de substances cardiotoniques dans les cultures in vitro de tissue de *Digitalis purpurea* L. CR Acad Sci Ser D 274: 846–847
41. Pilgrim H (1972) Cholesterol side-chain cleaving enzyme. Phytochemistry 11:1725–1728

42. Reinhard E, Boy M, Kaiser F (1975) Umwandlung von *Digitalis*-Glykosiden durch Zellsuspensionskulturen. Planta Med Suppl 1975:163–168
43. Rücker W (1982) Kombinierter Einfluß von Indolessigsäure, Gibberellin und Benzylaminopurin auf Kallus- und Organdifferenzierung an Blattexplantaten von *Digitalis purpurea*. Z Pflanzenphysiol 107:141–151
44. Rücker W, Jentzsch K, Wichtl M (1976) Wurzeldifferenzierung und Glykosidbildung bei in vitro kultivierten Blattexplantaten von *Digitalis purpurea* L.; Einfluß verschiedener Wuchsstoffe,
45. Rücker W, Jentzsch K, Wichtl M (1981) Organdifferenzierung und Glykosidbildung bei in vitro kultivierten Blattexplantaten von *Digitalis purpurea* L.; Einfluß verschiedener Wirkstoffe, Nährlösungen und Lichtverhältnisse. Z Pflanzenphysiol 102:207–220
46. Rücker W, Jentzsch K, Wichtl M (1983) Untersuchungen über Wachstum, Morphogenese und Glykosidbildung an Wurzelorgankulturen von *Digitalis purpurea* L. Biochem Physiol Pflanz 178:91–100
47. Schöner S, Reinhard E (1982) Clonal multiplication of *Digitalis lanata* by meristeme culture. Planta Med 45:155
48. Staba EJ (1962) Production of cardiac glycosides by plant tissue cultures. I. Nutritional requirements in tissue cultures of *Digitalis lanata* and *D. purpurea*. J Pharm Sci 51:249–254
49. Staba EJ, Lamba SS (1963) Production of cardiac glycosides by plant tissue cultures. II. Growth of *Digitalis lanata* and *Digitalis purpurea* in suspension cultures. Lloydia 26:29–35
50. Stohs SJ, Rosenberg H (1965) Steroids and steroid metabolism in plant tissue cultures. Lloydia 38:181–194
51. Tewes A, Wappler A, Peschke EM, Garve R, Nover L (1982) Morphogenesis and embryogenesis in long-term cultures of *Digitalis*. Z Pflanzenphysiol 106:311–324
52. Vogel E, Luckner M (1981) Distribution of cardenolides in *Digitalis lanata*. Planta Med 41: 161–165
53. Weiler EW, Westekemper P (1979) Rapid selection of strains of *Digitalis lanata* Ehrh. with high digoxin content. Planta Med 35:316–322
54. Weiler EW, Zenk MH (1976) Radioimmunoassay for the determination of digoxin and related compounds in *Digitalis lanata*. Phytochemistry 15:1537–1545
55. Wichtl M (1972) Über die Glykoside der Keimblätter von *Digitalis lanata* Ehrh. Sci Pharm 40: 242–247
56. Wichtl M, Freier R (1978) Veränderungen der Zusammensetzung des Glykosidgemisches in den Blättern von *Digitalis lanata* Ehrh. im Verlaufe der ersten Vegetationsperiode. Dtsch Apoth-Ztg 22:798–802
57. Wichtl M, Jentzsch K, Rücker W (1978) Wachstum und Glykosidbildung in Kalluskulturen und in Geweben verschiedener Organe bei *Digitalis purpurea* L. Pharmazie 33:229–233

Metabolism of Caffeoyl Derivatives in Plant Cell Cultures

B.E. ELLIS [1]

1 Introduction

Caffeic acid (3,4-dihydroxycinnamic acid) has been found in numerous plant species and tissues since its original isolation from coffee in 1867 [1–3]. Although the free acid occurs sometimes in small amounts, most commonly caffeic acid has been isolated as a conjugate (ester, amide or glycoside) and, as such, is a member of a large and varied family of hydroxycinnamoyl conjugates which also includes p-coumaric and ferulic acid derivatives. A unique feature of the caffeoyl derivatives is the ease with which the dihydroxyphenyl moiety can be oxidized to a reactive ortho-quinone species either through spontaneous air oxidation or enzyme (polyphenol oxidase) catalysis [4]. Ortho-quinones will react rapidly with nucleophiles such as thiol or amino functions on polypeptides, or with other dihydroxyphenyl units, to form covalently linked products. The impact of caffeoyl oxidation on protein structure and the ubiquitous distribution of hydroxycinnamoyl esters and amides in the plant kingdom have led to the suggestion that these compounds may be part of a pre-existing chemical defence system in plant cells. In the hypersensitivity response, mobilization of the system would involve either passive or active release of stored phenolic derivatives from the vacuolar storage compartment as a result of some challenge to the cell's integrity, followed by oxidation of the released compounds. While this response would ultimately be suicidal for the challenged cell, it is proposed that it would also deter advance of a pathogenic organism [5].

The potential importance of caffeoyl derivatives in plant defence and the paucity of information concerning the formation, storage, and mobilization of any representative except chlorogenic acid has led us to study a number of these caffeoyl compounds using plant cell suspension cultures as the experimental system. The specific goals have been the elucidation of the biogenesis of each component of the derivative and clarification of the regulation of its synthesis. In addition, we have begun evaluating the cell culture system for its physiological relevance to whole plant responses.

1 Department of Chemistry and Biochemistry, University of Guelph, Guelph, Ontario, N1G 2W1, Canada

Primary and Secondary Metabolism of Plant Cell Cultures, ed. by Neumann et al.

2 Caffeoyl Derivatives in Cultured Cells

Plant cell suspension cultures are notorious for failing to accumulate the metabolites for which the parent plant is noted [6, 7] but this tends not to be the case for caffeoyl derivatives. Indeed, some of the highest levels reported for secondary metabolite production in plant cell cultures are those for rosmarinic acid (Fig. 1) in *Coleus blumei* cul-

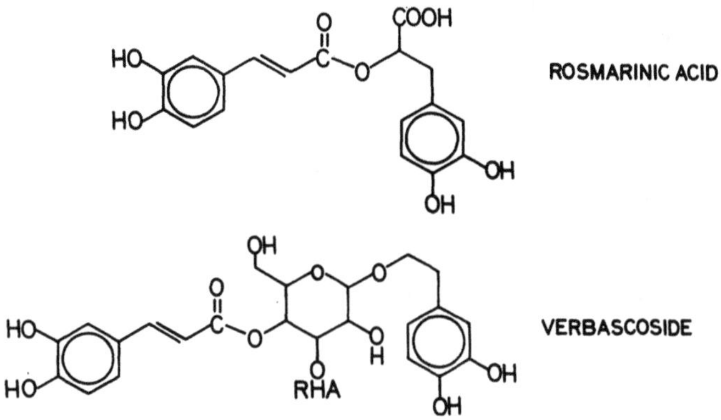

Fig. 1. Structures of rosmarinic acid and verbascoside

tures [8] and verbascoside (Fig. 1) in *Syringa vulgaris* [9]. Spontaneous overproduction of caffeoyl derivatives in cultures is not universally found, however, since cultures of *Trifolium pratense* do not accumulate any detectable clovamide [10] (N-caffeoyl DOPA) (B.E. Ellis, unpublished data) and cultures of *Daucus carota* and *Solanum tuberosum* accumulate relatively modest levels of chlorogenic acid [11, 12]. Culture lines of *Nicotiana tabacum* selected for resistance to p-fluorophenylalanine frequently display an apparently stable over-production of cinnamoyl putrescines, including caffeoyl putrescine [13, 14]. This is a particularly interesting response from a regulatory point of view because, although *Nicotiana* is capable of synthesizing both caffeoyl putrescine and chlorogenic acid, in this system phenylpropanoid overproduction is expressed only in the former end-product. In *Syringa* stem explants, the induction of overproduction of verbascoside appears to be dependent on the presence of 2,4-D in the culture medium, as shown in Fig. 2. Similar, although less dramatic, responses are observed for rosmarinic acid production in *Coleus* stem explants. In established suspension cultures, the production of these caffeoyl derivatives has been maintained without selection for over eight years. This stability is in contrast to the loss of production noted for some secondary metabolites in culture [15] and is maintained despite the existence within the suspension culture population of numerous cells with little or no intracellular accumulation of caffeoyl ester [16]. The continued presence of low-producing cells in a high-producing population implies either that, under in vitro culture conditions, the low-production characteristic carries little selective advantage within the population, or that the expression of caffeoyl derivative accumulation is a trait which varies considerably from one cell generation to the next in culture, or both. The results of single-

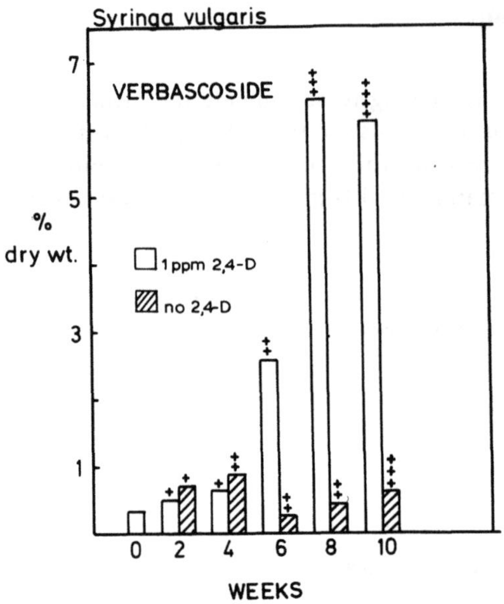

Fig. 2. Induction of verbascoside accumulation in stem explants of *Syringa vulgaris* on B5 medium with and without 2,4-D. *Crosses* indicate relative amount of callus formation

cell cloning of microspectrophotometrically-analyzed cells lend support to the idea that a high degree of cell-to-cell production variability is inherent in these cultures [17]. Detailed analysis of the dynamics of growth and rosmarinic acid production in suspension cultures of *Anchusa officinalis* showed that in B5 medium (1.0 mg l^{-1} 2,4-D) the synthesis of rosmarinic acid was initiated only after the cells had completed the exponential phase of their growth and their respiratory rate had declined markedly [18]. The synthesis of verbascoside in *Syringa* and of caffeoyl putrescine in *Nicotiana* appears to follow a similar pattern [9, 19].

3 Biogenesis of Caffeoyl Derivatives in Cultured Cells

The conversion of phenylalanine to caffeic acid comprises the general phenylpropanoid pathway [20] which is initiated by the action of phenylalanine ammonia-lyase (PAL). In addition to this common reaction sequence, however, the synthesis of each caffeoyl derivative requires a set of reactions which generates the non-caffeoyl moiety at an appropriate intracellular site and concentration, and couples it in ester or amide linkage with caffeic acid. In some instances, this moiety is a compound which has other metabolic roles in the cell (e.g. malic acid, glycerol, glucose, putrescine). It is not clear in these cases whether the caffeoyl-linked pool of the compound shares its biosynthetic origins with other cellular pools, or whether it is generated by a unique set of reactions solely for conjugate synthesis.

Clovamide (N-caffeoyl DOPA) and chlorogenic acid are examples of caffeoyl derivatives containing moieties which are biosynthetically only a single reaction removed from a primary metabolite (tyrosine and dehydroquinic acid, respectively), but the new reaction can now be considered an integral part of the synthesis of the pool of caffeoyl conjugate. This pattern becomes much more elaborate when we consider the biogenesis of rosmarinic acid and of the complex caffeoyl phenylethanol glycosides such as verbascoside. Both the 3,4-dihydroxyphenyllactic acid moiety of rosmarinic acid and the 3,4-dihydroxyphenylethanol moiety of verbascoside have been shown by radiotracer studies to have their origins in the aromatic amino acid tyrosine [9, 21], but several reactions are required in each case to modify the sidechain and add a 3-hydroxyl to the phenyl ring. The sequence of steps proposed to occur in these two pathways is shown in Fig. 3. The order presented for the final steps of verbascoside synthesis

TYROSINE
▼
TYRAMINE
▼
4-HYDROXYPHENYL ACETALDEHYDE
▼
4-HYDROXYPHENYL ETHANOL (TYROSOL)
▼
4-HYDROXYPHENYLETHANOL GLUCOSIDE (SALIDROSIDE)
▼
3,4-DIHYDROXYPHENYL ETHANOL GLUCOSIDE
▼
3,4-DIHYDROXYPHENYLETHANOL RHAMNOSYL GLUCOSIDE ▶ VERBASCOSIDE

TYROSINE
▼
4-HYDROXYPHENYLPYRUVATE
▼
4-HYDROXYPHENYLLACTATE
▼
3,4-DIHYDROXYPHENYLLACTATE
▼
ROSMARINIC ACID

Fig. 3. Proposed pathways for biogenesis of rosmarinic acid and verbascoside

is hypothetical since the substrate specificities of the relevant enzymes are still unknown. It is nevertheless striking that both in cultures producing rosmarinic acid and in those producing verbascoside, little if any of the proposed biosynthetic intermediates can be detected chromatographically. Whether the existence of some type of multienzyme assembly is responsible for this apparent channeling of phenylalanine and tyrosine metabolism in the cultured cells remains to be demonstrated.

4 Enzymology and Regulation

4.1 Phenylalanine Ammonia-Lyase (PAL)

Exactly how the metabolic flux through the general phenylpropanoid pathway in plant cells is regulated is still not clear. A number of studies, most notably those of Hahlbrock and Grisebach [20] have suggested that PAL catalyzes the rate-determining step between phenylalanine and the hydroxycinnamyl products. Consistent with this concept are the position of PAL at the beginning of the pathway and the ready inducibility of elevated PAL activity by stimuli which increase synthesis of phenylpropanoid metabolites [22, 23]. In flavonoid-producing tissues, the induced increase in PAL and, by inference, the increased flow in the pathway, has been shown to involve de novo synthesis of PAL enzyme [20].

In order to determine whether cultured cells which produce massive levels of caffeoyl derivatives possess any unique regulatory features at the PAL step, the enzyme from *Syringa vulgaris* cultures has been extensively characterized (N.B. Ramsey and B.E. Ellis, in preparation). Transfer of stationary-phase *Syringa* cells to fresh medium induces a rapid transient increase in extractable PAL, which declines to low levels by 72 h. The enzyme from 48-h cultures was purified by ammonium sulphate precipitation, chromatofocussing and preparative PAGE, and in most respects appears to be similar to PAL from other plant sources. In addition to being strongly inhibited by cinnamate, however, it is also sensitive to 1.0 mM phenylpyruvate (57% inhibition) and less so to 1.0 mM caffeic acid (26% inhibition), but not to verbascoside or to 4-hydroxyphenylpyruvate. The significance of inhibition by phenylpyruvate is not clear but is reminiscent of the strong inhibition of *Coleus blumei* PAL by 4-hydroxyphenylpyruvate [21]. In the latter case, the inhibitory compound is an intermediate in the biosynthesis of rosmarinic acid in the *Coleus* cells. Neither *Coleus* PAL nor *Syringa* PAL was stimulated in vitro by any of the phenylalanine or tyrosine metabolites tested which suggests that no positive effector arrangement exists between the tyrosine-derived pathways and the first step of the phenylalanine-derived general phenylpropanoid pathway.

Another approach to uncovering interactions between these pathways has been to attempt to block one pathway by use of specific enzyme inhibitors [21]. Inhibition of *Syringa* PAL in vivo with the potent competitive inhibitor L-α-aminooxyphenylpropionate (AOPP) [24] did effectively block synthesis of both aromatic moieties of verbascoside from phenylalanine and tyrosine. However, this was found to be due not to a metabolic interlock between the phenylalanine and tyrosine pathways but rather to a strong inhibition of the first enzyme in the tyrosine pathway by AOPP (see below).

4.2 Formation of Non-Caffeoyl Moieties

1. Caffeoyl Putrescine. The diamine component of caffeoyl putrescine has been shown by radiotracer studies in *Nicotiana* cell cultures to be formed from ornithine [13]. This conversion could potentially be accomplished either directly by ornithine decarboxylase (ODC), or by the sequence ornithine → arginine → agmatine → putrescine, in which arginine decarboxylase (ADC) catalyzes the committed step. Both decarboxylases can be detected in cultured *Nicotiana* cells but in a caffeoyl putrescine overproducer cell line, it is ADC activity which is most enhanced relative to that in the parent line (M. Walker, R. Downer, E. Dumbroff, and B.E. Ellis, in preparation). Treatment of the cells with the ODC suicide inhibitor α-difluoromethylornithine also failed to block incorporation of labelled ornithine into the putrescine moiety of caffeoyl putrescine [25]. It would be useful, however, to also examine in the effects of the analogous ADC inhibitor before drawing any final conclusions as to the relative responsibilities of these two enzymes for supporting caffeoyl putrescine production. Both enzymes have been characterized from other plant sources [26] but the properties of the *Nicotiana* culture enzymes have not been determined.

2. Chlorogenic Acid. The quinate component of chlorogenic acid is produced by reduction of dehydroquinate, an intermediate in the shikimate pathway. The enzyme responsible, quinate: NAD^+ oxidoreductase, has been extensively characterized in *Daucus carota* cell cultures, where it displays very interesting regulatory features. Not only can it exist in both a phospho- and dephospho-form, with the former more active than the latter [27], but the protein kinase responsible for the phosphorylation is in turn apparently controlled by interaction with CA^{2+} calmodulin, or some other type of calciprotein. Recently, it has also been reported that the regulatory characteristics of this system vary with the illumination status of the cultures [28]. Considering the numerous reports of light effects on phenylpropanoid metabolism, it will be very interesting to see how the impact of light on the quinate: NAD^+ oxidoreductase system is mediated and how these patterns are integrated with control of the general phenylpropanoid pathway in the overall biosynthesis of chlorogenic acid.

3. Rosmarinic Acid. The formation of 3,4-dihydroxyphenyllactic acid from tyrosine requires 3-hydroxylation of the ring at some point in the pathway. However, based on the specific accumulation of label from tyrosine in 4-hydroxyphenyllactate during in vivo AOPP inhibition of PAL in *Coleus blumei* cultures [21], we assume that the ring hydroxylation occurs late in rosmarinic acid biosynthesis. The initial conversion of tyrosine to 4-hydroxyphenylpyruvate has proven to be catalyzed by an aminotransferase, and full characterization of this enzyme from *Anchusa officinalis* cultures is now underway.

4. Verbascoside. Syringa vulgaris cell cultures accumulate both verbascoside, with its 3,4-dihydroxyphenylethanol moiety, and salidroside, the glucoside of 4-hydroxyphenylethanol (tyrosol) [9]. Both phenylethanol units rapidly become labelled when the cells are supplied with radioactive tyrosine and it seems likely that at least one pool of salidroside is actually an intermediate in verbascoside biosynthesis. There are seven enzymes required for conversion of tyrosine to part of the verbascoside structure and several of these are presently being characterized.

The first step is catalyzed by tyrosine decarboxylase (TDC), a pyridoxal phosphate-requiring enzyme whose activity in *Syringa* cells is induced by subculturing in a manner analogous to that of PAL. Earlier in vivo attempts to selectively inhibit the phenylalanine-derived pathway with AOPP and the tyrosine-derived pathway with the potent aromatic amino acid decarboxylase inhibitor α-fluoromethyltyrosine had resulted in both pathways being blocked by the AOPP treatment but neither being affected by the tyrosine analogue (C.C. Chapple and B.E. Ellis, in preparation). In vitro studies showed that AOPP was a powerful inhibitor of *Syringa* tyrosine decarboxylase (app. $Ki = 1.1 \times 10^{-8}$ M), whereas α-fluoromethyltyrosine was a relatively weak inhibitor of this plant enzyme. A comparison of the characteristics (pH optimum, thiol sensitivity, substrate specificity and inhibitor profile) of the *Syringa* enzyme with those of the analogous enzyme from three other plant species (*Nicotiana, Hordeum*, and *Sanguinaria*), *Streptococcus* and hog kidney showed that there were surprisingly few common features among these various decarboxylases. For example, tyrosine decarboxylase from *Hordeum* was sensitive to AOPP, while the enzyme from *Sanguinaria* was fully active at 200 μM AOPP.

Syringa TDC was slightly sensitive to inhibition by tyrosol (48% inhibition at 1.0 mM) and verbascoside (30% at 1.0 mM) but not to any other intermediates in the pathway. It was, however, more strongly inhibited by p-coumaric (43% at 0.1 mM) and caffeic (31% at 0.1 mM) acid, a feature also reported for the *Hordeum* enzyme [29]. No positive effectors were found from either pathway.

The product of tyrosine decarboxylation is converted to 4-hydroxyphenylacetaldehyde by a soluble tyramine oxidase which appears to have an inhibitor profile and substrate specificity setting it apart from either the well-characterized mammalian monoamine oxidases or the few plant amine oxidases which have been studied (Rancourt and Ellis, unpublished observations). Reduction of the 4-hydroxyphenylacetaldehyde to tyrosol is an NADPH-requiring step which has not yet been characterized further. Glucosylation of tyrosol is accomplished by a novel glucosyl transferase which has complete specificity for glucose transfer to the alkyl hydroxyl of tyrosol and for UDPG as the sugar donor substrate (Parissenti and Ellis, unpublished observations).

The progress made thus far on the enzymes of tyrosine metabolism in verbascoside synthesis suggests that they are a unique series of catalysts with specificity for the reactions needed to generate the non-caffeoyl portion of the verbascoside molecule. As isolated by cell disruption, they appear to be freely soluble proteins but the possibility of loose multienzyme associations existing within the cell cannot be eliminated. The lack of classical feed-back modulation on the initial steps of verbascoside synthesis, and the inducibility of PAL and tyrosine decarboxylase, both suggest that the main long-term control over the flux through the pathways may be the level of the active enzymes within the cytoplasm, much as it is in flavonoid biosynthesis in *Petroselinum* cultures [23]. Short-term control mechanisms must also exist, however, as exemplified by the NAD^+:quinate oxidoreductase system [27], and we can expect rapid progress in this area.

4.3 Caffeoyl Transferases

The exact point at which the caffeoyl unit becomes covalently coupled to its polar partner has been a point of debate for many years. Studies of chlorogenic acid biosynthesis suggested that the primary sequence followed was conversion of p-coumarate to p-coumaroyl quinate and then to chlorogenate [30], although in *Solanum tuberosum* cultures caffeic acid was efficiently incorporated directly into chlorogenate [12]. The latter result implied the existence of both a carboxyl-activating system and an appropriate acyltransferase. The hydroxycinnamoyl-CoA: quinate hydroxycinnamoyl transferase has since been characterized from *Nicotiana alata* and *Stevia rebaudiana* cell cultures [31]. The enzyme from both sources was capable of transferring not only caffeoyl units but also p-coumaroyl units. Interestingly, the chlorogenate-synthesizing activity could also be detected in a number of species which are known to produce caffeoyl derivatives other than chlorogenic acid (e.g. *Coleus, Cichorium, Trifolium, Catalpa*). At the same time, in cultures such as *Coleus* or *Nicotiana*, it has proven remarkably difficult to detect caffeoyl transferase activity which will utilize caffeoyl-CoA as a donor and 3,4-dihydroxyphenyllactate or putrescine as an acceptor (B.E. Ellis, unpublished results; J. Berlin, personal communication). Perhaps a more indirect series of transesterification steps is involved, or some regulatory factor is required which has not been satisfied by the incubation conditions used thus far.

5 Defence Responses in Caffeoyl Derivative-Accumulating Cultures

The question of the physiological relevance of observations made in cultured cell systems is always difficult to resolve. The cultures described above are clearly specialized in their secondary metabolism relative to many of the tissue types in the intact plant but they are presumably very similar to the phenolic-rich parenchymatous cells which occur widely in plants. As such, cultured cells accumulating caffeoyl derivatives also might be expected to respond to challenges to their integrity in the same fashion as phenolic-rich tissue in a traumatized plant [e.g. release and oxidation of stored caffeoyl derivative; increased polyphenoloxidase (PPO) activity]. Preliminary experiments with *Anchusa officinalis* cultures have confirmed that this is the case. An example of the impact of both an abiotic (4% DMSO) and a biotic (brief exposure to fungal cellulase) challenge on the metabolism of *Anchusa* cells is shown in Table 1. The dramatic rise in PPO activity observed after a brief hydrolysis of cell walls suggests that these cultures may well serve as useful model systems both for probing the biosynthesis of caffeoyl conjugates and for clarifying the mechanisms by which a threat to a plant's cellular integrity results in phenolic mobilization.

Table 1. Effect of dimethyl sulfoxide and cellulase on phenolic metabolism in *Anchusa officinalis* suspension cultures

Culture status 24 h post-treatment	Treatment of cells		
	4% DMSO for 1 h	2% Cellulase for 5 min	Control
Polyphenol oxidase ΔOD_{490}/min/ mg prot.[a]	0.00	0.58	0.01
PAL pkats mg^{-1} prot.	37.5	30.6	30.2
Rosmarinic acid content $\mu mol\ g^{-1}$ fr.wt.	12.3	14.0	11.0

[a] Measured with L-DOPA as substrate

Acknowledgements. This work has been made possible by the support of the Natural Sciences and Engineering Research Council of Canada.

References

1. Bate-Smith EC (1962) The phenolic constituents of plants and their taxonomic significance. J Linn Soc London Bot 58:95–173
2. Harborne JB (1966) Caffeic acid ester distribution in higher plants. Z Naturforsch 21b:604–605

3. Sondheimer E (1964) Chlorogenic acids and related depsides. Bot Rev 30:667–712
4. Pierpoint WS (1969) *o*-Quinones formed in plant extracts: their reactions with amino acids and peptides. Biochem J 112:609–617
5. Tomiyama K, Sakai R, Sakuma T, Ishizaka N (1967) The role of polyphenols in the defense reaction in plants induced by infection. In: Mirocha CJ, Uritani I (eds) The dynamic role of molecular constituents in plantparasite interaction. Am Phytopathol Soc, St Paul, Minn, p 165
6. Kartnig T (1977) Cardiac glycosides in cell cultures of *Digitalis*. In: Barz W, Reinhard E, Zenk MH (eds) Plant tissue culture and its bio-technological application. Springer, Berlin Heidelberg New York, p 17
7. Staba EJ, Chung AC (1981) Quinine and quinidine production by cinchona leaf, root and unorganized cultures. Phytochemistry 20:2495–2498
8. Zenk MH, El-Shagi H, Ulbrich B (1977) Production of rosmarinic acid by cell suspension cultures of *Coleus blumei*. Naturwissenschaften 64:585
9. Ellis BE (1983) Production of hydroxyphenylethanol glycosides in suspension cultures of *Syringa vulgaris*. Phytochemistry 22:1941–1943
10. Yoshihara T, Yoshihara H, Kunimatsu S, Sadamura S, Sakuma T (1977) New amino acid derivatives conjugated with caffeic acid and DOPA from red clover *(Trifolium pratense)*. Agric Biol Chem 41:1679–1684
11. Stärk D, Alfermann AW, Reinhard E (1976) Verlauf von Phenylalaninammonium-lyase-aktivität Anthocyan- und Chlorogensäurebildung in verschiedenen Zellstämmen von *Daucus carota*. Planta Med 30:104–117
12. Gamborg OL (1967) Aromatic metabolism in plants V. The biosynthesis of chlorogenic acid and lignin in potato cell cultures. Can J Biochem 45:1451–1457
13. Berlin J (1981) Formation of putrescine and cinnamoyl putrescines in tobacco cell cultures. Phytochemistry 20:53–55
14. Berlin J, Kukoschke KG, Knobloch K-H (1981) Selection of tobacco cell lines with high yields of cinnamoyl putrescines. Planta Med 42:17–180
15. Zenk MH (1978) The impact of plant cell culture on industry. In: Thorpe TA (ed) Frontiers of plant tissue culture 1978. Int Assoc Plant Tissue Cult, Calgary, p 1
16. Chaprin N, Ellis BE (1984) Microspectrophotometric evaluation of rosmarinic acid accumulation in single cultured plant cells. Can J Bot (in press)
17. Ellis BE (1982) Cell-to-cell variability in secondary metabolite production within cultured plant cell populations. In: Fujiwara A (ed) Plant tissue culture 1982. Jpn Assoc Plant Tissue Cult, Tokyo, p 395
18. De-Eknamkul W, Ellis BE (1984) Rosmarinic acid production and growth characteristics of *Anchusa officinalis* cell suspension cultures. Planta Med (in press)
19. Berlin J, Knobloch K-H, Höfle G, Witte L (1982) Biochemical characterization of two tobacco cell lines with different levels of cinnamoyl putrescines. J Nat Prod 45:83–87
20. Hahlbrock K, Grisebach H (1979) Enzymic controls in the biosynthesis of lignin and flavonoids. Annu Rev Plant Physiol 30:105–130
21. Ellis BE, Remmen S, Goeree G (1979) Interactions between parallel pathways during biosynthesis of rosmarinic acid in cell suspension cultures of *Coleus blumei*. Planta 147:163–167
22. Dixon RA, Lamb CJ (1979) Stimulation of de novo synthesis of L-phenylalanine ammonia-lyase in relation to phytoalexin accumulation in *Colletotrichum lindemuthianum* elicitor-treated cell suspension cultures of French bean *(Phaseolus vulgaris)*. Biochim Biophys Acta 586:453–463
23. Betz B, Schäfer E, Hahlbrock K (1978) Light-induced phenylalanine ammonia-lyase in cell suspension cultures of *Petroselinum hortense*. Quantitative comparison of rates of synthesis and degradation. Arch Biochem Biophys 190:126–135
24. Amrhein N, Gödeke K-H (1977) α-Aminooxy-β-phenylpropionic acid – a potent inhibitor of L-phenylalanine ammonia-lyase in vitro and in vivo. Plant Sci Lett 8:313–317
25. Berlin J, Forche E (1981) DL-α-Difluoromethyl ornithine causes enlargement of cultured tobacco cells. Z Pflanzenphysiol 101:277–282
26. Altman A, Friedman R, Levin N (1982) Arginine and ornithine decarboxylases, the polyamine biosynthetic enzymes of mung bean seedlings. Plant Physiol 69:876–879

27. Refeno G, Ranjeva R, Boudet AM (1982) Modulation of quinate: NAD^+ oxidoreductase activity through reversible phosphorylation in carrot cell suspensions. Planta 154:193–198
28. Graziana A, Ranjeva R, Salimath BP, Boudet AM (1983) The reversible association of quinate: NAD^+ oxidoreductase from carrot cells with a putative regulatory subunit depends on light conditions. FEBS Lett 163:306–310
29. Hosoi K (1972) Effects of aromatic acids on tyrosine carboxy-lase from barley roots. Plant Cell Physiol 13:179–181
30. Steck W (1968) Metabolism of cinnamic acid in plants: chlorogenic acid formation. Phytochemistry 7:1711–1717
31. Ulbrich B, Zenk MH (1979) Partial purification and properties of hydroxycinnamoyl-CoA: quinate hydroxycinnamoyl transferase from higher plants. Phytochemistry 18:929–933

Metabolic Relationships of Putrescine, GABA and Alkaloids in Cell and Root Cultures of Solanaceae

H.E. FLORES[1] and P. FILNER[2]

1 Introduction

The diamine putrescine (1,4-diaminobutane) is the precursor of the polyamines spermidine (Spd) and spermine (Spm) in all living systems studied to date. In view of the functions that polyamines may have in plant growth and development (Smith 1985), it is of importance to understand the mechanisms that regulate the availability of precursor pools for the synthesis of these compounds. In most, if not all higher plants, putrescine is synthesized via two alternate routes: (1) decarboxylation of ornithine, which gives rise to putrescine directly; or (2) decarboxylation of arginine, with agmatine and N-carbamylputrescine as intermediates. The arginine decarboxylase pathway appears to be important under conditions of stress (Flores et al. 1985), whereas the ornithine decarboxylase pathway is closely correlated with rapid cell division (Cohen et al. 1982). Spd and Spm are formed by the transfer of one or two aminopropyl moieties, respectively, derived from decarboxylated S-adenosylmethionine (Tabor and Tabor 1984). There are, however, other possible metabolic fates of putrescine besides being a polyamine precursor.

Several years ago, Berlin and Widholm (1977) selected a variant cell line of tobacco XD cells that was resistant to p-fluorophenylalanine (pfPhe). These cells had unusually high levels of phenolic derivatives which were identified as cinnamoyl putrescines (Berlin 1981). The putrescine conjugates reached levels of up to 10% (dry wt.) in the p-fPhe-resistant cell lines. In this paper we report physiological conditions, namely changes in the nitrogen source, which also lead to accumulation of cinnamoyl putrescines in tobacco XD cells.

In cereals and legumes (Flores and Filner 1985), putrescine can be metabolized via a two-step oxidation to γ-aminobutyric acid (GABA). A putrescine-utilizing mutant (PUT) was selected in this lab from XD cells. PUT cells differ from XD cells in that PUT cells efficiently convert putrescine to GABA in vivo. We report on the characterization of PUT cells, and compare its putrescine metabolism with that of wild-type cells, using various inhibitors as probes.

Putrescine, possibly in a bound form, is also the precursor for various alkaloids found in the Solanaceae, such as nicotine and the tropane alkaloids. An alternative system to cell suspensions is described below for studying the regulation of secondary metabolites related to putrescine and cinnamic acids; it involves the culture of roots transformed

1 Department of Plant Pathology, Louisiana State University, Baton Rouge, Louisiana 70803, USA
2 ARCO Plant Cell Research Institute, 6560 Trinity Court, Dublin, California 94568, USA

Primary and Secondary Metabolism of Plant Cell Cultures, ed. by Neumann et al.

with *Agrobacterium rhizogenes*. We report that transformed root clones grow as fast as cell suspensions, while producing the same amount of alkaloids as in the intact plant; cell suspensions on the other hand synthesize little or none of the alkaloids made by root cultures.

2 Nitrogens Sources for Tobacco Cell Suspensions: Effect on Free and Bound Putrescine Pools

Tobacco XD cells, a cell suspension line developed from *Nicotiana tabacum* cv. Xanthi, are able to grow in a variety of organic nitrogen sources in addition to nitrate or ammonium (Filner 1978). The inorganic sources support a doubling time of 2 days. On some organic nitrogen sources such as urea, growth at the same rate as nitrate-grown cells can be obtained only after selection of cells which metabolize the nitrogen source at a higher rate (i.e., they have higher urease activity) (Skokut and Filner 1980). The XD cells grow on arginine, but not on ornithine. They breakdown arginine to ornithine and urea by the action of arginase. The urea nitrogen is assimilated via urease action. These cells have ornithine decarboxylase activity, so ornithine can be converted to putrescine. Growth continues at the same rate without prior selection when cells are transferred from nitrate to GABA. Since GABA can be derived from putrescine in some higher plants (Flores and Filner 1985), it was of interest to know if tobacco cells could utilize putrescine as sole nitrogen source. Although growth was very slow during the initial transfers, a putrescine-utilizing variant (PUT cells) was eventually obtained, and has now been grown in N-MID medium + 1.5 mM putrescine + 3 mM succinate through over 50 serial subcultures made at 2- to 3-week intervals. The ability to utilize putrescine is retained by PUT cells after growth on nitrate for several months before transferring back to putrescine. Free and bound putrescine content was analyzed in cells grown under various nitrogen regimes using previously published methods (Flores and Galston 1982; Flores and Filner 1985; Ponchet et al. 1982).

As would be expected, the nitrogen source had a significant effect on the amino acid composition of XD cells (data not shown). Most dramatic, however, were the effects on the amount of free putrescine and cinnamoyl putrescines (Fig. 1). Nitrate-grown cells have very low levels of free and bound putrescine. In contrast, cinnamoyl putrescine represent a major organic nitrogen compound in the high-urease (14U) cell line grown on urea, or in XD cells grown on GABA. The total putrescine pool is higher than 5% dry wt. under these conditions. PUT cells grown on putrescine show intermediate levels of free and bound putrescine.

The accumulation of cinnamoyl putrescines can be induced by a simple switch in the nitrogen source, as shown in Fig. 2A for coumaroyl putrescine. When XD cells are transferred from KNO_3 to putrescine, a rapid accumulation of coumaroyl putrescine is observed after a 2- to 4-h lag. The presence of KNO_3 with putrescine somewhat enhances the level of bound putrescine obtained after 48 h, but transfer from KNO_3 to KNO_3 does not result in accumulation of cinnamoyl putrescine. Similar results have been observed upon transfer to urea or GABA. Even PUT cells, when freshly transferred to putrescine, are able to accumulate cinnamoyl putrescines, although at a slower rate

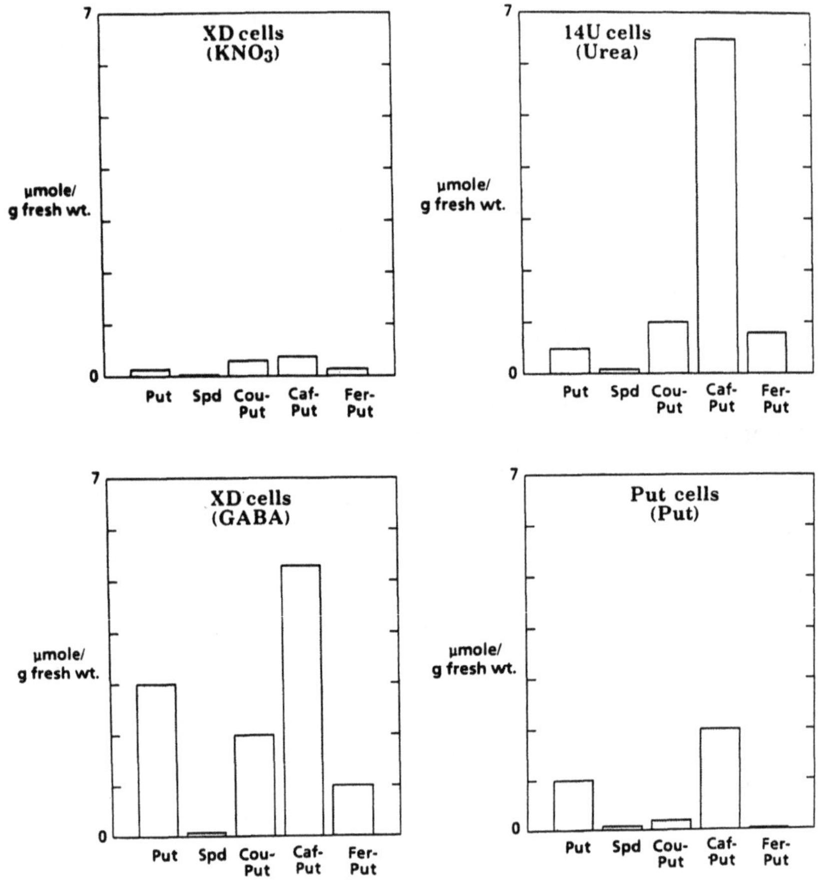

Fig. 1. Polyamine contents of XD tobacco cell lines grown on various nitrogen sources. Analysis performed on late exponential phase cultures, according to Flores and Filner (1985) and Ponchet et al. (1982)

than XD cells (Fig. 2B). As will be shown in Sect. 3, however, the main metabolic fate of putrescine in PUT cells is strikingly different than in wild-type cells. The accumulation of cinnamoyl putrescine can be reversed upon transfer back to KNO_3 (Fig. 3). It is not clear, however, whether the decrease of putrescine levels is due to turnover of the conjugates, or to dilution upon subsequent growth.

3 Putrescine Metabolism in PUT Cells

Putrescine metabolism in wild type and PUT cells was studied by the combined use of specific inhibitors and labeling with [1,4-^{14}C]-putrescine. The first step in the metabolism of GABA by plant cells is its transamination to pyruvate, yielding alanine and succinic semialdehyde, which is in turn oxidized to succinate. Gabaculine (GCU), a specific

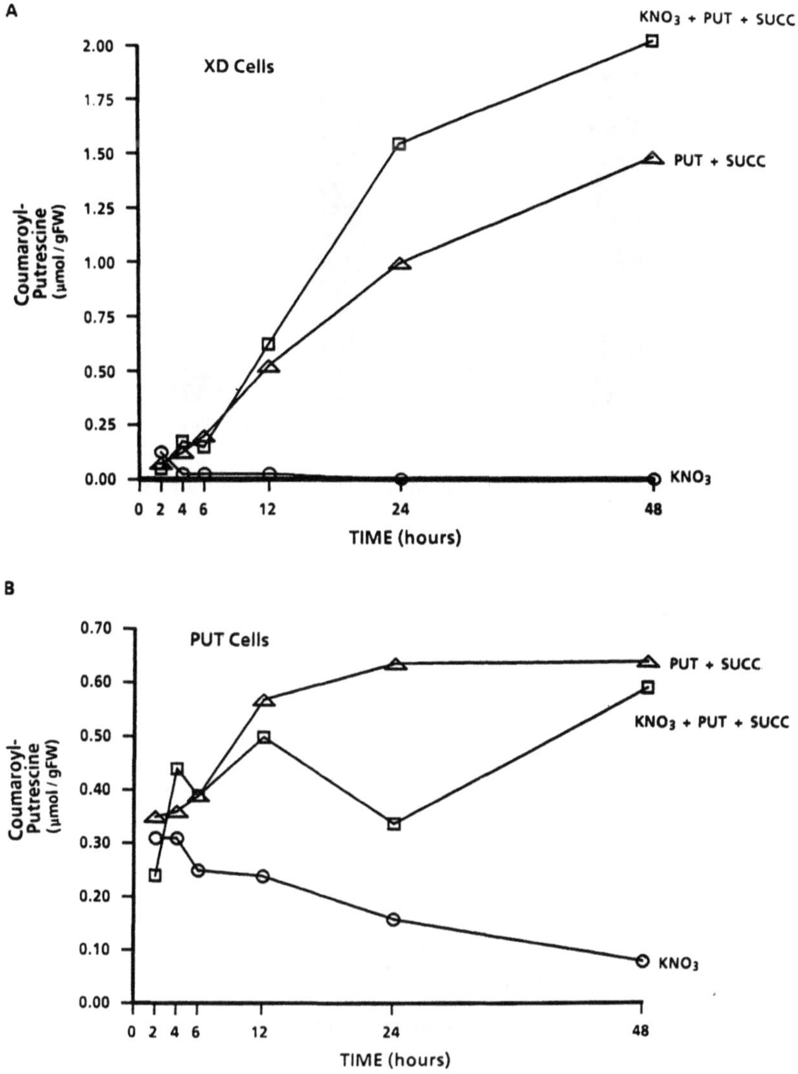

Fig. 2A,B. Time course of change in coumaroyl putrescine levels in XD cells (**A**) and PUT cells (**B**). Late exponential phase cells were washed with nitrogen-free media and resuspended in the indicated media

irreversible inhibitor of GABA-transaminase isolated from *Streptomyces toyocaensis* (Kobayashi et al. 1977), inhibits the tobacco enzyme (Filner, unpublished observation) and growth of the cells in GABA at 1–10 nM (Fig. 4). PUT cells growing on putrescine are also inhibited by GCU in the same concentration range, suggesting that putrescine is metabolized via its conversion to GABA. In contrast, growth of XD on nitrate or urea is inhibited by µM amounts of GCU. This 1,000-fold specificity "window" makes GCU a useful metabolic probe. Gabaculine at much higher concentrations has recently

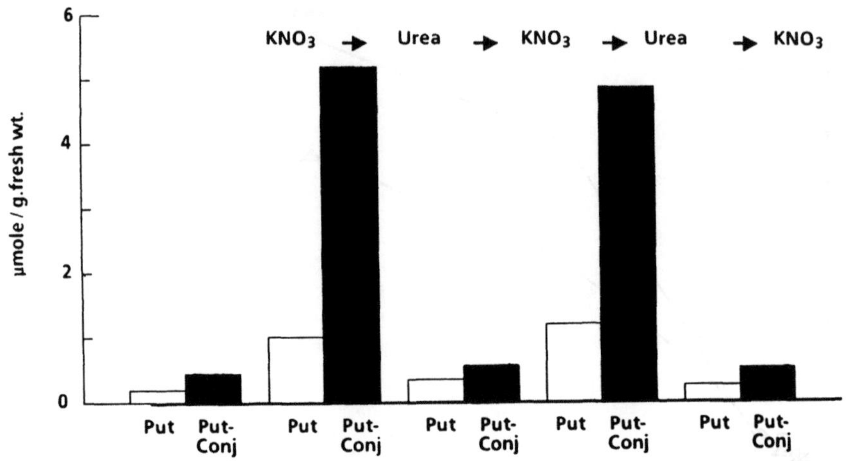

Fig. 3. Reversibility of cinnamoyl putrescine accumulation in tobacco XD cells. Analyses were done on exponentially growing cells transferred at 2-week intervals

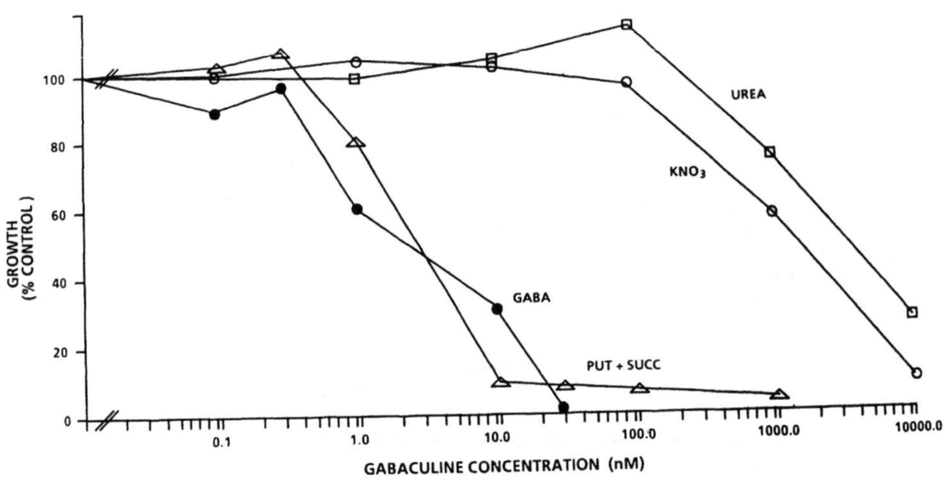

Fig. 4. Growth inhibition by gabaculine of tobacco cells on various nitrogen sources. Fresh weight determined after 2 weeks of culture. On NO_3^-, 100% = 26 g fresh wt l^{-1}; on urea 32; on GABA 29; on putrescine + succinate, 23

been reported to inhibit synthesis of δ-aminolevulinic acid from glutamate (Flint 1984). Because this blocks chlorophyll synthesis, GCU may be useful as a herbicide.

Both 14U and PUT cells take up labeled putrescine at the same rate. However, while the main metabolites in 14U cells are the cinnamoyl putrescines, PUT cells convert over 60% of the label into GABA within 2–4 h (Fig. 5A). GCU at 1 μM has no effect on the labeling pattern in 14U cells growing on urea, but in PUT cells growing on putrescine it causes a dramatic (8–10 fold) increase in the label found in GABA. This finding supports the contention that a pathway from putrescine to GABA is operating in PUT cells, but is absent in the wild type XD line.

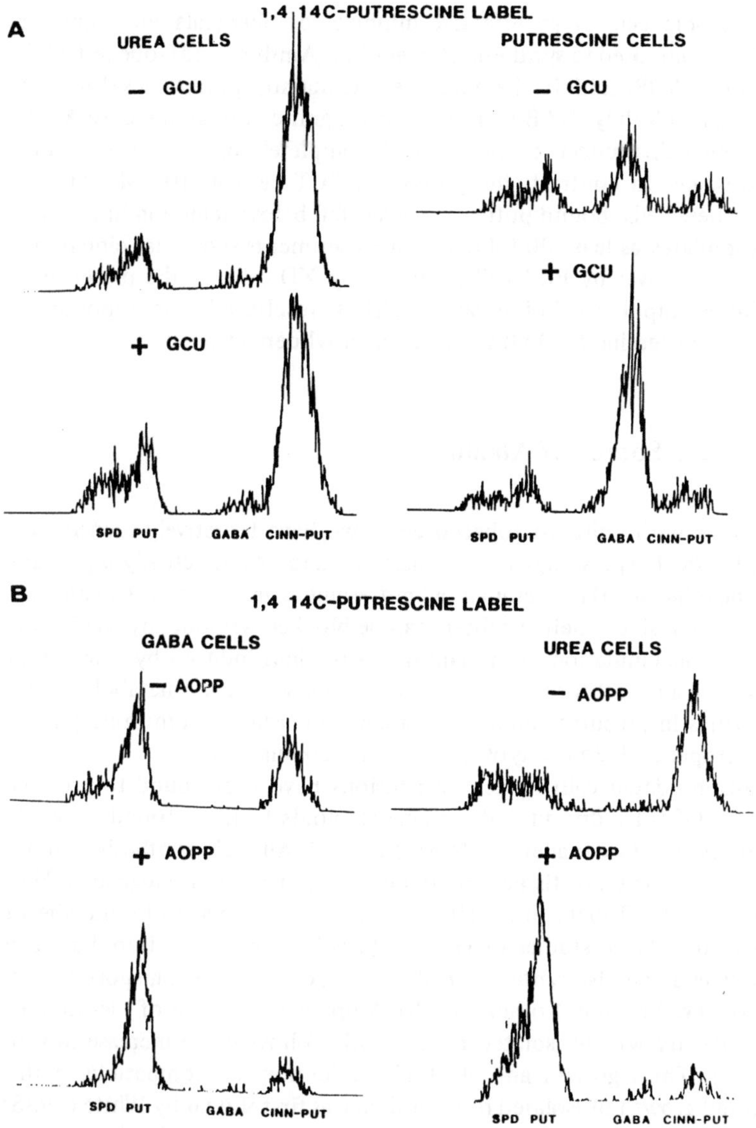

Fig. 5A,B. Labeling patterns in tobacco cells incubated with [1,4-^{14}C]-putrescine for 4 h in the absence or presence of GCU (1 μM) or AOPP (10 μM)

How is putrescine metabolized to GABA by PUT cells? We have recently established (Flores and Filner 1985) that diamine oxidase and pyrroline dehydrogenase extracted from pea can catalyze the conversion of putrescine to GABA, but we have not been able to detect such enzyme activities in extracts from PUT cells. In animal systems, it is well known that monoacetylated derivatives are better substrates for the di- or polyamine oxidases than the free amines (Tabor and Tabor 1984). It may be possible that in PUT cells a conjugated form, such as a cinnamoyl putrescine, is serving as a substrate.

Aminooxyphenylpropionic acid is an effective inhibitor of phenylalanine ammonia lyase, the first enzyme commited to synthesis of phenolics (Amrhein and Gödeke 1977). When added to XD cells, AOPP inhibits the synthesis of cinnamoyl putrescines (Fig. 5B). The growth of XD cells on KNO_3, GABA or urea is not significantly affected by AOPP (1–10 μM), even though this concentration of AOPP completely blocks the accumulation of bound putrescine. In contrast, the growth of PUT cells in 10 μM AOPP is inhibited > 80%; in these cells, bound putrescine levels fall below detection limits, and free putrescine accumulates as least 20-fold. Tracer experiments show that cinnamoyl putrescines can be turned over by PUT cells, but not by XD cells, in the presence of GCU and AOPP (Flores, unpublished observations). These results led us to hypothesize that PUT cells convert putrescine to GABA via a cinnamoyl derivative.

4 Cultured Roots as a Source of Alkaloids

From our observations with cultured tobacco cells, we have tentatively identified a system (PUT cells) in which a presumptive secondary metabolite has actually a primary metabolic role. In the original XD line, cinnamoyl putrescines seem to fit in the category of secondary compounds, since their synthesis can be blocked without any significant effect on growth. The accumulation of cinnamoyl putrescines induced by a switch in nitrogen regime may be a model for the regulation of other secondary metabolites. Because of the similarities in precursor molecules (a cinnamic acid and ornithine/putrescine), we chose the tropane alkaloids, hyoscyamine and scopolamine.

In general, undifferentiated calli or cell suspensions have been found to produce much lower amounts (< 0.1% dry wt.) of tropane alkaloids than are found in whole plants (0.4%–1.0% dry wt.) (see Waller and Novacki 1978). Alkaloid synthesis resumes either when growth slows down or tissues are induced to undergo morphogenesis (Hashimoto and Yamada 1983; Tabata et al. 1972). Roots were shown to be the site of nicotine synthesis in the classic studies of Dawson (1942). The same pattern has been established for tropane alkaloids: synthesis in the root, followed by transport and accumulation in leaves (Waller and Novacki 1978). Moreover, the rate of nicotine accumulation (up to 3% dry wt.) in isolated roots closely followed the increase in root dry wt. (Solt 1957). Thus, growth and alkaloid production are compatible in this system. The unlimited growth of isolated roots in vitro was first shown by White (1938) in tomato, and has since been extended to numerous species. However, the slow growth rates of root cultures has discouraged their use in metabolic studies.

Agrobacterium rhizogenes, is the cause of the "hairy root" disease of dicotyledonous plants. As in the case of *A. tumefaciens*, expression of the disease involves the stable integration of part of the Ri (for root-inducing) plasmid into the plant genome (Chilton et al. 1982). Roots arising at the site of infection can be cultured aseptically, and the transformed root clones obtained may be subcultured indefinitely on basal medium, growing at rates several times faster than normal roots. "Hairy root" clones of several Solanaceae were thus established for the study of tropane alkaloid production.

5 Development of Transformed Root Clones of Hyoscyamus

Seeds of *Hyoscyamus muticus* and *H. niger* were surface-sterilized (0.1% $HgCl_2$ for 5 min) and germinated on basal White's medium modified according to Street and Henshaw (1966). When seedlings were about 3 weeks old, shoots and the midveins of the most expanded leaves were punctured with a 3- to 5-day-old culture of *A. rhizogenes* (strain A4). Three weeks after infection, the adventitious roots formed at the inoculation sites (Fig. 6A) were removed, and excess bacteria were eliminated by either heat (37 °C for 2 weeks) or antibiotic (100 $\mu g\ ml^{-1}$ carbenicillin) treatment. Transformed root clones were subsequently established and maintained in rotary shakers in darkness (150 rpm at 30 °C) with transfer every 4 weeks.

The presence of a much higher number of meristematic points in transformed roots accounts for the faster rate of growth compared to normal root clones (Fig. 6B, Table 1). Numerous root hairs are found along the roots above the meristems. Although growing faster than normal roots, all the *Agrobacterium*-transformed root clones produce about the same amount of alkaloids and, in the case of *H. muticus*, the alkaloid pattern is also similar to that of normal roots and intact plants, i.e. hyoscyamine is the predominant compound. Transformed roots were also established from *Nicotiana tabacum* seeds, also showing rapid growth and high nicotine levels.

Table 1. Growth and alkaloid production by normal and *Agrobacterium rhizogenes* transformed root clones and callus cultures

Root clone	Fresh weight yield g/100 ml	Total alkaloid (% dry wt.)	Main alkaloid
Hyoscyamus muticus			
Normal C1 roots	0.51 ± 0.06	0.61	Hyoscyamine (> 80%)
H. muticus A4 RC1 roots	4.65 ± 0.19	0.55	Hyoscyamine (> 75%)
H. muticus A4 RC2 roots	4.81 ± 0.23	0.50	Hyoscyamine (> 85%)
H. muticus A4 RC3 roots	5.05 ± 0.22	0.55	Hyoscyamine (> 85%)
H. muticus A4 RC4 roots	4.71 ± 0.15	0.48	Hyoscyamine (> 90%)
H. muticus A4 RC5 roots	5.55 ± 0.28	0.50	Hyoscyamine (> 80%)
H. muticus A4 RC1 callus[a]	11.25 ± 0.50	N.D.	None
H. muticus A4 RC14 roots[b]	5.10 ± 0.20	0.58	Hyoscyamine (> 75%)
H. niger Normal C1 root	0.45 ± 0.09	0.55	Scopolamine (> 70%)
H. niger A4 RC1 roots	4.55 ± 0.21	0.40	Unknown (< 20% Scopolamine)
H. niger A4 RC2 roots	4.91 ± 0.30	0.45	Unknown (< 15% Scopolamine)
Nicotiana tabacum Normal roots	0.35 ± 0.50	>2.00	Nicotine
N. tabacum A4 RC1 roots	2.05 ± 0.10	>3.00	Nicotine

Roots harvested after 4 weeks of growth on White's basal medium (see text for details). Callus harvested after 6 weeks of growth on Murashige and Skoog's medium + 1.0 mg l^{-1} kinetin + 0.1 mg l^{-1} NAA

N.D. = Not detectable (< 0.025% dry wt.)

[a] Callus derived from A4 RC1 roots

[b] Second generation root clone derived from A4 RC1 callus

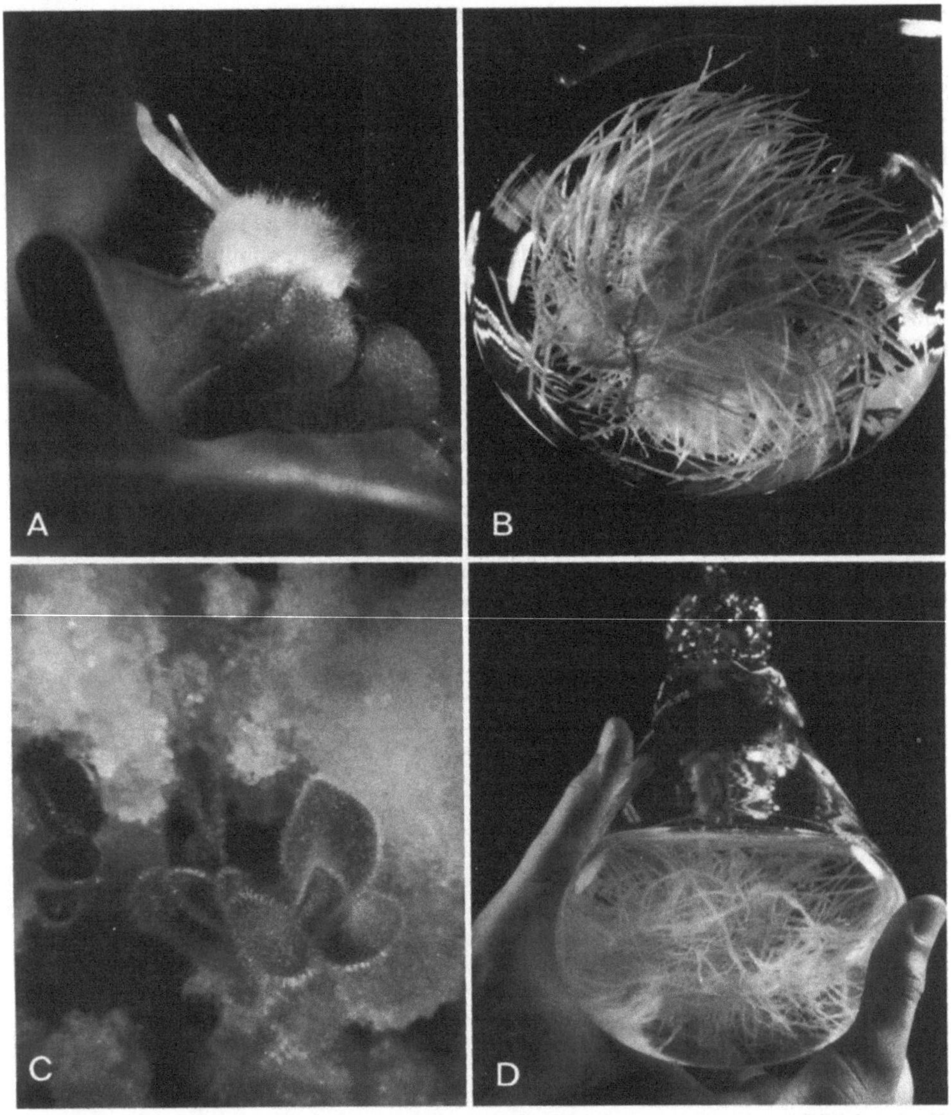

Fig. 6A–D. Establishment of hairy root cultures of *Hyoscyamus muticus*. **A** Roots emerging after 2 weeks of inoculation of leaf midvein with *A. rhizogenes* strain A4. **B** Primary transformed root clone (4 weeks of growth from a single root tip). **C** Shoot regeneration in callus derived from hairy roots grown on Murashige and Skoog medium + 1.0 mg l^{-1} kinetin + 0.1 mg l^{-1} NAA). **D** Second generation hairy root clone grown for 4 weeks from a single tip (5 mm) in a 1-l Erlenmeyer flask

In contrast with crown gall tumors, cells transformed by *A. rhizogenes* are capable of differentiating shoots spontaneously when transferred to solid medium (Tepfer 1984). Callus can be induced in the presence of exogenous plant hormones (Fig. 6C), from which shoots and roots can also be recovered upon transfer to basal medium. Alkaloids cannot be detected in transformed callus, but appear in regenerated roots, confirming

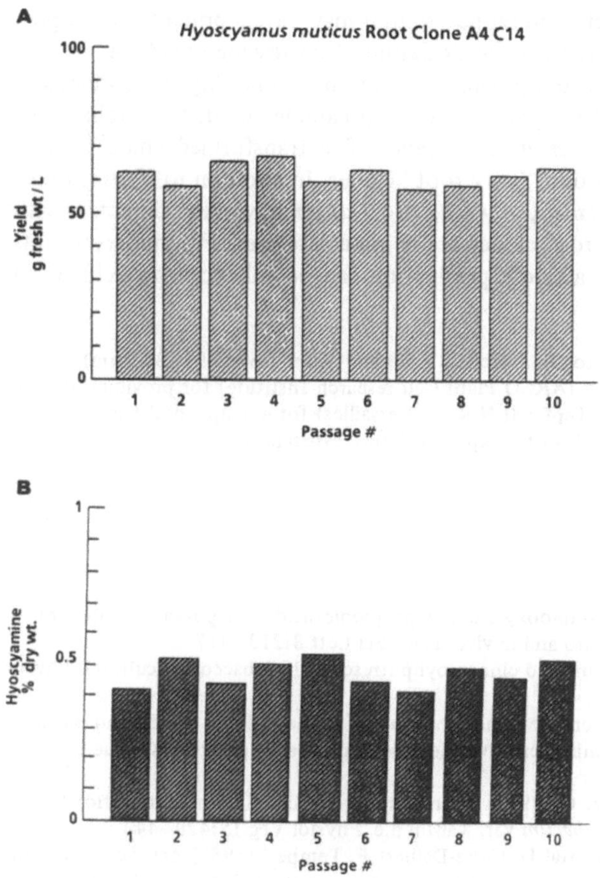

Fig. 7A,B. Stability of growth and alkaloid production in a second-generation hairy root clone of *Hyoscyamus muticus* (A4 C14). Yield and alkaloid content determined after each 4-week growing period from a single tip

that alkaloid production is dependent on organized growth. These second-generation roots can also be established as transformed clones, where the fast growth and alkaloid production have been stably maintained for at least 10 passages (Figs. 6D, 7).

6 Conclusions

The results discussed here show that putrescine in plant cells is more than just a precursor for polyamine synthesis. In fact, putrescine in the form of a cinnamoyl derivative can account for over 90% of the total diamine pool, especially in flower parts (Martin-Tanguy et al. 1978). A function in flower development has been proposed for bound polyamines (Cabanne et al. 1977). In tobacco XD cells we have established that the accumulation of cinnamoyl putrescine to levels similar to those found in flower parts can be induced by a simple switch in nitrogen source. This cell suspension system is thus a model to study the regulation of putrescine metabolism, and to probe the function of the bound forms. In addition we have preliminary evidence for at least one system, putrescine-utilizing cells, where cinnamoyl putrescines may have a primary role in putrescine metabolism, serving as carrier for the oxidation of putrescine to GABA.

The fact that putrescine is also the precursor of the pyrroline ring in nicotine, and probably of the tropine ring in hyoscyamine and scopolamine, led us to establish root clones transformed with *Agrobacterium rhizogenes*. The transformed character and alkaloid production are stable in our "hairy root" clones. In addition to *Hyoscyamus*, other *Solanaceae* such as *Brugmansia, Atropa,* and *Datura* have been infected with *A. rhizogenes*, and transformed root clones are being established. Preliminary results indicate than in this system too, alkaloid production can be enhanced by a change in nitrogen source.

Acknowledgements. We are grateful to Dr. Nikolaus Amrhein (Ruhr Universität-Bochum) for his gift of AOPP; to Dr. Robert Simpson (ARCO Plant Cell research Institute) for providing the A4 strain of *A. rhizogenes*; to Dr. David Tepfer (I.N.R.A., Versailles) for a sample of A4 transformed tobacco seeds; and to Mr. Mark Staebell for his expert technical assistance.

References

Amrhein N, Gödeke K-H (1977) α-aminooxy-β-phenylpropionic acid – a potent inhibitor of L-phenylalanine ammonia lyase in vitro and in vivo. Plant Sci Lett 8:313–317

Berlin J (1981) Formation of putrescine and cinnamoyl putrescines in tobacco cell cultures. Phytochemistry 20:53–55

Berlin J, Widholm JM (1977) Correlation between phenylalanine ammonia lyase activity and phenolic biosynthesis in *p*-fluorophenyl-alanine-sensitive and -resistant tobacco and carrot tissue cultures. Plant Physiol 59:550–553

Cabanne F, Martin-Tanguy J, Martin C (1977) Phénolamines associées 'à l'induction florale et à l'état reproducteur du *Nicotiana tabacum* var. Xanthi n.c. Physiol Veg 15:429–443

Chilton MD, Tepfer DA, Petit A, Chantal D, Casse-Delbart F, Tempé J (1982) *Agrobacterium rhizogenes* insert T-DNA into the genomes of the host plant root cells. Nature (London) 295: 432–434

Cohen E, (Malis) Arad E, Heimer YM, Mizrahi Y (1982) Participation of ornithine decarboxylase in early stages of tomato fruit development. Plant Physiol 70:540–543

Dawson RF (1942) Nicotine synthesis in excised tobacco roots. Am J Bot 29:813–815

Filner P (1978) Regulation of inorganic nitrogen and sulfur assimilation in cell suspension cultures. In: Thorpe TA (ed) Frontiers of plant tissue culture 1978. Int Assoc Plant Tissue Cult, Calgary, pp 437–442

Flint D (1984) Gabaculine inhibits δ-ALA synthesis in chloroplasts. Plant Physiol (Suppl) 75:170 (Abstr)

Flores HE, Filner P (1985) Polyamine catabolism in higher plants: Characterization of pyrroline dehydrogenase. Plant Growth Regul 3:275–289

Flores HE, Galston AW (1982) Analysis of polyamines in higher plants by high performance liquid chromatography. Plant Physiol 69:701–706

Flores HE, Young ND, Galston AW (1985) Polyamine metabolism and plant stress. In: Key JL, Kosuge T (eds) Cellular and molecular biology of plant stress. Liss, New York, in press

Hashimoto T, Yamada Y (1983) Scopolamine production in suspension cultures and redifferentiated roots of Hyoscyamus niger. Planta Med 47:195–199

Kobayashi K, Miyasawa S, Endo A (1977) Isolation and inhibitory activity of gabaculine, a new potent inhibitor of γ-aminobutyrate aminotransferase produced by a *Streptomyces*. FEBS Lett 76:207–210

Martin-Tanguy J, Cabanne F, Perdrizet E, Martin C (1978) The distribution of hydroxycinnamic acid amides in flowering plants. Phytochemistry 17:1927–1928

Ponchet M, Martin-Tanguy J, Poupet A, Marais A, Beck D (1982) Separation and quantification of basic hydroxycinnamic amides and hydroxycinnamic acids by reversed-phase high-performance liquid chromatography. J Chromatogr 240:397–404

Skokut TA, Filner P (1980) Slow adaptive changes in urease levels of tobacco cells cultured on urea and other nitrogen sources. Plant Physiol 65:995–1003

Smith TA (1985) Polyamines. Annu Rev Plant Physiol 36:117–143

Solt ML (1957) Nicotine production and growth of excised tobacco root cultures. Plant Physiol 32:480–488

Street HE, Henshaw GG (1966) Introduction and methods employed in plant tissue culture. In: Willmer EN (ed) Cells and tissues in culture. Methods, biology and physiology, vol III. Academic Press, London New York, p 459

Tabata M, Yamamoto H, Hiraoka N, Konoshima M (1972) Organization and alkaloid production in tissue cultures of *Scopolia parviflora*. Phytochemistry 11:949–955

Tabor CW, Tabor H (1984) Polyamines. Annu Rev Biochem 53:749–790

Tepfer D (1984) Transformation of several species of higher plants by Agrobacterium rhizogenes: Sexual transmission of the transformed genotype and phenotype. Cell 37:959–967

Waller GR, Novacki EK (1978) Alkaloid biology and metabolism in Plants. Plenum Press, New York London

White PR (1938) Cultivation of excised roots of dicotyledonous plants. Am J Bot 25:348–356

Metabolism and Degradation of Nicotinic Acid in Plant Cell Cultures

W. BARZ[1]

1 Introduction

In primary and secondary metabolism of higher plants, nicotinic acid (pyridine-3-carboxylic acid) plays an important role.

On the one hand, nicotinic acid in form of its amide constitutes the functional moiety of the pyridine nucleotide coenzymes, NAD and NADP. These nucleotides occur in all organisms which are capable of energy metabolism. The coenzymes appear to be among the most versatile molecules with respect to their biological functions. In addition to their well-documented role in a large number of oxidation-reduction reactions, these coenzymes are also involved in many aspects of metabolic regulation (Everse et al. 1982).

On the other hand, nicotinic acid is the building unit of many simple pyridine compounds (i.e. trigonelline, nicotine, anabasin, ricinine) and other more complex alkaloids which occur as secondary compounds in numerous plants of widely different taxonomic position (Mothes and Schütte 1969; Leete 1980).

Among the various biosynthetic routes known to exist in nature for nicotinic acid biosynthesis (Fig. 1), higher plants use the dihydroxyacetonephosphate/aspartate pathway which is also known from some microorganisms (White 1982). The first isolatable intermediate is quinolinic acid (QA). This 2,3-pyridine-dicarboxylic acid is also involved both in the oxidative catabolic route starting from tryptophane (fungi, several prokaryotes, vertebrates) and in the aspartate/formate pathway (anaerobic route in *Clostridium* species) leading to the pyridine ring structure. Quinolinic acid, as the major precursor of pyridine nucleotides, enters a biosynthetic and subsequently a salvage pathway which involve NAD (together with NADP) and several breakdown products of the pyridine nucleotide coenzyme (Fig. 1). These salvage pathways (pyridine nucleotide cycles) regenerate NAD from nicotinamide mononucleotide (NMN), nicotinamide (Nam) or nicotinic acid (Na). Pyridine nucleotide cycles (PNC) have been shown to be operative in microorganisms (Foster and Moat 1980), plants (Waller and Nowacki 1978), and animals (White 1982). Though plant PNC have not been studied very thoroughly, it is assumed that such cycles provide the main cellular pool of nicotinic acid used for secondary biosynthesis (Waller and Nowacki 1978).

Our studies on nicotinic acid metabolism in plant cell cultures mainly concentrated on the question whether plants possess the ability to degrade the pyridine ring of

1 Lehrstuhl für Biochemie der Pflanzen, Westfälische Wilhelms-Universität, 4400 Münster, FRG

Primary and Secondary Metabolism of Plant Cell Cultures, ed. by Neumann et al.

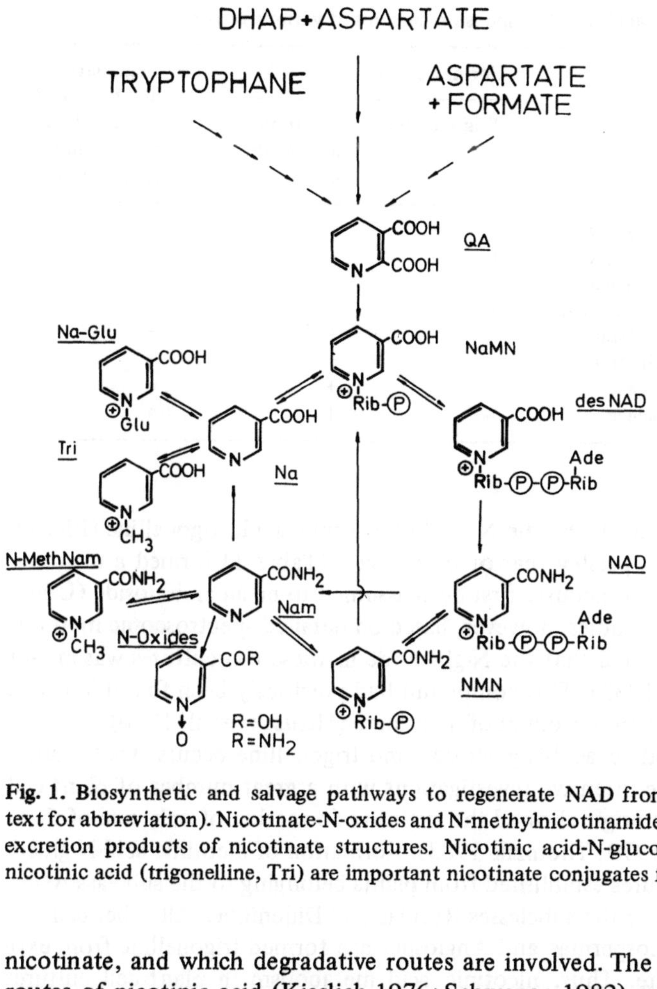

Fig. 1. Biosynthetic and salvage pathways to regenerate NAD from its breakdown products (see text for abbreviation). Nicotinate-N-oxides and N-methylnicotinamide (N-MethNam) are mammalian excretion products of nicotinate structures. Nicotinic acid-N-glucoside (Na–Glu) and N-methyl-nicotinic acid (trigonelline, Tri) are important nicotinate conjugates found in plants

nicotinate, and which degradative routes are involved. The various microbial catabolic routes of nicotinic acid (Kieslich 1976; Schwenen 1982), which all involve 6-hydroxy-nicotinic acid as first intermediate, have been taken as an example. It is of interest here to state that mammalian organisms cannot degrade nicotinate but rather excrete pyridine-N-oxides, trigonelline or N-methyl-nicotinamide (Fig. 1). Our studies were also designed to elucidate the enzymology and the metabolic fate of plant nicotinic acid conjugates such as nicotinic acid-N-glucoside (Na-Glu) or N-methylnicotinic acid (trigonelline, Tri) which are also shown in Fig. 1.

2 Conversion of Nicotinic Acid to Conjugates

Application of nicotinic acid or nicotinamide to plant cell suspension cultures always resulted in very rapid uptake of the substrates by the cells and the almost instantaneous conversion to conjugates (Heeger et al. 1976; Schwenen 1982). In case of soybean, mung bean, chickpea or *Chenopodium rubrum* cell cultures (Table 1), this primary

Table 1. Metabolism of nicotinic acid in cell suspension cultures of various plants

Cell culture species	Family	Formation of		Degradation of nicotinic acid by decarboxylation and ring fission
		Trigonelline	Nicotinic acid-N-glucoside	
Glycine max Merr.	Papilionaceae	+	–	–
Phaseolus aureus Roxb.	Papilionaceae	+	–	–
Cicer Arietinum L.	Papilionaceae	+	–	–
Chenopodium rubrum	Chenopodiaceae	+	–	–
Petroselinum hortense	Umbelliferae	–	+	+
Daucus carota	Umbelliferae	–	+	+
Nicotiana tabacum	Solanaceae	–	+	+
Nicotiana glauca	Solanaceae	–	+	+

conversion product was found to be the N-methyl nicotinic acid, trigonelline (Fig. 1). Other cell cultures, such as parsley, carrot or tobacco (Table 1) formed a nicotinic acid-N-glycoside which had erroneously first been assumed to be an arabinoside (Leienbach and Barz 1976). Later studies, however, based on detailed spectroscopic methods and chemical syntheses revealed that the N-glycoside in these cell cultures was in fact the N-glucoside (Thomzik 1983). This compound had previously been found in leaves of tobacco plants as a conversion product of nicotinate (Mizusaki et al. 1970).

This formation of nicotinic acid-N-glucoside and trigonelline occurs strictly alternatively (Table 1), because intensive investigations with a great number of plant cell cultures of wide taxonomic origin showed the exclusive formation of only one of these conjugates (Willeke et al. 1979; Thomzik 1983). Formation of nicotinic acid-N-glucoside was found in cell cultures established from plants belonging to the subclass Asteridae and the higher orders of the subclasses Rosidae and Dilleniidae. All other cell cultures from both the Gymnospermae and Angiospermae formed trigonelline from exogenously applied nicotinate. Thus, nicotinic acid metabolism in plant cell cultures could be of chemotaxonomic value as a marker at the level of order or subclass.

A further difference in nicotinic acid metabolism between these two groups of plant cell cultures concerns their ability to degrade this pyridinecarboxylic acid by either decarboxylation or ring fission (Table 1). As judge by the formation of $^{14}CO_2$ from either [^{14}C-carboxyl]- or [6-^{14}C]-nicotinic acid, the nicotinic acid-N-glucoside-producing cell cultures only showed the capacity for degradation of this substrate (Willeke et al. 1979; Schwenen 1982; Thomzik 1983).

Several investigations (Willeke et al. 1979; Thomzik 1983; Groß 1984; Upmeier and Barz, unpublished) have meanwhile shown that trigonelline and nicotinic acid-N-glucoside are also endogenous constituents of relevant plant cell cultures. Thus, cell cultures of soybean or mung bean when grown on nicotinate-free growth media for at least 20 passages contained significant amounts (0.01–0.03 μmol g^{-1} fr.wt.) of trigonelline as determined by chromatographic or HPLC techniques. Furthermore application of [U-^{14}C]-aspartate to soybean cell suspension cultures led to labelled ^{14}C-trigonelline (see Fig. 1). Similar experiments with parsley cell cultures and ^{14}C-aspartate clearly revealed endogenous synthesis of nicotinic acid-N-glucoside because label was readily detected in chromatographically purified conjugate.

The ability of certain plant cell cultures to form and accumulate trigonelline might be of physiological interest because this compound has been shown to function as a hormone in various legumes (Evans and Tramontano 1984). In primary root tissue of *Pisum sativum, Phaseolus vulgaris* or *Glycine max* trigonelline, which stems from the seeds, promotes preferential cell arrest in G2 of the cell cycle. The various investigations by Evans and coworkers (Tramontano et al. 1982, 1983) clearly demonstrated that N-methylnicotinic acid has a regulatory role in cell proliferation. It is presently not known whether trigonelline may be assigned a similar role in leguminous plant cell cultures.

3 Enzymology of Nicotinic Acid Conjugation Reactions

Recent studies have further clarified the metabolism of trigonelline and nicotinic acid-N-glucoside in plant cell cultures by purification and characterization of the responsible nicotinic acid N-methyltransferase and the N-glucosyltransferase. Early reports by Joshu and Handler (1960) and Antony et al. (1975) using crude protein preparations from *Pisum sativum* and *Trigonella foenum graecum* L. plants had provided evidence for nicotinic acid-N-methylation. S-Adenosylmethionine (SAM) or ATP in the presence of methionine were found to be essential cosubstrates.

In our own studies the S-adenosylmethionine: nicotinic acid-N-methyltransferase was purified from soybean cell suspension cultures (Groß 1984). The enzyme accepted SAM as sole methyl donor and in agreement with many N- and S-methyltransferases the S-adenosyl-L-homocysteine was found to be a strong competitive inhibitor of the enzyme reaction. The molecular weight of the protein was measured to be 110,000. The apparent K_m values for nicotinic acid and SAM were determined to be 120 μM and 50 μM, respectively. The enzyme reaction is shown in Fig. 2.

In soybean cell cultures the specific enzyme activity of the N-methyltransferase was found to be rather constant during a complete growth cycle; this observation and the

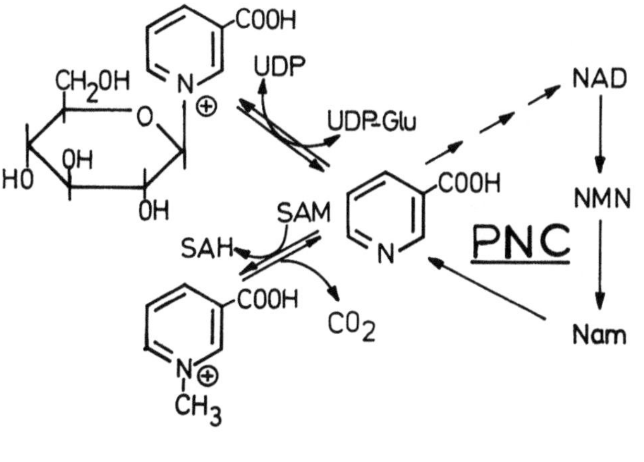

Fig. 2. Enzyme reactions for the formation of nicotinic acid-N-glucoside and trigonelline in plant cell suspension cultures in relation to the pyridine nucleotide cycle (PNC). The glucosyltransferase reaction is freely reversible whereas N-methylation and N-demethylation require different enzymes

rather constant level of trigonelline in such cultures (Upmeier and Barz unpublished) help to explain other findings that permanent trigonelline synthesis seems to occur in soybean cell cultures.

Present investigations in soybean cell suspension cultures involving pulse-labelling studies on trigonelline turnover (Upmeier and Barz, unpublished) seem to indicate that trigonelline N-demethylation proceeds rather slowly. In these rapidly dividing cells trigonelline appears to be diluted out of the cells by cell division.

Exogenous application of N-^{14}C-methyl-nicotinic acid to various leguminous cell cultures had shown oxidative N-demethylation reactions leading to nicotinic acid and other constituents of the PNC (Heeger et al. 1976). Recent enzymatic studies by Taguchi and Shimabayashi (1983) on trigonelline metabolism detected a demethylating enzyme activity in various organisms including higher plants so that an enzymic basis for the N-demethylase could be shown (Fig. 2). More work, however, is required to elucidate the precise role of N-demethylation and N-methylation of nicotinate in plant cell cultures.

The enzymology of nicotinic acid-N-glucoside formation has thoroughly been studied using protein preparations from parsley cell suspension cultures (Thomzik 1983). The final preparations were free from other glucosyltransferase activity and also did not show any hydrolysis of UDP-glucose. Essential properties of the enzyme are shown in Table 2.

A remarkable property of this N-glucosyltransferase reaction is its free reversibility. Thus, incubation of nicotinic acid-N-glucoside and UDP with the purified glucosyltransferase leads to the formation of nicotinic acid and UDP-glucose. Nicotinic acid-N-glucoside glucohydrolase activity could not be detected at any level of protein purification. Such data support earlier observations that nicotinic acid-N-glucosides, such as the riboside, are energy-rich compounds (Zatman et al. 1953).

The glucosyltransferase reaction (Fig. 2) can furthermore be assumed to provide nicotinic acid from the N-glucoside reservoir for the PNC if sufficient levels of UDP are available; this assumption requires further investigation. The level of N-glucosyltransferase activity during a growth cycle of parsley cell cultures was found to be rather constant, indicating that the enzyme is not related to any particular phase of the growth cycle.

Table 2. Properties of UDP-glucose: nicotinic acid-N-glucosyl-transferase from parsely cell suspension cultures. (Thomzik 1983)

Molecular weight	46,000
pH Optimum	7.8–8.2
Temperature optimum	28 °C
Sole glucosyl donor	UDP-Glucose
Best glucosyl accepter[a]	Nicotinic acid
K_m Nicotinic acid	0.17 mM
K_m UDP-Glucose	1.2 mM

[a] Nicotinic acid-methylester and 3-hydroxypyridine were found to be the only other, rather poor, glucosyl acceptors

4 Comparative Aspects of Nicotinic Acid Conjugation in Plants and Plant Cell Cultures

In plant cell cultures, trigonelline and nicotinic acid-N-glucoside occur as strictly alternative compounds. Furthermore, they are not metabolically exchangeable, because crossfeeding experiments in cell cultures showed that nicotinic acid-N-glucoside cannot be metabolized by trigonelline-producing cell cultures and that trigonelline is not metabolized in N-glucoside-forming cultures (Willeke et al. 1979). On the other hand, trigonelline, which has been detected in many plants (Evan's and Tramontano 1984), seems to occur in some plants which are known to produce at the level of cell cultures, nicotinic acid-N-glucoside, only (Thomzik 1983). Our own investigations on leaves and roots of potato, tobacco, and *Dahlia variabilis* (Table 3) have conclusively shown in such differentiated tissue the co-occurrence of both nicotinic acid conjugates (Thomzik 1983). Pulse-labelling experiments in potato tissue seemed to indicate that the nicotinic acid-N-glucoside is the first conjugate formed from exogenous nicotinic acid and that trigonelline finally accumulates at the expense of the N-glucoside (Thomzik 1983).

The data in Table 3 and our earlier observations with cell cultures (Willeke et al. 1979) have been corroborated by further investigations on the formation of trigonelline and/or nicotinic acid-N-glucoside and the occurrence of the corresponding enzymes (nicotinic acid-N-methyltransferase and nicotinic acid-N-glucosyltransferase) in differentiated tissue and cell cultures of *Solanum tuberosum* (Köster and Barz, unpublished). In the sequence: differentiated tissue → callus → cell suspension → callus → regenerated plant, remarkable changes in the occurrence of the two enzymes were found. While differentiated tissues clearly possessed significant levels of both enzymes, the N-methyltransferase could no longer be measured at the level of cell cultures. The ability of cell cultures for nicotinic acid-N-glucoside formation was, however, undoubtedly measured at the level of enzyme and by feeding experiments with ^{14}C-nicotinate. Though the N-glucoside was heavily labelled, formation of trigonelline could not be shown in these studies.

We interpret such results as indicating that in certain cases trigonelline is a true secondary plant compound which fails to be formed at the level of cell cultures. On the other hand, nicotinic acid-N-glucoside possibly fulfills a more primary metabolic function (see Sects. 3 and 5) which is essential to the cultured cells so that the ability to form this compound must also be preserved under cell culture conditions.

Table 3. Formation of trigonelline (Trig) and Nicotinic acid-N-glucoside (NIC-N-GLC) in various plants in comparison to literature data. (Thomzik 1983)

Plant	Organ	Literature data	Metabolites of nicotinate isolated
Solanum tuberosum	Leaves	Trig	NIC-N-GLC, Trig
(Solanaceae)	Tubes	Trig	NIC-N-GLC, Trig
	Roots	–	NIC-N-GLC, –
Nicotiana tabacum (Solanaceae)	Leaves	NIC-N-GLC	NIC-N-GLC, Trig
Dahlia variabilis	Leaves	Trig	NIC-N-GLC, Trig
(Compositae)	Roots	Trig	NIC-N-GLC, Trig

5 Evidence for the Pyridine Nucleotide Cycle in Plant Cell Cultures

Various lines of evidence have demonstrated that the pyridine nucleotide cycle (see Fig. 1) is also functioning in higher plant systems (Waller and Nowacki 1978; Martelli et al. 1984; Schwenen 1982). Though the individual enzymes of the cycle, together with the pool sizes of the various components, have not been measured in plant cell cultures it is justified to assume that such a cycle is also operating in plant cell cultures (i.e. parsley).

Analytical investigations (Schwenen 1982; Komoßa 1984) showed significant levels of NAD, NMN, nicotinamide-N-riboside, nicotinamide, and nicotinic acid to occur in such cell cultures. Application in parallel experiments of ^{14}C-labelled samples of nicotinic acid, nicotinamide, nicotinic acid-N-glucoside and finally [^{14}C-nicotinamide]-NAD (Schwenen 1982; Neuhann et al. 1979; Thomzik 1983) to parsley cells always led to the same qualitative and quantitative distribution of label over all pyridine compounds mentioned above. Such results are only possible if efficient salvage pathways for the conversion of nicotinic acid into NAD and NMN exist. The results further demonstrate that nicotinic acid-N-glucoside appears to be a metabolically active compound which, by loss of glucose, seems to enter the pyridine nucleotide cycle; the N-glucoside is considered to be a reservoir form of nicotinic acid.

6 Nicotinic Acid Degradation

Cell cultures which possess the ability to produce or which contain nicotinic acid-N-glucoside also show the potential for nicotinic acid degradation (Table 1). Cell cultures efficient in nicotinic acid degradation have been obtained from parsley, *Galium verum, Daucus carota, Catharanthus roseus, Tectona grandis* and other plants (Willeke et al. 1979; Schwenen 1982).

After application of various labelled pyridine derivatives such as [carboxyl-^{14}C]-nicotinic acid, [6-^{14}C]-nicotinic acid, [carbonyl-^{14}C]-nicotinamide, [carboxyl-^{14}C]-nicotinic acid-N-glucoside or [^{14}C-nicotinamide]-NAD to parsley cell suspension cultures substantial $^{14}CO_2$-formation could be measured (see Schwenen 1982). The rate of $^{14}CO_2$ evolution was found to occur in a long-term constant and linear fashion indicating steady-state concentrations in the catabolic pathway. Formation of $^{14}CO_2$ from the carboxyl group of nicotinic acid was always found to a higher extent than from ring positions. Thus, during nicotinate degradation in such plant systems decarboxylation of nicotinic acid obviously precedes degradation of the ring carbon chain.

In analogy to microbial systems (Kieslich 1976) 6-hydroxynicotinic acid had been assumed as first intermediate in nicotinic acid degradation in plant cell cultures. Earlier feeding experiments with [6-^{14}C]-hydroxynicotinic acid in various nicotinate-degrading cell cultures had failed to demonstrate its role in nicotinic acid degradation (Willeke et al. 1979). Later experiments showed that exogenously applied 6-hydroxynicotinic acid was converted to conjugated forms which were very slowly funnelled into catabolic routes (Schwenen 1982). Endogenously formed 6-hydroxynicotinic acid, however, occurs as free entity and appears to be involved in nicotinic acid degradation (Schwenen 1982) (Fig. 3).

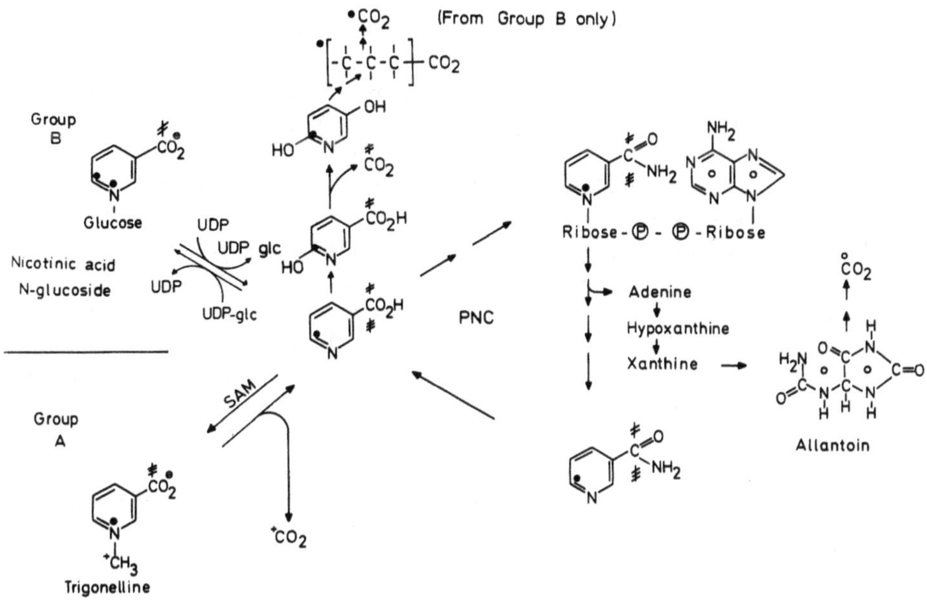

Fig. 3. Metabolism and degradation of nicotinic acid, nicotinamide and NAD in plant cell cultures. Conjugate formation in group A and group B of cell cultures refers to the mutually exclusive occurrence of trigonelline and nicotinic acid-N-glucoside (Willeke et al. 1979; Thomzik 1983). Symbols •, ‡, ⧧, and + refer to position-specific ^{14}C-labelled substrates

In a series of analytical studies involving trapping experiments with various dihydroxypyridines, a subsequent catabolite of nicotinic acid was determined to be 2,5-dihydroxypyridine (Fig. 3) (Schwenen 1982). Present investigations deal with the structural elucidation of a further catabolite of nicotinate which possesses a C_4 carbon skeleton and which under very mild conditions gives rise to a C_3 compound. This C_4 compound still shows label from position 6 of nicotinate and the structure of this catabolite will undoubtedly reveal important features of how 2,5-dihydroxypyridine is further catabolized (Fig. 3). In summary, these investigations demonstrate that at least in a restricted group of plants nicotinic acid degradation occurs and that cell suspension cultures are very suitable systems for the elucidation of catabolic routes.

7 Conclusions

The present knowledge of nicotinic acid metabolism and degradation in plant cell cultures is summarized in Fig. 3. In addition to the mutually exclusive formation of the conjugates trigonelline (group A) and nicotinic acid-N-glucoside (group B) and the presently known catabolic sequence of nicotinate, Fig. 3 also shows the degradative pathway of the adenine moiety of NAD (Neuhann et al. 1979). With regard to adenine catabolism in cell cultures, the same pathway was found to be operative regardless of the taxonomic position of the experimental plant and its cell culture. Future studies are required to elucidate the metabolic significance of nicotinic acid conjugates (group A and B) in cell cultures.

Our results have undoubtedly established that in plants nicotinic acid as part of the PNC can be catabolized and that the degradative route appears to be similar to what is known from *Pseudomonas fluorescens* (Behrman 1976). If the degradative route is further elucidated in more detail, nicotinic acid degradation can be seen as one of the first examples of an established catabolic pathway for an aromatic compound in a plant system. The advantages of plant cell suspension cultures for such investigations are very evident.

Future studies should also be directed at elucidating the very interesting aspects of regulation of nicotinic acid metabolism in plant cells. The different routes, as shown in Fig. 3 (conjugation, degradation, pyridine nucleotide resynthesis) require stringent measures of metabolic regulation which most likely will also involve different sites of cellular and organel localization of the pathways involved.

Acknowledgements. Our investigations were generously supported by Deutsche Forschungsgemeinschaft and Fonds der Chemischen Industrie (literature provision). Sincere thanks are expressed to all co-workers for their enthusiasm and help. Dr. G. Wenzel, Grünbach, kindly provided potato plants and callus cultures.

References

Antony A, Gopinathan KP, Vaidyanathan CS (1975) Biosynthesis of trigonelline in root callus cultures of Fenugreek. Indian J Exp Biol 13:39–41

Behrman EJ (1976) The bacterial oxidation of nicotinic acid, N-formylmaleamic and N-formylfumaramic acids. Arch Microbiol 110:87–90

Evans LS, Tramontano WA (1984) Trigonelline and promotion of cell arrest in G2 of various legumes. Phytochemistry 23:1837–1840

Everse J, Anderson B, You K (eds) (1982) The pyridine nucleotide coenzymes. Academic Press, London, New York

Foster JW, Moat AG (1980) Nicotinamide adenine dinucleotide biosynthesis and pyridine nucleotide cycle metabolism in microbial systems. Microbiol Rev 44:83–105

Groß W (1984) Untersuchungen zum Stoffwechsel und zur Enzymologie von Nikotinsäure-Konjugation in pflanzlichen Zellkulturen. Diplomarbeit, Münster Univ

Heeger V, Leienbach KW, Barz W (1976) Stoffwechsel von Nikotinsäure in pflanzlichen Zellsuspensionskulturen. III. Bildung und Stoffwechsel von Trigonellin. Hoppe-Seyler's Z Physiol Chem 357:1081–1087

Joshu JG, Handler P (1960) Biosynthesis of trigonelline. J Biol Chem 235:2981–2983

Kieslich K (ed) (1976) Microbial transformations of non-steroid cyclic compounds. Thieme, Stuttgart, pp 198–199

Komoßa D (1984) Zum Stoffwechsel und Abbau von Nikotinsäure in Zellkulturen und Pflanzen der Petersilie. Diplomarbeit, Münster Univ

Leete E (1980) Encycl Plant Physiol 8:79–91

Leienbach KW, Barz W (1976) Stoffwechsel von Nikotinsäure in pflanzlichen Zellsuspensionskulturen. II. Zur Isolierung, Charakterisierung und Enzymologie von Nikotinsäure-N-α-Arabinosid. Hoppe-Seyler's Z Physiol Chem 357:1069–1080

Martelli P, Bovalini L, Ferris S, Franchi GG (1984) High performance liquid chromatographic separation of the intermediates of pyridine coenzymes metabolism in *Ricinus communis* L. seedlings. Ital J Biochem 33:98–105

Mizusaki S, Tanabe Y, Kisaki T, Tamaki E (1970) Metabolism of nicotinic acid in tobacco plants. Phytochemistry 9:549–554

Mothes K, Schütte HR (eds) (1969) Biosynthese der Alkaloide. VEB Verlag, Berlin

Neuhann H, Leienbach KW, Barz W (1979) Degradation of nicotinamide adenine dinucleotide in cell suspension cultures. Phytochemistry 18:61–64

Schwenen L (1982) Untersuchungen über den Stoffwechsel von Nikotinsäure in pflanzlichen Zellsuspensionskulturen. Thesis, Münster Univ

Taguchi H, Shimabayashi Y (1983) Finding to trigonelline demethylating enzyme activity in various organisms and some properties of the enzyme from hog liver. Biochem Biophys Res Commun 113:562–568

Thomzik JE (1983) Untersuchungen zum Stoffwechsel der Nikotinsäure und zur enzymatischen Bildung ihrer Konjugate in Pflanzen und pflanzlichen Zellkulturen. Thesis, Münster Univ

Tramontano WA, Hartnett CM, Lynn DG, Evans LS (1982) Relationship between trigonelline concentration and promotion of cell arrest in G2 in cultured roots of *Pisum sativum*. Phytochemistry 21:1201–1206

Tramontano WA, Lynn DG, Evans LS (1983) Trigonelline, nicotinic acid and nicotinamide in seedlings of *Pisum sativum*. Phytochemistry 22:673–678

Waller GR, Nowacki EK (eds) (1978) Alkaloid biology and metabolism in plants. Plenum Press, New York London

White HB (1982) Biosynthetic and salvage pathways of pyridine nucleotide coenzymes. In: Everse J, Anderson B, You K (eds) The pyridine nucleotide coenzymes. Academic Press, London New York, pp 225–248

Willeke U, Heeger V, Meise M, Neuhann H, Schindelmeiser J, Vordemfelde K, Barz W (1979) Phytochemistry 18:105–110

Zatman LJ, Kaplan NO, Colowick SP (1953) Inhibition of spleen diphosphopyridine nucleotidase by nicotinamide, an exchange reaction. J Biol Chem 200:197–212

Plant Cell and Tissue Culture of *Cinchona* Species

R. VERPOORTE[1], R. WIJNSMA[1], Th. MULDER-KRIEGER[1], P.A.A. HARKES[2], and A. BAERHEIM SVENDSEN[1]

1 Introduction

The genus *Cinchona*, belonging to the family Rubiaceae, is still esteemed for its antimalaria activity. Several species, e.g. *C. pubescens* (syn. *C. succirubra*) and *C. ledgeriana*, have been cultured in plantations during the past 130 years for the production of *Cinchona* bark, the raw material for the alkaloids quinine and quinidine. *Cinchona* plantations are mainly found in Middle-America, Central Africa, India, Phillipines, and Indonesia. After 7–12 years of growth the bark of the trees is harvested. The alkaloid content is then about 10%–15% (Smit 1984).

Quinine has for many years been the major antimalaria drug, however, the advent of the (semi)-synthetic drugs reduced the role of quinine considerably. However, there is a renewed interest in quinine, as malaria parasites have developed resistance against other known drugs. Quinidine is used as an antiarrhythmic drug. It has recently been shown to be also a potent antimalaria agent (Warhurst 1981). The bitter taste of the *Cinchona* alkaloids has led to their widespread use in beverages.

Besides the alkaloids quinine and quinidine (see Scheme 1 for structures of the alkaloids), their dihydro derivatives and their non-methoxylated analogues cinchonidine and cinchonine are found in the bark. Small amounts of indole alkaloids can also be found in the bark, however, these alkaloids are more abundant in the leaves (Zèches et al. 1980; Keene et al. 1983). Particularly the semidimeric alkaloids of the cinchophylline type (Scheme 1) are found in the leaves. The biosynthesis of the alkaloids occurs in the leaves according to Klein Horsman-Relijk (1960), from which they are transported to the bark and stored. The biosynthesis of all the *Cinchona* alkaloids follows the usual pathway for the indole alkaloid biosynthesis, starting with strictosidine (Scheme 1). However, not all steps have been proven yet.

The cell and tissue culture of *Cinchona* species has several interests:

Vegetative Propagation. As *Cinchona* trees are still the source for the production of the alkaloids, methods for rapid in-vitro propagation of high-producing or disease resistant varieties is of great interest. Several authors reported methods for the in-vitro

1 Center of Bio-Pharmaceutical Sciences, Division of Pharmacognosy, State University of Leiden, Biotechnology Delft Leiden (BDL), P.O. Box 9502, 2300 RA Leiden, The Netherlands

2 Department of Plant Molecular Biology, State University of Leiden, Biotechnology Delft Leiden (BDL), P.O. Box 9502, 2300 RA Leiden, The Netherlands

Primary and Secondary Metabolism of Plant Cell Cultures, ed. by Neumann et al.

propagation of *Cinchona* species (Chatterjee 1974; Hunter 1979; Krikorian et al. 1981; Noerhadi 1981; Koblitz et al. 1983a).

Studies of Production of Alkaloids by Cell Cultures. Cinchona seems to be an interesting model system for the production of alkaloids by cell culture systems. In the plant apparently two types of cells are involved in the alkaloid metabolism. There are cells where biosynthesis and alkaloid excretion occurs and cells that absorb and store the alkaloids. In fact also in suspension culture of *Cinchona* several types of cells can be distinguished, some clearly containing large amounts of blue fluorescent compounds similar to those reported by Neumann et al. (1983) for *Catharanthus roseus* cell suspension cultures. Thus for studying the processes that regulate the excretion and the storage of the alkaloids, *Cinchona* cell cultures seem to be a promising model system. Furthermore the *Cinchona* cells can produce two different types of secondary products at the same time, i.e. alkaloids and anthraquinones (see below). A model system therefore exists for studying possible competitive effects in the biosynthesis of these products. In our studies it was found that the biosynthetic pathways leading to the various alkaloids and anthraquinones are, at least partly, independently influenced by media components. Several groups have already reported on the production of alkaloids in *Cinchona* cell and tissue cultures (Whitten and Dougall 1981; Anderson et al. 1982; Hunter et al. 1982; Mulder-Krieger et al. 1982a,b,c, 1984; Staba and Chung 1981; Koblitz et al. 1983b).

Studies of the Biosynthesis of Cinchona Alkaloids. Still some steps in the *Cinchona* alkaloid biosynthesis are not completely known. The cell culture system may offer an excellent possibility to study alkaloid biosynthesis in more detail. Here we will briefly survey our studies in the field of *Cinchona* cell and tissue culture.

2 Analytical Procedures

For the analysis of the *Cinchona* alkaloids liquid chromatographic methods are more suitable than GLC, due to the low volatility of the alkaloids (Verpoorte and Baerheim Svendsen 1984). The closely related structures of the quinoline alkaloids, make it rather difficult to separate these alkaloids. TLC always requires a combination of solvents to identify the various alkaloids. No single TLC system is capable of separating the eight naturally occurring alkaloids (Verpoorte et al. 1980). An HPLC reversed-phase ion-pair system was developed that could separate all quinoline alkaloids (Verpoorte et al. 1983, 1984). Recently McCalley (1983) reported a straight-phase HPLC system for the analysis of *Cinchona* alkaloids, which we have adapted for our latest studies as it is more convenient to operate.

Immunoassay methods have been effectively used for the rapid and sensitive analysis of various compounds in plant material (Weiler 1982). Immunoassay kits are commercially available for the analysis of quinidine in plasma or serum. The cross-reactivity of various related *Cinchona* alkaloids was tested (see Table 1), the sensitivity obtained was not better than that with HPLC using 280 nm UV detection. Due to a higher

Table 1. Cross-reactivity of *Cinchona* alkaloids in immunoassay methods. (Wijnsma et al., unpublished results)

Alkaloid	EMIT[a]	FIA[b]
Quinidine	100%	100%
Dihydroquinidine	46	91
Cinchonine	22	93
Cupreidine	20	74
Epiquinine	5	9
Cinchoninon	–	8
Quinidinon	–	4
Cinchonamine	–	3
Cinchonidine	–	–
Quinine	–	–
Epiquinidine	9	–
Hydroquinine	–	–
Detection limit Qd	70 ng ml^{-1}	40 ng ml^{-1}

[a] Enzyme immunoassay (SYVA)
[b] Fluorescent immunoassay (Ames TDA)

selectivity an EMIT assay was found the method of choice, although alkaloids other than quinidine contributed to the results. The sensitivity is comparable with HPLC. It has the advantage of a short time for analysis when compared to HPLC and TLC, and can be used for larger numbers of samples.

For the analysis of the indole alkaloids TLC has proven to be a useful method. Using a ferric (III) chloride in perchloric acid spray reagent a selective and sensitive detection of the indole alkaloids is possible on TLC plates (Mulder-Krieger et al. 1982c). Recently Robins et al. (1984) reported the determination of quinine in cell culture material by means of a radioimmunoassay (RIA) method.

For the qualitative analysis of the anthraquinones both TLC and HPLC were found to be the methods of choice (Wijnsma et al. 1984), while for quantitative analysis a modification of the spectrophotometric method as described by Zenk et al. (1975) and Schulte et al. (1984) has been used.

3 Plant Tissue Culture

3.1 Growth

Callus cell lines of both *C. pubescens* and *C. ledgeriana* were obtained from various parts of the plant. However, calli growth was found to be slow often with browning and a decrease of growth or even death of the calli. Therefore extensive studies were made to optimizethe factors in the medium that affect the growth, i.e. growth hormones and salts. The first experiments concerned the salt concentration in six otherwise different media, all derived from the Murashige-Skoog (MS) medium. The salt concentration was varied from 1/8, 1/4, 1/2, and 1/1 of the original MS medium. The results are summarized in Table 2. Between the various media (1–6) clearly differences in growth are observed, for each medium also the concentration of the salts clearly affects the

Table 2. Tissue growth and percentage of water in callus cultures of *Cinchona pubescens* after 13 weeks

Medium	Salt concentration[a]	Total fresh weight (g)[b]	Dry weight (g)[b]	% Water
1	MS • 1/8	3.10	0.44	86
1	MS • 1/4	3.33	0.45	86
1	MS • 1/2	1.38	0.21	85
1	MS • 1	0.94	0.13	86
2	MS • 1/8	1.82	0.27	85
2	MS • 1/4	1.69	0.25	85
2	MS • 1/2	1.20	0.16	87
2	MS • 1	0.37	0.05	86
3	MS • 1/8	2.14	0.24	89
3	MS • 1/4	4.60	0.49	89
3	MS • 1/2	9.97	0.73	93
3	MS • 1	8.95	0.84	91
4	MS • 1/8	2.44	0.28	89
4	MS • 1/4	4.50	0.47	90
4	MS • 1/2	8.69	0.85	90
4	MS • 1	5.67	0.58	90
5	MS • 1/8	1.19	0.19	84
5	MS • 1/4	0.76	0.12	84
5	MS • 1/2	0.18	0.03	83
5	MS • 1	0.36	0.05	86
6	MS • 1/8	1.04	0.17	84
6	MS • 1/4	0.96	0.16	86
6	MS • 1/2	0.77	0.12	84
6	MS • 1	0.53	0.07	87

Organic constituents present in the media (mg l^{-1})

Constituents	Medium					
	1	2	3	4	5	6
Adenine	–	–	–	–	1	1
Glycine	–	–	–	–	2	–
Cysteine HCl	50	50	50	50	10	10
Thiamine HCl	10	10	10	10	1	1
Casein-hydrolysate (vitamin-free)	–	100	–	100	100	100
Meso-Inositol	100	100	100	100	100	100
BA	2	2	2	2	1	1
IBA	8	8	–	–	5	5
2,4-D	–	–	1	1	–	–
Sucrose	30,000	30,000	30,000	30,000	30,000	30,000
Agar	7,000	7,000	7,000	7,000	7,000	7,000

[a] MS = salt concentration as in original MS medium

[b] Average of 4–9 explants, weight starting material 8 ± 1 mg

growth. In the media which resulted in low growth yields, the lower salt concentrations (MS 1/8, 1/4, and 1/2) gave relatively the best results. For the media that resulted in higher yields, MS 1/2 salt concentration was best, being nearly as good as MS 1/1 (Mulder-Krieger et al. 1982a). Using the medium which gave the best growth in the first experiments (medium 3, Table 2) further studies were made on the influence of the phytohormones and culture condition on the growth and the alkaloid production (Mulder-Krieger et al. 1982b).

Regarding phytohormones it was found that high auxin levels caused root formation [NAA at 10.8 (medium C) or 16.2 μMol (medium D)] (Fig. 1). Best results were obtained with an auxin concentration of 1 μMol, resulting in friable calli and relatively rapid growth. Comparison of BA and zeatin showed that the latter cytokinin gave better results (Fig. 1). Direct comparison of the auxins IBA, IAA, 2,4-D, and NAA (media F, G, H, and K in Fig. 1), using similar media containing 1 μMol of these auxins, showed no significant differences for growth yields. Considering the alkaloid content, however, 2,4-D and IBA were found to increase the alkaloid levels. From Fig. 1 it can also be observed that light improved the growth yields (Mulder-Krieger et al. 1982b). Based on these results larger amounts of calli were grown on medium H for further studies of the alkaloids present (see below).

Recently a further study has been started for the factors affecting the growth of the *Cinchona* cell cultures using the method reported by de Fossard et al. (1974). In this experiment 81 different media are tested, in which four groups of constituents: mineral salts, auxins, cytokinins, and organic growth factors are varied in three different concentrations being low (L), medium (M), and high (H). The experiment was performed

Table 3. Relative amounts of alkaloids present in extracts of freshly harvested leaves and stems of *C. pubescens* and callus cultures initiated from leaves and stems of the same specimen. No separation of cinchoninone and quinidinone could be obtained by TLC. (Mulder-Krieger et al. 1984)

Alkaloid	Leaves living plant	Stems living plant	Leaf callus culture	Stem callus culture
Cinchonine	++	+++	+++	+++
Cinchonidine	++	+++	+++	+++
Quinine	++	+++	++	+
Quinidine	++	+++	++	++
Dihydrocinchonine	–	++	+++	+++
Dihydrocinchonidine	–	++	++	+++
Dihydroquinine	–	++	+	+
Dihydroquinidine	–	++	++	++
Cinchonamine	–	+++	+++	++
10-Methoxycinchonamine	+	+	+++	++
Quinamine	+++	++	+	+
3α,17β-Cinchophylline	+++	+++	–	–
3α,17α-Cinchophylline	++	++	–	–
Cinchoninone/quinidinone	–	++	–	–

+++: major compound
+: minor compound
–: not detected

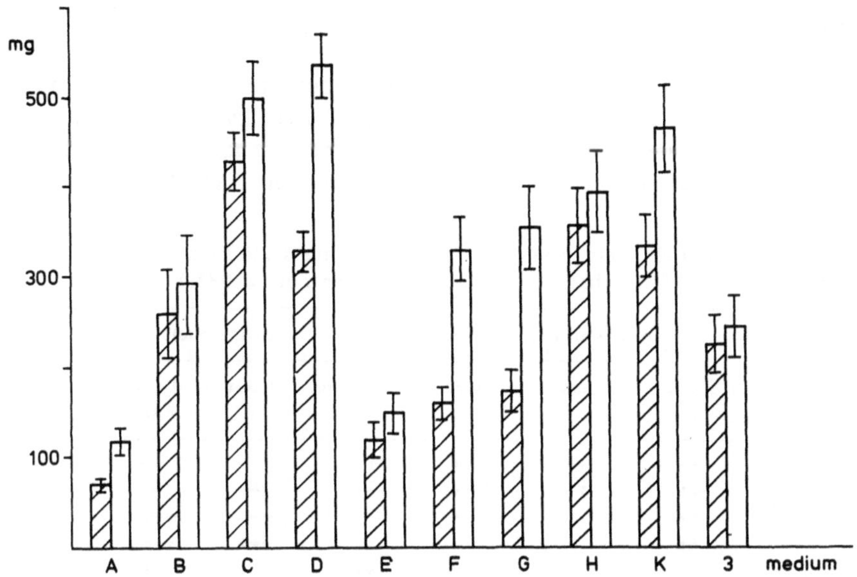

Fig. 1. Effect of combinations of various auxins and cytokinins on the growth of *C. pubescens*. After 2 months the calli were harvested and the mean fresh weights were determined. The 95% confidence intervals are indicated. The cultures were either kept in the dark (*shaded*) or grown with 12 h artificial light followed by a 12 h dark period (*not shaded*). The composition of the media is with exception of the growth regulators the same as in medium 3 (with 1/2 of the MS salt-concentration (see Table 2). The auxins and cytokinins are listed below

Medium	Auxin	Auxin concentration (μM)	Cytokinin	Cytokinin concentration (μM)
A	IAA	1.0	BA	5.0
B	2,4-D	4.5	Kinetin	0.5
C	NAA	10.8	Kinetin	0.9
D	NAA	16.2	BA	4.4
E	IBA	1.0	BA	5.0
F	IBA	1.0	Zeatin	1.0
G	IAA	1.0	Zeatin	1.0
H	2,4-D	1.0	Zeatin	1.0
K	NAA	1.0	Zeatin	1.0
3	2,4-D	4.5	BA	8.1

IAA: indoleacetic acid, 2,4-D: 2,4-dichlorophenoxyacetic acid
NAA: naphthaleneacetic acid
IBA: indolebutyric acid, BA: benzyladenine

both in the light and the dark, thus resulting in altogether 162 different sets of parameters. Growth was found to be improved by higher auxin levels. Best growth was observed in the light series, although many media combinations did not result in growth at all in the light series, e.g. all media with low auxin concentrations (Harkes et al. 1984).

3.2 Alkaloids

In the callus cultures only small amounts of alkaloids were found to be present, typically 0.01% on dry weight basis. By means of preparative TLC, the alkaloids of the various cell lines could be isolated and characterized by their spectral data and TLC comparison with reference compounds. The results are summarized in Table 3 and Table 4 for the

Table 4. Relative amounts of alkaloids present in some tissue cultures and living plant material of *C. ledgeriana* obtained from the same batch of seeds. (Mulder-Krieger et al. 1982c)

Alkaloid	Plant tissue culture				Living plant		
	Root 1	Leaves 1	Seed 1	Seedling	Root	Leaves + stem	Leaves
Cinchonine	+++	+++	+++	+++	++	++	–
Cinchonidine	+++	++	++	+++	++	++	–
Quinine	++	+	+	++	+++	+++	+
Quinidine	++	+	+	++	+++	+++	+
Dihydrocinchonine	++	++	++	+++	+	+	–
Dihydrocinchonidine	+	–	+	++	+	+	–
Dihydroquinine	+	–	+	+	++	++	–
Dihydroquinidine	+	+	+	++	++	++	–
Quinamine	+	+	++	+	++	++	++
Cinchoninone	+	–	–	++	++	–	–
Cinchophyllines	–	–	–	–	–	+++	+++
Cinchonamines	–	–	–	–	–	–	+

+++: major component
+: minor component
–: not detected

two species studied. Besides the quinoline type of alkaloids also some indole alkaloids were identified, e.g. quinamine and cinchonamine (for structures see Scheme 1). However, the cinchophylline type of alkaloids could not be detected. One new alkaloid was isolated and characterized as 10-methoxy-cinchonamine by its spectral data (Mulder-Krieger et al. 1982c).

Quantitative analysis of the eight naturally occurring quinoline alkaloids in several cell lines of *C. pubescens* showed a clear increase of the alkaloid content for calli grown in the light (Verpoorte et al. 1984).

The calli grown in the above mentioned experiment according to de Fossard et al. (1974) were also analyzed for their alkaloid content. The quinoline alkaloids were analyzed by the method reported by McCalley (1983), the indoles, cinchonamine and quinamine, were only analyzed semiquantitatively by means of TLC. The major alkaloid in all experiments was quinidine and/or cinchonamine. The quinoline alkaloids were found in highest concentrations in media with low auxin levels, whereas indole alkaloids were most abundant in media with higher auxin levels. Indoles were also more frequently encountered in calli grown in the light than in those grown in the dark. Although the alkaloid levels in the calli were generally lower in the light series, the yield

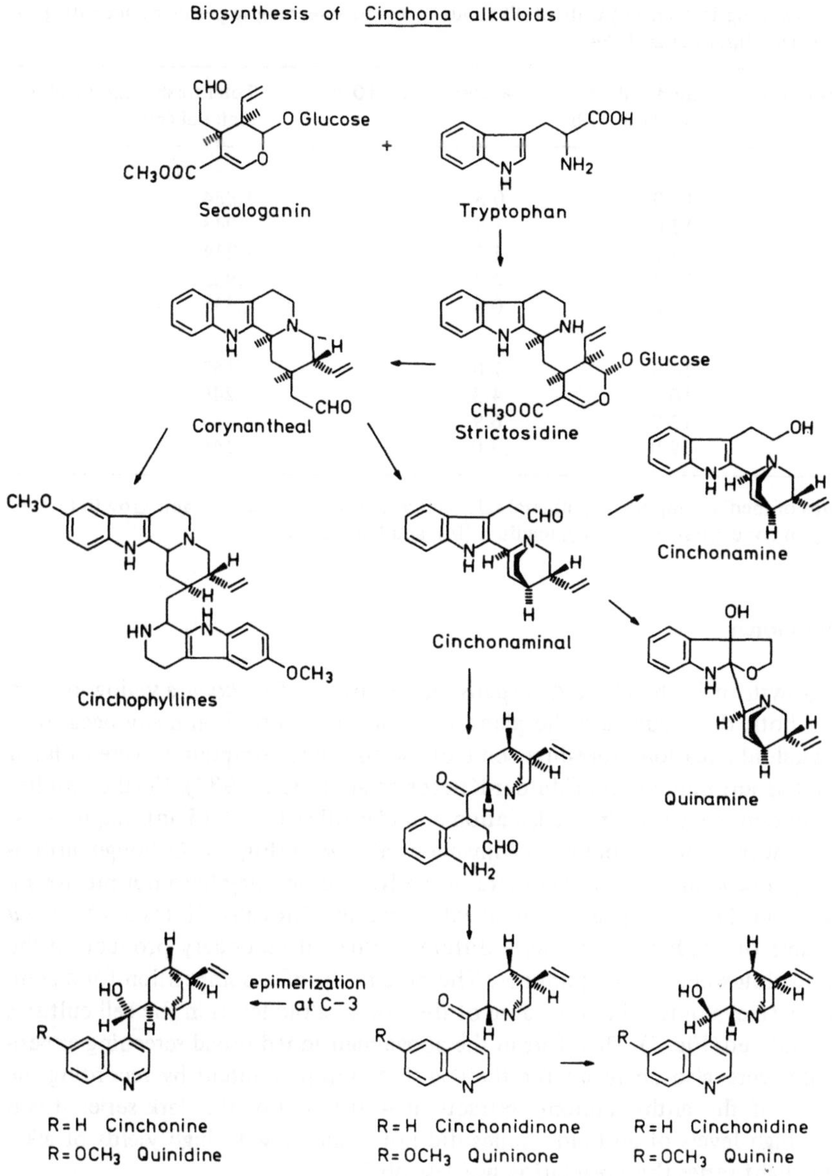

Scheme 1

per petridish was considerably higher if compared to the dark series, due to better growth in the light (Table 5).

Considering the highly variable yields of alkaloids, a two-step system, with a growth and a production medium, seems thus to be most promising for production purposes.

Table 5. Media resulting in highest yields of alkaloids in broad screening experiment according to De Fossard (1974). (Harkes et al. 1984)

Medium composition	μg Alkaloid per Petri dish	% Alkaloid (× 10^{-3})	Total fresh weight callus material (mg)
Light series			
MMHM[a]	10.0	0.5	1,884
HMLL	13.0	1.4	943
HMMM	26.1	1.2	2,214
HMHM	50.7	2.7	1,905
HHHM	10.0	0.8	1,335
Dark series			
MLMH	10.9	7.0	157
HLLL	10.3	4.3	240
HLMM	12.7	6.9	185
HLMH	19.0	13.0	146

[a] Four groups of media components, mineral salts, auxins, cytokinins, and organic growth factors respectively in concentration low (L), medium (M), and high (H)

3.3 Anthraquinones

During the growth of both calli and suspension cultures of *Cinchona* a yellow-orange coloration of both the media and the plant material was observed on many occasions. During the alkaloid isolation procedure some of the coloured compounds were isolated and identified as anthraquinones (Mulder-Krieger et al. 1982c, 1984). Further studies for these compounds resulted in the isolation and identification of 15 anthraquinones, eight of which were new compounds (Wijnsma et al. 1984) (Fig. 2). Although anthraquinones are quite common in several genera of the Rubiaceae, they have not previously been isolated from *Cinchona* plant material with certainly. The cell cultures of *Cinchona* are thus capable of producing two quite different types of secondary products at the same time: alkaloids and anthraquinones. The possibility of a competition for a common precursor (mevalonic acid) as a cause of the low alkaloid levels in the cell cultures should be considered (Fig. 3). Therefore in the above-mentioned broad screening experiment the calli were also analyzed for their anthraquinone content by recording the UV-absorption of the anthraquinone extracts at 490 nm. For the dark series it was obvious that high levels of anthraquinones did not coincide with high yields of alkaloids. For the light series this correlation was less obvious.

The anthraquinone levels were highest at the lower auxin concentration in this experiment. For those media that resulted in the highest anthraquinone levels in the dark series, no growth at all could be observed in the light series. The possibility of photodegradation of the anthraquinones to toxic products for the cells should be considered as a possible cause of the death of the cells. In general in the dark series higher anthraquinone levels were found. Calculation of the yield per petridish for the various media resulted in the interesting observation that the yield in the light series was similar for all media being at the average of 107 μg Petri dish (± 17 μg). The variation in the dark series was much larger. Again from these results it can be concluded that for production

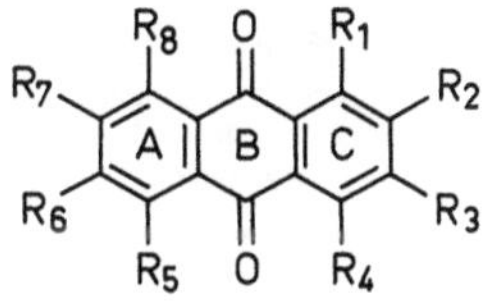

	R_1	R_2	R_3	R_4	R_5	R_6	R_7	R_8
Purpurin	OH	OH	H	OH	H	H	H	H
Anthragallol-1,2-dimethylether	OCH_3	OCH_3	OH	H	H	H	H	H
Anthragallol-1,3-dimethylether	OCH_3	OH	OCH_3	H	H	H	H	H
Rubiadin	OH	CH_3	OH	H	H	H	H	H
1-Hydroxy-2-hydroxy-methyl A[a]	OH	CH_2OH	H	H	H	H	H	H
1-Hydroxy-2-methyl A	OH	CH_3	H	H	H	H	H	H
Morindone-5-methylether or 1,7-dihydroxy-8-methoxy-2-methyl A	OH	CH_3	H	H	OCH_3 or H	OH or H	H or OH	H or OCH_3
5,6-Dimethoxy-1-(or -4-)hydroxy--2-(or -3-)hydroxy-methyl A	OH or H	CH_2OH or H	H or CH_2OH	H or OH	OCH_3	OCH_3	H	H
5-Methoxy-2-(or -3-)methyl-1,4,6-trihydroxy A	OH	CH_3 or H	H or CH_3	OH	OCH_3	OH	H	H
2-Hydroxy-1,3,4-trimethoxy A	OCH_3	OH	OCH_3	OCH_3	H	H	H	H
4-Methoxy-1,3,5-trihydroxy A	OH	H	OH	OCH_3	OH	H	H	H
1,4-Dimethoxy-2,3-methylene-dioxy A	OCH_3	$O-CH_2-O$		OCH_3	H	H	H	H
1,3-Dihydroxy-4-methoxy A	OH	H	OH	OCH_3	H	H	H	H
1,3-Dihydroxy-2,5-dimethoxy A	OH	OCH_3	OH	H	OCH_3	H	H	H
2,5-(or 3,5-)Di-hydroxy-1,3,4-(or -1,2,4-)tri-methoxy A	OCH_3	OH or OCH_3	OCH_3 or OH	OCH_3	OH	H	H	H

[a] A = Anthraquinone

Fig. 2. Structures of the anthraquinones isolated from callus cultures of *Cinchona ledgeriana*

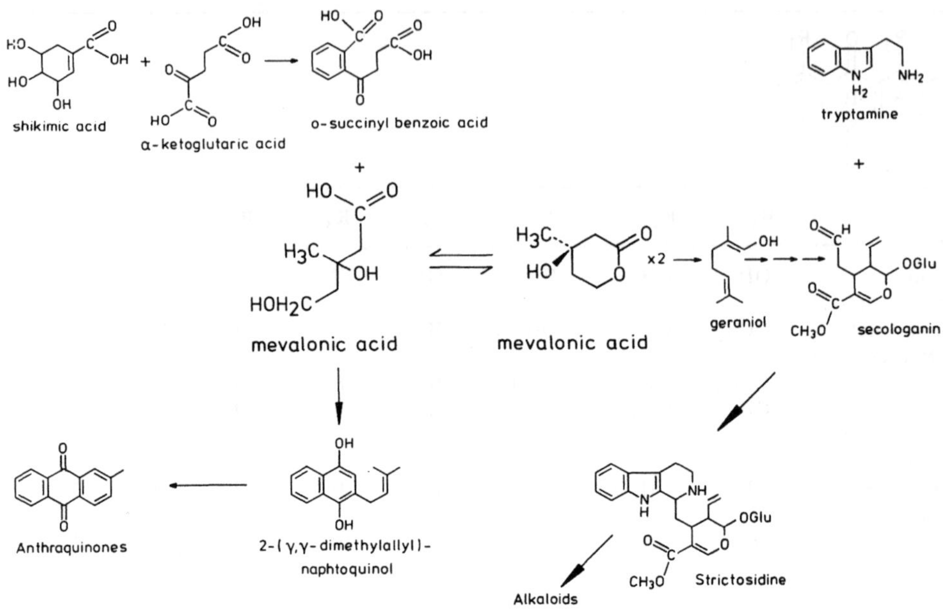

Fig. 3. The competition for mevalonic acid between anthraquinone- and alkaloid biosynthesis in tissue cultures of *Cinchona* sp.

purposes a two-stage process of growth and biosynthesis seems the most promising approach.

4 General Conclusions and Discussion

Although *Cinchona* plant cell and tissue cultures are difficult to grow, several investigators have been successful in obtaining both viable callus and cell suspension cultures.

In callus cultures usually low alkaloid levels were found, typically in the range of 0.01%; during years of subculturing a decrease of alkaloid levels was observed in the various cell lines. Preliminary studies of some cell suspension cultures showed more promising results: alkaloid levels up to 0.9% have been noted, without any cell line selection procedures or addition of precursors to the medium (BDL, unpublished results). In these cell suspension cultures alkaloid production is very low during the growth phase, however, with the onset of the stationary phase it clearly increases.

Although *Cinchona* species are well known for their alkaloid production, in plant cell and tissue culture they more readily produce anthraquinones than alkaloids. This may point to the possibility of production of the anthraquinones under stress conditions. The antimicrobial activity reported for these compounds may suggest that they play a role in the defense of wounded plant tissues against infections, i.e. the anthraquinones might be assumed to be phytoalexins. Studies to test this hypothesis are in progress.

Acknowledgements. A support from NATO (grant number 599/83) for collaborative studies with Prof. E.J. Staba is gratefully acknowledged. We are indebted to Prof. E.J. Staba for many fruitful discussions and critically reading of the manuscript.

References

Anderson LA, Keene AT, Phillipson JD (1982) Alkaloid production by leaf organ, root organ and cell suspension cultures of *Cinchona ledgeriana*. Planta Med 46:25–27

Chatterjee SK (1974) Vegetative propagation of high quinine yielding *Cinchona*. Indian J Hortic 31:174–177

Fossard RA de, Myint A, Lee ECM (1974) A broad spectrum tissue culture experiment with tobacco (Nicotiana tabacum) pith tissue culture. Physiol Plant 30:125–130

Harkes PAA, Kreijbolder L, Libbenga KR, Wijnsma R, Nsengiyaremge T, Verpoorte R (1984) Influence of various media constituents on the growth of *Cinchona ledgeriana* tissue cultures and the production of alkaloids and anthraquinones therein. (Submitted)

Hunter CS (1979) In vitro culture of *Cinchona ledgeriana* L. J Hortic Sci 54:111–114

Hunter CS, McCalley DV, Barraclough AJ (1982) Alkaloids produced by cultures of *Cinchona ledgeriana* L. In: Fujiwara A (ed) Proc 5th Int Congr Plant Tissue Cell Cult, Plant Tissue Culture. Tokyo, pp 317–318

Keene AT, Anderson LA, Phillipson JD (1983) Investigation of *Cinchona* leaf alakloids by High-performance liquid chromatography. J Chromatogr 260:123–128

Klein Horsman-Relijk J (1960) Onderzoek naar de biosynthese van enkele alkaloiden in *Cinchona succirubra*. Thesis, Amsterdam

Koblitz H, Koblitz D, Schmauder HP, Gröger D (1983a) Studies on tissue cultures of the genus *Cinchona* L. Plant Cell Rep 2:95–97

Koblitz H, Koblitz D, Schmauder HP, Gröger D (1983b) Studies on tissue cultures of the genus *Cinchona* L. Alkaloid production in cell suspension cultures. Plant Cell Rep 2:122–125

Krikorian AD, Singh M, Quinn CE (1981) Aseptic micropropagation of *Cinchona*: prospects and problems. In: Rao AN (ed) Proceedings of the COSTED Symposium on tissue culture of economically important plants. Singapore, pp 167–174

McCalley DV (1983) Analysis of *Cinchona* alkaloids by high-performance liquid chromatography. J Chromatogr 260:184–188

Mulder-Krieger Th, Verpoorte R, Baerheim Svendsen A (1982a) Tissue culture of *Cinchona pubescens*: Effects of media modifications on the growth. Planta Med 44:237–240

Mulder-Krieger Th, Verpoorte R, Graaf YP de, Kreek M van der, Baerheim Svendsen A (1982b) The effects of plant growth regulators and culture conditions on the growth and the alkaloid content of callus cultures of *Cinchona pubescens*. Planta Med 46:15–18

Mulder-Krieger Th, Verpoorte R, Water A de, Gessel M van, Oeveren BCJA van, Baerheim Svendsen A (1982c) Identification of the alkaloids and anthraquinones in *Cinchona ledgeriana* callus cultures. Planta Med 46:19–24

Mulder-Krieger Th, Verpoorte R, Kreek M van der, Baerheim Svendsen A (1984) Identification of alkaloids and anthraquinones in *Cinchona pubescens* callus cultures; the effect of plant growth regulators and light on the alkaloid content. Planta Med 50-1:17–23

Neumann D, Krauss G, Hieke M, Gröger D (1983) Indole alkaloid formation and storage in cell suspension cultures of *Catharanthus roseus*. Planta Med 48:20–23

Noerhadi E (1981) Plant tissue culture research in Indonesia. In: Rao AN (ed) Proceedings of the COSTED Symposium on tissue culture of economically important plants. Singapore, pp 290–292

Robins RJ, Webb AJ, Rhodes MJC, Payne J, Morgan MRA (1984) Radioimmunoassay for the quantitative determination of quinine in cultured plant tissues. Planta Med 50:235–238

Schulte U, El-Shagi H, Zenk MH (1984) Optimization of 19 rubiaceae species in cell culture for the production of anthraquinones. Plant Cell Rep 3:51–54

Smit EHD (1984) Verleden, heden en perspectieven van *Cinchona* spp. Pharm Weekbl 119:159–162

Staba EJ, Chung AC (1981) Quinine and quinidine production by *Cinchona* leaf, root, and unorganized cultures. Phytochemistry 20:2495–2498

Verpoorte R, Baerheim Svendsen A (1984) J Chromatogr Library 23B:Chromatography of alkaloids – GC and HPLC

Verpoorte R, Mulder-Krieger Th, Troost JJ, Baerheim Svendsen A (1980) Thin-layer chromatographic separation of *Cinchona* alkaloids. J Chromatogr 184:79–96

Verpoorte R, Mulder-Krieger Th, Verzijl MJ, Verzijl JM, Baerheim Svendsen A (1983) High-performance liquid chromatography of *Cinchona* alkaloids. I. Normal-phase HPLC. J Chromatogr 261:172–175

Verpoorte R, Mulder-Krieger Th, Wijnsma R, Verzijl JM, Baerheim Svendsen A (1984) HPLC analysis of alkaloids in extracts of callus cultures of *Cinchona* spp. Z Naturforsch 39c:680–682

Warhurst DC (1981) *Cinchona* alkaloids and malaria. The Lancet:1346

Weiler E (1982) Application of immunoassay techniques in pharmacognosy. Pharm Weekbl 117: 477–479

Whitten GH, Dougall DK (1981) Quinine and quinidine accumulation by root, callus and suspension cultures of *Cinchona ledgeriana*. In vitro 17:220

Wijnsma R, Verpoorte R, Mulder-Krieger Th, Baerheim Svendsen A (1984) Anthraquinones in callus cultures of *Cinchona ledgeriana*. Phytochemistry 23:2307–2311

Zèches M, Richard B, Thepenier P, Le Men-Olivier L, Le Men J (1980) Alcaloides des feuilles du *Cinchona ledgeriana*. Phytochemistry 19:2451–2454

Zenk MH, El-Shagi H, Schulte U (1975) Anthraquinone production by cell suspension cultures of *Morinda citrifolia*. Planta Med Suppl:78–101

The Production of Pyrethrins by *Chrysanthemum cinerariaefolium* (Trev) Bocc.

E.J. STABA [1] and S.W. ZITO [2]

1 Introduction

The genus *Chrysanthemum* in the Compositae includes over 200 species, of which a very few are known for their insecticidal activity. The term "pyrethrum flower" refers most often to the perenial *Chrysanthemum cinerariaefolium* (Trev) Bocc., which is the commercially important species for pyrethrins, and occasionally to *C. coccineum* Willd. (garden pyrethrum) and *C. marschalli* Aschers. The annual garden chrysanthemums (*C. morifolium* Ramat.) do not contain pyrethrins. Pyrethrins have been reported present in the tissue cultures of the *Compositae* plant *Tagetes erecta* L. and *T. minuta* [1, 2].

Pyrethrins are produced commercially by harvesting the flower heads of plants cultivated throughout the world, but particularly in Kenya and Ecuador. The in vitro micropropagation of *C. cinerariaefolium* (pyrethrum) from capitulum or shoot meristem explants is reported from the Netherlands [3], India [4], Ecuador [5], and Kenya [6]. Thousands of plants from pyrethrum clones have been successfully transplanted into the fields of Ecuador [5]. The tissue culture micropropagation method is sometimes preferred because it is more rapid than field splitting and insures nematode free plants for a period of time [6]. Aseptic shoots may be initiated on media containing benzyladenine (BA) from 0.2 mg l^{-1} [6] to 3.0 mg l^{-1} [7], and roots in media containing 0.1 mg l^{-1} [8] to 0.2 mg l^{-1} naphthaleneacetic acid [6].

Pyrethrins are a collection of six structurally related insecticidal esters found in pyrethrum. The pyrethrins I are esters formed between chrysanthemic acid and three structurally related rethrolones known individually as pyrethrolone, cinerolone, and jasmolone. Thus the term pyrethrins I refers to a mixture of pyrethrin I, cinerin I, and jasmolin I, which is as they occur in the plant [9]. Also found in the plant are pyrethrins II which are esters formed between pyrethric acid and the same three rethrolones as in the pyrethrins I, thus the term pyrethrins II refers to a mixture of pyrethrin II, cinerin II, and jasmolin II. These compounds are effective, non-persistent insecticides used commercially in many applications, particularly in pediculocides and in preparations used for insect control in industrial kitchens.

1 Department of Medical Chemistry and Pharmacognosy, University of Minnesota, Minneapolis, MN 55455, USA

2 College of Pharmacy, St. John's University, Jamaica, NY 11439, USA

Primary and Secondary Metabolism of Plant Cell Cultures, ed. by Neumann et al.

2 Pyretheum Tissue Culture Production of Pyrethrins

Chrysanthemum plants that are not obtained from commercial pyrethrum growers often contain low amounts of pyrethrins. In 1975 it was reported that the callus and root organ cultures of such non-commercial "pyrethrums," and other selected chrysanthemums, did not produce pyrethrins. Some of the chrysanthemum cultures did contain trace amounts of pyrethrins when examined by gas chromatography and a mosquito larvae bioassay [10]. In 1976 a Japanese patent reported that extracts of pyrethrum callus, green callus, and redifferentiated plants contained pyrethrins. The pyrethrins were identified by gas chromatography and were active in a mosquito larvae test [8].

Cashyap et al. [11] could not detect pyrethrins in pyrethrum callus cultures. Zieg et al. [7] reported that callus from 54% of 133 cultures established from pyrethrum flower and leaf explants contained 35 mg-% or less pyrethrins. The highest pyrethrin containing callus culture was 113 mg-%.

As early as 1966 Head [12] realized that the vegetative tissues of pyrethrum plants as well as the flowers contained pyrethrins. Zieg et al. [7] reported that only 2 of 55 aseptic pyrethrum shoot cultures did not contain detectable levels of pyrethrins. Yields of 50mg-% were common from the shoot cultures and included one that contained 341 mg-%. The pyrethrins are present in each leaf layer, but more appears present in the epidermal layers of young leaves, the mesophyll cells of more mature leaves, and where a high oil gland concentration appears [13].

3 Light Effects on Pyrethrum Tissue Culture Production of Pyrethrins

The pyrethrum plant is a quantitative short-day plant requiring photoperiods of about 12 h. In addition, Roest [14] recommends a low temperature (9 °C) for 6 weeks to initiate flowering and higher temperatures (17°–25 °C) for flower and vegetative development.

Flowering in the ornamental chrysanthemum occurs best with 8 h of light followed by 16 h of darkness [15]. Long light days of 16 h will keep the plant in the vegetative stage [16]. Many ornamental chrysanthemums are vernalized by 3–4 weeks of chilling, and devernalized by low light intensity at a temperature of 28 °C for 4 weeks [17]. Gibberellin treatment of unvernalized or partially vernalized plants induced growth extension and flowering but not in fully vernalized plants [18]. Devernalization seems to preserve the perennial character of the chrysanthemum.

Our efforts to induce flowers on pyrethrum shoot cultures were unsuccessful. The shoot cultures received varying treatments of temperature, photoperiods and giberrellins (unpublished data). It was observed that light significantly enhances the production and/or accumulation of pyrethrins in shoot cultures that received 400 fc for 16 h at 25 °C. The light can be removed for 1 or 2 weeks without significantly effecting the yield of pyrethrins [18].

4 The Biosynthesis of Pyrethrins

The chrysanthemic acid portion of the pyrethrins I is known to be derived biosynthetically from acetate via mevalonic acid by the well-known isoprene pathway. Incorporation of radiolabeled acetate, mevalonate, and chrysanthemate into the pyrethrins I has been accomplished by using cut flower stems and isolated achenes. Further, radiolabeled chrysanthemate has also been incorporated into pyrethric acid and pyrethrins II, thus establishing the relationship between the two types of pyrethrins. The rethrolone portion of the pyrethrins has been shown to be derived from acetate and since no intermediates were detected it is presumed that the acetate is incorporated via the polyketide pathway. This work was also done using isolated ovules (achenes) from the flower heads [9].

5 Incorporation of Radioactivity into Pyrethrins I by a Pyrethrum Cell-Free System

In this report we further extend our knowledge of the biosynthetic pathway to the pyrethrins by establishing the incorporation of radioactivity from 1-(^{14}C)-isopetenyl pyrophosphate into pyrethrins I and chrysanthemyl alcohol (Fig. 1) by cell-free homo-

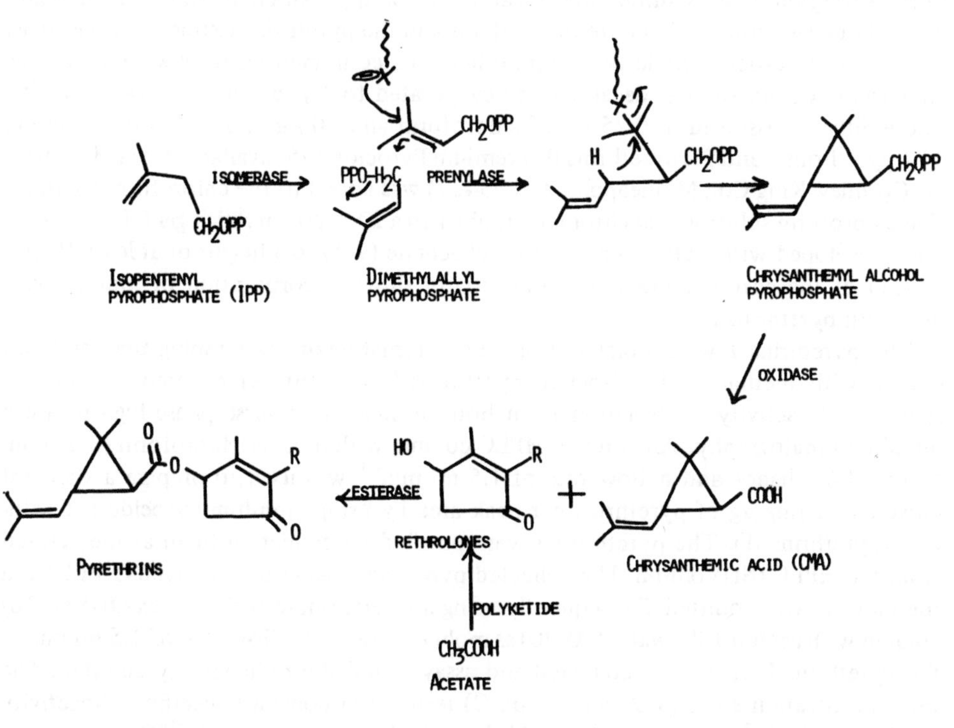

Fig. 1. Incorporation of radioactivity from 1-(^{14}C)-isopentenyl pyrophosphate into pyrethrins I and chrysanthenyl alcohol

genate prepared from the flower buds of pyrethrum. The cell-free procedure described, with shoot cultures, resulted in an activity equally to approximately 10% of the buds.

5.1 Preparation of Cell-Free Homogenate

Fresh unopened flower buds are mixed with an equal amount of powdered insoluble polyvinylpolypyrrolidone in a glass mortar and pestle. Homogenizing buffer, consisting of 0.1 M MES (pH 7.0), 0.5 M sucrose, 0.1 M sodium ascorbate, and 0.1 M β-mercaptoethanol is added in a ratio of 5 ml g^{-1} of fresh plant material. Grinding in the mortar and pestle is continued until a smooth, free-flowing paste is achieved. The mixture is centrifuged at 7,000 rpm for 10 min and the supernatant is recovered as the cell-free homogenate. These procedures are done at 4 °C and the cell-free homogenate is stable for at least 4 weeks when stored in a closed glass container at that temperature. Boiling or freezing of the cell-free homogenate results in loss of activity.

5.2 Incorporation of 1-(^{14}C)-isopentenyl Pyrophosphate into Pyrethrins I

A freshly prepared cell-free system (1 ml) containing 0.5–1.0 mg ml^{-1} of protein as determined by the Bradford method is incubated with 0.2 μCi of 1-(^{14}C)-isopentenyl pyrophosphate ammonium salt having specific activity of 53 mCi $mmol^{-1}$ (Amersham Corp., Arlington Heights, Illinois 60005) at 35 °C for approximately 18 h. The incubate is then boiled in order to denature the proteins and the pyrethrins extracted three times with 5 ml of pesticide grade petroleum ether. The combined extracts were dried over anhydrous sodium sulfate, decanted and evaporated to dryness on a rotovap at 40 °C. The residue is dissolved in 0.25 ml of chloroform and 10 μg of a standard containing 11% pyrethrins I and 9% pyrethrins II (Premium Pyrocide 175, available from McLaughlin Gormley King Co., Minneapolis, MN 55427) was added to the chloroform solution. The chloroform solution was chromatographed on a 20 × 20 cm Silica gel GF analytical plate developed with petroleum ether:ethyl acetate (9:1) to a height of at least 10 cm. A scan of the chromatogram for radioactivity shows incorporation of radioactivity into both pyrethrins I and II.

The pyrethrins I were isolated from the chromatogram by scraping the band and eluting with methanol. The isolated pyrethrins I were further purified to constant specific radioactivity by isolation from both normal and reverse phase high pressure liquid chromatography. An Altex Si HPLC column with hexane:chloroform (3:2) containing 1% ethanol and a flow rate of 1.5 ml min^{-1} was used to prepare a standard curve comparing μg of pyrethrins per peak area by using Premium Pyrocide 175 solution (pyrethrins II). The pyrethrins I were purified by collecting them as they eluted from the HPLC (Si) column. The collected pyrethrins I were then assayed (HPLC) and the radioactivity counted. Subsequently using a reverse phase HPLC Altex-ODS (C-18) column with acetonitrile:water (80:20) as mobile phase and a flow rate of 1.5 ml min^{-1} the pyrethrins I were again collected and assayed and the radioactivity counted. The three purification steps (TLC and 2 HPLC) resulted in constant specific radioactivity ($2.37 \pm 0.1 \times 10^7$ dpm $mmol^{-1}$) establishing the incorporation of 1-(^{14}C)-isopentenyl pyrophosphate into the pyrethrins as 0.9%.

5.3 Incorporation of 1-(^{14}C)-Isopentenyl Pyrophosphate into Chrysanthemyl Alcohol

A freshly prepared cell-free system (CFS) (2.5 ml) containing 0.5–1.0 mg ml^{-1} of protein was incubated with 3.0 μCi of 1-(^{14}C)-isopentenyl pyrophosphate having a specific activity of 53 mCi $mmol^{-1}$ at 35 °C for 18 ± 2 h. The incubate was then boiled to denature protein and the pyrethrins and other free terpenes were extracted with petroleum ether three times. The aqueous phase was adjusted to pH of 9 with 0.1 N KOH and 1 mg of alkaline phosphatase was added. The mixture was then incubated at 37 °C for approximately 18 h. The incubation was stopped by the addition of an equal volume of 10% KOH. The mixture was extracted three times with petroleum ether, dried over anhydrous sodium sulfate, decanted and evaporated to dryness on a rotovap at 40 °C. The residue is dissolved in 250 μl of chloroform.

The chloroform solution is chromatographed on a 20 × 20 cm Silica gel GF analytical plate and developed with petroleum ether:ethyl acetate (8:2) to a height of at least 10 cm. The band corresponding to chrysanthemyl alcohol is scraped and eluted with methanol. An aliquot was removed for counting radioactivity and to the remainder is added 20 mg of unlabeled chrysanthemyl alcohol as carrier. The methanol was evaporated and the residue dissolved in 2 ml of anhydrous pyridine and mixed with 240 mg of 3,5-dinitrobenzoyl chloride and 15 ml of 5% potassium bicarbonate solution. The mixture was heated on a steam bath for 1.5 h followed by extraction three times with 15 ml of chloroform. The extract was washed twice with 30 ml of 0.5 N HCl followed by two washes with 30 ml of distilled water. The chloroform is dried over anhydrous sodium sulfate and evaporated to dryness. The residue (3,5-dinitrobenzoate of chrysanthemyl alcohol) was recrystallized to constant specific radioactivity from 95% ethanol ($7.1 \pm 0.1 \times 10^4$ dpm $mmol^{-1}$) establishing the incorporation of 1-(^{14}C)-isopentenyl pyrophosphate into chrysanthemyl alcohol as 4.4%.

6 Conclusions

The incorporation of radioactivity from 1-(^{14}C)-isopentenyl pyrophosphate into the pyrethrins I confirms its intermediacy in the biochemical pathway which heretofore was assumed, based upon previous work which utilized radioactive mevalonic acid. Our work further establishes the presence of chrysanthemyl alcohol pyrophosphate as a most probable intermediate in the biosynthesis of the pyrethrins. Therefore this research is entirely consistent and supportive of the proposed biochemistry of the pyrethrins. The importance of the CFS should be highlighted since it enables the use of putative precursors which do not ordinarily pass membrane barriers (i.e., pyrophosphates).

Acknowledgements. Support for this study was received from McLaughlin Gormley King Company, Minneapolis, MN 55427. Selected pyrethrum plant materials were received from Industria Extractora, CA, Quito, Ecuador.

References

1. Khanna P, Khanna R (1976) Endogenous free ascorbic acid and effect of exogenous ascorbic acid on growth and production of pyrethrins from in vitro culture of *Tagetes erecta* L. Indian J Exp Biol 14:630–631
2. Jain SC (1977) Chemical investigation of *Tagetes* tissue cultures. Planta Med 31:68–70
3. Roest S, Bokelmann GS (1973) Vegetative propagation of *Chrysantheum cinerariaefolium* in vitro. Sci Hortic 1:120–122
4. Grewal S, Sharma K (1978) Indian J Exp Biol 16:1119
5. Levy LW (1981) A large-scale application of tissue culture: the mass propagation of pyrethrum clones in Ecuador. Environ Exp Bot 21:389–395
6. Wambugu FM, Rangan TS (1985) In vitro clonal multiplication of pyrethrum (*Chrysanthemum cinerariaefolium* Vis) by micropropagation. Plant Sci Lett (in press)
7. Zieg RG, Zito SW, Staba EJ (1983) Selection of high pyrethrin producing tissue cultures. Planta Med 48:88–91
8. Aoki S, Kaneto K, Hashimoto S, Oogai H (1976) Production of pyrethrin by tissue culture. Jpn Appl Sho 51:97868 (Patent 78:24097, March, 1978)
9. Cassada JE (ed) (1973) Pyrethrum – the natural insecticide. Academic Press, London New York
10. Chumsri P, Staba EJ (1975) Pyrethrins content and larvicidal activity of Chrysanthemum plants and tissue cultures. Acad Pharm Sci Abstr 5:169
11. Cashyap MM, Jueh JSH, MacKenzie IA, Pattenden C (1978) In vitro synthesis of pyrethrins from tissue cultures of *Tanacetum cinerariifolium*. Phytochemistry 17:544–545
12. Head SW (1966) A study of the insecticidal constituents in *Chrysanthemum cinerarizefolium*. Pyrethrum Post 8:32–37
13. Zito SW, Zieg RG, Staba EJ (1983) Distribution of pyrethrins in oil glands and leaf tissue of *Chrysanthemum cinerariaefolium*. Planta Med 47:205–207
14. Roest S (1976) Flowering and vegetative propagation of pyrethrum (*Chrysantheum cineraria-efolium* Vis) in vivo and in vitro. Centre Agric Pub and Doc, Wageningen, pp 104
15. Jeffcoat B, Cockshull KE (1972) Changes in the levels of endogenous growth regulators during development of the flowers of *Chrysanthemum morifolium*. J Exp Bot 23:722–732
16. Harada H (1967) Flower induction in excised shoot apices of *Pharbitis* and *Chrysanthemum* cultured in vitro. Nature (London) 214:1027–1028
17. Schwabe WW (1957) Factors controlling flowering in the Chrysanthemum. VI De-vernalization by low light intensity in relation to temperature and carbohydrate supply. J Exp Bot 8: 220–234
18. Staba EJ, Nygaard BG, Zito SW (1984) Light effects on pyrèthrum shoot cultures. Plant Cell Tissue Org Cult 3:211–214

Biosynthesis of Chorismate-Derived Quinones in Plant Cell Cultures[1]

E. LEISTNER[2]

1 Introduction

Although quinones are derived from a variety of different precursors (for reviews see Leistner 1981, 1984), they belong to those secondary plant products that are produced in cell cultures in rather large amounts. Thus acetate derived anthraquinones occur in tissue cultures of *Cassia* (Tabata et al. 1975; Takahashi et al. 1978; Noguchi and Sankawa 1982) and *Rheum* (Furuya et al. 1975; Rai 1978). Selection of strains yielded a *Nicotiana tabacum* cell culture which produced rather large amounts of ubiquinone-10 (Matsumoto et al. 1980). This quinone is derived from phenylpropanoids and mevalonic acid, as are the naphthoquinones alkannin and shikonin occurring in *Echium* (Fukui et al. 1983) and *Lithospermum* (Fukui et al. 1984) cell suspension and callus cultures. The present review concentrates on quinones derived from chorismate[1] via o-succinylbenzoic acid.

2 Quinones in Morinda and Galium Plant Cell Cultures

Tissue cultures producing large amounts of natural products are derived from plants belonging to the genera *Morinda* and *Galium*, which are members of the family Rubiaceae. The cultures have been established from aerial parts of the plants. The pigments, however, are localized in the roots and therefore these cell cultures represent an impressive example of the totipotency of every single cell in an intact organism.

Experiments on the biosynthesis of chorismate-derived quinones were difficult to carry out as long as tissue cultures were not available because roots are rather unsuitable systems for precursor feeding experiments or even enzyme extractions. The use of tissue culture in the study of quinone biosynthesis has resulted in high incorporation rates (Zenk et al. 1975), which have seldom been experienced with intact plants. Moreover, there is hope that tissue cultures of the plants mentioned above may serve as a source for enzymes catalyzing steps in the pathway leading from chorismic acid to anthraquinones (Heide and Leistner 1983; Heide et al. 1982a,b).

1 Note added in proof: See page 224

2 Institut für Pharmazeutische Biologie, Nußallee 6, 5300 Bonn 1, FRG

Primary and Secondary Metabolism of Plant Cell Cultures, ed. by Neumann et al.

2.1 Morinda Cell Cultures

A cell suspension culture of *Morinda citrifolia* L. has been established by Zenk et al. (1975). The anthraquinones occurring in this culture were isolated and identified (Inoue et al. 1981a; Leistner 1973a, 1975). Droplet countercurrent chromatography (DCCC) proved to be particularly suitable in the isolation of aglycones and glycosides (Inoue et al. 1981a). Some of the constituents isolated from the culture were known compounds which also occur in the intact plant. However, seven new pigments have also been isolated from the culture. Six of these compounds were identified (Inoue et al. 1981a). The tissue culture materials used in both isolations (Inoue et al. 1981a; Leistner 1975) were grown in media containing either naphthalene acetic acid (Leistner 1975) or 2,3-dimethoxy phenoxyacetic acid (Inoue et al. 1981a). In one of the isolations (Leistner 1975), no attempt was made to isolate glycosides.

The publication of Zenk et al. (1975) on this *Morinda* tissue culture is one of the most cited papers in the field of secondary metabolites and plant cell cultures. This may be due to the fact that it represents a study on a genetically stable line in which the influence of various factors such as hormones, carbohydrates, amino acids, inorganic ions, vitamins, pH, temperature, and biosynthetic precursors on growth and anthraquinone formation were investigated. It was repeatedly noted that pigment formation and growth of the tissue culture were not correlated. It was also found that a biosynthetic precursor of anthraquinones viz. o-succinylbenzoic acid (Leistner 1973a) was very efficiently incorporated into anthraquinones (Zenk et al. 1975). Indeed 70% of the precursor administered to the culture was converted to anthraquinones. One of the striking features of this culture is that both 2,4-dichlorophenoxyacetic acid (2,4-D) and naphthalene acetic acid (NAA) support the growth of cultures but that only NAA stimulates anthraquinone formation, whereas 2,4-D completely suppresses pigment production. This phenomenon has been further investigated using 40 different synthetic phenoxyacetic acids (Zenk et al. 1984). Phenoxyacetic acids substituted in para position with halogen all support growth, whereas anthraquinone formation increases with the change of the halogen atom in the 4 position from 4-F via 4-Cl and 4-Br to 4-I-phenoxyacetic acid. Highest stimulation of anthraquinone formation was seen when 2,3-dimethyl- or 2-Br-phenoxyacetic acid was used. On the contrary, a marked inhibition of anthraquinone biosynthesis was observed when L-tryptophan or its precursors were administered to the culture (El-Shagi et al. 1984). 2-Aminobenzothiazole, however, was most effective in the inhibition of anthraquinone biosynthesis. This sulfur- and nitrogen-containing compound is not known to be a naturally occurring compound. Precursor feeding experiments on the *Morinda* culture in combination with tryptophan showed that the amino acid interfered with the late steps in the biosynthetic pathway leading to anthraquinones. It was concluded that accumulation of primary metabolites (e.g. L-tryptophan) can shut down a secondary pathway (anthraquinone biosynthesis). While these experiments were carried out using a cell culture strain derived from a *Morinda citrifolia* plant, we have established a cell culture from *Morinda lucida* (Igbavboa, Sieweke, Leistner, Röwer, Hüsemann, Barz 1985, Planta, in press). This plant also contains anthraquinones and anthraquinone glycosides (Demagos et al. 1981), as does the heterotrophic cell culture obtained from this plant. The main pigment in the culture was found to be the primveroside of lucidin, whereas in the intact root the primveroside of morindone prevails (Röwer 1983). When the freshly established callus was exposed to continuous light (6,000–8,000 lx), green cell lumps developed which were transferred to a liquid culture. Repeated transfers and continuous illumination gave a green suspension culture which finally grew in medium A (Bauch and Leistner 1978a) without sucrose. Determination of chlorophyll and lipoquinones [i.e. phylloquinone (vitamin K_1), plastoquinone, tocopherol, and ubiquinone] showed that the amounts of plastoquinone, ubiquinone, and chlorophyll present in the cells matched those of the intact plant when compared on a dry weight basis. On the other hand, phylloquinone and α-tocopherol were only 28 and 44% of those encountered in the plant (Igbavboa 1981). Anthraquinones were not detectable (Röwer 1983). Similar characteristics were exhibited by a culture which was kept under illumination (6,000–8,000 lx) in the same medium but with sucrose. When an inoculum of this cultures was transferred into new medium (containing sucrose) and subsequently kept on a shaker in the dark, lipoquinones, and chlorophyll disappeared from the cells and anthraquinones developed as shown in Fig. 1. This contrasts with the situation in the heterotrophic culture of *Morinda citrifolia*. When a heterotrophic culture was exposed to continuous illumination (600 lx) or darkness, pigment production was not affected (Zenk et al. 1975). However, plant cells of *Morinda lucida* carrying out photosynthesis undergo a drastic metabolic change when transferred to darkness and kept in a medium containing sucrose. The intact plant behaves in a similar way: lipoquinones are confined to the photosynthetic tissue (leaves), anthraquinones occur only in the heterotrophic nonphotosynthetic tissue (roots). The metabolic switch observed with the *M. lucida* culture is also interesting because of the following reason: both the heterotrophic culture and the culture carrying out photosynthesis produced quinones which are derived from chorismic acid[1], o-succinylbenzoic acid (1) with 1,4-dihydroxy-2-naphthoic acid (3) representing the branch point for both pathways leading to anthraquinones or phylloquinone (Fig. 2).

The biosynthesis of phylloquinone proceeds in the chloroplasts (Kaiping et al. 1984), but the site of synthesis of anthraquinones within the cell is still unknown.

2.2 Galium Cell Cultures

Although *Morinda* and *Galium* belong to the same family (Rubiaceae), the former comprises woody tropical plants, while the latter species are weeds indigenous to Europe. Plants of both genera contain the same type of o-succinylbenzoic acid-derived anthraquinones in their roots (for a review see Leistner 1981, 1984). Plant cell cultures derived from *Morinda* and *Galium* plants produce large amounts of anthraquinones. Those present in the cell suspension culture of *Galium mollugo* have been identified (Bauch and Leistner 1978a). Lucidin primveroside is the main pigment. The growth response and anthraquinone production of the *Galium* cell culture was investigated. It was found that different components (hormones, nutrients) in the media influence not only the total amount of secondary products, but also the absence or presence of individual compounds. Absence of an anthraquinone from the cultured cells is not due to degradation of the pigment because they have been shown to be metabolically stable (Zenk et al. 1975; Bauch and Leistner 1978b).

Analysis of the factors that influence growth and anthraquinone formation in a batch culture is rather difficult because growth rate and environment change. A chemostat

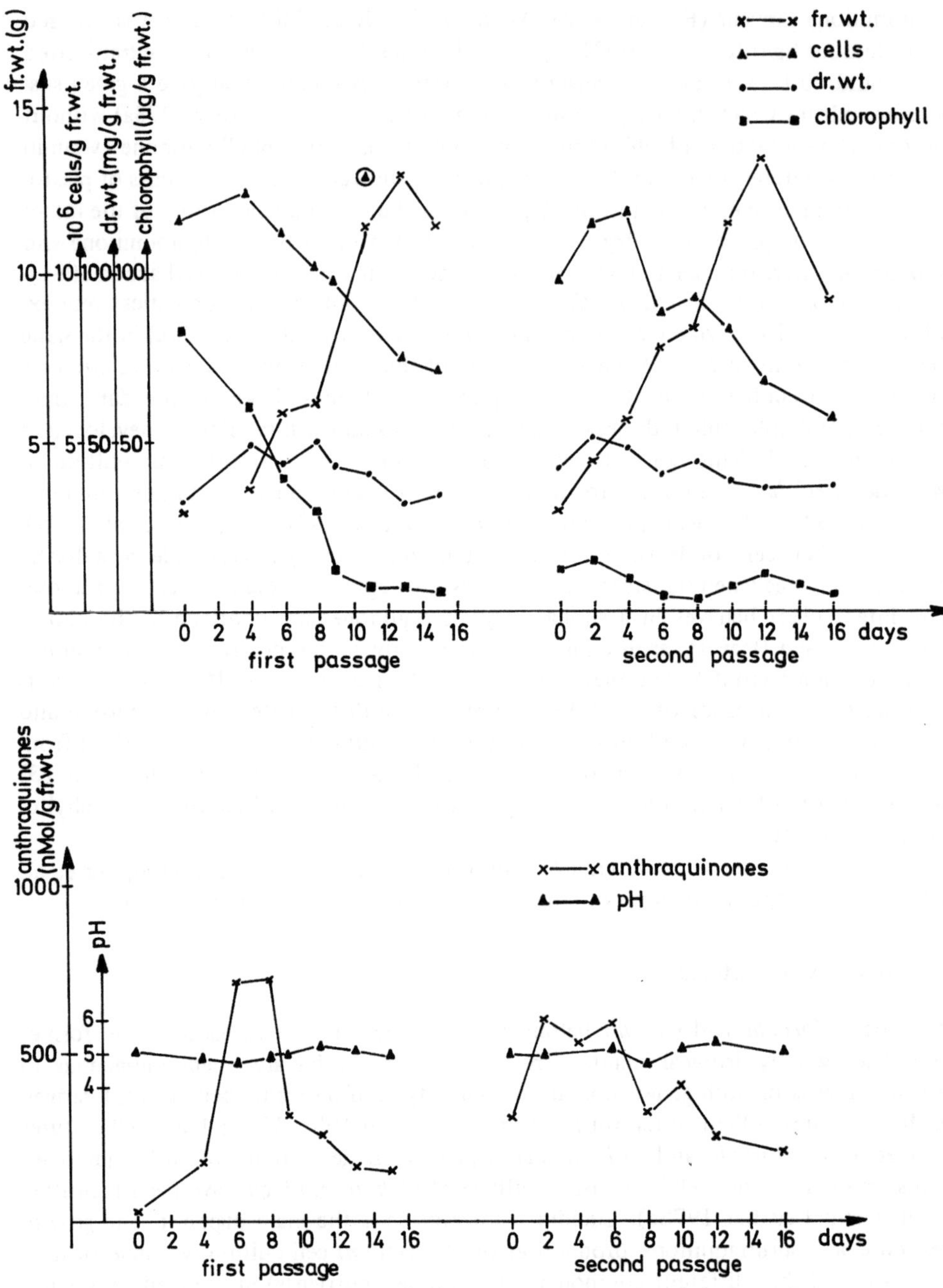

Fig. 1. Transfer of photosynthesizing cells of *Morinda lucida* grown in Gamborg's B 5-medium (40 ml; including sucrrose) from light (6,000–8,000 lx) into darkness

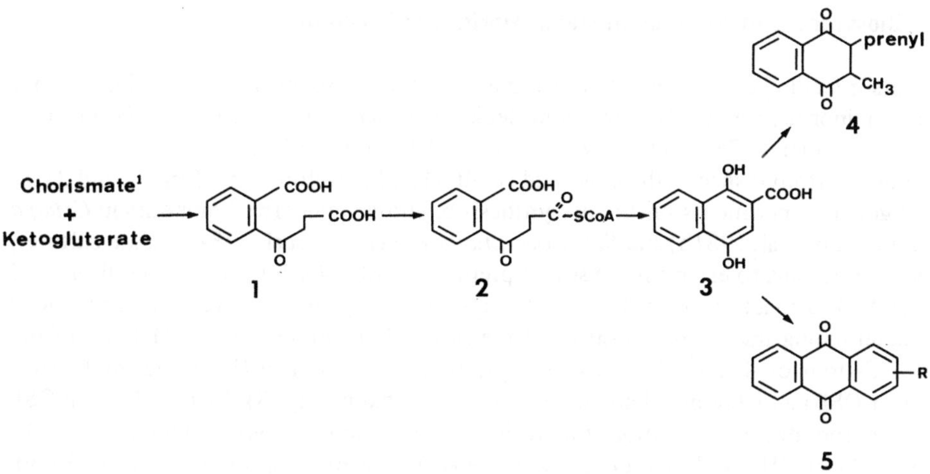

Fig. 2. Branched pathway leading from chorismate[1] to vitamin K (3) or anthraquinones

culture, however, provides for the control of cell division and pigment production (Wilson and Marron 1978) when phosphate is used as a growth-limiting factor. It was shown that cell proliferation and anthraquinone production coincide. Thus primary and secondary metabolism are not separated in a *Galium mollugo* cell culture.

Plant species may produce different types of secondary metabolites. An example is provided by *Galium mollugo* plants which contain not only quinones but also iridoids (e.g. asperuloside) (Briggs and Nickolls 1954). The fact that anthraquinones are formed abundantly in the cell suspension culture does not necessarily mean that iridosis are also present. We failed to detect these metabolites in the cultured cells, although we were able to confirm their presence in the intact plant (Bauch and Leistner 1978a). Iridoids are derived from mevalonic acid and sugars. Although anthraquinones are derived from o-succinylbenzoic acid, mevalonic acid and sugars, the production of anthraquinones in the *Galium mollugo* cell suspension culture can be stimulated by the addition of o-succinylbenzoic acid alone. This shows that mevalonic acid and sugars are available within the cultured cells at a non-limiting level. In spite of this observation, iridoids which are derived from mevalonic acid and sugars could not be detected. The absence or presence of precursors is therefore not necessarily the key for the induction of secondary metabolites in a cell suspension culture.

Batch cultures were also established from 19 different Rubiaceae species *(Galium, Rubia, Asperula, Sherardia)* (Schulte et al. 1984). Although the cultures were derived from plants belonging to one family and producing secondary metabolites by the same biosynthetic pathways (Fig. 2), conditions for the optimum growth and pigment formation were different in every single case. Thus the regulatory principles underlying secondary metabolism in cell suspension cultures are not understood. A different ability of different cell strains to produce natural effectors (hormones) may be a reason for the differences observed. A deeper insight into the regulation of secondary product formation may be possible if the enzymes involved in anthraquinone biosynthesis are known.

2.3 Biosynthesis of Anthraquinones in Morinda and Galium

Precursor feeding experiments to intact plants and cell suspension cultures showed that anthraquinones in *Rubia* (Leistner and Zenk 1967; Leistner 1973b), *Morinda* (Leistner 1973a; Leistner 1975), and *Galium* (Bauch and Leistner 1978a,b; Inoue et al. 1984a) are derived from o-succinylbenzoic acid (OSB, 1) and mevalonic acid (Figs. 2 and 3). In the light of experiments on the biosynthesis of chorismatederived quinones in *Catalpa ovata* (Inoue et al. 1981b) and *Streptocarpus dunnii* (Inoue et al. 1984b) tissue cultures, it was important to establish the site of prenylation of 1,4-dihydroxy-2-naphthoic acid (DHNA, 3) a reaction occurring in anthraquinone biosynthesis (Fig. 3). This was done by determining the relative position of the carbon in lucidin (7 in Fig. 3) derived from the 2'C-atom of 4-(2'-^{13}C carboxyphenyl)-4-oxobutanoic acid (i.e. o-succinylbenzoic acid or OSB) and the β-carbon (i.e. CH_2OH) in lucidin (Fig. 3). Since ^{13}C from OSB was incorporated into position 9 of lucidin, prenylation is likely to take place in the 3-position of DHNA (Inoue et al. 1979. 1984a). This prenylation of DHNA in 3 position is also suggested by the finding that 2-methoxycarbonyl-3-prenyl-1,4-naphthoquinone (8) (Heide and Leistner 1981, 1982), 1,4-dihydroxy-3-prenyl-2-naphthoic acid methyl ester diglucoside (9) (Inoue et al. 1984a), and mollugin (10) (Schildknecht et al. 1976) are natural products occurring in the intact plants or cell suspension cultures of *Galium mollugo*. The experiments in which ^{13}C labelled o-succinylbenzoic

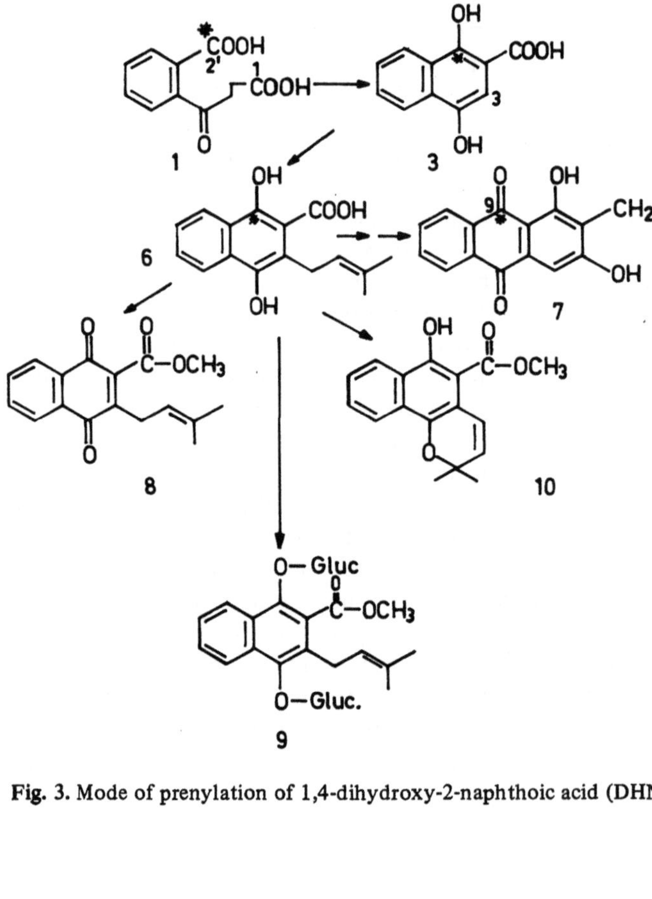

Fig. 3. Mode of prenylation of 1,4-dihydroxy-2-naphthoic acid (DHNA, 3)

acid was administered to these cell suspension cultures were remarkable, because for the first time in precursor feeding experiments chemostat cultures were used (Inoue et al. 1979, 1984a). While precursor feeding experiments with intact plants normally result in an incorporation rate of 1% or less, in these cases an incorporation rate of 83.9% of o-succinylbenzoic acid into lucidin primveroside was observed. ^{13}C NMR analysis demonstrated that a specific and non-random incorporation took place. It is obvious that the high incorporation rate observed is also due to the metabolic stability of anthraquinones in the cell suspension cultures (vide supra).

The first step in anthraquinone biosynthesis is assumed to be the nucleophilic attack of succinic semialdehyde thiamine pyrophosphate on chorismic acid (for a review see Leistner 1984). The enzyme (o-succinylbenzoic acid synthase) has hitherto only been demonstrated in bacterial vitamin K_2 (menaquinone) biosynthesis (Meganathan and Bentley 1983). The enzyme requires thiamine pyrophosphate as a cofactor. This can be demonstrated by chromatography of the enzyme on an ion exchanger. The synthase will then decompose and two protein fractions are detectable, one that catalyzes the decarboxylation of α-ketoglutarate in the presence of thiamine pyrophosphate and the other one that contains the intact OSB synthase. When U-^{14}C ketoglutarate is incubated with the former enzyme fraction in the presence of thiamine pyrophosphate, $^{14}CO_2$ and a second radioactive material, which was assumed to be the succinic semialdehyde adduct of thiamine pyrophosphate, are formed. Indeed this adduct decomposes to give succinic semialdehyde as one of the products. The presence of thiamine pyrophosphate in the adduct was demonstrated using a thiamine auxotrophic strain of *Lactobacillus viridescens* (Weische and Leistner, unpublished). Although the reaction mechanism postulated for the decarboxylation of ketoglutarate by OSB synthase (Meganathan and Bentley 1983) is identical with that of ketoglutarate dehydrogenase, both enzyme systems represent distinct catalytic entities (Weische and Leistner, unpublished[1]). The next reaction in anthraquinone biosynthesis is the activation of OSB (Fig. 2). This reaction has been demonstrated in cell free extracts of *Galium mollugo* cell suspension cultures (Heide et al. 1982b) after a procedure had been worked out for the successful removal of anthraquinones from cell free extracts (Heide and Leistner 1983). The activated o-succinylbenzoic acid was demonstrated to be identical with the corresponding compound involved in bacterial menaquinone biosynthesis (Heide et al. 1982a). Label from ^{32}P or ^{3}H-adenine labelled ATP were not incorporated into the activated OSB but ^{3}H-coenzyme A was. A molar ratio of 1:1 for OSB and Coenzyme A was observed. Thus only one of the two carboxyl groups of OSB were activated (Heide et al. 1982a). It had been suggested (Bentley and Meganathan 1982) and experimental evidence seemed to indicate (Kolkmann et al. 1982; Heide et al. 1982a) that the "aromatic" carboxyl group [i.e. the 2'C-atom of 4-(2'-carboxyphenyl)-4-oxo-butanoic acid, see 1 in Figs. 2 and 3] was activated. Synthesis of samples of OSB activated either in the position 2' or 1 (Fig. 3), however, showed clearly that it is the position 1 in OSB (see 2 in Fig. 2) that is biologically activated (Kolkmann and Leistner 1985). Previous experiments (Kolkmann et al. 1982) led to a wrong conclusion. Only this activated species of OSB (2 in Fig. 2) is metabolized to 1,4-dihydroxy-2-naphthoic acid (3 in Fig. 2) by bacterial enzyme preparations. The ring closure reaction of 2 to DHNA (3) has not yet been demonstrated in cell fre extracts of *Galium* cell cultures.

3 Biosynthesis of Quinones in Cell Cultures of Juglans, Cinchona, Tectona, Streptocarpus, and Catalpa

Chorismate[1]-derived quinones from plants which belong neither to the genus *Morinda* nor to *Galium* have also been described. The biosynthesis of naphthalene derivatives were investigated in *Juglans* plants and callus cultures (Müller and Leistner 1978a,b). Quinones occurring in *Cinchona* (Mulder-Krieger et al. 1982) and *Tectona* (Dhruva et al. 1972) tissue cultures are also likely to be derived from chorismic acid via o-succinylbenzoic acid (Leistner 1981, 1984). The biosynthesis of quinones occuring in *Streptocarpus* (Inoue et al. 1984b) and *Catalpa* (Inoue et al. 1981a,b) has already been mentioned above (vide supra).

Acknowledgement. The authors work reported herein was supported by the Deutsche Forschungsgemeinschaft and Fonds der Chemischen Industrie.

References

Bauch HJ, Leistner E (1978a) Aromatic metabolites in cell suspension cultures of *Galium mollugo* L. Planta Med 33:105–123

Bauch HJ, Leistner E (1978b) Attempts to demonstrate incorporation of labelled precursors into aromatic metabolites in cell suspension cultures of *Galium mollugo* L. Planta Med 33:124–127

Bentley R, Meganathan R (1982) Biosynthesis of vitamin K (menaquinone) in bacteria. Microbiol Rev 46:241–280

Briggs LH, Nickolls GA (1954) Chemistry of the *Coprosma* Genus. Part VIII. The occurrence of asperuloside. J Chem Soc:3940–3943

Demagos GP, Baltus W, Höfle G (1981) New anthraquinones and anthraquinone glycosides from *Morinda lucida*. Z Naturforsch 86b:1180–1184

Dhruva BR, Rama Rao AV, Srinivasan R, Venkataraman K (1972) Structure of a quinone from teak tissue culture. Indian J Chem 10:683–685

El-Shagi H, Schulte U, Zenk MH (1984) Specific inhibition of anthraquinone formation by amino acid compounds in *Morinda* cell cultures. Naturwissenschaften 71:267

Fukui H, Tsukada M, Mizukami H, Tabata M (1983) Formation of stereoisomeric mixture of naphthoquinone derivatives in *Echium lycopsis* callus cultures. Phytochemistry 22:453–456

Fukui H, Yoshikawa N, Tabata M (1984) Induction of benzoquinone formation by activated carbon in *Lithospermum erythrorhizon* cell suspension cultures. Phytochemistry 23:301–305

Furuya T, Ayabe S, Noda K (1975) Chrysophanol and emodin from callus tissue of *Rhubarb (Rheum palmatum)*. Phytochemistry 14:1457

Heide L, Leistner E (1981) 2-Methoxycarbonyl-3-prenyl-1,4-naphthoquinone, a metabolite related to the biosynthesis of mollugin and anthraquinones in *Galium mollugo* L. JCS Chem Commun 334–336

Heide L, Leistner E (1982) Versuche zur Synthese natürlicher vorkommender prenylierter Naphthalinderivate. Nachweis eines neuen Prenylchinonderivates in *Galium mollugo*. Z Naturforsch 37c:354–362

Heide L, Leistner E (1983) Enzyme activities in extracts of anthraquinone-containing cells of *Galium mollugo*. Phytochemistry 22:659–662

Heide L, Arendt S, Leistner E (1982a) Enzymatic synthesis, characterization and metabolism of the coenzyme A ester of o-succinylbenzoic acid, an intermediate in menaquinone (vitamin K_2) biosynthesis. J Biol Chem 257:7396–7400

Heide L, Kolkmann R, Arendt S, Leistner E (1982b) Enzymic synthesis of o-succinylbenzoyl-CoA in cell-free extracts of anthraquinone producing *Galium mollugo* L. cell suspension cultures. Plant Cell Rep 1:180–182

Igbavboa U (1981) Diplomarbeit, Inst Pharm Biol Phytochem, Westfälische Wilhelms-Univ, Münster, FRG

Inoue K, Shiobora Y, Nayeshiro H, Inouye H, Wilson G, Zenk MH (1979) Site of prenylation in anthraquinone biosynthesis in cell suspension cultures of *Galium mollugo*. JCS Chem Commun: 957–959

Inoue K, Nayeshiro H, Inouye H, Zenk MH (1981a) Anthraquinones in cell suspension cultures of *Morinda citrifolia*. Phytochemistry 20:1693–1700

Inoue K, Ueda S, Shiobara Y, Kimura I, Inouye H (1981b) Quinones and related compounds in higher plants. Part 11. Role of 2-carboxy-2,3-dihydro-1,4-naphthoquinone and 2-carboxy-2-(3-methyl-but-2-enyl)-2,3-dihydro-1,4-naphthoquinone in the biosynthesis of naphthoquinone congeners of *Catalpa ovata* callus tissue. JCS Perkin I:1246–1258

Inoue K, Shiobara Y, Nayeshiro H, Inouye H, Wilson G, Zenk MH (1984a) Biosynthesis of anthraquinones and related compounds in *Galium mollugo* cell suspension cultures. Phytochemistry 23:307–311

Inoue K, Ueda S, Nayeshiro H, Moritone N, Inouye H (1984b) Biosynthesis of naphthoquinones and anthraquinones in *Streptocarpus dunnii* cell cultures. Phytochemistry 23:313–318

Kaiping S, Soll J, Schulz G (1984) Site of methylation of 2-phytyl-1,4-naphthoquinol in phylloquinone (vitamin K_1) synthesis in spinach chloroplasts. Phytochemistry 23:89–91

Kolkmann R, Knauel G, Arendt S, Leistner E (1982) Site of activation of o-succinylbenzoic acid during its conversion to menaquinones (vitamin K_2). FEBS Lett 137:53–56

Kolkmann R, Leistner E (1985) Synthesis and revised structure of the o-succinylbenzoic acid coenzyme A ester, an intermediate in menaquinone biosynthesis. Tetrahedron L 26:1703–1704

Leistner E (1973a) Biosynthesis of morindone and alizarin in intact plants and cell suspension cultures of *Morinda citrifolia*. Phytochemistry 12:1669–1674

Leistner E (1973b) Mode of incorporation of precursors into alizarin (1,2-dihydroxy-9,10-anthraquinone). Phytochemistry 12:337–345

Leistner E (1975) Isolierung, Identifizierung und Biosynthese von Anthrachinonen in Zellsuspensionskulturen von *Morinda citrifolia*. Planta Med Suppl:214–224

Leistner E (1981) Biosynthesis of plant quinones. In: Conn EE (ed) The biochemistry of plants, vol VII. Academic Press, London New York, pp 403–423

Leistner E (1985) Occurrence and biosynthesis of quinones in woody plants. In: Higuchi T (ed) Biosynthesis and biodegradation of wood components. Academic Press, London New York, pp 273–290

Leistner E, Zenk MH (1967) Ein neuer Biosyntheseweg für Anthrachinone: Der Einbau von Shikiminsäure in 1,2-Dihydroxy-anthrachinon (Alizarin) und 1,2,4-Trihydroxy-anthrachinon (Purpurin) in *Rubia tinctorum* L. Z Naturforsch 22b:865–868

Matsumoto T, Ikeda T, Kanno N, Kisaki T, Noguchi M (1980) Selection of high ubiquinone 10-producing strain of tobacco cultured cells by cell cloning technique. Agric Biol Chem 44:967–969

Meganathan R, Bentley R (1983) Thiamine pyrophosphate requirement for o-succinylbenzoic acid synthesis in *Escherichia coli* and evidence for an intermediate. J Bacteriol 153:739–746

Müller WU, Leistner E (1978a) Metabolic relation between naphthalene derivatives in *Juglans*. Phytochemistry 17:1735–1738

Müller WU, Leistner E (1978b) Aglycones and glycosides of oxygenated naphthalenes and a glycosyltransferase from *Juglans*. Phytochemistry 17:1739–1742

Mulder-Krieger Th, Verpoorte R, Water A de, Gesel M von, Oeveren BCJA van (1982) Identification of the alkaloids and anthraquinones in *Cinchona ledgeriana* callus cultures. Planta Med 46: 19–24

Noguchi H, Sankawa U (1982) Formation of germichrysone by tissue cultures of *Cassia torosa*: Induction of secondary metabolism in the lag phase. Phytochemistry 21:319–323

Rai PP (1978) The production of anthraquinones in callus cultures of *Rheum palmatum*. J Nat Prod 41:114–116

Röwer I (1983) Diplomarbeit, Lehrstuhl Biochemie der Pflanzen, Westfälische Wilhelms-Univ, Münster, FRG

Schildknecht H, Straub F, Scheidel B (1976) Mollugin, eine neuer Farbstoff aus Rhizomen des Wiesen-Labkrautes *Galium mollugo* L. Liebigs Ann Chem:1295–1306

Schulte U, El-Shagi H, Zenk MH (1984) Optimization of 19 *Rubiaceae* species in cell culture for the production of anthraquinones. Plant Cell Rep 3:51–54

Tabata M, Hiraoka N, Ikenoue M, Sano Y, Konoshima M (1975) The production of anthraquinones in callus cultures of *Cassia tora*. J Nat Prod 38:131–134

Takahashi S, Kitanaka S, Takido M, Ebizuka Y, Sankawa U, Hoson M, Koboyashi M, Shibata S (1978) Formation of anthraquinones by the tissue culture of *Cassia obtusifolia*. Planta Med 33: 389–392

Wilson G, Marron P (1978) Growth and anthraquinoine biosynthesis by *Galium mollugo* L. cells in batch and chemostat culture. J Exp Bot 29:837–851

Zenk MH, El-Shagi H, Schulte U (1975) Anthraquinone production by cell suspension cultures of *Morinda citrifolia*. Planta Med Suppl:79–101

Zenk MH, Schulte U, El-Shagi H (1984) Regulation of anthraquinone formation by phenoxy acetic acids in *Morinda* cell cultures. Naturwissenschaften 71:266

Note added in proof. Recent results show that chorismate is converted to isochorismate prior to incorporation into quinones [Weische A, Leistner E (1985) Cell free synthesis of o-succinylbenzoic acid from *iso*-chorismic acid, the key reaction in vitamin K_2 (menaquinone) biosynthesis. Tetrahedron L 26:1487–1489]

The Role of Leucine in Terpenoid Metabolism: Incorporation of Leucine into Sesquiterpenoids and Phytosterols by Andrographis Tissue Cultures

K.H. OVERTON[1]

1 Introduction

The possible involvement of the amino acid leucine in the biosynthesis of isoprenoids has long been the subject of conjecture. The catabolic pathway for leucine in mammals was established in the early 1950's, largely by the work of M.J. Coon et al. [15, 24, 36], as outlined in Scheme 1. It was subsequently shown that the *same* (3S)-antipode of 3-hydroxy-3-methylglutaryl-CoA (HMG-CoA) (5) lies on the catabolic pathway of leucine [20, 37] (Scheme 1, a) and on the anabolic pathway from acetate to (3R)-mevalonic acid (MVA) (6) [16, 17] (Scheme 1, b). The latter is, of course, the well-established biosynthetic route to terpenoids and steroids, but the interesting question arises whether an alternative, if less important route to (3R)-MVA (6) might start from leucine, rather than acetate and proceed via HMG-CoA, incorporating the carbon skeleton of leucine intact [but for loss of C–1 in the conversion of (1) to (2)].

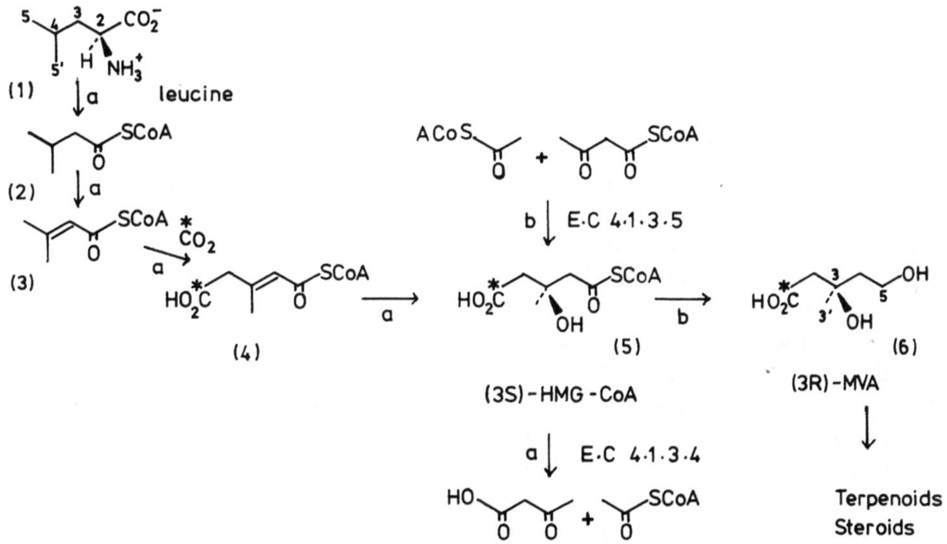

Scheme 1. Cross-over of the catabolic pathway for leucine and the anabolic pathway for terpenoids at HMG-CoA

1 Department of Chemistry, University of Glasgow, Glasgow G12 8QQ, United Kingdom

Primary and Secondary Metabolism of Plant Cell Cultures, ed. by Neumann et al.

There have been reports of relevant work in the literature for more than 30 years. Thus Bloch showed in 1944 [8], well before MVA was first isolated from distillers' solubles and identified as the key intermediate in terpenoid biosynthesis [41, 43], that deuteriated leucine and isovaleric acid [see (2)] are incorporated by rats into cholesterol via acetate. Bloch, Clark, and Harary later showed [9] that acids corresponding to (3), (4), and (5) (Scheme 1) are converted in rats to cholesterol without, apparently, prior breakdown to acetate or acetoacetate, while Brady and Gurin found [11] that isovalerate and acetoacetate were similarly incorporated into cholesterol intact. There have been more recent attempts to detect the incorporation of leucine into terpenoids, including gibberrellins [25], hop resins [26], the sesquiterpenoid ipomeamarone [27], cantharidin [34], and the indole alkaloid vindoline [42]. Suga and his colleagues have studied [38, 39] the incorporation of (S)-[U-^{14}C]-leucine into linalool by *Cinnamomum camphora* and into geraniol in *Pelargonium roseum*. More recently they have shown that germinating seeds of *Pisum sativum* incorporate label from [U-^{14}C]-leucine uniformly into squalene and β-amyrin [40].

Two studies that precede our own work merit serious attention. The first concerns incorporation of leucine into mevalonic acid and β-carotene by the fungus *Phycomyces blakesleeanus*. Goodwin and Lijinski [19] and MacKinney et al. [23] independently reported in 1952 that *P. blakesleeanus* incorporates $^{14}CO_2$ into β-carotene only if leucine is present in the medium. A more recent study by Lowry and Chichester [21] demonstrates that 88% of the radioactivity thus incorporated is located at C–1 of MVA. Since C–1 is lost in the conversion of MVA into isoprenoids, only 12% of label, located elsewhere in the MVA molecule, gets through to β-carotene. These results and earlier experiments by the Chichester group [14], which investigated the fate of leucines labelled with ^{14}C at C–1, C–2, C–3, C–4, and C–5 on their conversion into β-carotene, are readily explained in terms of the transformations indicated in Scheme 1. Thus $^{14}CO_2$ probably carboxylates ββ-dimethylacrylyl-CoA (3) as in the course of leucine catabolism in mammals and thus labels C–1 successively of HMG-CoA (5) and MVA (6). If this were the only source of MVA, the β-carotene would be totally unlabelled (see above). The 12% of MVA molecules labelled at positions other than C–1 (predictably at C–3/C–3′ and C–5, although this was not established) could arise from cleavage of HMG-CoA to acetyl-CoA and acetoacetate, followed by re-assembly. Chichester's earlier results with specifically labelled leucines can be accommodated on the same basis: C–1 of leucine is not incorporated; C–2 and C–3 are weakly incorporated since they become acetyl-CoA in equilibrium with a large unlabelled pool, while C–4 and C–5 are strongly incorporated since the acetoacetate formed from them presumably encounters a smaller pool. It is particularly significant that *P. blakesleeanus* seems to transform HMG-CoA formed from leucine degradation *to a major extent* (88%) directly into MVA. Unfortunately, however, the evidence for this is indirect.

Clearly, what is required to establish beyond doubt the mode of incorporation of leucine into isoprenoids, is to determine the fate of ^{13}C-labelled leucines in the metabolites formed by NMR spectroscopy. This approach has been used in only one instance apart from our own work. *Aspergillus amstelodami* produces two metabolites, echinuline (7) and flavoglaucine (8), substances of mixed biogenesis which contain respectively three and one meroterpenoid (C_5) unit. Fuganti et al. have studied [12, 18] the incorporation of (4S)- and (4R)-[5-^{13}C]-leucine into echinuline and flavoglaucine and find

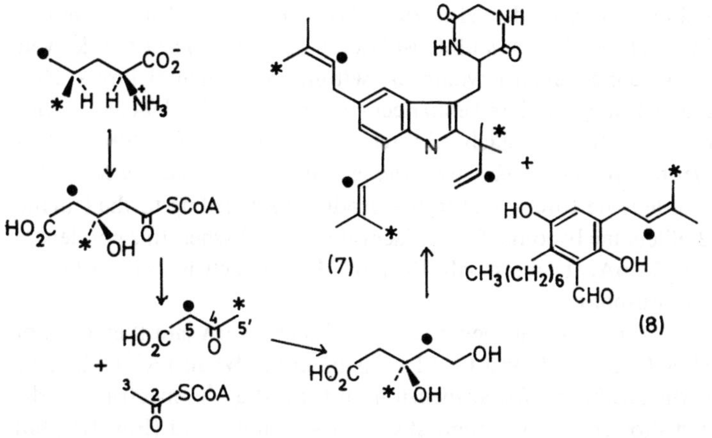

Scheme 2. Incorporation of leucine into echinuline and flavoglaucine

that a major mode of incorporation, although certainly not the only mode, is in accordance with cleavage of the intermediate HMG-CoA to acetoacetate (C–4, C–5, and C–5′) and acetyl-CoA (C–2 and C–3), and intact incorporation of the C_4 fragment (Scheme 2). These experiments do not provide evidence as to the fate of C–2 and C–3, nor does the evidence exclude the possible direct incorporation of HMG-CoA to a significant extent.

2 Incorporation of Leucine into Sesquiterpenoids by Andrographis Tissue Cultures

We have investigated the *general* problem of the incorporation of leucine into isoprenoids in a range of biological systems, bearing in mind the two pathways revealed by the earlier work and the possibility that the balance of these pathways may well vary in different systems. We were above all concerned to establish the fate of *each* carbon atom of leucine when the amino acid is incorporated into an isoprenoid. The method of choice must be to incorporate leucines specifically labelled with ^{13}C and to locate the labelled carbon atoms in the product by NMR. For this one requires a system that efficiently incorporates the labelled precursor into the product and our best system so far has been a callus culture of *Andrographis paniculata*. This culture makes a family of sesquiterpenoid lactones whose biosynthesis we have studied in detail [2, 3, 22, 28–32]. The major lactone is paniculide B (paniculide) (9) and we discovered some time ago that *Andrographis* suspension cultures incorporate [U-^{14}C]-leucine into paniculide rather well [I_{spec} 15%–20%]. This therefore appeared to be an ideal system for investigating *how* leucine is incorporated.

Our first experiments showed quite clearly that *leucine is not incorporated intact* via HMG-CoA. Thus, leucine labelled at C–2 will label C–1 of HMG-CoA and, if this is incorporated intact, C–5 of MVA and hence C–4, C–8, and C–12 of paniculide.

When (2RS)-[2-^{13}C]-leucine [35] was fed to *Andrographis* suspension cultures together with (2S)-[U-^{14}C]-leucine, neither the paniculide isolated (I_{spec} 15.5%) not the derived diacetate showed any peak enhancement whatever, within experimental limits. Had direct incorporation occurred, the calculated enhancement from [U-^{14}C]-leucine should have been approximately three-fold for each of three peaks. *Clearly therefore, in this experiment HMG-CoA from leucine breakdown does not enter paniculide intact.* But how is label from C–2 of leucine totally lost in the product? According to the leucine catabolic pathway, C–2 of leucine becomes C–1 of acetyl-CoA and when this condenses with unlabelled acetoacetyl-CoA, it labels only C–1 of MVA which is lost as CO_2 on forming isopentenyl pyrophosphate.

In the same way, C–3 of leucine will become C–2 of acetyl-CoA and then C–2 of MVA, which should label C–3, C–7, and C–11 of paniculide. When [3-^{13}C]-leucine [4] was incubated with the *Andrographis* suspension culture, the ^{13}C spectrum of the paniculide formed indeed showed enhancement at C–3, C–7, and C–11 (Fig. 1B), but these were not the only enriched signals. There were six additional enriched peaks, corresponding to C–1, C–5, C–9, C–13, C–14, and C–15. Moreover, the intensity of labelling was absolutely uniform across the nine enriched peaks and this must mean that the [2-^{13}C]-acetyl-CoA formed condenses with itself and then with a third equivalent to label MVA at C–2, C–3′, and C–4 (Scheme 3). When [2-^{13}C]-leucine was

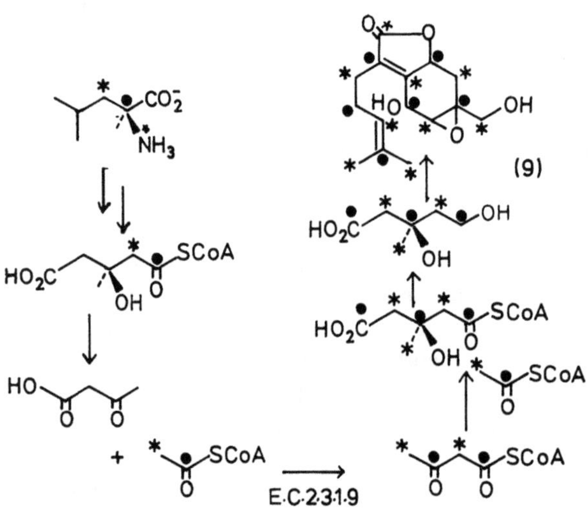

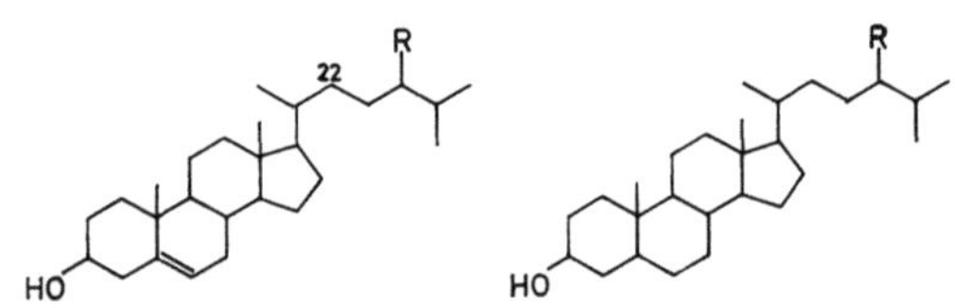

(9) $R = C_2H_5$
(10) $R = C_2H_5$; Δ^{22}
(11) $R = CH_3$

(12) $R = C_2H_5$
(13) $R = CH_3$
(14) $R = H$

Scheme 3. Incorporation of [2-^{13}C]- and [3-^{13}C]-leucines into paniculide B via acetyl CoA

metabolised by the cultures used in the above experiment and the product harvested under the same conditions, the complementary result was obtained (see Scheme 3). Thus, *six* positions were enriched, C–2, C–4, C–6, C–8, C–10, and C–12, as expected (Fig. 1A). The apparent discrepancy between this and the earlier experiment in which *no* label was incorporated from [2-^{13}C]-leucine, can be explained if the condensing enzyme, acetyl-CoA acetyltransferase (EC 2.3.1.9), functions at a time when *labelled*

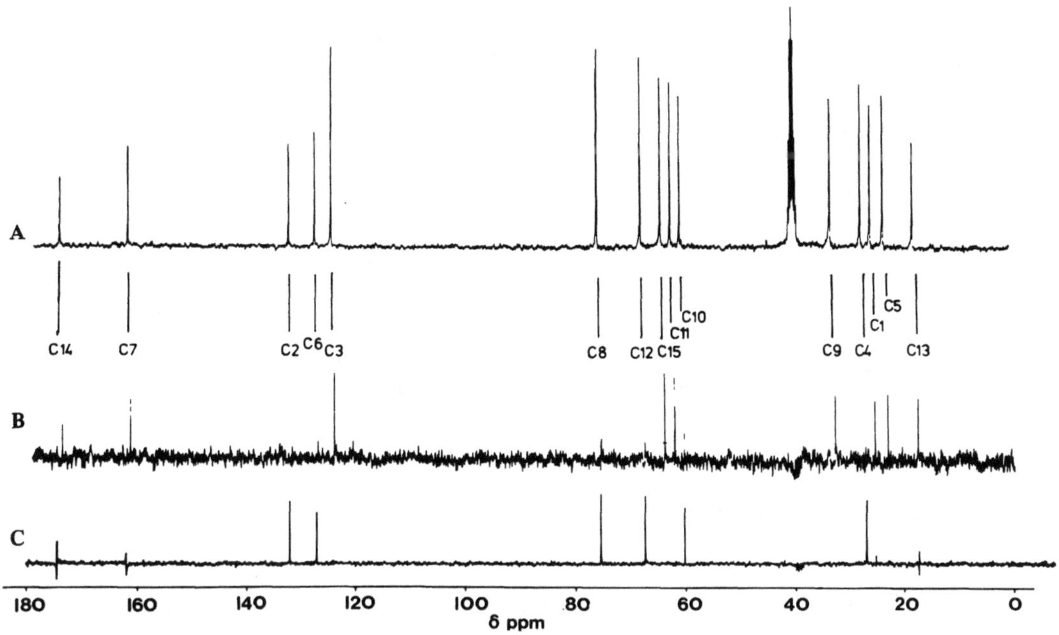

Fig. 1A–C. Labelling pattern of paniculide B derived from [2-^{13}C]- and [3-^{13}C]-leucines. **A** ^{13}C n.m.r. difference spectrum of paniculide B labelled from [2-^{13}C]-leucine – paniculide B at natural abundance. **B** ^{13}C n.m.r. difference spectrum of paniculide B labelled from [3-^{13}C]-leucine – paniculide B at natural abundance. **C** Natural abundance spectrum of paniculide 0.25 mmol in $(CD_3)_2SO$; 4,000 scans; 90.65 MHz

acetyl-CoA is available in the later but not the earlier experiment. The two sets of experiments were in fact separated by some 3 years, so that the cultures as well as the feeding conditions may have differed significantly.

The experiments described so far disclose the fate of C–2 and C–3 of leucine that become C–1 and C–2 of acetyl-CoA. What of C–4, C–5, and C–5′ of leucine that become acetoacetate? Is this incorporated at all and, if it is, is it incorporated intact? To answer this question, we labelled the two methyl groups attached to C–4 of leucine individually with ^{13}C. Although chemically apparently identical, they are spatially distinct; they are *prochiral* and hence distinguishable by an enzyme. One can work out that each will label one carbon of acetoacetate and, if this is incorporated intact, three carbons of paniculide. *Assuming* that isovaleryl-CoA dehydrogenase [the enzyme that converts (2) into (3), Scheme 1] functions with *anti*-stereochemistry, (4S)-[5-^{13}C]-leucine (•) should label C–3, C–7, and C–11 of paniculide and (4R)-[5-^{13}C]-leucine (*) should label C–13, C–14, and C–15 (Scheme 4). The two isotopomers were synthesised [33] and separately incubated with the *Andrographis* culture. The ^{13}C spectra of the paniculides thus produced were at first sight alarming; they were apparently indistinguishable from each other and also from the spectrum of paniculide biosynthesised from [3-^{13}C]-leucine. The *major* pathway must therefore correspond to cleavage of either [2-^{13}C]- (from (4S)-[5-^{13}C]-leucine) or [4-^{13}C]-acetoacetate (from (4R)-[5-^{13}C]-leucine) to [2-^{13}C]-acetyl-CoA and elaboration as before (Scheme 4, A). However,

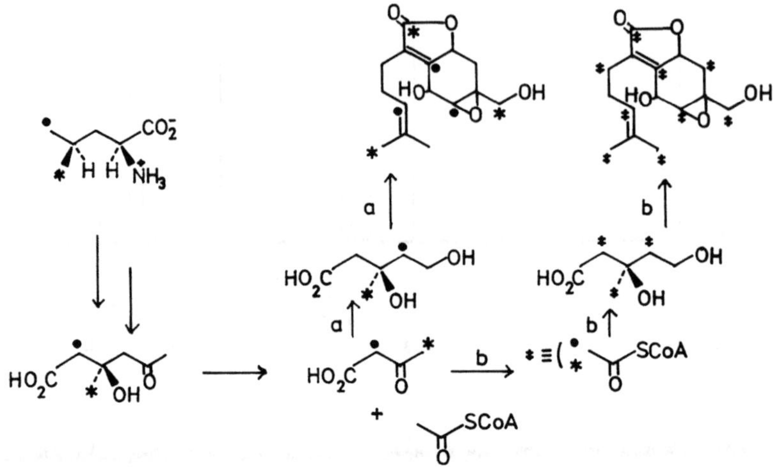

Scheme 4. Incorporation of (4R)- and (4S)-[5-^{13}C]-leucine into paniculide B

careful scrutiny of the ^{13}C NMR spectra revealed about 20% additional enrichment, in each case, of the peaks that should be labelled (Scheme 4, B, C) by direct incorporation of acetoacetate (Fig. 2).

Thus we have shown conclusively that, in this system, at any rate, leucine

a) is not incorporated direct via HMG-CoA;
b) is incorporated via the two end products of the leucine catabolic pathway, both of which are incorporated,

thus establishing for the first time the fate of each carbon atom when leucine is incorporated into a terpenoid.

3 Incorporation of Leucine and Dimethylacrylic Acid into the Phytosterols of Andrographis Tissue Cultures

In view of Bloch's pioneering work on leucine incorporation into cholesterol in rats and present concern about a possible connection between dietary cholesterol and heart disease, the possible role of leucine in steroid biosynthesis is a subject of considerable interest. With this in mind we had examined the much-studied [5–7] steroid-producing tissue cultures of tobacco and bramble but found them to be uncooperative for our purpose, yielding highly intractable mixtures of steroids and levels of incorporation that were too low to be useful. Cultures of *Andrographis*, on the other hand, produced a simple mixture of β-sitosterol (9), stigmasterol (10), and compesterol (11). The proportions varied, but the mixture was easily isolated and the dilution values were workable. For NMR analysis, the mixture was hydrogenated to produce stigmastanol (12) and campestanol (13). A portion was acetylated, the ratio of acetates determined by GLC, and a mixture of stigmastanol and campestanol of the same composition prepared

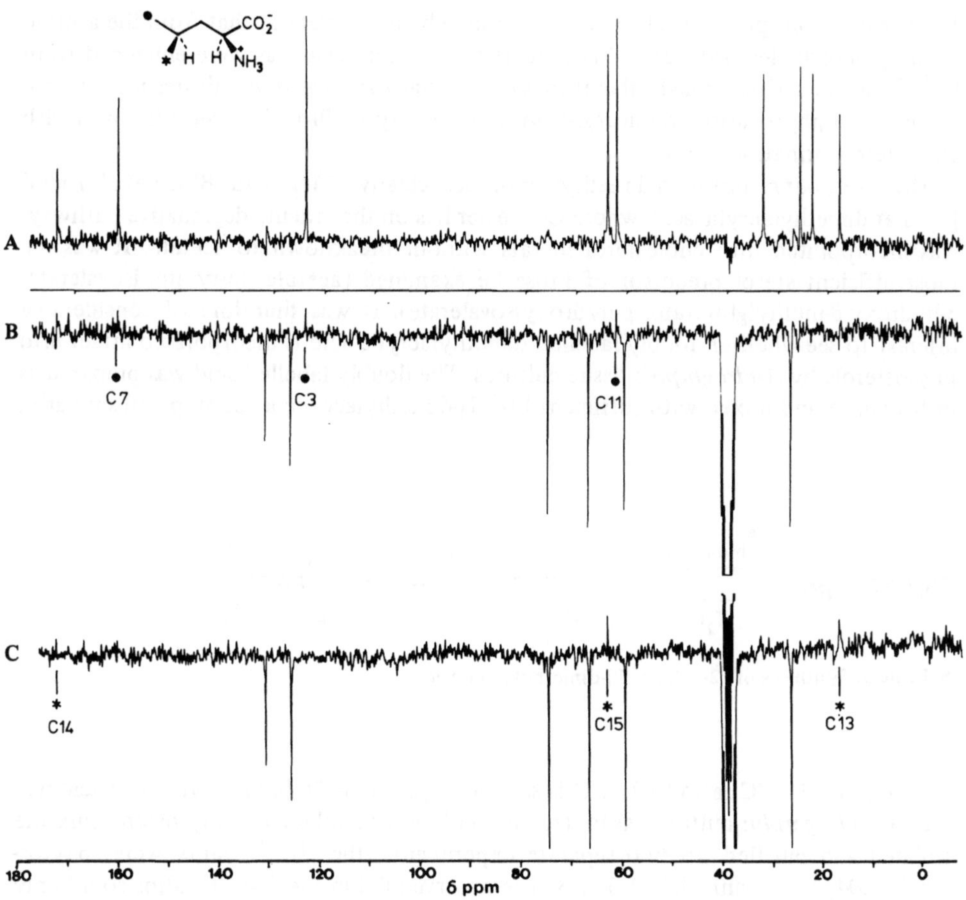

Fig. 2. A Labelling pattern of paniculide B derived from (4S)-[5-^{13}C]-leucine (•) and (4R)-[5-^{13}C]-leucine (*). **B** Difference spectrum showing peaks labelled via [2-^{13}C]-acetoacetate. **C** Difference spectrum showing peaks labelled via [4-^{13}C]-acetoacetate. Conditions as for Fig. 1

to provide the ^{13}C standard at natural abundance. The NMR assignments were checked out against the published values [10] for cholestanol (14), using the stigmastanol-campestanol mixture obtained from $^{13}CH_3CO_2Na$ feeding. A expected, there were only very minor differences for the carbon resonances of the tetracyclic nucleus and the attached methyls (C–18, C–19) as between cholestanol (14), stigmastanol (12), and campestanol (13). The signals for C–18 and C–19 which have been the subject of controversy in the spectrum of cholestanol, can be confidently assigned in our spectrum biosynthetically enriched from [2-^{13}C]-acetate: thus, the signal at δ 12.30 (C–18) is a triplet (singlet + doublet from coupling with labelled C–13), while the signal at δ 12.05 (C–19) is a singlet; C–3 is, as required, a quintuplet and C–17 a triplet. This agrees with the latest revision of the ^{13}C spectrum of cholestanol, based on analysis of the Yb-shifted spectrum [13]. We next fed [3-^{13}C]-leucine and [3-^{14}C]-leucine under the same conditions as for [2-^{13}C]-acetate and processed the phytosterol mixture as before.

The substraction spectrum, although unfortunately not as clear as that from the acetate feeding, nevertheless showed enhancement of the same peaks as were enhanced from [2-^{13}C]-acetate. This strongly therefore suggests that *Andrographis* cultures incorporate leucine into phytosterols via acetyl-CoA, analogously to Bloch's earliest findings with cholesterol formation in rats.

However, our results with dimethylacrylic acid clearly differ. Thus Bloch et al. found [9] that dimethylacrylic acid, whose CoA-ester lies on the leucine degradative pathway, was incorporated into cholesterol in rats without breakdown to acetate. It was the most efficient sterol precursor of those he examined (acetate, butyrate, isovalerate, β-hydroxy-β-methylglutarate, β-hydroxyisovalerate). It was therefore of considerable interest to see whether doubly labelled dimethylacrylic acid is incorporated intact into phytosterols by *Andrographis* tissue cultures. The doubly labelled acid was prepared as in Scheme 5 and mixed with sufficient [2-^{3}H]-dimethylacrylic acid, prepared similarly,

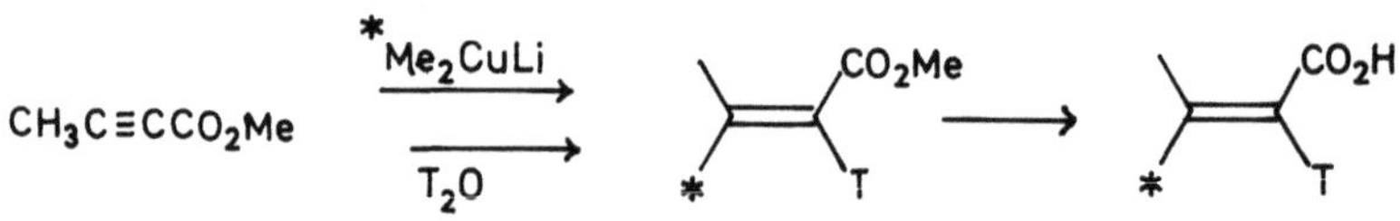

Scheme 5. Synthesis of [2-^{3}H,4-^{14}C]-dimethylacrylic acid

to bring the ^{3}H/^{14}C ratio to 7.8. This sample of [2-^{3}H, 4-^{14}C]-dimethylacrylic acid was fed to *Andrographis* cultures, as for leucine, and the paniculide and phytosterol mixture isolated and purified. In two separate experiments, the ^{3}H/^{14}C ratios were: phytosterols 1.04, 1.74; paniculide 1.5, 1.3. The observed fall in the ^{3}H/^{14}C ratios could only to a minor extent have been the consequence of secondary deuterium isotope effects. Clearly, dimethylacrylic acid has not been incorporated intact into either phytosterols or paniculide in these experiments. Incorporation via acetoacetate and acetyl-CoA would readily accommodate the observed results.

Acknowledgements. It is a pleasure to acknowledge Sobak Singh and Pete Anastasis, who did the chemistry, Isabel Freer, who tended the cultures, David Rycroft, who procured the ^{13}C NMR spectra and the SERC, who paid the bill.

References

1. Aberhart DJ, Tann C-H (1981) Substrate stereochemistry of isovaleryl-CoA dehydrogenase. Elimination of the 2-pro-R hydrogen in biotin-deficient rats. Bioorg Chem:200–205
2. Allison AJ, Butcher DN, Connolly JD, Overton KH (1968) Paniculides A, B, C. Bisabolenoid lactones from tissue cultures of *Andrographis paniculata*. Chem Commun:1493–1494
3. Anastasis P, Freer I, Gilmore C, Mackie H, Overton K, Swanson S (1982) Cyclisation of farnesylpyrophosphate to γ-bisabolene in tissue cultures of *Andrographis paniculata*. Chem Commun:268–270

4. Anastasis P, Overton K, Singh SB (1984) Synthesis of (2RS)-[3-^{13}C]- and (2RS)-[3-^{14}C]-leucine. J Lab Compds Radiopharm:393–400
5. Benveniste P, Heintz R (1974) Plant sterol metabolism. Enzymatic cleavage of the 9 β, 19 β cyclopropane ring of cyclopropyl sterols in bramble tissue cultures. J Biol Chem 249:4267–4274
6. Benveniste P, Hirth L, Ourisson G (1966) Biosynthesis of sterols in tobacco tissue culture II. Details of biosynthesis of phytosterols. Phytochemistry 5:45–58
7. Benveniste P, Hewlins MJT, Fritig B (1969) Biosynthesis of sterols in cultured tobacco tissues. Kinetics of formation of sterols and their precursors. Eur J Biochem 9:526–533
8. Bloch K (1944) Some aspects of the metabolism of leucine and valine. J Biol Chem:255–263
9. Bloch K, Clark LC, Harary I (1954) Utilization of branched chain acids in cholesterol synthesis. J Biol Chem:687–699
10. Blunt JW, Stothers JB (1977) ^{13}C Nmr spectra of steroids – a survey and commentary. Org Mag Res 9:439–464
11. Brady RO, Gurin S (1951) The synthesis of radioactive cholesterol and fatty acids in vitro. J Biol Chem 189:371–377
12. Cardillo R, Fuganti C, Ghiringhelli D, Grasselli P (1977) Pattern of incorporation of leucine samples asymmetrically labelled with ^{13}C in the isopropyl unit into the C_5-isoprenoid units of echinuline and flavoglaucine. Chem Commun:474–476
13. Chadwick DJ, Williams DH (1974) The full assignment of the carbon-13 nuclear magnetic resonance spectrum of 5α-cholestan-3β-ol with the aid of the lanthanide shift reagent $Yb(dpm)_3$. J Chem Soc Perkin II:1903–1908
14. Chichester CO, Yokoyama H, Nakayama TOM, Lukton A, Mackinney G (1959) Leucine metabolism and carotene biosynthesis. J Biol Chem 234:598–602
15. Coon MJ, Kupiecki FP, Decker EE, Schlesinger MJ, Del Campillo A (1959) The enzymic synthesis of branchedchain acids. In: Wolstenholme GEW, O'Connor M (eds) Ciba Foundation Symposium biosynthesis of terpenes and sterols. Churchill, London, pp 62–74
16. Cornforth JW (1968) Terpenoid biosynthesis. Chem Br 4:102–106
17. Cornforth JW (1974) Enzymes and stereochemistry. Tetrahedron 30:1515–1524
18. Fuganti C, Grasselli P, Pedrocchi-Fantoni G (1979) On the mode of incorporation of leucine samples labelled at positions 5 and 2 with ^{3}H and ^{14}C into the isoprenylated tryptophan derivative echinuline from *Aspergillus amstelodami*. Tetrahedron Lett:2453–2454
19. Goodwin TW, Lijinski W (1952) Carotene production by *Phycomyces blakesleeanus*. The effect of different amino acids when used in media containing low concentrations of glucose. Biochem J 50:268–273
20. Higgins MJP, Kornblatt JA, Rudney H (1972) Acyl-CoA ligases. In: Boyer PD (ed) The enzymes, vol VII, 3rd edn. Academic Press, London New York, pp 407–434
21. Lowry LK, Chichester CO (1971) The role of thiamine, leucine and CO_2 in the biosynthesis of carotenes by the mould *Phycomyces blakesleeanus*. Phytochemistry 10:323–333
22. Mackie H, Overton KH (1977) Hydrolysis and isomerization of trans,trans-farnesyl diphosphate by *Andrographis* tissue-culture enzymes. Eur J Biochem:101–106
23. Mackinney G, Nakayama TOM, Buss CD, Chichester CO (1952) Carotenoid production in *Phycomyces*. J Am Chem Soc:3456–3457
24. Meister A (1965) Intermediary metabolism of amino acids. Biochemistry of the amino acids, vol II, 2nd edn. Academic Press, London New York, pp 742–747
25. Mertz D (1970) Light-stimulated incorporation of L-leucine into the gibberellin of *Gibberella fujikuroi*. Plant Cell Physiol 11:273–279
26. Mori Y (1961) Incorporation of leucine into hop resins. Bull Brew Sci 6:28–31 (see Chem Abstr (1964) 60:7150 b)
27. Oshima-Oba K, Sugiura I, Uritani I (1969) The incorporation of leucine [U-^{14}C] into ipomeamarone. Agric Biol Chem 33:586–591
28. Overton KH, Picken DJ (1976) Biosynthesis of bisabolene by callus cultures of *Andrographis paniculata*. Chem Commun:105–106
29. Overton KH, Roberts FM (1973) Biosynthesis of 2-trans, 6-trans- and 2-cis,6-trans-farnesols by soluble enzymes from tissue cultures of *Andrographis paniculata*. Chem Commun:378–379

30. Overton KH, Roberts FM (1974) Stereochemistry of C–1 hydrogen exchange in the interconversion of trans,trans- and cis,trans-farnesols by soluble enzymes from tissue cultures of *Andrographis paniculata*. Chem Commun:385–386
31. Overton KH, Roberts FM (1974) Biosynthesis of trans,trans- and cis,trans-farnesols by soluble enzymes from tissue cultures of *Andrographis paniculata*. Biochem J:585–592
32. Overton KH, Roberts FM (1974) Interconversion of trans,trans- and cis,trans-farnesol by enzymes from *Andrographis*. Phytochemistry:2741–2743
33. Overton KH, Singh SB (1984) Synthesis of (4R)- and (4S)-[5-^{13}C]leucine. Unpublished results
34. Peter HG, Woggon W-D, Schlatter C, Schmid H (1977) Einbauversuche mit Geraniol und Farnesol in Cantharidin. Helv Chim ACta 60:844–866
35. Picken DJ (1977) Biosynthesic studies in plant tissue culture. PhD Thesis, Glasgow, pp 71, 79–82
36. Rodwell VW (1969) Carbon catabolism of amino acids. In: Greenberg DM (ed) Metabolic pathways, part II, vol III, 3rd edn. Academic Press, London New York, pp 191–235
37. Stegink LD, Coon MJ (1968) Stereospecificity and other properties of highly purified β-hydroxy-β-methylglutaryl coenzyme A. Cleavage enyzme from bovine liver. J Biol Chem 243: 5272–5279
38. Suga T, Hirata T, Shishibori F, Tange K (1974) First proof of the biosynthesis of isoprenoid from amino acid in higher plant. The incorporation of leucine into linalool. Chem Lett:189–192
39. Suga T, Hirata T, Tange K (1975) Biosynthesis of isoprenoid from amino acid in higher plant. Incorporation of L-leucine and L-valine into geraniol and citonellol. Chem Lett:243–246
40. Suga T, Tange K, Iccho K, Hirata F (1980) Biosynthesis of triterpenoids from amino acids in *Pisum sativum*. The distribution of the radioactivity in squalene biosynthesised from radioisotopically labelled L-leucine and L-valine. Phytochemistry 19:67–70
41. Tavormina PA, Gibbs MH (1956) The metabolism of β,γ-dihydroxy-β-methylvaleric acid by liver homogenates. J Am Chem Soc 78:6210
42. Wigfield DC, Wren BP (1977) Indole alkaloid biosynthesis. An investigation into leucine as a possible precursor of the monoterpene portion of vindoline. Bioorg Chem 6:511–519
43. Wolf DE, Hoffman CH, Aldrich PE, Skeggs HR, Wright LD, Folkers K (1957) Determination of structure of β,δ-dihydroxy-β-methylvaleric acid. J Am Chem Soc 79:1486–1487

Accumulation of Antineoplastic Agents by Plant Tissue Cultures

M. MISAWA [1], M. HAYASHI [2], and S. TAKAYAMA [2]

1 Introduction

It is self-evident that the ultimate objective of the research on production of useful metabolites by plant tissue cultures is the industrial manufacturing of the compounds on the large scale in a way similar to microbial fermentation. In spite of remarkable advances in this technology, the production cost of metabolites is still high especially for the production of food stuffs, food additives or pharmaceutical which can be more easily produced by chemical synthesis or by fermentation. Therefore, it is very important to choose what kind of products we should make through tissue culture technology. In this sense, efficient production of antineoplastic agents might be a very attractive research target because of their high added value. The plant investigation program of the National Cancer Institute in the U.S. for the discovery of new anticancer agents is the largest in the world and the number of plant species tested to date in this program is estimated at approximately 30,000 [1].

As a result of these screening programs, various interesting antineoplastic agents have been isolated and some of them are at the stage of clinical trials. However, in general, growth of the plants is slow and concentration of the compounds is extremely low. Furthermore, in some cases, it is difficult to harvest the plants because of their low propagation rate and of a drastic decrease in their availability. The production of these compounds by extraction from the intact plants on a large scale is very difficult. On the other hand, plant tissue culture could be the means of producing these compounds more efficiently.

In order to produce high added value compounds, the author and his group at Kyowa Hakko Kogyo Co. Ltd. in Japan have studied the production of proteinase inhibitors, plant virus inhibitors, camptothecine etc. using plant suspension cultures since 1970 [2]. During the course of these investigations, the studies on more efficient production of new promising antineoplastic agents have been carried out by a research contract of the N.C.I. [3]. Since some of the compounds are still in clinical trials, they have not yet been put into commercial production, but a few compounds are very promising.

1 Allelix Inc., 6850 Goreway Drive, Mississauga, Ontario, Canada L4V 1P1
2 Research Institute for Biological Sciences, Kyowa Hakko Kogyo Co. Ltd., Ami-machi, Inashiki-gun, Ibraki-Ken, Japan

Primary and Secondary Metabolism of Plant Cell Cultures, ed. by Neumann et al.

2 Results

2.1 Collection of Plants and Induction of Callus Tissues

Since most antineoplastic producing plants were not native to Japan, their seeds, plantlets and/or bulbs were imported from U.S.D.A., Beltsville, Maryland or from the Botanical Research Institute, Republic of South Africa. The seeds were sown in soil and also in sterilized agar media, and the plantlets were transplanted and cultivated in a greenhouse. Only *Tripterygium wilfordii* and *Cephalotaxus harringtonia* were available in Japan.

Callus tissues were induced from the stems of these plants on 10 ml of Murashige and Skoog's (MS) medium containing sucrose and different concentrations of phytohormones as shown in Table 1 in a test tube at 25 °C in the light (ca. 1,000 lx) for approximately 1 month. After 5 or 6 subcultures, each about 30 days on suitable media, friable callus tissues were obtained.

To obtain suspension cultures, approximately 5 g of the callus inoculated into 100 ml of MS liquid containing 3% sucrose and phytohormones in a 300 ml Erlenmeyer flask and cultivated at 25 °C for 10–20 days on a rotary shaker operated at 180 rpm in the dark. After successive transfers to the fresh medium, most of the cells were now in suspension and thus pipettable.

Table 1. Antineoplastic agent-producing plants and phytohormones for their callus induction

Plants	Antineoplastic agents	Phytohormones ($mg\ l^{-1}$)
Baccharis megapotamica	Baccharin	Kin. 1, 2,4-D 0.5
		Kin. 1, NAA 10
Brucea antidysenterica	Bruceantin	Kin. 1, 2,4-D 6
Caesalpinia gilliesii	Cesalin	Kin. 1, 2,4-D 0.5
Cephalotaxus harringtonia	Harringtonine, Homoharringtonine	NAA 3
Colchicum speciosum	3-Deoxycolchicine	not induced
Fagara zanthoxyloides	Fagaronine	not induced
Heliotropium indicum	Indicine-N-oxide	Kin. 1, 2,4-D 0.5
		Kin. 1, NAA 1
Ochrosia moorei	Ellipticine, 9-Methoxyellipticine	Kin. 1, 2,4-D 0.5
Putterlickia verrucosa	Maytansine	Kin. 1, 2,4-D 6
Taxus brevifolia	Taxol	not induced
Thalictrum dasycarpum	Thalicarpine	Kin. 0.1, 2,4-D 0.5
Tripterygium wildfordii	Triptolide, Tripdiolide	Kin. 1, 2,4-D 6
		Kin. 1, NAA 1
		Kin. 1, NAA 10

Kin.: Kinetin

2.2 Maytansine

Maytansine (Fig. 1) is an ansa macrolide compound which was isolated from several *Maytenus* species and *Putterlickia verrucosa* [4]. One of the *Maytenus* species has been

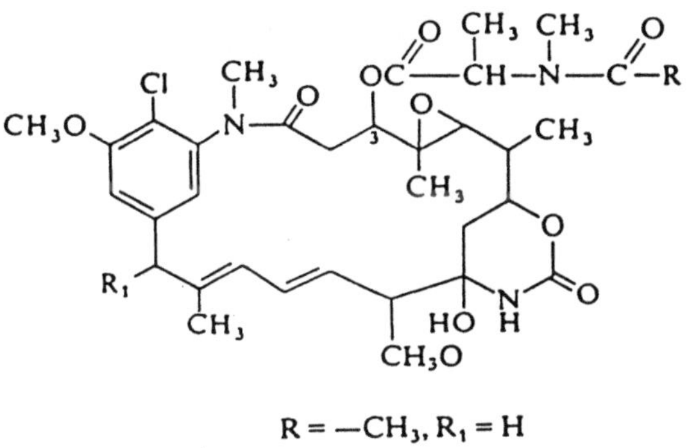

Fig. 1. Chemical structure of maytansine

used in South Africa in a herbal mixture for carcinoma and sarcoma. It is of interest that maytansine has a very potent activity in animal tumor systems at the level of $\mu g\ kg^{-1}$ of animal body weight and particularly active in P388 leukemia, L1210 leukemia, and B16 melanoma. According to recent information, however, the compound did not show the potent activity in clinical trials which had been expected.

The growth of callus tissues and suspension cultures of *Putterlickia verrucosa* was improved by optimization of environmental conditions. The medium supplemented with 1 mg l^{-1} kinetin and 1 mg l^{-1} NAA supported the highest both in callus and in suspension cultures. Alcoholic extracts of the callus and of the suspension cultured cells showed cytotoxicity against KB cells obtained from the Cancer Institute, the Japan Foundation for Cancer Research.

A dose response curve of the effect of authentic maytansine on KB cells was shown in Fig. 2. In order to identify the cytotoxic compounds(s), the callus tissues of *P. ver-*

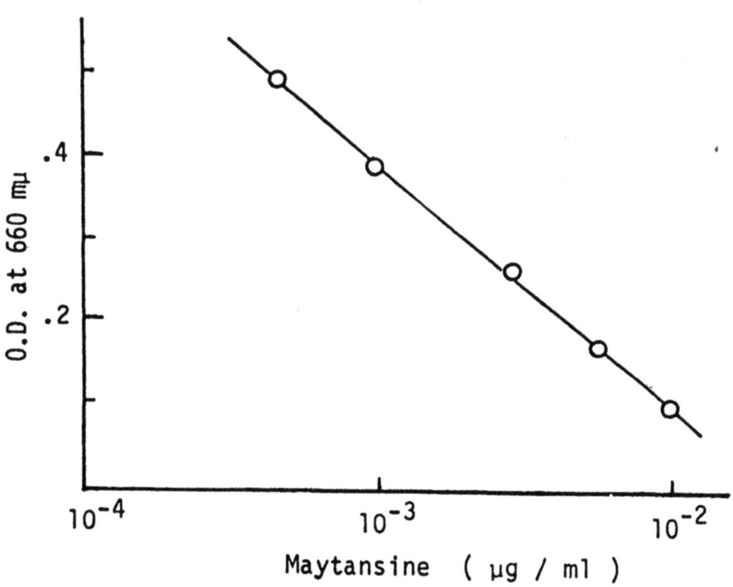

Fig. 2. Dose response of maytansine to KB cells

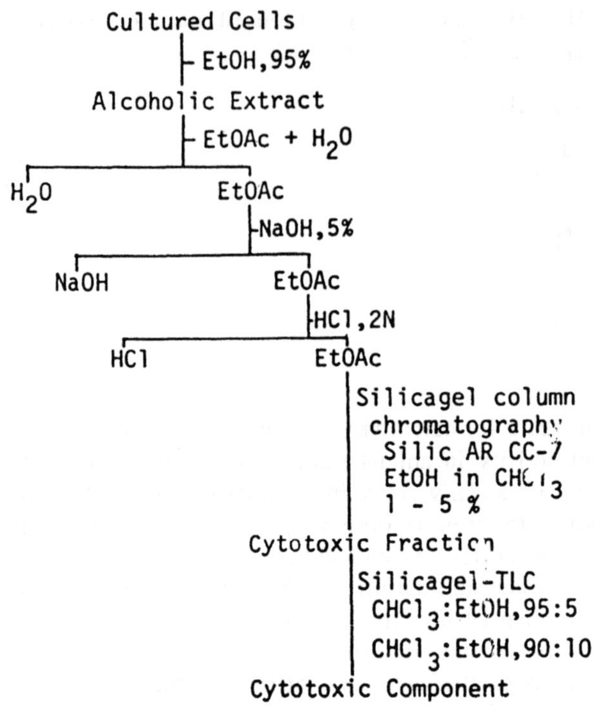

Fig. 3. Purification procedure of *P. verrucosa* product

rucosa were harvested after 50 days of cultivation on MS medium containing 3% sucrose, 1 mg l^{-1} kinetin, and 1 mg l^{-1} NAA. The purification procedure of the compound is shown in Fig. 3, in which partition between ethyl acetate and water, chromatography on silic ARCC-7 and TLC on silica gel plates were carried out. Furthermore, chromatography using aluminium oxide II–III columns and HPLC on silica and reversed columns were employed using the procedure shown in this figure. The results from TLC (Fig. 4)

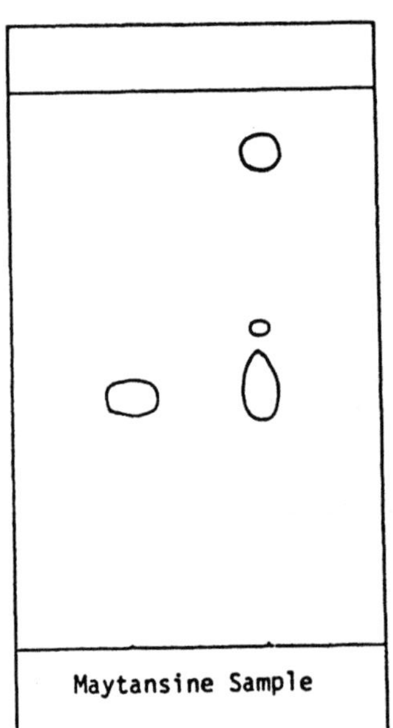

Fig. 4. TLC of products in *P. verrucosa* cultured cells.
TLC: Preparative silicagel F_{254}
Solvent system: 10% C_2H_5OH in $CHCl_3$

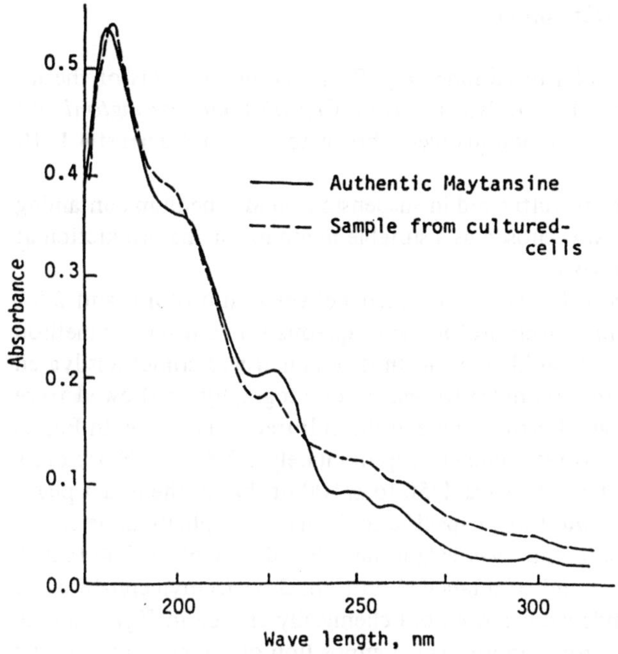

Fig. 5. UV spectra of authentic maytansine and of isolated product from cultured cells of *P. verrucosa*

and from UV spectrum (Fig. 5) showed that the isolated compound was very similar to the authentic maytansine and the retention times of the compound on HPLC with a silica gel nucleosil column (3% methanol in methylene chloride as an eluate) and with a silanized unisil DC-18 reversed column (65% methanol as an eluate) corresponded to those of maytansine, respectively. The result from HPLC with a silica gel nucleosil column is shown in Fig. 6. The fraction thus obtained also showed potent cytotoxicity against KB cells. However, the amount was only 15 μg of maytansine in 300 g of the dried callus, based on its cytotoxicity, which was about 1/240 of that in the mother plant.

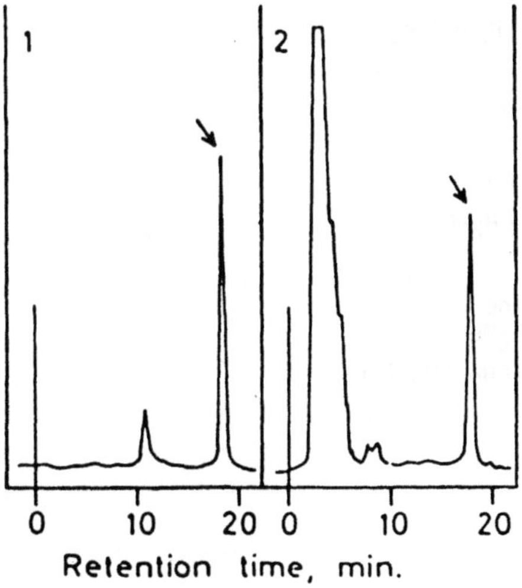

Fig. 6. HPLC of authentic maytansine and of isolated product from cultured cells of *P. verrucosa*. *1* authentic maytansine; *2* sample → maytansine

2.3 Homoharringtonine and Harringtonine

These alkaloids, together with isoharringtonine (Fig. 7) are complex esters of the inactive alcohol cephalotaxine which were isolated from *Cephalotaxus harringtonia* by Powell et al. in 1969 [5]. These compounds were shown to be active against L1210 leukemia, P388 and KB cells.

The cells of *C. harringtonia* were cultivated in suspension on MS medium containing 1% sucrose and 3% NAA which was chosen as a suitable medium for the production at 25 °C on a rotary shaker for 15 days.

The alcoholic extract of the cells was partitioned between chloroform and 2.5% aqueous tartaric acid solution and then chloroform and aqueous ammonia by the method of Powell et al. [5]. The crude alkaloid fraction thus obtained was trimethylsilyated with [bis(trimethylsilyl)-trifluoroacetamide] for gas chromatography. It showed trace amounts of homoharringtonine and harringtonine in the cultured cells as seen in Fig. 8. The cytotoxicity of this fraction corresponded to approximately 1.7×10^{-5} % as homoharringtonine in the dried cells whose level was 1/50 to 1/100 of that in the intact plant.

Since the KB cell inhibitory method is not specific to determine cephalotaxine esters, a radioimmunoassay for cephalotaxine was established. Syntheses of cephalotaxine-bovine serum albumin conjugate and of a labelled antigen, 3-o-succinylcephalotaxine [^{125}I-tyrosine methyl ester] amide were carried out chemically as seen in Figs. 9 and 10 and it's antiserum was obtained from rabbits. The composition of the reaction mixture is shown in Table 2 and thus standard curves of RIA with authentic cephalotaxine is shown in Fig. 11. It is necessary to hydrolyze homoharringtonine, harringtonine, and other antitumor esters, with 2N-HCl for a couple of minutes at 100 °C for the deter-

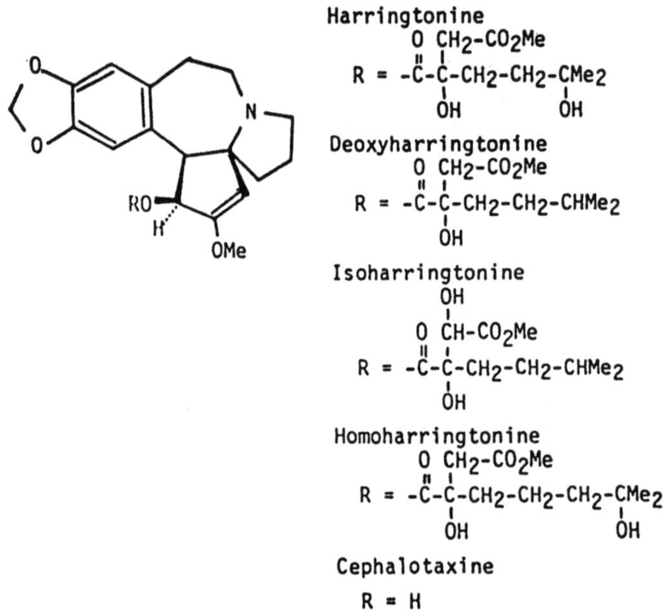

Fig. 7. Structures of harringtonine and its derivatives

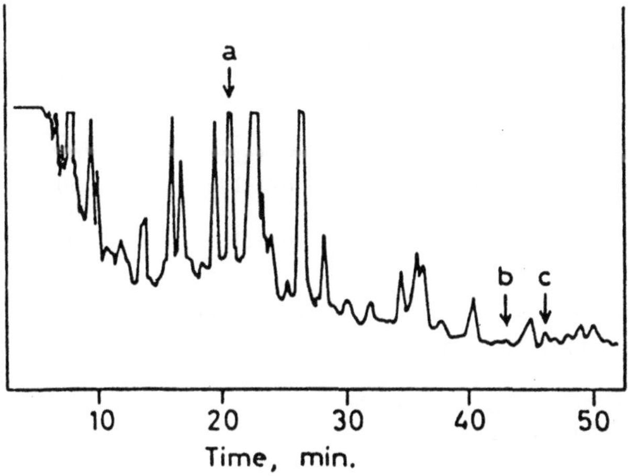

Fig. 8. Gas chromatography of a crude alkaloid fraction obtained from suspension cultured cells of *C. harringtonia*. Silicon OV-1 on Gaschrom Q; glass column (1 m); temperature: 150° to 300 °C at 2 °C min^{-1}; H_2 : 0.6 kg cm^{-2}; detector: FID. *a* Methyllignocerate (C_{24}:O), internal standard; *b* Harringtonine; *c* Homoharringtonine

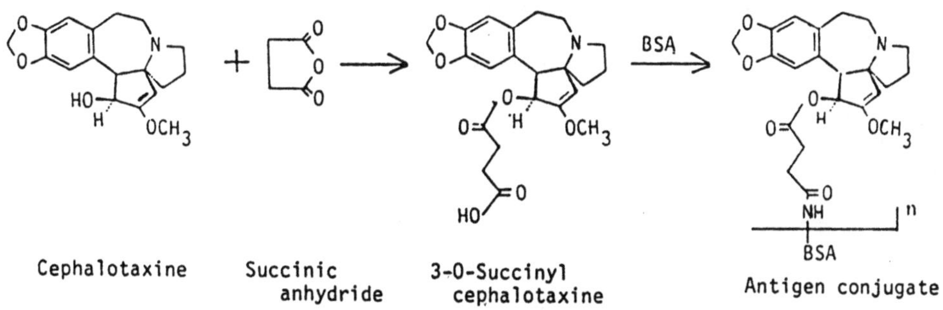

Fig. 9. Synthesis of cephalotaxine-bovine serum albumine conjugate

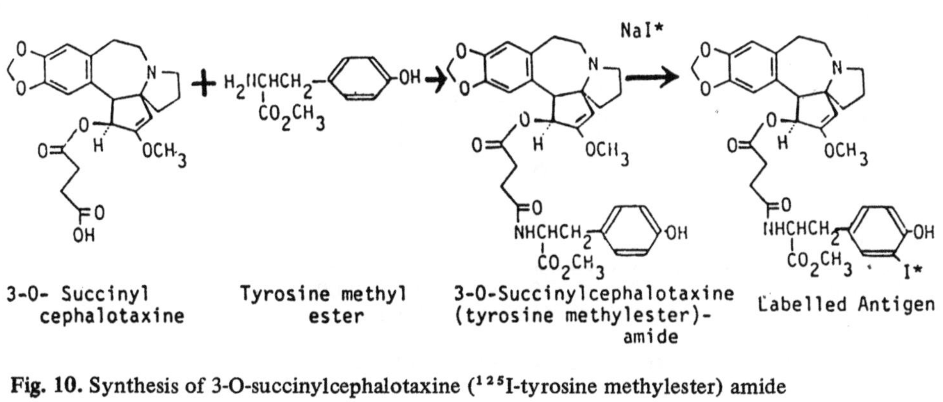

Fig. 10. Synthesis of 3-O-succinylcephalotaxine (^{125}I-tyrosine methylester) amide

mination because the method is specific for cephalotaxine as seen in the cross reactivity (Table 3).

Table 4 shows the levels of cephalotaxin and its esters as homoharringtonine in cultured cells, plantlets grown in vitro and in plants grown in the field. Although the level

Table 2. Reaction mixtures for RIA

Component	T[d]	Blank	Bo	Sample
PBS[a]	200 μl	200 μl	200 μl	100 μl
0.2% Gelatin	200 μl	200 μl	100 μl	100 μl
Antiserum[b]	–	–	100 μl	100 μl
Sample	–	–	–	100 μl
^{125}I-Antigen[c]	100 μl	100 μl	100 μl	100 μl
Total				500 μl

[a] 0.01 M-Phosphate buffer (pH 7.5), 0.15 M NaCl, 0.1% NaN_3
[b] 1/5,000 Solution of Antiserum A
[c] 3-O-Succinylcephalotaxine-(^{125}I-typrsine methyl ester) amide
[d] Total counts, without treatment of $(NH_4)_2SO_4$ or DCC

Table 3. Cross-reactivity with cephalotaxine esters

Compounds	Cross-reactivity (%)
Cephalotaxine	100
Harringtonine	9.1
Homoharringtonine	6.0
Isoharringtonine	0.7

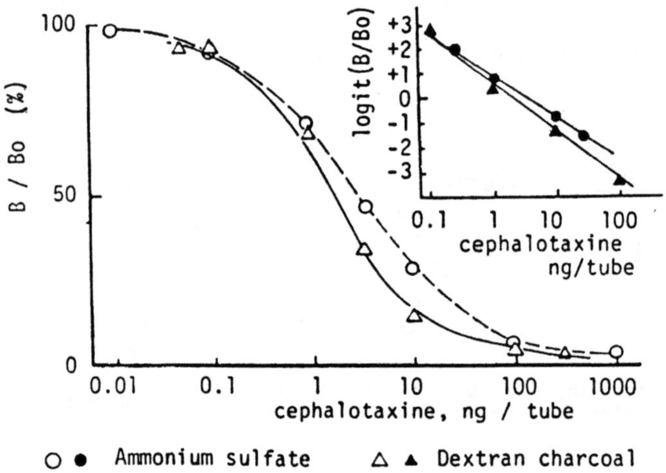

Fig. 11. Standard curves of RIA with cephalotaxine. *B* binding of ^{125}I-cephalotaxine to antiserum in the presence of cephalotaxine standard; *Bo* binding in the absence of cephalotaxine. logit (B/Bo) = ln [(B/Bo)/(1 – B/Bo)]

Table 4. Levels of cephalotaxine and its esters in *Cephalotaxus harringtonia*

Source	Concentration (μg kg^{-1} dry wt.)
Suspension cultured cells	0.55
Plantlet grown in vitro	36.0
Plants grown in the field	180.0

in the cells was only 1/300 of the intact plants, plantlets regenerated from the cells in vitro accumulated a higher amount of the alkaloids than the undifferentiated cells.

2.4 Triptolide and Tripdiolide

These antileukemia diterpenoid triepoxides (Fig. 12) were isolated by Kupchan et al. from *Tripterygium wilfordii* in 1972 [6]. They were found to show significant activity

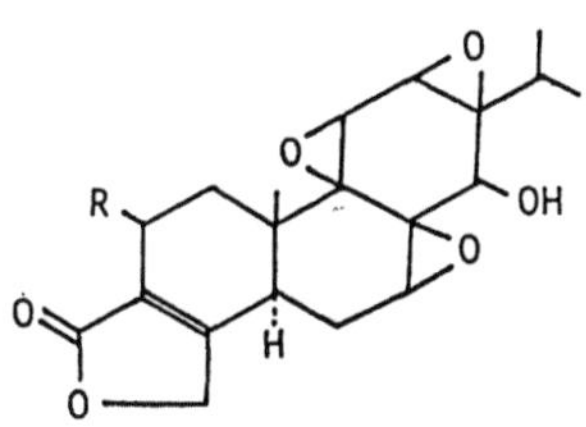

Fig. 12. Chemical structures of triptolide and tripdiolide. *R* H triptolide; *R* OH tripdiolide

against the L1210 leukemia in mice, P388 and Lewis lung carcinoma. A medium containing 3% sucrose, 1 mg l^{-1} kinetin and 1 mg l^{-1} NAA was best for growth of *T. wilfordii* callus. A cytotoxic fraction was extracted from 150 g dried cells with 95% ethanol (800 and 450 ml) at room temperature for 24 h and with 450 ml of 95% ethanol at 50 °C for 5 h. The extract was concentrated in vacuo and suspended in 45 ml and then washed in about 100 ml of n-hexane after the pH was adjusted to 5.5. The cytotoxic component in the aqueous solution was extracted with a 3 × 30 ml portions of ethyl acetate. The combined extracts were concentrated and chromatographed by a silica gel column with ethyl acetate as an eluant and with acetone-n-hexane (30:70), respectively. The fraction corresponding to tripdiolide was applied on HPLC (reverse phase) twice with ethanol-water (30:70 and 25:75, respectively) as an eluant. A cytotoxic compound (33 μg as tripdiolide) obtained by preparative TLC was analyzed by TLC, UV, and mass spectra. A spot on TLC was detected by an exposure of iodine vapor or by a spray of 2% ceric ammonium sulfate in 12% sulphuric acid. Both the retention time on HPLC and Rf value on TLC showed that the isolated compound was identical with authentic tripdiolide. Ultraviolet and mass spectra also showed the identical patterns with those of tripdiolide: UV λ_{max}^{EtOH} 217 mm; MS m l^{-1} (Rel. Int.); 592(2), 505(100), 433(15), 419(8), and 413(8). From these results, the isolated compound was identified as tripdiolide.

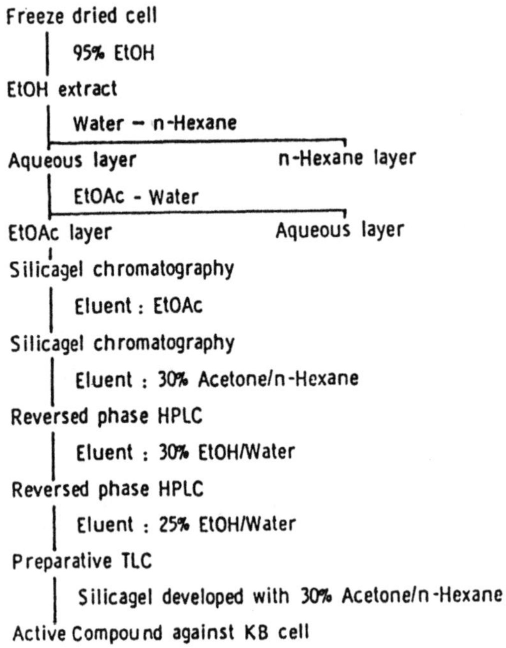

Fig. 13. Purification procedure of tripdiolide

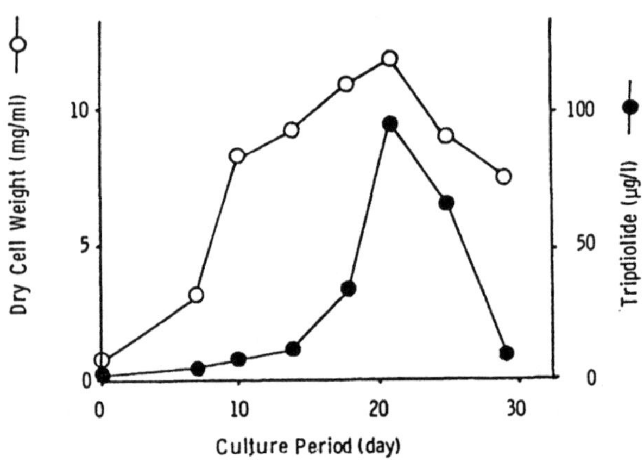

Fig. 14. A typical time course of *Tripterygium wilfordii* suspension cultures

Figure 14 shows a typical time course of *T. wilfordii* suspension culture in MS medium containing 3% sucrose, 1 mg l^{-1} kinetin and 0.1 mg l^{-1} NAA. The intracellular level of tripdiolide and the cell mass reached the maximum, 96 $\mu g\, l^{-1}$ and 11.8 mg ml^{-1}, respectively on the 21st day of the cultivation. The relationship between the growth and the production was typical for a secondary metabolite fermentation.

Using suspension cultures, a large number of experiments were carried out in order to optimize the culture conditions. Only a few of them are briefly described here.

Among several phenylurea derivatives synthesized chemically by Okamoto et al. at the University of Tokyo, N-phenyl-N′-(2-chloro-4-pyridyl) urea, so-called 4-PU-30, showed a potent cytokinin-like activity to *T. wilfordii* cells. The level of tripdiolide was increased by addition of 0.1 mg l^{-1} 4-PU-30 as shown in Table 5.

Table 5. Stimulative effect of 4-PU-30 on production of tripdiolide by *Tripterygium wilfordii* suspension culture

	Level	Dry cell wt.	Tripdiolide
4-PU-30	0.01 mg l^{-1}	12.5 mg ml^{-1}	72.5 $\mu g\ l^{-1}$
	0.1	11.7	153.5
	1.0	6.8	45.0
Kinetin	1.0	10.8	91.1

Tripdiolide is assumed to be synthesized biologically via a mevalonic acid pathway. Since farnesylpyrophosphate is located in a key point on the branching pathway, farnesol was added to the medium. Cells which could grow in a medium containing 100 μg of farnesol per litre produced higher amounts of tripdiolide (Table 6). Roots were observed in these cells.

The levels of the alkaloids in the callus seemed to be higher than those in the suspension cultured cells. As seen in Table 7, 100.6 μg of tripdiolide and 69.8 μg of triptolide per gram of the dried cells were detected, which were about 10 and 7 times higher than those in the intact plants, respectively.

Table 6. Effect of farnesol on growth and production of tripdiolide by *Tripterygium wilfordii* cells in suspension

Farnesol added $\mu g\ ml^{-1}$	Growth mg ml^{-1}	Tripdiolide $\mu g\ g^{-1}$ cell dry wt.
0	14.3	6.3
30	12.7	13.0
100	10.1	20.4

Cultivated for 3 weeks

Table 7. Levels of tripdiolide and triptolide in the callus of *Tripterygium wilfordii*

Tripdiolide	Triptolide
100.6 $\mu g\ g^{-1}$ dried cells	69.8 $\mu g\ g^{-1}$ dried cells

Determined by HPLC method

2.5 Other Antineoplastic Agents

Crude extracts of cultured cells of *Baccharis megapotamica, Holacantha emoryi*, and of *Brucea antidysenterica* were found to have cytotoxicity against KB cells. The preliminary experiments showed that the products of these cultured cells might be baccharine, holacanthone, and bruceantine, respectively.

Acknowledgements. This study was supported in part by the contract of N.C.I. NO1-CM-87910. We are grateful to Dr. J. Douros, N.C.I. in the U.S. for very useful suggestions and for providing authentic samples of antineoplastic agents and plants.

References

1. Suffness M, Douros J (1980) Anticancer agents based on natural product models. Academic Press, London New York, pp 465–487
2. Misawa M, Suzuki T (1982) Recent progress in plant cell culture, research on the production of useful plant metabolites in Japan. Appl Biochem Biotechnol 7:205–216
3. Misawa M, Hayashi M, Takayama S (1983) Production of antineoplastic agents by plant tissue cultures. (1) Induction of callus tissues and detection of the agents in cultured cells. Planta Med 49:115–119
4. Kupchan SM, Komoda Y, Court WA et al. (1972) Maytansine, a novel antileukemia ansa macrolida from *Maytenus ovatus*. J Am Chem Soc 95:1354–1356
5. Powell RG, Weisleder D, Smith CR Jr et al. (1969) Structure of cephalotaxine and related alkaloids. Tetrahedron Lett 46:4081–4084
6. Kupchan SM, Court WA, Dailey RG Jr (1972) Triptolide and tripdiolide. Novel antileukemia diterpenoid triepoxides from *Tripterygium wilfordii*. J Am Chem Soc 94:7194–7195

Induction of Enzymes of Phytoalexin Synthesis in Soybean Cells by Fungal Elicitor

J. EBEL, M.R. STÄB, and W.E. SCHMIDT [1]

1 Introduction

Higher plants have aquired effective defence mechanisms during evolution, which secure their survival in the presence of a large variety of infective microorganisms. Resistance mechanisms of plants are expressed at different levels in host-parasite interactions including preformed physical and chemical defence barriers as well as defences triggered by the invader [16]. One type of active response of plants to attempted infection is the production of low molecular weight antimicrobial compounds called phytoalexins. Considerable evidence supports the view that the phytoalexin response is one mechanism by which plants resist diseases [6, 12, 14, 21].

Phytoalexins are hardly detectable in healthy, noninduced plants. Phytoalexin synthesis may be caused by infection with potential pathogens or treatment of various plant tissues, including suspension-cultured cells [10], with pathogen-derived elicitors [6,9,12]. Our studies are aimed at the elucidation of the mechanisms regulating phytoalexin accumulation in plant tissues.

2 Phytoalexin Synthesis in Soybean

Inoculation with the fungus *Phytophthora megasperma* f.sp. *glycinea*, a pathogen causing stem and root rot of soybean *(Glycine max)*, or treatment with a β-glucan derived from the cell walls of the fungus (glucan elicitor) [3] causes various tissues of soybean to synthesize large amounts of the phytoalexin glyceollin (Fig. 1). Glyceollin was identified as a mixture of structurally related pterocarpans [5, 19, 20]. Labeling studies in vivo have demonstrated that, in challenged tissue, the induced level of glyceollin is controlled mainly by its rate of synthesis [22, 23].

The pterocarpanoid phytoalexin, glyceollin, belongs to the large group of isoflavonoids which are common natural constituents of the legumes. As outlined in Fig. 1, glyceollin biosynthesis includes a number of reactions catalyzed by enzymes of general phenylpropanoid metabolism (phenylalanine ammonia-lyase, cinnamate 4-hydroxylase, and 4-coumarate:CoA ligase) and of flavonoid biosynthesis (acetyl-CoA carboxylase and chalcone synthase). The chalcone synthase identified so far from different soy-

1 Biologisches Institut II der Universität, 7800 Freiburg, FRG

Primary and Secondary Metabolism of Plant
Cell Cultures, ed. by Neumann et al.

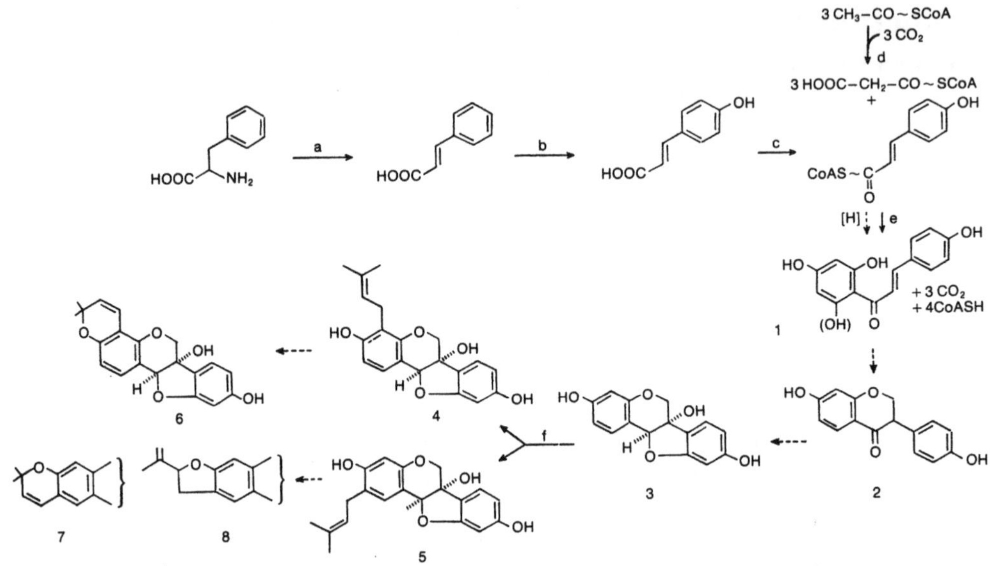

Fig. 1. Proposed biosynthetic pathway of the soybean phytoalexin, glyceollin. *1* 4,2′,4′-trihydroxychalcone; *2* dihydroxyisoflavanone; *3* glycinol (3,6a,9-trihydroxypterocarpan); *4, 5* glyceollidin I and II (4- and 2-dimethylallyl-trihydroxypterocarpan); *6* glyceollin I; *7* glyceollin II; *8* glyceollin III. Enzymes: *a* phenylalanine ammonia-lyase; *b* cinnamate 4-hydroxylase; *c* 4-coumarate:CoA ligase; *d* acetyl-CoA carboxylase; *e* chalcone synthase; *f* prenyltransferase

bean tissues catalyzes the formation of 4,2′,4′,6′-tetrahydroxychalcone. It is not yet known how the synthesis of 4,4′,6′-trihydroxychalcone, the putative precursor of glyceollin, is accomplished in vivo [11]. Recent investigations demonstrated a dimethylallylpyrophosphate: 3,6a9-trihydroxypterocarpan dimethylallyltransferase in a particulate fraction from elicitor-treated or infected tissues of soybean [18, 26]. The products of dimethylallyltransfer in vitro are likely intermediates in the biosynthesis of the various structurally related soybean phytoalexins.

3 Model System: Cell Cultures and Elicitors

The plant's resistance response to a pathogen represents a complex interaction of two living organisms which, at least in the initial phase of infection, is restricted to a few cells of the challenged plant organ, for example a root. Although a complete picture of the resistance responce can only be obtained with the intact system, the sensitivity of many of the available analytical methods is too low to enable the detection of induced biochemical changes within a few cells. This is not the case with a simplified model system, in which cultured soybean cells are effectively induced by an elicitor of microbial origin. Some of the advantages of using the elicitor in combination with cultured plant cells are that large batches of sterile, relatively uniform, and undamaged cells are simultaneously exposed to the same trigger of known amount and composition. For the purpose outlined below, the model system is ideally suited for exclusively measuring responses of the plant cells. The glucan elicitor of *P. megasperma* which is used to induce soybean cells is a structural polysaccharide found in the cell walls of the fungus. Detailed analyses demonstrated that the polysaccharide elicitor is a β-1,3-glucan with glucosyl branches at position 6 of some of the glucose residues [1–4, 25].

4 Induction of Enzymes of Phytoalexin Synthesis in Soybean

The process of induction of phytoalexin synthesis in soybean cells can be viewed as a signal-response event in which the elicitor generates an external signal that finally is transformed into a greatly increased rate of phytoalexin synthesis (Fig. 2). Rapid and large transient increases in the catalytic activities of the enzymes related to glyceollin biosynthesis (Table 1 and Fig. 1) can account for the phytoalexin production.

Little is known at present about the initial event(s) of elicitor recognition (Fig. 2) and the possible involvement of (a) second messenger(s) in signal transmission [9]. We have obtained preliminary evidence that a partially purified and radioactively labelled elicitor fraction binds to soybean protoplasts (W. Schmidt, unpublished results) and thus may generate responses of the soybean cells which ultimately will lead to increased phytoalexin synthesis. It appears unlikely that adenosine-3′,5′-cyclic monophosphate which serves as a second messenger in animal systems plays an analogous role in the response of soybeans to infection with *P. megasperma* or the response of cultured cells

Elicitor
↓
Elicitor recognition
↓
Enzyme synthesis
↓
Phytoalexin synthesis

Fig. 2. Proposed steps of the elicitor-mediated phytoalexin response in soybean cells

Table 1. Effects of elicitor treatment of cultured soybean cells on catalytic activities of enzymes of general phenylpropanoid metabolism, the flavonoid pathway, and pterocarpan metabolism

Enzyme	Treatment	
	None	Elicitor
	µkat kg^{-1}	
Phenylalanine ammonia-lyase	8	37
4-Coumarate:CoA ligase		
Isoenzyme 1	3.3	4.2
Isoenzyme 2	3.5	5.8
Acetyl-CoA carboxylase	4.8	9.6
Chalcone synthase	1.7	10
Prenyltransferase	2	12

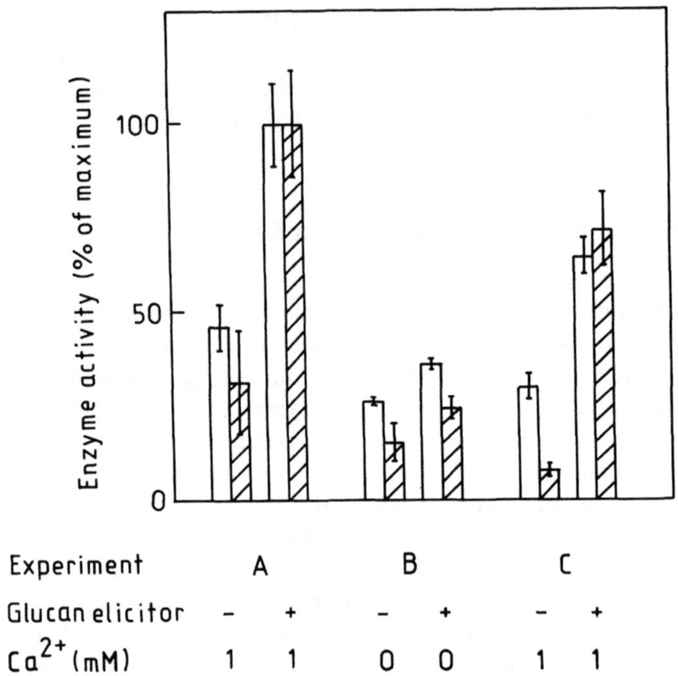

Fig. 3. Effect of Ca^{2+} ions on the elicitor-mediated enhancement of phenylalanine ammonia-lyase (□) and chalcone synthase (▨) catalytic activities of soybean cells. Extracellular Ca^{2+} was removed by EGTA treatment of the batch culture followed by washing with Ca^{2+}-free culture medium. Ca^{2+} (1 mM) was added either 10 h prior to *A* or at the onset of elicitor treatment *C*, or was omitted from the culture medium *B*. Treatment with glucan elicitor was for 10 h (+); maximal activities were 17.3 μkat/kg for phenylalanine ammonia-lyase and 3.3 μkat kg^{-1} for chalcone synthase

to the elicitor [13]. However, Ca^{2+} ions could have some role in the transmission of the primary elicitor signal into the plant cell. To support this hypothesis, the Ca^{2+} ion concentration in the cell culture medium of a soybean batch culture was varied by removal of extracellular Ca^{2+} through an EGTA treatment of the culture followed by a washing procedure and by replenishment of Ca^{2+}. The results in Fig. 3 indicate that removal of Ca^{2+} ions (Fig. 3,B) almost completely abolished the elicitor-mediated induction of two of the enzymes of glyceollin biosynthesis in soybean cells, phenylalanine ammonia-lyase and chalcone synthase, whereas re-addition of Ca^{2+} ions to the culture medium either 10 h prior to (A) or at the onset of treatment with a crude elicitor preparation (C) restored the enzyme induction. Transport studies in vitro demonstrated (Fig. 4) that the same crude elicitor preparation strongly inhibited active Ca^{2+} uptake by membrane vesicles of soybean cells. Partially purified elicitor fractions had surprisingly little effect on Ca^{2+} uptake. Although these partially purified elicitor fractions were highly effective in inducing glyceollin accumulation in the cotyledon assay [2], they poorly induced phenylalanine ammonia-lyase and chalcone synthase in cultured soybean cells. Further investigations are required to establish whether the ability to induce enzymes of phytoalexin synthesis in soybean cells and to inhibit Ca^{2+} uptake in vitro are related to the same function of the glucan elicitor.

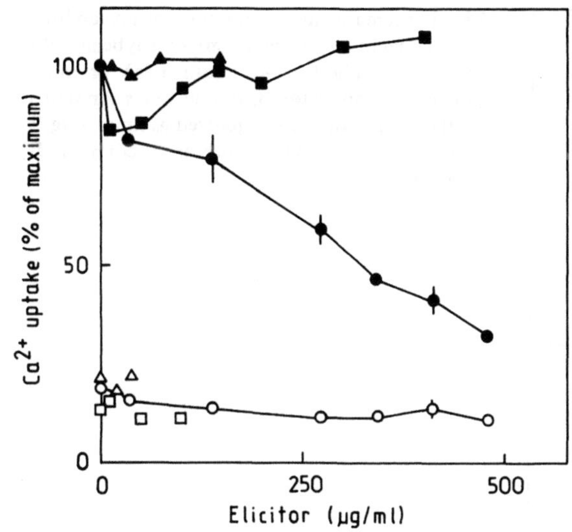

Fig. 4. Effects of different *P. megasperma* elicitor fractions on the ATP-mediated Ca^{2+} uptake by microsomal vesicles from soybean cells. Uptake was studied [8] in the presence of crude cell wall released elicitor (●, ○) or elicitor fractions released by partial acid hydrolysis of mycelial cell walls and purified by gel filtration (▲, △; ■, □); *open symbols* uptake in the absence of ATP; glucose equivalents were used as the measure of elicitor concentration

The rapid elicitor-induced increases in the catalytic activities of two of the enzymes, phenylalanine ammonia-lyase and chalcone synthase, were correlated with high rates of enzyme synthesis, as measured by immunoprecipitation of the in vivo and in vitro labeled proteins [11, 15]. Following the onset of induction, the mRNA activities for both enzymes increased significantly at 3 h, reached highest levels at 5–7 h and subsequently returned to low values at 10 h. More recently, RNA blot hybridization with cDNA complementary to chalcone synthase mRNA [17] was used to demonstrate that the timing is the same for changes in amount and activity of chalcone synthase mRNA with highest values again occurring at about 5 h after the onset of elicitor induction (Fig. 5). These and similar results obtained for chalcone synthase mRNA following infection of soybean seedlings with *P. megasperma* suggest that phytoalexin synthesis in soybean is regulated by temporary gene activation [24].

Increases in the catalytic activities of phenylalanine ammonia-lyase and chalcone synthase have also been observed when cultured soybean cells are incubated with two

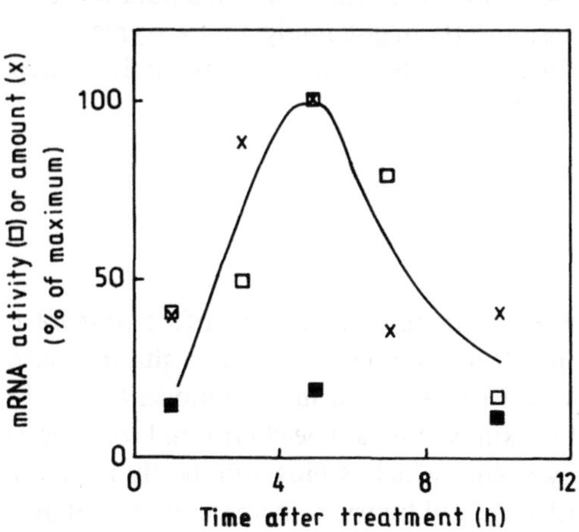

Fig. 5. Time course of chalcone synthase mRNA induction in soybean cells after treatment with *P. megasperma* glucan elicitor. Cell culture samples were taken at the indicated times after onset of treatment and used to measure translational activity in vitro (□, ■) and hybridizable amount (x) of chalcone synthase mRNA; (■) translational activity of untreated cells

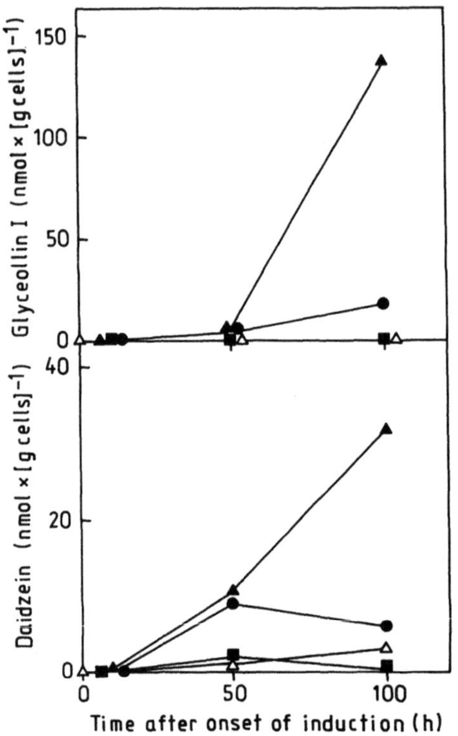

Fig. 6. Increases in the amounts of glyceollin I and daidzein after treatment of soybean cells with glucan elicitor (▲), xanthan (■), or endo-polygalacturonase (●) for various times. Amounts of the compounds are expressed as nmol × [g^{-1} fresh wt. of cells]; (△) amounts of the compounds in untreated cells

other microbial compounds, xanthan, an extracellular polysaccharide from *Xanthomonas campestris*, and endopolygalacturonase from *Aspergillus niger*. As in the induction with the *P. megasperma* glucan elicitor, the changes in enzyme activities can be accounted for by corresponding changes in the rates of de novo synthesis of the enzymes [11]. However, whereas the glucan elicitor induced the accumulation of large amounts of glyceollin in soybean cells, endopolygalacturonase induced only low, albeit significant amounts, but xanthan did not enhance glyceollin accumulation under the conditions used (Fig. 6) [11]. A similar pattern was observed for daidzein (Fig. 6), a putative intermediate in glyceollin biosynthesis [7]. These results might imply that enzymes other than phenylalanine ammonia-lyase and chalcone synthase catalyze rate-limiting steps of glyceollin biosynthesis in soybean cells [11].

5 Conclusions

One type of active defence response of plants to attempted microbial infection involves the production of phytoalexins. Suspension-cultured soybean cells and a glucan elicitor from *P. megasperma*, a soybean pathogen, can be used as a suitable model system for studies of the mechanisms controlling phytoalexin synthesis. The elicitor-induced formation of the pterocarpanoid phytoalexin, glyceollin, which is biosynthetically related to the group of isoflavonoids, is correlated with large and transient increases in the catalytic activities of several enzymes of the biosynthetic pathway. A role of Ca^{2+} ions in the initial elicitor-plant cell interaction is indicated because the presence of Ca^{2+} ions in the culture medium of soybean cells enhances the elicitor-mediated enzyme induction.

The transient increases in the catalytic activities of at least some of the enzymes are accounted for by their de novo synthesis. Since we have shown that the induced changes in chalcone synthase mRNA activity coincide with changes in amount of mRNA we suggest that phytoalexin synthesis in soybean cells is controlled at the level of gene transcription. The entire biosynthetic pathway of the soybean phytoalexins involves probably more than ten enzymatic steps. Its regulation may be exerted by an interplay of a number of the enzymes.

Acknowledgements. This work was supported by Deutsche Forschungsgemeinschaft (SFB 206) and by BASF (Ludwigshafen). The authors thank Dr. D. Marmé, Freiburg, for valuable help with Ca^{2+} uptake studies and Dr. P. Albersheim, Boulder, for a sample of partially purified *P. megasperma* glucan elicitor.

References

1. Albersheim P, Valent BS (1978) Host-pathogen interactions in plants. Plants, when exposed to oligosaccharides of fungal origin, defend themselves by accumulating antibiotics. J Cell Biol 78:627–643
2. Ayers AR, Ebel J, Finelli F, Berger N, Albersheim P (1976) Host-pathogen interactions. IX. Quantitative assays of elicitor activity and characterization of the elicitor present in the extracellular medium of cultures of *Phytophthora megasperma* var *sojae*. Plant Physiol 57:751–759
3. Ayers AR, Ebel J, Valent B, Albersheim P (1976) Host-pathogen interactions. X. Fractionation and biological activity of an elicitor isolated from the mycelial walls of *Phytophthora megasperma* var *sojae*. Plant Physiol 57:760–765
4. Ayers AR, Valent B, Ebel J, Albersheim P (1976) Host-pathogen interactions. XI. Composition and structure of wall-released elicitor fractions. Plant Physiol 57:766–774
5. Burden RS, Bailey JA (1975) Structure of the phytoalexin from soybean. Phytochemistry 14: 1389–1390
6. Darvill AG, Albersheim P (1984) Phytoalexins and their elicitors – A defense against microbial infection in plants. Annu Rev Plant Physiol 35:243–275
7. Dewick PM (1982) Isoflavonoids. In: Harborne JB, Mabry TJ (eds) The flavonoids: Advances in research. Chapman and Hall, London, pp 535–640
8. Dieter P, Marmé D (1980) Ca^{2+} transport in mitochondrial and microsomal fractions from higher plants. Planta 150:1–8
9. Ebel J (1984) Induction of phytoalexin synthesis in plants following microbial infection or treatment with elicitors. In: Ory RL, Rittig FR (eds) Bioregulators, chemistry and uses, ACS Symp Ser, vol 257. Am Chem Soc, Washington DC, pp 257–271
10. Ebel J, Ayers AR, Albersheim P (1976) Host-pathogen interactions. XII. Response of suspension-cultured soybean cells to the elicitor isolated from *Phytophthora megasperma* var *sojae*, a fungal pathogen of soybeans. Plant Physiol 57:775–779
11. Ebel J, Schmidt WE, Loyal R (1984) Phytoalexin synthesis in soybean cells: Elicitor induction of phenylalanine ammonia-lyase and chalcone synthase mRNAs and correlation with phytoalexin accumulation. Arch Biochem Biophys 232:240–248
12. Grisebach H, Ebel J (1978) Phytoalexins, chemical defense substances of higher plants? Angew Chem 17:635–657 (Int Ed Engl)
13. Hahn MG, Grisebach H (1983) Cyclic AMP is not involved as a second messenger in the response of soybean to infection by *Phytophthora megasperma* f sp *glycinea*. Z Naturforsch 38c: 578–582

14. Hahn MG, Bonhoff A, Grisebach H (1985) Quantitative localization of the phytoalexin glyceollin I in relation to fungal hyphae in soybean roots infected with *Photophthora megasperma* f sp *glycinea*. Plant Physiol 77:591–601
15. Hille A, Purwin C, Ebel J (1982) Induction of enzymes of phytoalexin synthesis in cultured soybean cells by an elicitor from *Phytophthora megasperma* f sp *glycinea*. Plant Cell Rep 1: 123–127
16. Horsfall JC, Cowling EB (eds) (1980) Plant disease, vol V. Academic Press, London New York
17. Kreuzaler F, Ragg H, Fautz E, Kuhn DN, Hahlbrock K (1983) UV-induction of chalcone synthase mRNA in cell suspension cultures of *Petroselinum hortense*. Proc Natl Acad Sci USA 80:2591–2593
18. Leube J, Grisebach H (1983) Further studies on induction of enzymes of phytoalexin synthesis in soybean and cultured soybean cells. Z Naturforsch 38c:730–735
19. Lyne RL, Mulheirn LJ (1978) Minor pterocarpinoids of soybean. Tetrahedron Lett 3127–3128
20. Lyne RL, Mulheirn LJ, Leworthy DP (1976) New pterocarpinoid phytoalexins of soybean. J Chem Soc Chem Commun:497–498
21. Mansfield JW (1982) The role of phytoalexins in disease resistance. In: Bailey JA, Mansfield JW (eds) Phytoalexins. Blackie, Glasgow, pp 253–288
22. Moesta P, Grisebach H (1981) Investigation of the mechanism of phytoalexin accumulation in soybean induced by glucan or mercuric chloride. Arch Biochem Biophys 211:39–41
23. Moesta P, Grisebach H (1981) Investigation of the mechanism of glyceollin accumulation in soybean infected by *Phytophthora megasperma* f sp *glycinea*. Arch Biochem Biophys 212: 462–467
24. Schmelzer E, Börner H, Grisebach H, Ebel J, Hahlbrock K (1984) Phytoalexin synthesis in soybean *(Glycine max)*. Similar time courses of mRNA induction in hypocotyls infected with a fungal pathogen and in cell cultures treated with fungal elicitor. FEBS Lett 172:59–63
25. Sharp JK, McNeil M, Albersheim P (1984) The primary structures of one elicitor-active and seven elicitor-inactive hexa-β-D-glucopyranosyl-D-glucitols isolated from the mycelial walls of *Phytophthora megasperma* f sp *glycinea*. J Biol Chem 259:11321–11336
26. Zähringer U, Ebel J, Mulheirn LJ, Lyne RL, Grisebach H (1979) Induction of phytoalexin synthesis in soybean. Dimethylallylpyrophosphate:trihydroxypterocarpan dimethylallyl transferase from elicitor-induced cotyledons. FEBS Lett 101:90–92

Protoplast Fusion of Secondary Metabolite-Producing Cells

Y. YAMADA and H. MORIKAWA [1]

1 Introduction

Great interest in the production of such useful substances as alkaloids, steroids, vitamins, and pigments has arisen with the development of plant cell technology. Some attempts to produce metabolites on an industrial scale have been reported [17].

Secondary metabolites are produced in cultured plant cells [47]. The metabolite productivity of the cultured cells, however, is very low compared with the productivity of intact plants. Recently, there have been several reports of the successful establishment of stable cell lines that produce large amounts of secondary metabolites. For example, Yamamoto et al. [114] reported that *Euphorbia millii* cell lines produce seven times as much anthocyanin as the original calluses. Also, Yamada and Sato [80, 112] reported the establishment of *Coptis japonica* cell lines that produce more berberine alkaloid than the original plants.

Many researchers have pointed out the great usefulness of protoplast culture for plant breeding [88, 96, 101], for obtaining drug-resistant mutants and auxotrophs [6, 49, 98], and for somatic hybridization [28, 53, 83, 115]. Most of the plant genera studied so far, however, have been limited to solanaceous plants. Also the usefulness of protoplast culture for the production of secondary metabolites has yet to be deeply explored; only a few reports of successful studies of the culture of protoplast from cultured cells that produce large amounts of secondary metabolites have been presented. Protoplast technology based on the culture of protoplasts from selected high metabolite-producing cell lines undoubtedly will be an important field of plant technology.

2 Selection of High Metabolite-Producing Cell Lines

Several reports have been published of cell lines that contain large amounts of primary and secondary metabolites. They include vitamin B_6 [113], ubiquinone 10 [51], anthocyanin [114], and alkaloids [66, 80, 109, 112, 116]; these cell lines have been isolated by cell selection methods.

1 Research Centre for Cell and Tissue Culture, Faculty of Agriculture, Kyoto University, Sakyo-ku, Kyoto 606, Japan

Primary and Secondary Metabolism of Plant Cell Cultures, ed. by Neumann et al.

How then do we select cells that are stable and good producers of secondary metabolites? We must determine which of the two selection methods in use for single cell (protoplast) cloning and small cell aggregate cloning, is the more effective and efficient for this purpose.

We will focus on the results [80, 112] with selected cells of *Coptis japonica* that produce several times more berberine alkaloid than the parent plant. Berberine, a useful antibacterial and stomachic agent, is a compound whose antiinflammatory effect recently has been reported. It takes 5-6 years to produce *Coptis* roots when contain berberine, the raw material used in berberine production. Primary research [112] in the laboratory has proved that cultured *Coptis* cells can produce the same amounts of berberine as *Coptis* roots (2%-4% on a dry weight basis) within a short period (3 weeks). But for industrial purposes this amount is too little; thus, the productivity of cultured *Coptis* cells must be improved. A clue to improving berberine yields comes from fluoromicroscopy which clearly shows that individual cultured *Coptis* cells differ in their abilities to produce berberine.

Therefore, Sato and Yamada [80] selected high berberine-producing cell lines by cloning small aggregates that contained large amounts of berberine as indicated by the intensity of fluorescence present. By repeated cloning of these high berberine-producing aggregates over about ten generations, they have obtained cell lines that consistently produce large amounts of berberine and whose berberine production is stable. Using this small cell aggregate selection method they have obtained [80] one cell line whose average productivity is 8.2% (dry wt.) of the cells and 0.90 g l^{-1} of culture medium. The highest production of this cell line is 13.2% (dry wt.) and 1.39 g l^{-1} of culture medium [80]. The former value is much higher than the amount of berberine obtained from intact *Coptis* roots (ca. 5% dry wt.).

Yamada has cultured protoplasts (single cell cloning) from the selected *Coptis* cell lines that produce large amounts of berberine. Mino and Yamada recently succeeded [57] in forming callus colonies from the protoplasts of high berberine-producing *Coptis* cells and have analyzed the berberine alkaloid contents of these protoplast-derived cell lines. They found that although the berberine alkaloid spectrum of the protoplast-derived cell lines is similar to that of the parental cell line, almost all the protoplast-derived cell lines had a berberine content that was lower than the high berberine-producing parental cell line, though many protoplast-derived cell lines grow more rapidly than the parental cell line. In culturing *Coptis* protoplasts, white cells which did not produce berberine were formed from yellow cells which did, frequently resulting in mosaic colonies composed of yellow and white cells derived from a single protoplast. Interestingly, these protoplast-derived cell lines showed chromosomal variations even though each clone was derived from a single protoplast [57]. Constabel et al. [14] also have reported variations in metabolic production in cell lines derived from protoplasts of a single leaf of periwinkle *(Catharanthus roseus)*.

Higher plants are multicellular organisms that consist of differentiated cells (tissues). Special differentiated cells have the potential to produce specific substances in quantity and in a stable manner. In this regard, higher plants differ from single cell organisms such as bacteria for which single cell cloning is very effective for obtaining a specific character cell. For the differentiation of secondary metabolites, we must, therefore, use a method for selecting cells that have a specific character that differs from the selection method used with single cell organisms.

Protoplast culture technique undoubtedly has importance for the production of secondary metabolites. The productivity of metabolites produced by cell culture is a function of the amounts of the metabolite in the cells and of the growth rate of the cells. Using protoplast culture, it is possible to obtain cell lines that grow very rapidly. Interestingly, intraspecific protoplast fusion of a high metabolite-producing cell line and a rapidly growing cell line produces a "plant hybridoma".

3 Protoplast Isolation

The enzymatic isolation of plant protoplasts was developed by Cocking [10] in 1960, who succeeded in isolating protoplasts from tomato roots using a culture filtrate of the fungus *Myrothcium verrucaria*. His enzyme method was further developed by Takebe et al. [90] who isolated large numbers of viable protoplasts from tobacco leaves using cell-wall-degrading enzymes produced by the Japanese Fermentation Industry.

During the last decade, a number of cell-wall-degrading enzyme preparations have become available commercially. All these preparations are crude mixtures of various hydrolyzing enzymes. The most commonly used enzyme preparations are Cellulase Onozuka R-10 and Cellulase RS from *Trichoderma viride*, which is rich in cellulase, Driselase from a basidiomycete rich in cellulase and hemicellulase, Macerozyme R-10 from *Rhizopus* sp., which is rich in pectinase and hemicellulase, Pectolyase Y-23 from *Aspergillus japonicus*, which is rich in pectinase, and other pectinases and hemicellulases [21, 95].

The plant cell wall is a composite material in which crystalline cellulose microfibrils are embedded in matrix substances composed of pectic substances, hemicelluloses, and proteins. The average composition of the primary cell walls is 30% cellulose, 30% pectic substances, 30% hemicellulose, and 10% proteins [77]. These cell wall polysaccharides and proteins are organized in a nascent morphology formed during the biosynthesis of polymers [77]. Their detailed chemical structures are not known. Consequently, when isolating protoplasts it is necessary to use a mixture of crude enzymes rather than purified enzymes.

Protoplasts have now been isolated from various plant tissues [21, 95] that include leaves, roots, petioles, coleoptiles, pods, and nodules, as well as from cultured cells in vitro. At present, however, the isolation of protoplasts from certain plant tissues remains difficult. For example, protoplasts have not been isolated from the mesophyll tissue of sweet potato although a number of viable free cells have been isolated from this tissue [4]. Our research has shown that the yield of viable protoplasts from cultured photoautotrophic or photomixotrophic tobacco (*Nicotiana tabacum* var. Samsun) cells is still very low with any enzyme combination, osmotic variation or pH range. Actually, large amounts of protoplasts have been obtained from photomixotrophic cells using a combination of Driselase and Pectolyase Y-23; but these protoplasts died within 1 h during the enzyme treatment. We have partially purified the crude enzyme preparations with Sephadex G-25, but obtained only poor results in photomixotrophic cells. The addition of 5% bovine serum albumin to the enzyme solution [4] greatly increased the yield of viable protoplasts from photomixotrophic tobacco cells.

4 Culture of Protoplasts

So far, the successful culture of protoplasts mainly has been limited to those from solanaceous plants, particularly the mesophyll cells of these plants. For many species, the protoplasts isolated from cultured cells do not divide easily and form calluses. The current techniques used for the culture of protoplasts and for obtaining protoplast fusion products are reviewed.

4.1 Plating on Agar Medium

Protoplasts suspended in a liquid medium at a density of 10^4 ml^{-1} are mixed gently, but quickly, with an equal volume of medium prepared in agar (1.2%) which has been kept at 45 °C. Small volumes (5 ml per Petri dish), of the protoplast-containing agar medium are dispensed into Petri dishes (6 cm in diameter), which then are sealed with Parafilm and incubated. This method has been used to culture a variety of protoplasts from cells of tobacco [12, 18, 63] and petunia [12, 23] leaves, from *Vigna* stem and hypocotyl [3] and from callus tissues derived from *Citrus* ovules [94] and *Lycoperisicon* petioles [58].

This method has also been used to culture the fusion products of protoplasts of *Petunia hybrida* and *P. parodii* [11] and of *Arabidopsis thaliana* and *Brassica campestris* [35]. The advantage of this technique is that the protoplasts remain in a fixed position and the development of individual protoplasts can be followed with ease.

4.2 Liquid Culture

Protoplasts suspended in a liquid medium at a density of 10^4 to 10^5 ml^{-1} are poured directly into a petri dish (1 ml per Petri dish, 35 mm in diameter) or onto a layer of agar medium of the same composition as the liquid medium. The former method is referred to as the liquid culture method and the latter as the liquid-over-agar culture method.

The liquid culture method has been used in the culture of protoplasts from mesophyll cells of petunia [33], cucumber [15], tobacco [64], haploid *Hyoscyamus muticus* [102], *Solanum melongena* [81], and haploid *Brassica napus* [45], and those from suspension cells of *Atropa belladonna* [30], *Sorghum bicolor* [5], and *Trifolium arvense* [103]. This method also has been used to culture the fusion products of Solanaceous plants [32, 79, 82, 83, 93], Cruciferous plants [68], and rapeseed and soybean [43].

The liquid-over-agar culture method has been used to culture protoplasts from mesophyll cells of *N. tabacum* [69] and *Rehmania glutinosa* [105], and protoplasts from root cells of *Phaseolus aureus* [106], also for the culture of fusion products of *P. hybrida* and *P. parodii* [76].

Recently Maeda et al. [48] reported successful callus formation from protoplasts of cultured *Lythospermum erythrorhizon* cells that produce large amounts of shikonin. They showed that factors affecting protoplast division are the use of glucose as the osmoticum, the liquid-over-agar plating method and the culture of protoplasts at an osmolarity lower than that of the isolation solution. They also showed [48] that when the mannitol or sorbitol in the protoplast-culture medium is replaced with glucose, and coconut milk is added to the medium, the frequency of colony formation increases markedly (more than 1%).

In the liquid-over-agar culture the agar medium serves as a reservoir of nutrients, and the liquid layer should be kept thin (ca. 1 mm thick) to facilitate aeration. Usually cell colonies formed in the liquid medium are transferred to a soft agar (0.4%) medium or to an agar medium, after which they are cultured.

4.3 Droplet Culture

Protoplasts suspended in liquid medium at a density of 10^4 to 10^5 ml^{-1} are transferred in 50–200 μl droplets to plastic Petri dishes, microculture chambers or Cuprak dishes, that then are sealed and incubated. The development of individual protoplasts is followed through an inverted microscope. This method has been used for the culture of protoplasts from suspension cells of soybean [39], *Vicia hajastana* [41], carrot [20], haploid *Datura innoxia* [24] and *N. tabacum* [25], and for the culture of protoplasts from mesophyll cells of *P. hybrida* [71], *N. tabacum* [26], *Pisum sativum* [13], and *D. innoxia* [46]. It also has been used to culture the fusion products of soybean and *N. glauca* [38], *Arabidopsis thaliana* and *Brassica campestris* [27], *N. rustica* and *N. tabacum* [19], and *Atropa belladonna* and *N. chinensis* [29].

4.4 Feeder Layers and Nurse Culture

Nondividing but metabolically active, X-ray irradiated protoplasts or suspension cells embedded in agar medium support the growth of protoplasts from the same source plated above them at a density as low as 5–50 ml^{-1} [8, 78].

This cross-feeding effect also has been reported between protoplasts of different species. Nurse cultures have been used by Menczel et al. [54] to clone somatic hybrids from *Nicotiana* heterokaryons.

We also have been using a nurse culture chamber to culture heterokaryons (Fig. 1): a ring made of nylon mesh (8 mm in diameter), is placed in the center of a plastic petri dish (35 mm in diameter) then liquid medium (1 ml) of the same composition as the agar medium is added to the dish [110]. Heterokaryons are picked up from the site of cell fusion by micro-pipette (Fig. 2) and transferred to the area inside the nylon-mesh

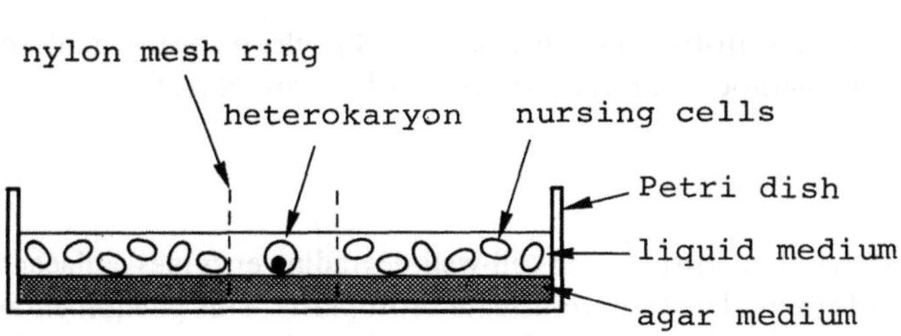

Fig. 1. A microculture chamber used for the culture of heterokaryons with nursing cells (see text for details)

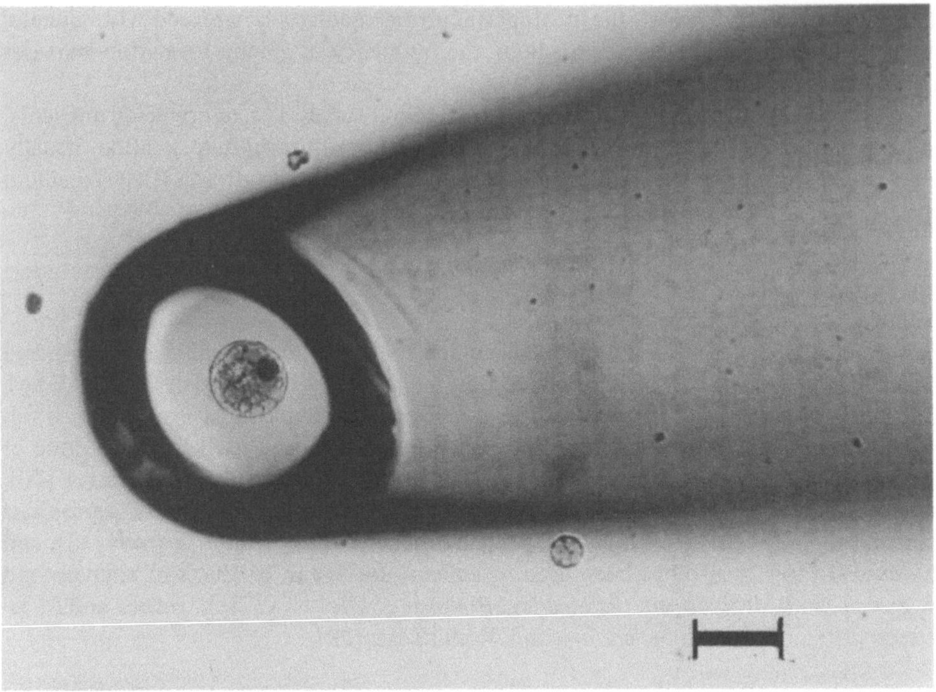

Fig. 2. A micropipette picking up a heterokaryon cell formed from *C. japonica* and *E. millii* protoplasts. *Bar* 50 μm

ring, after which nursing cells are added to the area of the dish around the ring. The dish then is sealed and incubated. The advantages of this chamber are that one can (1) culture heterokaryons at a density as low as one to ten heterokaryons per milliliter, (2) easily change culture conditions such as the basal medium, osmotic pressure of the medium and nursing cells, and (3) follow the process of growth and cell division of individual heterokaryons in the ring within the chamber.

5 Fusion of Protoplasts

The general methods for protoplast fusion and a discussion of the electrofusion method are given below then the PEG method and electrical method for protoplast fusion.

5.1 Spontaneous Fusion

During the treatment of some plant tissues with cell-wall-degrading enzymes, adjacent cells within the tissue fuse forming large multinucleate protoplasts. This phenomenon has been termed spontaneous fusion and is found in a number of tissues that include soybean callus [56], *Datura* and *Papaver* callus [111], oat root tip [104] and tobacco leaf [73, 104].

Withers and Cocking [104] have presented electron microscopic evidence that, in the case of oat root and tobacco leaf, spontaneous fusion is caused by the symplastic nature of plant tissues, i.e. by the plasmodesmatal connections in the tissues. Ito [36], and Ito and Maeda [37] have reported that cell fusion in meiotic protoplasts of *Lilium longiflorum* and *Trillium kamtschaticum* is induced by physical contact between the protoplasts during isolation. This phenomenon, termed spontaneous fusion, differs from the spontaneous fusion described earlier because no plasmodesmatal connections are involved in the fusion of meiotic protoplasts. Up to 90% of these meiotic protoplasts are bi- or multinuclear. The ability of the protoplasts to fuse is quickly lost (within 30 min of isolation). Using the method, Ito and Meada [37] obtained about 11% interspecific hybrid protoplasts from meiotic protoplasts of *L. longiflorum* and *T. kamtschaticum*.

Usually, direct contact of isolated protoplasts is not sufficient to produce fusion, treatment with a fusogen is needed. But, protoplasts isolated from cells cultured at a low temperature, whose cell membranes are rich in phospholipids with a low phase transition point, readily fuse on simple physical contact or easily fuse when treated with a fusogen [108].

5.2 Fusion Produced by High Ionic Concentrations

Power et al. [74] succeeded in the intra- and interspecific fusion of isolated protoplasts of oat and maize roots by treating them with 0.25 M sodium nitrate. Carlson et al. [7] have reported the successful parasexual production of a hybrid between *N. glauca* and *N. langsdorffii* by this sodium-nitrate technique. This has been confirmed by Smith et al. [89] and by Chupeau et al. [9].

During their study of membrane fusion by Sendai virus, Toister and Loyter [92] found that avian erythrocytes fused when there was a high concentration of $CaCl_2$, a high pH and a high temperature, i.e. the addition of 40 mM $CaCl_2$ caused agglutination of the erythrocytes at pH 10.5 in the cold, and fusion occurred when the agglutinated cells were transferred to 37 °C. Using this method Melchers and his co-workers [44, 52] produced somatic hybrids from mesophyll protoplasts of haploids of two chlorophyll-deficient, light-sensitive varieties of *N. tabacum*. Recently Power et al. [72] reported the successful somatic hybridization of sexually incompatible petunias, *P. parodii* and *P. parniflora*, by the same method.

The surface of the protoplast membrane is negatively charged [31, 59, 62]. High concentrations of cations such as Na^+ and Ca^{2+} reduce the negative charges on the surface of the protoplasts membrane, and this may facilitate agglutination and the fusion of protoplasts. Ca^{2+} ion is known to be adsorbed on the protoplast surface membrane which decreases the surface charge [62]. Ca^{2+} ion also induces a phase change in lipid membranes that is considered an important event in cell fusion. A high pH value induces the lysis of phospholipids in the membrane which results in the formation of lyso compounds such lysolecitin [44, 92]. Furthermore, a high temperature increases the fluidity of the membrane.

5.3 Fusion Produced by High Polymers

Kao and Michayluk [40] and Wallin et al. [99] have found that protoplast fusion is induced by treatment with high concentrations (40%–50%) of polyethylene glycol (PEG) (MW 1,500–8,000). This PEG method is the most commonly used technique for fusion of plant protoplasts, mammalian cells and fungal protoplasts, as well as for the interkingdom fusion of plant and animal cells [75, 84]. There are reports of a number of somatic hybrid plants [22, 75, 84] obtained by this fusion method, but the genera that undergo successful somatic hybridization are still limited.

It has been suggested [40] that PEG acts as a molecular bridge between adjacent membranes, either by hydrogen bonding or indirectly together with Ca^{2+} ion. PEG also has been reported to decrease the surface potential of lipid monolayers and to have a dehydration effect [75].

Polyvinyl alcohol (D.P. = ca. 500) also induces protoplast fusion [61, 85]. Kameya et al. have reported that dextran with a high molecular weight (ca. 5×10^5) also is fusogenic [42]. They obtained hybrids of *Daucus* plants using dextran.

5.4 Temperature Effects

Raising the temperature in fusion experiments facilitates the fusion of animal cells [92] and the fusion of plant protoplasts [44]. Based on the kinetic analysis of the PEG-induced cell fusion process with *Rauwolfia serpentina* protoplasts, the values for the activation energy for the fusion process are comparable to the flow activation energy (7–15 kcal mol^{-1}) of the liposomes and biological membranes [85].

Cell fusion is considered [1, 16] to involve the (a) aggregation of membrane proteins which results in the exposure of areas of (protein free) lipid bilayers, (b) close contact of the exposed bilayers, and (c) phase changes in the two outer monolayers of adjacent lipid bilayers. The action of fusogenic compounds such as PEG may be involved in stage (a) in this fusion mechanism. The presence of Ca^{2+} and the increase in temperature probably are effective at each stage.

On the basis of the chemical analysis of the distribution of phospholipid types and of the phospholipid molecular species in the cell membranes of various cultured cells, Yamada et al. [107] have suggested that membrane fluidity, as determined by the phase transition of phospholipids, has an important function in the cell fusion process. Interestingly, there is a difference in the relative distribution of the phospholipid molecular species in cells cultured at different temperatures (10° and 25 °C), and protoplasts whose membranes have a large number of unsaturated fatty acids show increased speed of the protoplast fusion induced by PEG [108].

5.5 Electrofusion

Protoplast fusion by electrical stimulation, the socalled electrofusion method, was first developed by Senda et al. [87] in 1979. They induced fusion of plant protoplasts by electrical stimulation applied by microelectrodes. Neumann et al. [65] reported the electrofusion of mold cells and Weber et al. [100] of yeast protoplasts in the presence of PEG; both used pairs of metal electrodes (parallel electrodes). Zimmermann and his co-workers [97, 117-119] introduced an improvement for high fusion frequency electrofusion by combining the electrofusion method and Pohl's dielectrophoresis method [70]. At present, there are three types of electrofusion methods; the microelectrode method [86, 87], the parallel electrode method [65, 86, 91, 100], and the dielectrophoresis chamber method [97, 117-119].

5.5.1 Microelectrode Method [86, 87]

In the microelectrode method the tips of two glass capillary microelectrodes are brought close to the surface of protoplasts adhering at two points, and a DC pulse is applied acrossed the two electrodes. Usually, the amplitude of the DC pulse is in the range of 1-50 μA and the duration of this pulse is between 0.25-50 ms when protoplasts are suspended in 0.6 M sorbitol or mannitol containing 1-10 mM $CaCl_2$, with or without 3 mM MES buffer salts.

The advantage of this method is that protoplast fusion is induced from one to the other of a pair of protoplasts and the fusion process can be followed in detail. Direct contact of the microelectrode tips on the surface of the protoplasts, so often described by other researchers, however, is not needed to induce fusion.

5.5.2 Parallel Electrode Method [86]

A pair of platinum electrodes are aligned parallel to each other and glued onto a glass plate so that the distance between the electrodes is ca. 0.5-2 mm. Both ends of the electrodes are sealed with pieces of glass. The electrodes are connected through an electric switch to a capacitor. A mixture of protoplasts suspended in solution as described above (see Sect. 5.5.1) is placed between the electrodes in the fusion chamber. The capacitor first is charaged to the desired voltage; then, by resetting the position of the switch, it is discharged through the protoplast suspension. Usually the voltage of the capacitor is changed from 500 to 1,000 V, the capacity being from 0.01 to 100 μF, and the resistance of the protoplast suspension in the fusion chamber from 0.5 to 2 kΩ.

This method produces large numbers of viable, electrically fused cells by one electrical stimulation. One of its advantages is that almost no direct contact occurs between the protoplasts and electrodes during the induction of fusion. This is in sharp contrast to the method that uses the dielectrophoresis chamber. We know from experience that upon electrical stimulation the surface of the protoplast becomes very "sticky" and that direct contact of the protoplast surface with the electrode results in serious damage to the protoplast.

5.5.3 Dielectrophoresis Chamber Method [97, 117-119]

The fusion chamber and electric set-up are constructed in a manner similar to that used in the parallel electrode method, but the electrodes are connected to an oscillator which provides AC fields and to a pulse generator which provides the DC pulse. Protoplasts first are subjected to a high-frequency (500 kHz) AC field so that individual protoplasts can be collected and aligned to form a "pearl chain" on the surface of the electrodes. This phenomenon, called dielectrophoresis, first was found by Pohl [70, 117]. A DC pulse is applied to the aligned "protoplast chains" to induce electrofusion.

One advantage of this method is that it induces a number of point-adhering protoplasts which results in a very high fusion frequency (50%–80%). Vienken and Zimmermann [97] reported a new technique for producing high yields of a desired combination of two heterokaryon cells. Recently, Bates et al. [2] reported that mesophyll protoplasts that have been subjected to electric fields in the dielectrophoresis chamber are still capable of growth and cell division. We have succeeded in forming callus colonies from a mixture of *C. japonica* and *E. millii* protoplasts that was subjected to electrofusion by the parallel electrode method.

6 Comparison of the PEG and Electrical Methods of Fusion

Heteroplasmic fusion of protoplasts from high berberine-producing *C. japonica* and of high anthocyanin producing *E. millii* cell lines (determined by characteristic cytoplasmic marker for heterokaryons) is induced at almost the same fusion frequency (0.5%–3% of the total protoplasts) by both the PEG and parallel electrode methods [110].

About half of the heterokaryons formed from *C. japonica* and *E. millii* usually die [110] during the first night after fusion treatment with these two methods. We succeeded in forming callus colonies from a mixture of protoplasts from *C. japonica* and *E. millii* that had undergone fusion treatment with the PEG or parallel electrode method. In the colonies formed, some cells are red (anthocyanin-producing), others yellow (berberine-producing) [110]. We have separated a number of cell lines from these colonies and are currently analyzing the metabolites in the cells to determine whether somatic hybrid cells are present.

Protoplasts treated with the PEG technique often adhere to each other or to the bottom of the Petri dish, whereas electrically stimulated protoplasts do not. This non-adhering nature of the electrically fused protoplasts is of advantage in the mechanical isolation of fused cells by micropipette.

7 Selection of Fused Hybrid Cells

There is no general method applicable to the separation of fused hybrid cells from unfused cells. The selection methods for fused hybrid cells in use at present can be divided into three groups: (1) selection before culture of the fusion products, (2) selection after callus formation from fusion products, and (3) selection after regeneration of plants from the callus formed from fusion products. The first method employs the direct, mechanical pick-up of fused heterokaryons with a micropipette operated by a micromanipulator (pick-up method). The latter two methods are based on the use of various genetic markers, physiological growth response and/or drug resistance.

7.1 Selection by Genetic and Biochemical Markers

The first somatic hybrid plants formed between *N. glauca* and *N. langsdorffii* were recovered through selection for auxin autotrophy [7]. Power et al. [76] used different responses of *P. hybrida* and *P. parodii* to growth medium and to actinomycin D to select interspecific hybrid plants. Zelcer et al. [115] reported the successful selection of fusion products based on suppression of cell division by X-ray irradiation of cytoplasmic sterile male *N. tabacum* protoplasts and by the use of a mannitol medium which is unfavorable to the growth of *N. sylvestris* protoplasts.

Correction of albino lesions between complementary chlorophyll-deficient mutants is the most widely used technique for the selective identification of somatic hybrids from species in which plant regeneration from protoplasts is possible; these include *Nicotiana* [19, 52], *Datura* [82, 83], and *Petunia* [11].

Mutants induced in vitro also have been used for the selection of somatic hybrids. The kanamycin resistance of a cell line of *N. sylvestris*, KR103, that had been isolated from cultured cells, was used as a genetic marker to recover hybrid cells of *N. sylvestris* and *N. knightiana* [50]. A streptomycin-resistant mutant of *N. tabacum* SR1 also has been used to isolate several hybrids formed from *N. tabacum* and *N. knightiana* [55].

7.2 Pick-Up Method

In contrast to the cell selection by genetic and biochemical markers described above, the pick-up, or mechanical isolation, method is applicable to any type of protoplast combination. Undoubtedly, this method will find wider use as microculture techniques for single protoplasts develop further.

Kao [38] has used a disposable micropipette to isolate small samples (25 μl) of hybrids of soybean-*Nicotiana glauca* and transfer them to the wells of Cuprak dishes for culture. Heterokaryons in the wells are distinguished from the parental protoplasts because they contain green plastids, a characteristic of *N. glauca* protoplasts; but these heterokaryons resemble soybean morphologically in that they have dense cytoplasmic strands. Using green plastids and dense cytoplasmic strands as markers for heterokaryons, other researchers have mechanically isolated heterokaryons and cultured them: these include heterokaryons of *N. knightiana* and *N. sylvestris* which, after mechanical isolation, were transferred to nurse cultures of albino cells [54]; those of *Brassica campestris* and *Arabidopsis thaliana* which were cultured in microdroplets of ca. 1 μl [27]; those of *N. tabacum* and *P. hybrida* which were cultured in microdroplets [67]; and those of *N. paniculata* and *N. tabacum* cnx-68 and of *N. sylvestris* and *N. tabacum* cnx-68 which were incubated in co-culture with their auxotrophic parental protoplasts [34].

During our cell fusion study of high berberine-producing *C. japonica* and high anthocyanine-producing *E. millii* cells, we found dark red "droplets" surrounded by a membraneous structure formed in the heteroplasmically fused protoplasts induced by the PEG and electric methods [110]. Using these droplets as a marker, we picked up the heterokaryons and transferred them to a nurse culture in order to continue their culture. As a result, we observed that one heterokaryon cell divided to become ten new cells.

8 Conclusions

Single cell cloning by means of protoplast culture often has been postulated as the best way to isolate genetically homogeneous cell clones. There is evidence, however,that protoplast cloning produces variations both in genotype (chromosomal number) and in phenotype (alkaloid production) in cultured *C. japonica* cells. Why do these protoplast-derived clones often show a heterogeneous character? Probably because of the "singleness" of the isolated protoplasts; liberation of individual cells from the organized structure of the cell aggregates in which they interacted with neighboring cells. The reason for the variations found in protoplast culture undoubtedly is related to the inter-recognition mechanism that functions in the multicellular tissues of eukaryotes. Recently, Murphy, who based his studies on the Luria-Delbruck test for the trait of chlorite resistance in cultured rose cells, reported that the organization of plant cells into cell aggregates modifies the expected distribution of mutations in clones [60].

The instability of protoplast culture is very useful for metabolite production, because increased variability is a prerequisite for breeding. For example, Shepard et al. [88] reported that among regenerants from leaf protoplasts of the commercial tetraploid potato, there is marked variation in several agronomically useful characteristics, including disease resistance.

The introduction of foreign genetic information into protoplasts by somatic cell fusion and by micro-injection is important for both plant breeding and for the production of secondary metabolites for commercial use. Most studies of somatic hybridization, however, so far have been limited to several model plants [22]; therefore, before new types of cells that can be used for metabolite production can be obtained by cell fusion techniques, several difficulties must be overcome – particularly those of the selection and culture of fusion products as described above.

The importance of the cultured cells produced by cell fusion techniques is that:

1. If a cell line can be established from hybrid plant cells obtained from protoplast fusion, the hybrid cells may not only be able to produce the two different species of metabolites of the parent cells simultaneously, but may also yield one or more entirely new substances. The productivity of metabolites in the hybrid cell also may be increased by the conjunction of intermediates in the biosynthetic pathways.
2. If homo- or heterokaryotic cells can be obtained from the fusion of high metabolite-producing cells with rapidly dividing cells, metabolite production by "plant hybridoma" has the potential to become a very efficient experimental and industrial method.

References

1. Ahkong QF, Fisher D, Tampion W, Lucy JA (1975) Mechanism of cell fusion. Nature (London) 253:194–195
2. Bates GW, Gaynor JJ, Shekhawat NS (1983) Fusion of plant protoplasts by electric fields. Plant Physiol 72:1110–1113
3. Bharal S, Rashid A (1980) Isolation of protoplasts from stem and hypocotyl of the legume *Vigna sinensis* and some factors affecting their regeneration. Protoplasma 102:307–313

4. Bidney DL, Shepard JF (1980) Colony development from sweet potato petiole protoplasts and mesophyll cells. Plant Sci Lett 18:335–342
5. Brar DS, Rambold S, Constabel F, Gamborg OL (1980) Isolation, fusion and culture of *Sorghum* and corn protoplasts. Z Pflanzenphysiol 96:269–275
6. Carlson PS (1970) Induction and isolation of auxotrophic mutants in somatic cell cultures of *Nicotiana tabacum*. Science 168:487–489
7. Carlson PS, Smith HH, Dearing RD (1972) Parasexual interspecific plant hybridization. Proc Natl Acad Sci USA 69:2292–2294
8. Cella R, Galun E (1980) Utilization of irradiated carrot cell suspensions as feeder layer for cultured *Nicotiana* cells and protoplasts. Plant Sci Lett 19:243–252
9. Chupeau Y, Missonier C, Hommel M-C, Goujaud J (1978) Somatic hybrids of plants by fusion of protoplasts. Observations on the model system *Nicotiana glauca – Nicotiana langsdorffii*. Mol Gen Genet 165:239–245
10. Cocking EC (1960) A method for the isolation of plant protoplasts and vacuoles. Nature (London) 187:962–963
11. Cocking EC, George D, Price-Jones MJ, Power JB (1977) Selection procedures for the production of inter-species somatic hybrids of *Petunia hybrida* and *Petunia parodii*. II. Albino complementation selection. Plant Sci Lett 10:7–12
12. Cocking EC, Power JB, Evans PK, Safwat F, Frearson EM, Hayward C, Berry SF, George D (1974) Naturally occurring differential drug sensitivities of cultured plant protoplasts. Plant Sci Lett 3:341–350
13. Constabel F, Kirkpatrick JW, Gamborg OL (1973) Callus formation from mesophyll protoplasts of *Pisum sativum*. Can J Bot 51:2105–2106
14. Constabel F, Rambold S, Chatson KB, Kurz WGW, Kutney JP (1981) Alkaloid production in *Catharanthus roseus* (L.) G. Don. VI. Variation in alkaloid spectra of cell lines dericed from one single leaf. Plant Cell Rep 1:3–5
15. Coutts RHA, Wood KR (1975) The isolation and culture of cucumber mesophyll protoplasts. Plant Sci Lett 4:189–193
16. Cullis PR, Hope MJ (1978) Effects of fusogenic agent on membrane structure of erythrocyte ghosts and the mechanism of membrane fusion. Nature (London) 271:672–674
17. Curtin ME (1983) Harvesting profitable products from plant tissue culture. Bio/technology 1:650–657
18. Davey MR, Frearson EM, Withers LA, Power JB (1974) Observations on the morphology, ultrastructure and regeneration of tobacco leaf epidermal protoplasts. Plant Sci Lett 2:23–27
19. Douglas GC, Keller WA, Setterfield G (1981) Somatic hybridization between *Nicotiana rustica* and *N. tabacum*. I. Isolation and culture of protoplasts and regeneration of plants from cell cultures of wild-type and chlorophyll-deficient strains. Can J Bot 59:208–219
20. Dudits D, Kao KN, Constabel F, Gamborg OL (1976) Embryogenesis and formation of tetraploid and hexaploid plants from carrot protoplasts. Can J Bot 54:1063–1067
21. Eriksson T (1977) Technical advances in protoplast isolation and cultivation. In: Barz W, Reinhard E, Zenk MH (eds) Plant tissue culture and its biotechnological application. Springer, Berlin Heidelberg New York, pp 313–322
22. Evans DA (1983) Agricultural applications of plant protoplast fusion. Bio/technology 1: 253–261
23. Frearson EM, Power JB, Cocking EC (1973) The isolation, culture and regeneration of *Petunia* leaf protoplasts. Dev Biol 33:130–137
24. Furner IJ, King J, Gamborg OL (1978) Plant regeneration from protoplasts isolated from a predominantly haploid suspension culture of *Datura innoxia* (Mill.). Plant Sci Lett 11:169–176
25. Gamborg OL, Shyluk JP, Fowke LC, Wetter LR, Evans D (1979) Plant regeneration from protoplasts and cell cultures of *Nicotiana tabacum* sulfur mutant (Su/su). Z Pflanzenphysiol 95: 255–264
26. Gleba YY (1978) Microdroplet culture. Tobacco plants from single mesophyll protoplasts. Naturwissenschaften 65:158–159
27. Gleba YY, Hoffmann F (1978) Hybrid cell lines *Arabidopsis thaliana + Brassica campestris*: no evidence for specific chromosome elimination. Mol Gen Genet 165:257–264

28. Gleba YY, Hoffmann F (1979) "Arabiobrassica": plant-genome engineering by protoplast fusion. Naturwissenschaften 66:547–554
29. Gleba YY, Momot VP, Cherep NN, Skarzynskaya MV (1982) Intertribal hybrid cell lines of *Atropa belladonna* (x) *Nicotiana chinensis* obtained by cloning individual protoplast fusion products. Theor Appl Genet 62:75–79
30. Gosch G, Bajaj YPS, Reinert J (1975) Isolation, culture, and induction of embryogenesis in protoplasts from cell-suspensions of *Atropa belladonna*. Protoplasma 86:405–410
31. Grout BWW, Willison JHM, Cocking EC (1973) Interactions at the surface of plant cell protoplasts: an electrophoretic and freeze-etch study. J Bioenerg 4:311–328
32. Gupta PP, Gupta M, Schieder O (1982) Correction of nitrate reductase defect in auxotrophic plant cells through protoplast-mediated intergeneric gene transfers. Mol Gen Genet 188:378–383
33. Hayward C, Power JB (1975) Plant production from leaf protoplasts of *Petunia parodii*. Plant Sci Lett 4:407–410
34. Hein T, Przewozny T, Schieder O (1983) Culture and selection of somatic hybrids using an auxotrophic cell line. Theor Appl Genet 64:119–122
35. Hoffmann F, Adachi T (1981) "Arabidobrassica": Chromosomal recombination and morphogenesis in asymmetric intergeneric hybrid cells. Planta 153:586–593
36. Ito M (1973) Studies on the behavior of meiotic protoplasts II. Induction of a high fusion frequency in Protoplasts from liliaceous plants. Plant Cell Physiol 14:865–872
37. Ito M, Maeda M (1973) Fusion of meiotic protoplasts in liliaceous plants. Exp Cell Res 80: 453–456
38. Kao KN (1977) Chromosomal behavior in somatic hybrids of soybean-*Nicotiana glauca*. Mol Gen Genet 150:225–230
39. Kao KN, Gamborg OL, Miller RA (1971) Cell divisions in cells regenerated from protoplasts of soybean and *Haplopappus gracilis*. Nature (London) New Biol 232:124
40. Kao KN, Michayluk MR (1974) A method for highfrequency intergeneric fusion of plant protoplasts. Planta 115:355–367
41. Kao KN, Michayluk MR (1975) Nutritional requirements for growth of *Vicia hajastana* cells and protoplasts at a very low population density in liquid media. Planta 126:105–110
42. Kameya T, Horn ME, Widholm J (1981) Hybrid shoot formation from fused *Daucus carota* and *D. capillifolius* protoplasts. Z Pflanzenphysiol 104:459–466
43. Kartha KK, Gamborg OL, Constabel F, Kao KN (1974) Fusion of rapeseed and soybean protoplasts and subsequent division of heterokaryocytes. Can J Bot 52:2435–2436
44. Keller WA, Melchers G (1973) The effect of high pH and calcium on tobacco leaf protoplast fusion. Z Naturforsch 28c:737–741
45. Kohlenbach HW, Wenzel G, Hoffmann F (1982) Regeneration of *Brassica napus* plantlets in cultures from isolated protoplasts of haploid stem embryos as compared with leaf protoplasts. Z Pflanzenphysiol 105:131–142
46. Koop H-U, Weber G, Schweiger H-G (1983) Individual culture of selected single cells and protoplasts of higher plants in microdroplets of defined media. Z Pflanzenphysiol 112:21–34
47. Kurz WGW, Constabel F (1979) Plant cell cultures, a potential source of pharmaceutical. Adv Appl Microbiol 25:209–240
48. Maeda Y, Fujita Y, Yamada Y (1983) Callus formation from protoplasts of cultured *Lithospermum erythrorhizon* cells. Plant Cell Rep 2:179–182
49. Maliga P (1980) Isolation, characterization, and utilization of mutant cell lines in higher plants. Int Rev Cytol 11A:225–253
50. Maliga P, Lázár G, Joó FH, Nagy A, Menczel L (1977) Restoration of morphogenic potential in *Nicotiana* by somatic hybridization. Mol Gen Genet 157:291–296
51. Matsumoto T, Ikeda T, Kanno N, Kisaki T, Noguchi M (1980) Selection of high ubiquinone-10 producing strains of cultured tobacco cells by cell cloning technique. Agric Biol Chem 44: 967–969
52. Melchers G, Labib G (1974) Somatic hybridization of plants by fusion of protoplasts. I. Selection of light resistant hybrids of "haploid" light sensitive varieties of tobacco. Mol Gen Genet 135:277–294

53. Melchers G, Sacristan MD, Holder AA (1978) Somatic hybrid plants of potato and tomato regenerated from fused protoplasts. Carlsberg Res Commun 43:203–218
54. Menczel L, Lázár G, Maliga P (1978) Isolation of somatic hybrids by cloning *Nicotiana* heterokaryons in nurse cultures. Planta 143:29–32
55. Menczel L, Nagy F, Kiss ZsR, Maliga P (1981) Streptomycin resistant and sensitive somatic hybrids of *Nicotiana tabacum* + *Nicotiana knightiana*: correlation of resistance to *N. tabacum* plastids. Theor Appl Genet 59:191–195
56. Miller RA, Gamborg OL, Keller WA, Kao KN (1971) Fusion and division of nuclei in multinucleated soybean protoplasts. Can J Genet Cytol 13:347–353
57. Mino M, Yamada Y (1983) Genetic analysis of cultured cells that produce useful secondary metabolites: A theory. Chem Regul Plants 18:91–99 (In Japanese)
58. Morgan A, Cocking EC (1982) Plant regeneration from protoplasts of *Lycopersicon esculentum* Mill. Z Pflanzenphysiol 106:97–104
59. Morikawa H, Ichikawa Y, Senda M, Yamada Y, Sato F (1982) Surface charge characteristics of mesophyll and cultured cell protoplasts. In: Fujiwara E (ed) Proc 5th Int Plant Tissue Cell Cult, Plant Tissue Cult. Maruzen, Tokyo, pp 625–626
60. Murphy TM (1982) Analysis of distributions of mutants in clones of plant-cell aggregates. Theor Appl Genet 61:367–372
61. Nagata T (1978) A novel cell-fusion method of protoplasts by polyvinyl alcohol. Naturwissenschaften 65:263–264
62. Nagata T, Melchers G (1978) Surface charge of protoplasts and their significance in cell-cell interaction. Planta 142:235–238
63. Nagata T, Takebe I (1971) Plating of isolated tobacco mesophyll protoplasts on agar medium. Planta 99:12–20
64. Nagy JI, Maliga F (1976) Callus induction and plant regeneration from mesophyll protoplasts of *Nicotiana sylvestris*. Z Pflanzenphysiol 78:453–455
65. Neumann E, Gerisch G, Opatz K (1980) Cell fusion induced by high electric impulses applied to *Dictyostelium*. Naturwissenschaften 67:414–415
66. Ogino T, Hiraoka N, Tabata M (1978) Selection of high nicotine-producing cell lines of tobacco callus by single-cell cloning. Phytochemistry 17:1907–1910
67. Patnaik G, Cocking EC, Hamill J, Pental D (1982) A simple procedure for the manual isolation and identification of plant heterokaryons. Plant Sci Lett 24:105–110
68. Pelletier G, Primard C, Vedel F, Chetrit P, Remy R, Rousselle P, Renard M (1983) Intergeneric cytoplasmic hybridization in Cruciferae by protoplast fusion. Mol Gen Genet 191:244–250
69. Pental D, Cooper-Bland S, Harding K, Cocking EC, Müller AJ (1982) Cultural studies on nitrate reductase deficient *Nicotiana tabacum* mutant protoplasts. Z Pflanzenphysiol 105:219–227
70. Pohl HA (1978) *Dielectrophoresis*. Cambridge Univ Press, Cambridge
71. Potrykus I, Harms CT, Lörz H, Thomas E (1977) Callus formation from stem protoplasts of corn (*Zea mays* L.). Mol Gen Genet 156:347–350
72. Power JB, Berry SF, Chapman JV, Cocking EC (1980) Somatic hybridization of sexually incompatible petunias: *Petunia parodii, Petunia parviflora*. Theor Appl Genet 57:1–4
73. Power JB, Cocking EC (1970) Isolation of leaf protoplasts: Macromolecule uptake and growth substance response. J Exp Bot 21:64–70
74. Power JB, Cummins SE, Cocking EC (1970) Fusion of isolated plant protoplasts. Nature (London) 225:1016–1018
75. Power JB, Evans PK, Cocking EC (1978) Fusion of plant protoplasts. In: Poste G, Nicolson GL (eds) Surface reviews. Membrane fusion, vol V. Elsevier/North-Holland, Amsterdam New York, pp 370–385
76. Power JB, Frearson EM, Hayward C, George D, Evans PK, Berry SF, Cocking EC (1976) Somatic hybridization of *Petunia hybrida* and *P. parodii*. Nature (London) 263:500–502
77. Preston RD (1974) The physical biology of plant cell walls. Chapman and Hall, London
78. Raveh D, Huberman E, Galun E (1973) In vitro culture of tobacco protoplasts: use of feeder techniques to support division of cells plated at low densities. In vitro 9:216–222
79. Rennie PJ, Weber G, Constabel F, Fowke LC (1980) Dedifferentiation of chloroplasts in interspecific and homospecific protoplast fusion products. Protoplasma 103:253–262

80. Sato F, Yamada Y (1984) High berberine-producing cultures of *Coptis japonica* cells. Phytochemistry 23:281–285
81. Saxena PK, Gill R, Rashid A, Maheshwari SC (1981) Plantlet formation from isolated protoplasts of *Solanum melongena* L. Protoplasma 106:335–359
82. Schieder O (1977) Hybridization experiments with protoplasts from chlorophyll-deficient mutants of some *Solanaceous* species. Planta 137:253–257
83. Schieder O (1978) Somatic hybrids of *Datura innoxia* Mill + *Datura discolor* Bernh. and of *Datura innoxia* Mill + *Datura stramonium* L. var. *tatula* L. I. Selection and characterization. Mol Gen Genet 162:113–120
84. Schieder O, Vasil IK (1980) Protoplast fusion and somatic hybridization. Int Nat Rev Cytol 11B:21–45
85. Senda M, Morikawa H, Katagi H, Takada T, Yamada Y (1980) Effect of temperature on membrane fluidity and protoplast fusion. Theor Appl Genet 57:33–35
86. Senda M, Morikawa H, Takeda J (1982) Electrical induction of cell fusion of plant protoplasts. In: Fujiwara E (ed) Proc 5th Int Plant Tissue Cell Cult, Plant Tissue Cult. Maruzen, Tokyo, pp 615–616
87. Senda M, Takeda J, Abe S, Nakamura T (1979) Induction of cell fusion of plant protoplasts by electrical stimulation. Plant Cell Physiol 21:1441–1443
88. Shepard JF, Bidney D, Shahin E (1980) Potato protoplasts in crop improvement. Science 208:17–24
89. Smith HH, Kao KN, Combatti NC (1976) Interspecific hybridization by protoplast fusion in *Nicotiana*. Confirmation and extension. J Hered 67:123–128
90. Takebe I, Otsuki Y, Aoki S (1968) Isolation of tobacco mesophyll cells in intact and active state. Plant Cell Physiol 9:115–124
91. Teissie J, Knutson VP, Tsong TY, Lane MD (1982) Electric pulse-induced fusion of 3T3 cells in monolayer culture. Science 216:537–538
92. Toister U, Loyter A (1973) The mechanism of cell fusion. II. Formation of chicken erythrocyte polykaryons. J Biol Chem 248:422–432
93. Uchimiya H, Ohgawara T, Kato H, Akiyama T, Harada H (1983) Detection of two different nuclear genomes in parasexual hybrids by ribosomal RNA gene analysis. Theor Appl Genet 117–118
94. Vardi A, Spiegel-Roy P, Galun E (1975) Citrus cell culture: isolation of protoplasts, plating densities, effect of mutagens and regeneration of embryos. Plant Sci Lett 4:231–236
95. Vasil IK, Vasil V (1980) Ioslation and culture of protoplasts. Int Rev Cytol 11B:1–19
96. Vasil V, Vasil IK (1980) Isolation and culture of cereal protoplasts. Part 2: Embroygenesis and plantlet formation from protoplasts of *Pennisetum americanum*. Theor Appl Genet 56: 97–99
97. Vienken J, Zimmermann U (1982) Electric field-induced fusion: electrohydraulic procedure for production of heterokaryon cells in high yield. FEBS Lett 137:11–13
98. Vunsh R, Aviv D, Galun E (1982) Valine resistant plants derived from mutated haploid and diploid protoplasts of *Nicotiana sylvestris* and *N. tabacum*. Theor Appl Genet 64:51–58
99. Wallin A, Glimelius K, Eriksson T (1974) The induction of aggregation and fusion of *Daucus carota* protoplasts by polyethylene glycol. Z Pflanzenphysiol 74:64–80
100. Weber H, Förstser W, Jacob H-E, Berg H (1981) Microbiological implications of electric field effects. III. Stimulation of yeast protoplast fusion by electric field pulses. Z Allg Mikrobiol 21:555–562
101. Wenzel G, Schieder O, Przewozny T, Sopory SK, Melchers G (1979) Comparison of single cell culture-derived *Solanum tuberosum* L. Plants and a model for their application in breeding programs. Theor Appl Genet 55:49–55
102. Wernicke W, Lörz H, Thomas E (1979) Plant regeneration from leaf protoplasts of haploid *Hyoscyamus muticus* L. produced by anther culture. Plant Sci Lett 15:239–249
103. White DWR, Bhojwani SS (1981) Callus formation from *Trifolium arvense* protoplast-derived cells plated at low densities. Z Pflanzenphysiol 102:257–261
104. Withers LA, Cocking EC (1972) Fine-structural studies on spontaneous and induced fusion of higher plant protoplasts. J Cell Sci 11:59–75

105. Xu X-H, Davey MR (1983) Shoot regeneration from mesophyll protoplasts and leaf explants of *Rehmannia glutinosa*. Plant Cell Rep 2:55–57
106. Xu Z-H, Davey MR, Cocking EC (1981) Isolation and sustained division of *Phaseolus aureus* (mung bean) root protoplasts. Z Pflanzenphysiol 104:289–298
107. Yamada Y, Hara Y, Senda M, Nishihara M, Kito M (1979) Phospholipids of membranes of cultured cells and products of protoplast fusion. Phytochemistry 18:423–426
108. Yamada Y, Hara Y, Katagi H, Senda M (1980) Protoplast fusion. Effect of low temperature on the membrane fluidity of cultured cells. Plant Physiol 65:1099–1102
109. Yamada Y, Hashimoto T (1982) Production of tropane alkaloids in cultured cells of *Hyoscyamus niger*. Plant Cell Rep 1:101–103
110. Yamada Y, Hirabayashi Y, Morikawa H, Ohnishi N, Hayashi Y (1985) Callus formation from *Euphorbia millii* protoplasts and culture of heterokaryons formed from *E. millii* and *Coptis japonica* protoplasts (in preparation)
111. Yamada Y, Nakaminami M (1973) The isolation and fusion of protoplasts from alkaloid plant cells cultured. In vitro Colloq Int CNRS 212:373–383
112. Yamada Y, Sato F (1981) Production of berberine in cultured cells of Coptis *Coptis japonica*. Phytochemistry 20:545–547
113. Yamada Y, Watanabe K (1980) Selection of high vitamin B_6-producing strains in cultured green cells. Agric Biol Chem 44:2683–2687
114. Yamamoto Y, Mizuguchi R, Yamada Y (1982) Selection of a high and stable pigment-producing strain in cultured. *Euphorbia millii* cells. Theor Appl Genet 61:113–116
115. Zelcer A, Aviv D, Galun E (1978) Interspecific transfer of cytoplasmic male sterility by fusion between protoplasts of normal *Nicotiana sylvestris* and x-ray irradiated protoplasts of male-sterile *N. tabacum*. Z Pflanzenphysiol 90:397–407
116. Zenk MH, El-shagi H, Arens H, Stöckgt J, Weiler EW, Deus B (1977) Formation of the indole alkaloids serpentine and ajmalicine in cell suspension cultures of *Catharanthus roseus*. In: Barz W, Reinhard E, Zenk MH (eds) Plant tissue culture and its bio-technological application. Springer, Berlin Heidelberg New York, pp 27–43
117. Zimmermann U (1982) Electric field-mediated fusion and related electrical phenomena. Biochim Biophys Acta 694:227–277
118. Zimmermann U, Scheurich P (1981) High frequency fusion of plant protoplasts by electric fields. Planta 151:26–32
119. Zimmermann U, Scheurich P, Pilwat G, Benz R (1981) Cells with manipulated functions: new perspectives for cell biology, medicine, and technology. Angew Chem 20:325–344 (Int Ed Engl)

Conventional and New Approaches to Increase the Alkaloid Production of Plant Cell Cultures

J. BERLIN, H. BEIER, L. FECKER, E. FORCHE, W. NOÉ, F. SASSE, O. SCHIEL, and V. WRAY[1]

1 Introduction

The commercial production of natural compounds from plant cell cultures has repeatedly been envisaged [1, 14, 19, 23]. Recently the development of the first commercial process, the production of shikonin derivatives from suspension cultures of *Lithospermum erythrorhizon*, has been announced by Japanese scientists [7]. This first demonstration that plant cell cultures can indeed serve as a source of natural products must not conceal, however, the fact that we are still far away from a breakthrough in this area of biotechnology [2]. The conventional approaches (e.g. screening for highly productive variants, development of production media) were not really successful and thus not sufficient for many culture systems [2]. More knowledge on the molecular level about regulatory controls in the expression of secondary metabolism under cell culture conditions seems to be urgently needed. Our group has been involved in this area of biotechnology for 6 years. Here we would like to summarize briefly our experiences in increasing alkaloid formation of plant cell cultures by conventional methods. We also describe a new approach by which we hope to obtain a first insight into how to manipulate the expression of secondary pathways in plant cell cultures.

2 Isoquinoline Alkaloid Formation

The large group of isoquinoline alkaloids contains several alkaloids of great commercial importance. Thus we tried to establish morphinane producing cell cultures from various *Papaver* species. Roughly 200 callus cultures from individual plants of five morphinane-producing *Papaver* species were screened by a radioimmunoassay for morphine, by HPLC and capillary GC for morphinane-producing clones. Despite extensive screenings and variations of media, we failed to establish a cell line with significant levels of morphinanes. Most of the cell lines produced high levels of sanguinarines and related alkaloids. The development of one line was followed over a period of 3 years (Fig. 1). While growth rates increased during this time, the specific yields of alkaloids decreased from 6% to 0.2%. This phenomenon, that during long-term cultivation the productivity

1 GBF-Gesellschaft für Biotechnologische Forschung mbH, Mascheroder Weg 1, 3300 Braunschweig, FRG

Primary and Secondary Metabolism of Plant Cell Cultures, ed. by Neumann et al.

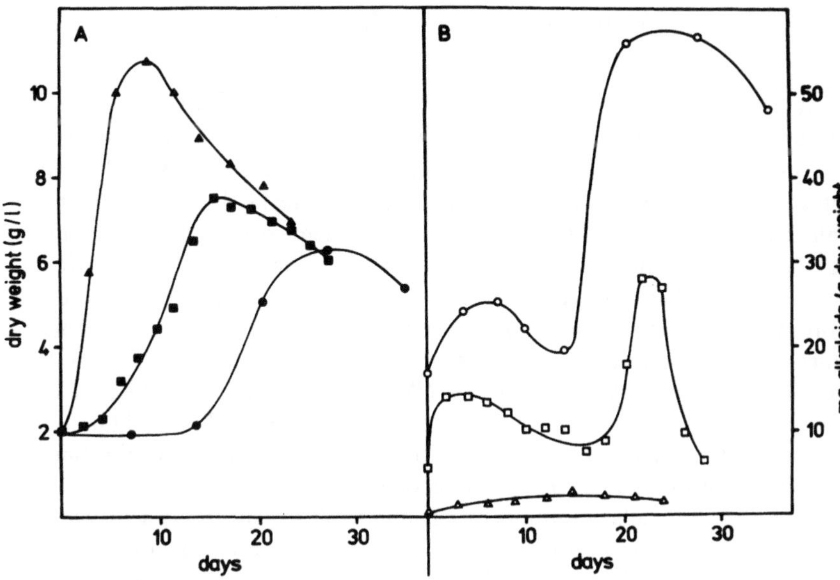

Fig. 1. Growth and alkaloid production of a suspension culture of *Papaver somniferum* after 5 (●, ○), 20 (■, □), and 35 (▲, △) growth cycles. While the suspension culture became finer with time, the culture cycles had to be reduced from 28 to 14 days. Sanguinarines (nor, dihydro, oxy) counted for more than 95% of total alkaloids, minor components were protopine, cryptopine, and salsoline

decreases in favor of better growth, is often observed. By continuously screening for producing cell clones one can overcome this problem of production instability. Another way is to transfer the low producing cells into production media. The production of benzophenanthridine alkaloids by *Eschscholtzia californica* is increased from 13 mg l^{-1} to 150 mg l^{-1} by transferring the cells from the growth medium to the production medium IM2 [6]. In some cases high production of these types of alkaloid is stable over long cultivation periods. Without any selection or optimization of media, a fine suspension culture of *Thalictrum rugosum* was established accumulating 750–1,000 mg protoberberine alkaloids l^{-1} (e.g. magnoflorine, columbamine, berberine) (Forche et al., in preparation). The productivity of this cell line with a growth cycle of 2 weeks has been remarkably stable for at least 18 months. Our data are in good agreement with those published by other authors. Some widely distributed isoquinoline alkaloids such as protoberberines and benzophenanthridines are formed spontaneously in reasonable or even high levels on typical growth media by many plant cell cultures [8, 9, 17]. The productivity of such spontaneously producing cell lines can be further optimized by screenings or media variation. Despite many attempts, cell lines containing special isoquinoline alkaloids which are restricted to very few plant species have rarely been established. There are a few reports claiming the presence of morphinane alkaloids in cultured cells. However, as none of the allegedly producing cell cultures [20] has been further optimized, it looks as though the morphinane production did not respond to the conventional techniques of stimulating the productivity of cell cultures.

3 Physiological State of Cells Determined by in Vivo 31-P-NMR

During our studies on the medium-induced stimulation of secondary metabolism in plant cell cultures we found that phosphate was a negative effector of secondary metabolism. The inducibility of tobacco cells for cinnamoyl putrescine biosynthesis depended on the internal levels of phosphate accumulated within the cells [12]. We have extended this observation in such a way that many media constituents can "induce" secondary metabolism when they are the growth-limiting factor [18]. The fact is that in most culture systems phosphate becomes the first growth-limiting constituent. Therefore the levels of inorganic phosphate and phosphate metabolites seem to be good indicators of the physiological state of cultured plant cells. A convenient and non-invasive method of monitoring such metabolites is in vivo 31-P-NMR spectroscopy [22]. We have used this technique also to determine the physiological state at which cells should be transferred to the production medium for optimal expression of secondary metabolism. As cells enlarge during a growth cycle and thus become more sensitive to shear stress, we usually did not grow the cells in bioreactors to the fresh weight levels achievable in shake flasks. Hence the accumulation of media constituents influencing expression of secondary metabolism, and thus inducibility, may be different in cells scaled up in various fermenters from those of shake flask at the same cell density (fresh wt l^{-1}). In Fig. 2A a typical growth curve of *Catharanthus roseus* cells is shown. After 8–9 days growth was only due to cell enlargement. The fresh weight curve was surely a poor indicator for determining the physiological state of the cells. When 10 ml of cell sus-

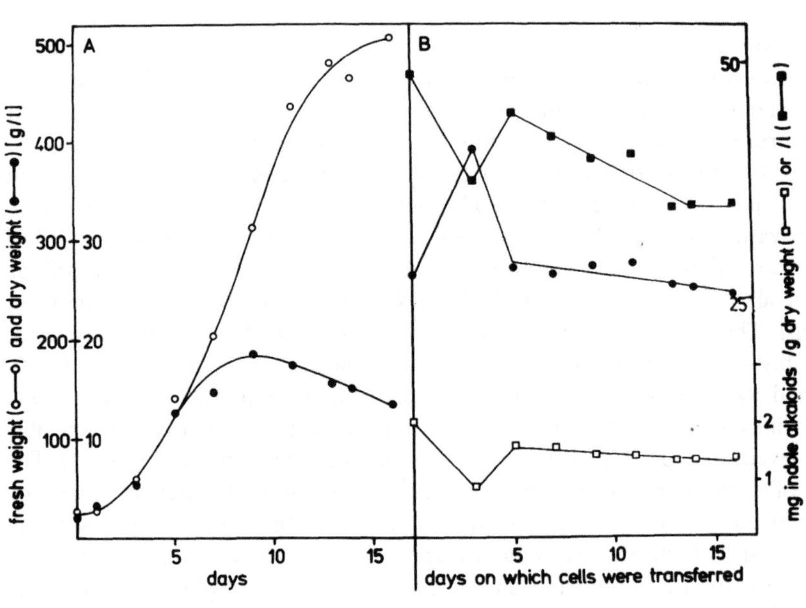

Fig. 2. Growth kinetic of *Catharanthus roseus* cells on MX medium (A). On the days indicated 10 ml suspensions were transferred to IM2-medium. After 8–16 days samples were analyzed for growth and alkaloid production. The values given (B) are the mean values of 4 samples when production reached a plateau

pensions were transferred from the growth medium to the induction medium the variously aged cells responded differently (Fig. 2B). Ten ml suspension taken on the third day showed the best growth in the production medium (although only 0.5 g cells were transferred) while the induction of indole alkaloid biosynthesis was poor (Fig. 2B). If one looks at the corresponding in-vivo 31-P-NMR spectrum (Fig. 3), one realizes that this was the only one of the samples transferred to the induction medium IM2 having accumulated high levels of internal Pi, which allowed the cells to grow best in the phosphate starved production medium. It is evident from this experiment that cells in which phosphate has just been depleted (Figs. 2, 3) are the best for optimal induction and production of indole alkaloids in *Catharanthus roseus*.

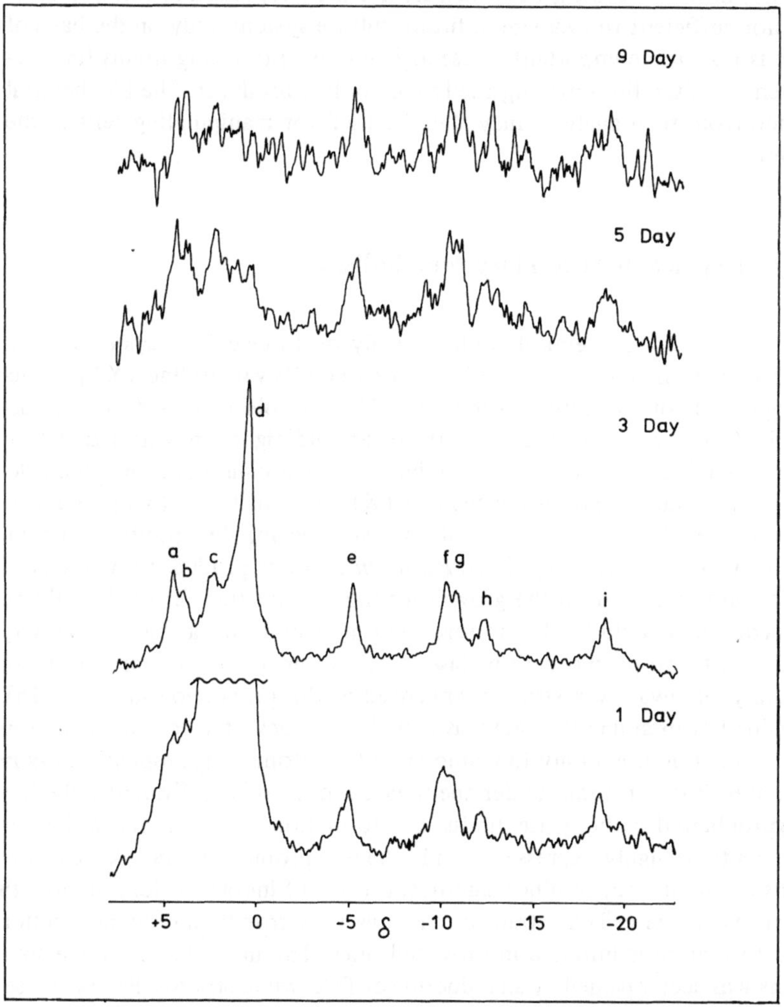

Fig. 3. In-vivo 31-P-NMR spectra, at 162 MHz with aeration, of *Catharanthus roseus* cells grown on MX medium for the times shown. The signals correspond to *a* glucose-6-phosphate; *b* sugar phosphates; *c* cytoplasmic inorganic phosphate; *d* vacuolar inorganic phosphate; *e* γ-ATP; *f* α-ATP; *g, h* nucleoside diphosphoglucose; *i* β-ATP. Note that cells of day 1 were not transferred to production media (Fig. 2). If possible, include the above sentence as it helps to avoid misunderstandings

When critically analyzing the state of the art of the biotechnological production of natural products by plant cell cultures, we have to admit that many important secondary compounds have at present not been found in morphologically undifferentiated cell cultures or were recalcitrant to any conventional approaches of productivity improvement. The conventional techniques seem to work mainly in systems where expression of secondary metabolism occurs spontaneously or can be induced by simple variation of the medium composition. The conventional techniques are evidently not sufficient to achieve a general breakthrough for the commercial production of natural products from plant cell cultures. As only a small portion of the total genetic information is expressed in cultured cells, we should develop techniques of manipulating gene expression and eventually influence those genes which are normally repressed in cultured cells. Therefore it is not sufficient to evaluate in future culture systems only on the basis of actual yields. It is now more important to use high and low producing strains for analyzing the biochemical alterations making a cell a high or low producer. The biochemical knowledge gained from such analyses may then be used for manipulating cells in the required direction.

4 Regulatory Enzymes of Secondary Metabolism

We have concentrated our biochemical studies mainly on three culture systems. First, we compared biochemically the tobacco cell lines TX1 and its variant line TX4 producing low and high levels of cinnamoyl putrescines. The ten-fold increased product accumulation of TX4 cells was mainly due to up to ten fold higher enzyme activities in the phenylpropanoid branch and three fold increase of arginine and ornithine decarboxylase [3]. When the low producing cell line TX1 was transferred to a production medium, alkaloid levels increased four- to sixfold and, again, the specific and total enzyme activities increased distinctly [10]. *Peganum* cultures produced low levels of harman alkaloids and serotonin on the growth medium. After transfer of the cells to a production medium, alkaloid levels and serotonin increased drastically. The activity of the first enzyme, tryptophan decarboxylase (TDC), diverting the primary precursor into the secondary pathway was strongly enhanced in the production medium [16]. Interestingly, added tryptamine (2 mM) was completely converted to serotonin even by cells which formed de novo only low amounts of serotonin. Tryptophan was only poorly incorporated into serotonin under the same conditions [16]. Evidently the low activity of tryptophan decarboxylase limits the formation of serotonin while the hydroxylase seems to be highly expressed even in the low producing state. Cell cultures of *Catharanthus roseus* do not produce significant levels of indole alkaloids in growth media. However, when the cells are transferred to various growth-limiting production media, alkaloid formation is initiated in most cell lines. The induction of indole alkaloid biosynthesis was accompanied by an induction of TDC while strictosidine synthase, the key enzyme combining tryptamine and secologanin, was found to be constitutively expressed [13]. The observations made for these three systems are summarized in Fig. 4. In all systems the activity of the enzyme diverting the primary metabolite into the secondary pathway had to be increased greatly before increased accumulation of second-

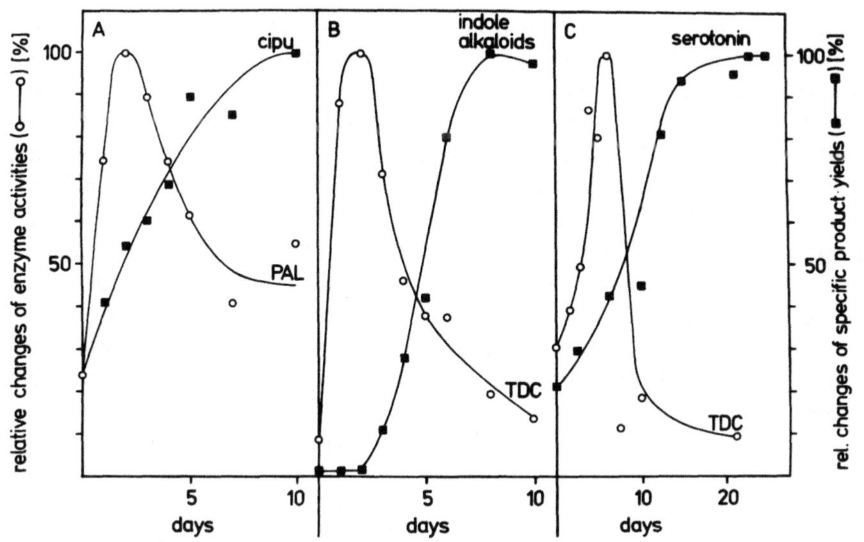

Fig. 4A–C. "Induction" of enzymes linking primary and secondary metabolism on growth-inhibiting media and accumulation of corresponding secondary metabolites. A tobacco cells, cipu = cinnamoyl putrescines, PAL = phenylalanine ammonia lyase; B *Catharanthus* cells; C Peganum cells

ary metabolites occurred. Thus these enzymes seem to be a bottleneck for optimal productivity, although we do not claim that increased activities of such enzymes are the only requirement for increased product formation [11]. However, we feel that these enzymes exert a decisive regulatory control in the accumulation of secondary metabolites and are therefore suitable targets for initial studies on how to manipulate the expression of secondary metabolism in cultured plant cells.

5 A Genetic Engineering Approach for Overcoming Regulatory Bottlenecks

The induction of these enzymes occurred on culture media in which growth was greatly reduced. The value of selecting variant cells in which these enzymes are expressed on growth media is clearly demonstrated by the TX4 cells producing up to 1 g cinnamoyl putrescines l^{-1} in the growth medium [3]. A system for the selection of such variants has been proposed [4]. Another possibility would be to modify cell lines, having such high enzyme activities only under growth-limiting conditions, with sufficient and constitutively expressed levels by the techniques of genetic engineering. The most suitable system for this approach seems to be the *Peganum* culture. Consequently, our goal is to endow *Peganum* cells with a constitutively expressed tryptophan decarboxylase. As indicated above, the second and final step of the biosynthetic sequence to serotonin is always present in the cells. This investigation is not complete, but we can report our results to date. Tryptophan decarboxylase of *Catharanthus roseus* cells was purified to homogeneity and characterized [15]. Antibodies against this enzyme have been prepared in rabbits. The antiserum recognized the induction of TDC in crude enzyme preparations of induced *Catharanthus* cells but did not bind to TDC of induced *Peganum* cells (not shown). It was shown by in vivo labelling with 35-S-methionine and immunoprecipitation that the increase of TDC activity in the production medium results from de novo synthesis (Fig. 5). The maximum of TDC synthesis precedes the maximum specific activity by 10–12 h. A half-life time of 21 h for TDC can be calculated from the kinetics. Poly(A)$^+$-RNA from induced and non-induced *Catharanthus roseus* cells have been translated in a wheat germ cell-free system. Only the mRNA preparations of induced cells synthesized proteins which were immunoprecipitable by the TDC antiserum, whereas non-induced cells showed almost no reaction (not shown). The construction of gene banks from induced and non-induced cells has been started. Screening of the gene bank for positive clones may cause some problems as the TDC represents only 0.03% of total protein even in induced cells [15]. However, this is perhaps a general problem one has to cope with when studying regulatory aspects of secondary metabolism in plant cells.

Parallel to this investigation we have initiated a similar approach for *Lupinus* cells. The tetracyclic structure of quinolizidine alkaloids is formed by two enzymatic steps from lysine [21]. The poor synthesis of quinolizidine alkaloids of *Lupinus* cells seems to be partly due to the very low activity of the first enzyme in this pathway, lysine decarboxylase. Therefore, we plan to transform *Lupinus* cells with a constitutively expressed bacterial lysine decarboxylase (LDC). LDC of the Gram-negative bacterium *Hafnia alvei* has been purified to homogeneity, a specific antiserum has been prepared and purified. Isolated DNA from *Hafnia alvei* was partially digested with the restriction endonuclease Sau 3a. Fragments of a size of 6–18 kB were ligated into the BamHI site of the plasmid pBR322 and *E. coli* strain HB101 was transformed with this re-

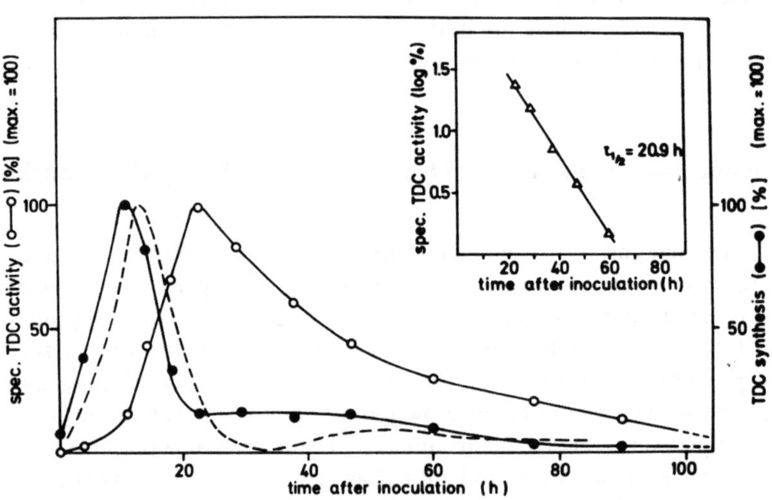

Fig. 5. Time course of TDC activity of *Catharanthus roseus* and its apparent synthesis on IM2 medium. The maximum specific activity was 76.6 pkat mg^{-1}. The rate of the synthesis is expressed as a percentage of the maximum immunoprecipitable specific incorporation of 35S-methionine. *Dotted line* represents the synthesis calculated according to Schimke (1974) (Int. Rev. Sci. Biochem. 9:1983)

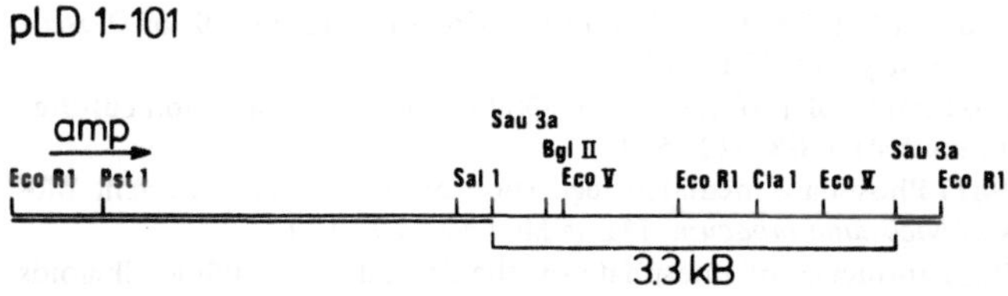

Fig. 6. Restriction map of the *Hafnia alvei* insert of the plasmid pLD1–101. Within this pattern the gene of LDC is located as all LDC-positive clones contain this pattern

combinant plasmids. The expression of Hafnia genes in isolated minicells of *E. coli* K12 made it possible to screen the gene bank for LDC-positive clones with the purified LDC antibodies [5]. One of the positive clones (462) had an insert of about 7.5 kB. According to the MW of the LDC subunits (75,000 MW) the size of the insert was three times too large. The plasmid pLD462 was then partially digested with Sau 3a and fragments of the size of 1.5–2.5 and 2.5–5 kB were isolated and ligated as before. From HB101 transformants two clones (1–101 and 2–19) were screened expressing large amounts of *Hafnia alvei* LDC. The size of the inserts are now 3.3 and 2.9 kB. No positive clones were found in transformed colonies with inserts smaller than 1.5 kB. In Fig. 6 the restriction map of that part of the *Hafnia alvei* DNA insert is given which is common to all LDC-positive clones. The DNA sequence analysis is under way. Thus the LDC gene of *Hafnia alvei* may soon be available for transforming *Lupinus* cells with Ti-plasmid vectors carrying the LDC gene. The answers one can expect from this type of experiment will evidently not be restricted to the question of whether one can stimulate quinolizidine or serotonin biosynthesis by this means.

Acknowledgements. We are grateful to the Commission of European Countries for supporting the genetic aspects of our research within the context of the "Biomolecular Engineering Programme".

References

1. Barz W, Ellis BE (1981) Plant cell cultures and their biotechnological potential. Ber Dtsch Bot Ges 94:1–26
2. Berlin J (1984) Plant cell cultures – a future source of natural products? Endeavour New Ser 8:5–8
3. Berlin J, Knobloch KH, Höfle G, Witte L (1982) Biochemical characterization of two tobacco cell lines with different levels of cinnamoyl putrescines. J Nat Prod 45:83–87
4. Berlin J, Witte L, Hammer J, Kukoschke KG, Zimmer A, Pape D (1982) Metabolism of p-fluorophenylalanine in p-fluorophenylalanine sensitive and resistant tobacco cell cultures. Planta 155:244–250
5. Berlin J, Beier H, Fecker L, Noe W, Schairer HU (1983) Transformation of plant cell cultures for the increased formation of secondary metabolites. Prog Rep 1983 Biomol Eng Programme CEC, Brussels, vol I, pp 126–127
6. Berlin J, Forche E, Wray V, Hammer J, Hösel W (1983) Formation of benzophenanthridine alkaloids by suspension cultures of *Eschscholtzia californica*. Z Naturforsch 38c:346–352
7. Fujita Y, Tabata M, Nishi A, Yamada Y (1982) New medium and production of secondary compounds with the two staged culture method. In: Fujiwara A (ed) Plant tissue culture 1982. Maruzen, Toyko, pp 399–400

8. Furuya T, Yoshikawa T, Kiyohara H (1983) Alkaloid production of cultured cells of *Dioscoreophyllum cuminsii*. Phytochemistry 22:1671–1673
9. Hinz H, Zenk MH (1981) Production of protoberberine alkaloids by cell suspension cultures of *Berberis* species. Naturwissenschaften 68:620–621
10. Knobloch KH, Berlin J (1981) Phosphate mediated regulation of cinnamoyl putrescine biosynthesis in cell suspensions of *Nicotiana tabacum*. Planta Med 42:167–172
11. Knobloch KH, Berlin J (1983) Influence of phosphate on the formation of indole alkaloids and phenolic compounds in cell suspension cultures of *Catharanthus roseus*. I. Comparison of enzyme activities and product accumulation. Plant Cell Tissue Org Cult 2:333–340
12. Knobloch KH, Beutnagel G, Berlin J (1981) Influence of accumulated phosphate on culture growth and formation of cinnamoyl putrescines in medium-induced cell suspension cultures of *Nicotiana tabacum*. Planta 153:582–585
13. Knobloch KH, Hansen B, Berlin J (1981) Medium induced formation of indole alkaloids and concomitant changes of interrelated enzyme activities in cell suspension cultures of *Catharanthus roseus*. Z Naturforsch 36c:40–43
14. Misawa M, Suzuki T (1982) Recent progress in plant cell culture. Research on the production of useful plant metabolites in Japan. Appl Biochem Biotech 7:205–216
15. Noe W, Mollenschott C, Berlin J (1984) Tryptophan decarboxylase from *Catharanthus roseus* cell suspension cultures: purification, molecular and kinetic data of the homogeneous protein. Plant Mol Biol 3:281–288
16. Sasse F, Heckenberg U, Berlin J (1982) Accumulation of β-carboline alkaloids and serotonin by cell cultures of *Peganum harmala*. II. Interrelationship between accumulation of serotonin and activities of related enzymes. Z Pflanzenphysiol 105:315–322
17. Sato F, Yamada Y (1984) High berberine producing cultures of *Coptis japonica* cells. Phytochemistry 23:281–285
18. Schiel O, Jarchow-Redecker K, Piehl GW, Lehmann J, Berlin J (1984) Increased formation of cinnamoyl putrescines by fedbatch fermentation of cell suspension cultures of *Nicotiana tabacum*. Plant Cell Rep 3:18–20
19. Shuler ML (1981) Production of secondary metabolites from plant cell cultures – problems and prospect. Ann NY Acad Sci 369:65–80
20. Tam WHJ, Constabel F, Kurz WGW (1980) Codeine from cell suspension cultures of *Papaver somniferum*. Phytochemistry 19:486–487
21. Wink M, Hartmann T (1979) Cadaverine-pyruvate transamination: The principal step of enzymatic quinolizidine alkaloid biosynthesis in *Lupinus polyphyllus* cell suspension cultures. FEBS Lett 101:343–346
22. Wray V, Schiel O, Berlin J (1983) High field phosphorus-31 nuclear magnetic resonance investigation of the phosphate metabolites in cell suspension cultures of *Nicotiana tabacum*. Z Pflanzenphysiol 112:215–220
23. Zenk MH (1978) The impact of plant cell cultures on industry. In: Thorpe T (ed) Frontiers in plant tissue culture 1978. The bookshop, Univ Calgary, pp 1–14

Multiple Shoot Cultures: A Viable Alternative in Vitro System for the Production of Known and New Biologically Active Plant Constituents

M.R. HEBLE [1]

1 Introduction

A major part of modern chemotherapeutics and traditional drugs are derived from higher plants [1]. While a few medicinal plant species are cultivated, a large number of them are procured from the naturally growing flora which has resulted in a gradual depletion of the resources. These considerations warrant methods for systematic cultivation and preservation of valuable genotypes. Alternatively methods for in vitro production of known drugs and new biologically active principles would be of particular value.

During the past decade considerable attention has been focussed on utilizing plant cell cultures as a source for secondary metabolites. Of the several cultures tested, a few appear to be promising for large-scale production. Examples of such systems are *Coleus blumei* (rosmarinic acid) [2], *Lithospermum erythrorhizon* (shikonin derivatives) [3], *Nicotiana tabacum* (ubiquinone) [4], and *Coptis japonica* (berberine) [5]. The biotransformation potential of *Digitalis lanata* cells [6] is also an important feature. On the other hand, a majority of cell cultures fail to synthesize the desired compounds or lose the capacity to synthesize during serial subcultures. This is possibly due to the instability of the cultured cells caused by nuclear irregularities [7]. In view of these limitations it is essential to explore additional in vitro methods for obtaining plant constituents. The inherent ability of a large number of medicinal plants to form multiple shoots in culture from axillary and shoot tip meristem has opened up avenues towards production of plant constituents of medicinal importance (Table 1), and preservation/propagation of elite genotype [8, 9].

2 Initiation and Growth of Multiple Shoot Cultures

The shoot tip and axillary meristem develop into shoots on a minimal inorganic basal medium of Murashige and Skoog [10] both on agar surface and liquid suspension. When the medium is supplemented with plant growth hormones such as BA, Kn, Zeatin, IAA and NAA singly or in combination, multiple shoots develop from their meristem and the basal ends. In certain cases the basal ends developed proliferating cells which

1 Bio-organic Division, Bhabha Atomic Research Centre, Bombay 400085, India

Primary and Secondary Metabolism of Plant Cell Cultures, ed. by Neumann et al.

Table 1. Secondary products isolated from multiple shoot cultures

Plant	Culture origin	Compound (S) isolated	Shoot culture yield mg %	Plant part yield mg %	Reference
Dioscorea Composita	Axillary meristem	Diosgenin	200	210 (leaf)	Heble and Staba [22]
Withania somnifera	Axillary meristem	Withanolides	320	–	Heble et al. (unpl.)
Digitalis lanata	Seedling	Digoxin	9	230 (leaf)	Lui and Staba [24]
Digitalis lanata	Cells	Cardenolides	15	–	Kuberski et al. [26]
Digitalis purpurea	Seedling callus	Digitoxin	2–4	–	Hagimori et al. [27]
Rauwolfia serpentina	Axillary meristem	Indole alkaloids Ajmaline Ajmalidine Yohimbine	710	540 (leaf)	Roja et al. [16]
Catharanthus roseus	Seedling	Indole alkaloids Vindoline	–	–	Krueger et al. [17]
Cinchona ledgeriana	Seedling	Cinchona alkaloids quinine quinidine	890	–	Staba and Chung [19]
Cinchona succirubra	Seedling	Cinchonine cinchonidine	796	–	Staba and Chung [19]
Papaver somniferum	Seedling	Morphine	–		Staba et al. [21]
Pyrethrum	Disc floret	Pyrethrins	342	1,215 (leaf)	Zeig et al. [29]

subsequently gave rise to shoot buds [11]. Multiple shoot cultures can be propagated on agar or liquid medium by routine periodic subcultures. The shoot cultures can also be grown in 2-l capacity flasks (Heble et al., personal communication) and multiliter fermentors [12]. The type of explant, the hormonal combinations and the conditions of growth play an important role in the establishment and maintenance of shoot cultures.

3 Alkaloid Formation in Shoot Cultures

3.1 Tropane Alkaloids

Shoot cultures of *Atropa belladonna* were initiated from shoot tip and nodal meristem on Wood and Braun medium [13] supplemented with NAA and kinetin. Active growth of the tissue was achieved on MS medium containing BA *(5 ppm)*. The shoot buds developed from the meristems and from the unorganized cells which had developed at

Table 2. Alkaloid content in shoot cultures and plant shoot of *Atropa belladonna*

Plant Type	Plant shoots Alkaloids mg %	Shoot cultures Alkaloids mg %	Culture condition
Control plants	150	10	MS + BA
Control plants	150	62	MS + BA + Ornithine + phenylalanine
Selected plant I	200	10	MS + BA
Selected plant II	300	10	MS + BA
Selected plant III	400	Traces	MS + BA

the base. In this system multiple shoots also developed from leaf explants (discs) in a similar manner [14]. On MS liquid medium the cultures gave a sixfold growth over a 4-week incubation period (Fig. 1). Shoot cultures of *A. belladonna* showed different morphogenetic responses to plant growth regulators such as root initiation on Kn + IAA medium, broadening of leaves on BA + NAA medium and unorganized cell proliferation on Kn + 2,4-D medium. The cultures were sensitive to exogenous growth hormones.

The tropane alkaloids are distributed in all parts of the plant and atropine forms the major constituent. The shoot cultures contained generally low levels of alkaloids as compared to the parent. Atropine was detected in the cultures by TLC only in trace quantities. Plant growth hormones did not influence the alkaloid production, but precursors of tropane alkaloids marginally augmented the alkaloid production [11]. Cultures established from high alkaloid-yielding selected plants also contained low levels of alkaloid (Table 2). Plants regenerated from the cultures and grown in the field contained alkaloids similar to the parent. The results indicated that the cultures have retained the genetic information for alkaloid biosynthesis but the juvenile stage of the cultures is not conducive to alkaloid production.

3.2 Indole Alkaloids

3.2.1 *Rauwolfia serpentina*

Shoot cultures of *R. serpentina* were initiated from axillary meristem of the plant on MS agar medium supplemented with NAA + BA, and established in MS liquid medium of the same composition [14]. The cultures gave a high biomass yield during a 5-week growth period (Fig. 1). The shoots developed from the nodal meristem and also the base without the intervention of callus.

Exogenous supply of growth hormones did not influence the morphogenetic pattern of the cultures except on medium containing IAA + Kn where sporadic rooting was observed. The hormones, however, influenced the growth characteristics.

The plant *R. serpentina* contains more than 50 alkaloids of which reserpine, rescinnamine, ajmaline, ajmalicine, yohimbine, and serpentine are the most important [15]. The root contains most of the alkaloids but the leaves contain a much simpler alkaloid profile. The shoot cultures contain yohimbine, ajmaline, ajmalidine, and other dihydroindoles as major components [16]. The alkaloid content of the cultures was more than

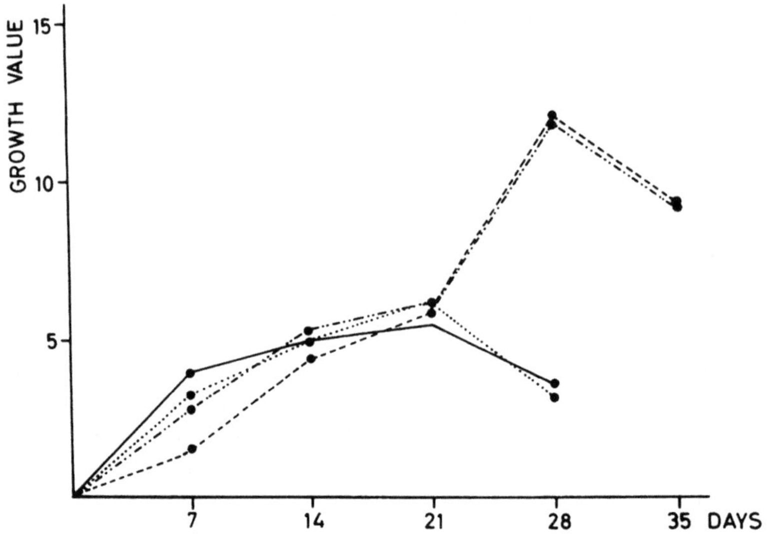

Fig. 1. Growth pattern of *Atropa belladonna* and *Rauwolfia serpentina* shoot cultures. *A. belladonna*; ●———● Fresh wt. ●·········● Dry wt. *R. serpentina*; ●– – – – – –● Fresh wt. ●–··–··–··● Dry wt. Growth value = Final wt. / Inoculum wt.

Table 3. Alkaloid profile of *Rauwolfia serpentina*

Alkaloids	Leaves	Shoot cultures	Roots
Ajmaline	–	+	+
Ajmalidine	+	+	–
Yohimbine	–	+	+
Reserpine	–	–	+
Ajmalicine	+	–	+
Serpentine	+	–	+
Total alkaloids	0.54%	0.71%	2.7%

the parent shoots, and the alkaloid profile resembled partly that of shoots and partly that of roots (Table 3). The active growth of the cultures yielding high biomass and consistent production of alkaloids over prolonged periods indicated that *Rauwolfia* shoot cultures could be useful in the production of indoles.

3.2.2 *Catharanthus roseus*

Krueger et al. [17] established shoot cultures from aseptic seedlings of *C. roseus* on MS medium supplemented with BA. The cultures gave nearly 12-fold growth in 35 days. The alkaloid pattern of the cultures was not identical with the parent. Vindoline was isolated and characterized. The presence of vindoline, monomer of the anticancer alkaloid vinblastine has not been reported in cell cultures.

Working with *C. roseus* cell cultures Constable et al. [19] regenerated shoots usually multiple shoots from a low percentage of young callus tissues. The shoots thus formed contained strictosidine, ajmalicine, hoerhammericine, vindolinine, vindoline, and catharanthine. The authors report that the accumulation of vindoline is favoured by structural organization.

3.3 Cinchona Alkaloids

Multiple shoot cultures were established from the seedling of cinchona *(Cinchona ledgeriana, Cinchona succirubra)*, on RT liquid medium supplemented with growth hormones [19]. The growth of leaf cultures was significantly more than the root cultures (*C. ledgeriana*, GI: 21.45, *C. succirubra*, GI: 20.1).

The shoot cultures yielded comparatively high levels of alkaloids (*C. ledgeriana* 0.89, *C. succirubra* 0.80 dry wt.). Both cultures contained cinchonine, cinchonidine, quinine, and quinidine.

In the plant the alkaloids accumulate in the trunk bark and heart wood in 8 to 20 years [20]. It is, therefore, of particular interest that 6-week-old leaf cultures could accumulate nearly 10% of the bark alkaloids and 50% of the 1-year-old *cinchona* plants.

Cultivation of tree species for medicinal purposes is an expensive proposition. There is also a wide variation in the alkaloid content of the *cinchona* trees, which could be realized only after several years. Shoot cultures from high alkaloid-yielding plants would be advantageous, for germplasm preservation, multiplication of elite genotypes and production of alkaloids.

3.4 Poppy Alkaloids

In vitro, shoots of *Papaver bracteatum* produced 0.007% of thebaine which is more than the corresponding cell cultures [22]. Production of thebaine and codeine by in vitro methods would be of advantage since these are the minor alkaloids in the plant as compared to morphine.

4 Shoot Cultures of Steroid-Yielding Plants

4.1 Dioscorea composita

Shoot cultures established from nodal meristem of the plant *D. composita* produced diosgenin in concentrations similar to the parent shoot [22]. Growth pattern and diosgenin synthesis were influenced by different levels of BA in the medium (Table 4). The results indicated that the shoot cultures exhibited, similar biosynthetic capacity as the parent explant and the synthesis of diosgenin is influenced by plant growth hormones.

Table 4. Diosgenin in *Dioscorea composita* shoot cultures

BA mg l^{-1}	Growth response	Diosgenin mg %
0.1	++	78
0.5	++++	203
0.75	+++	120
1.0	++	113
Plant shoots	----	210

4.2 Digitalis

Cardenolide formation in organised structures of *Digitalis* has received much attention. Hirotani and Furuya [23] reported that the leaves regenerated from *D. purpurea* callus produced pharmaceutically important cardenolides digitoxin and purpurea glycoside A. The observation was of significance since callus cultures of *Digitalis* failed to synthesize cardenolides. Lui and Staba established multiple shoot cultures of *D. lanata* on RT medium supplemented with BA. The cultures grew rapidly (GI: 20 + 2.2) and produced digoxin (9.0 + 1.6 mg:% dry wt.). Precursors of cardenolide biosynthesis such as sodium glycocholate, cholesterylacetate, and progesterone augmented the digoxin formation. The cardenolide pattern in shoot cultures was similar to the plant leaf but quantitative differences were observed. Hagimori et al. [25] established shoot forming cultures of *D. purpurea*. The authors observed that the light grown cultures accumulated higher levels of digitoxin than the dark grown cultures and suggested that the chloroplast is not essential for digitoxin synthesis but the proplastids which would develop into chloroplast upon illumination contain the cardenolide biosynthetic system. Kuberski et al. [26] observed that in *D. lanata* the cardenolide content of the irradiated embryos reached to 0.15 μg/mg during plantlet regeneration.

It is evident from the above data that structural differentiation is a prerequisite for cardenolide biosynthesis. The type of differentiation, and exogenous supply of growth hormones and precursors determine the level and pattern of cardenolide synthesis.

4.3 Withania somnifera

The production of withanolides a group of steroids possessing cytotoxic activity [27] has been investigated in cell and shoot cultures of *W. somnifera* (Heble et al., unpubl. data). Multiple shoot cultures were established from nodal meristem and cell cultures from internodal stem segments. Cell cultures did not contain detectable amounts of withanolides. The shoot cultures gave nearly four fold growth during a 7 day incubation period and contained 0.32% dry wt. of total withanolides. The pattern of withanolides in the shoot culture was different from that reported in the leaves of the plant [28]. The shoot cultures did not contain withaferin A the major component of the plant. Withanone, withanolide G, withanolide I, withanolide E, and trace amounts of 4-β-hydroxywithanolide E were identified by thin layer chromatography and high performance liquid chromatography (Fig. 2). The shoot culture extracts showed cyto-

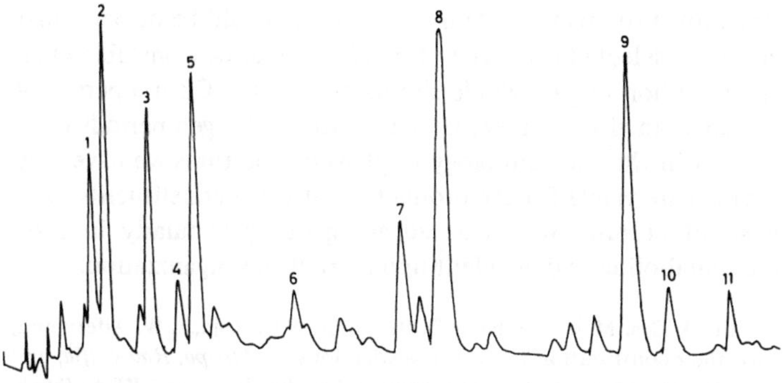

Fig. 2. HPLC profile of withanolides extracted from multiple shoot cultures of *Withania somnifera*: *7* withanolide E; *8* withanone + withanolide D; *9* withanolide G; *10* withanolide I

toxic activity against mouse fibroblast suspension culture L-929. Although the shoot cultures synthesized high levels of withanolides, qualitative differences between the plant and the cultures were significant. Such cultures might be useful in obtaining new biologically active components.

5 Pyrethrum

Pyrethrins, a group of insecticidal esters, have been isolated in substantial amounts (0.32% dry wt.) from shoot cultures of *Chrysanthemum cinerariaefolium* established from a CS - 44 Equadorian plant selection [29]. The authors suggested that the type of tissue organization appears to have a bearing on pyrethrin production. It is of particular interest that pyrethrum cultures derived from high yielding plant selection provided more high yielding lines.

6 Conclusion

In vitro culture of isolated shoot tip was attempted in the early 1930s by White [30]. Culturing *Asparagus* stem tips, Loo [31] obtained plantlets. Shoot tip and meristems cultures have been extensively, used in the area of horticulture, plant pathology, and clonal propagation [32].

It is now possible to cultivate multiple shoots in fermentors and obtain high biomass. Takayama and Misawa [12] obtained nearly 22-fold growth in bubblecolumn fermentor, and suggested that for the production of a large number of plantlets submerged culture techniques similar to those used for microbial cultivation are superior when the culture conditions are optimized. An added advantage in shoot cultures is that they give 10%-15% dry mass as compared to cell cultures, which give 4%-10% dry mass.

Since shoot cultures are known to maintain stem lineage they would be of particular advantage in the preservation of selected genotype of medicinal plants. Constable et al. [18] have reported the preservation of periwinkle shoots at 12°–15 °C for a period of 6 months. The shoot meristems can also be freeze-preserved for prolonged periods [33].

Despite the rapid advances in the biotechnology of plant cell cultures we have very few cell cultures systems which are viable for the production of plant constituents [34]. Alternative system like shoot cultures would be advantageous particularly in cases where the synthesis of the metabolites is dependent upon structural organization.

Acknowledgement. I wish to thank Drs. M.S. Chadha, B.D. Benjamin, P.C. Roja, A.W. Alfermann, E. Reinhard, and J. Berlin for their contribution towards investigations on *Atropa, Rauwolfia,* and *Withania*. I also wish to thank The Department of Atomic Energy, Bomby, India, and KFA, Jülich, Federal Republic of Germany for the support.

References

1. Zenk MH (1979) The impact of plant cell culture on industry. In: Thorpe TA (ed) Frontiers of plant tissue culture, Calgary, International Association for plant tissue culture, pp 1–13
2. Zenk MH, Deus B (1982) Natural product synthesis by plant cell cultures. In: Fujiwara A (ed) Plant Tissue Culture, Tokyo, Japan, pp 391–394
3. Fujita J, Maeda Y, Suga C, Morimota T (1983) Production of Shikonin derivatives by cell suspension cultures of *Lithospermum erythrorhizon*. III. Comparison of shikonin derivatives of cultured cells and ko-shikon. Plant Cell Reports 2:192–193
4. Matsumoto I, Kanno N, Ikeda T, Obi Y, Kisaki T, Noguchi M (1981) Selection of cultured tobacco cell strains producing high levels of ubiquinone – 10 by a cell cloning technique. Agric Biol Chem 45:1627–1633
5. Seeto F, Yamada Y (1984) High berberine producing cultures of *Coptis japonica*. Phytochemistry 23:281–285
6. Alfermann AW, Bergmann W, Figur C, Helmbold U, Schwantag D, Schuller I, Reinhard E (1983) In: Mantell SH, Smith H (ed) Plant Biotechnology Cambridge University press. Cambridge, London New York New Rochelle Melbourne Sydney, pp 67–74
7. Chadha MS, Heble MR (1980) Biotechnological applications of plant tissue and cell cultures: Problems and prospects. In: Srivastava US (ed) Golden Jubilee Commemoration Volume 1980. National Academy of Sciences, India, pp 681–708
8. Staba EJ (1982) Production of useful compounds from plant tissue culture. In: Fujiwara A (ed) Plant tissue culture 1982. Japanese Association for Plant Tissue Culture Tokyo, Japan, pp 25–30
9. Shabde MM, Murashige T (1974) Organ culture. In: Durbin R (ed) Nicotiana. Procedures for experimental use. Washington DC, US Department of Agriculture Tech. Bull. No. 1586, pp 40 51
10. Murashige T, Skoog F (1962) A revised medium for rapid growth and bioassays with tobacco tissue cultures. Physiol Plant 15:473–497
11. Benjamin BD, Roja PC, Heble MR, Chadha MS, Multiple shoot cultures of *Atropa belladonna*: Effect of physicochemical factors on growth and alkaloid formation (Z Pflanzenphysiol, Submitted to press)
12. Takayama S, Misawa M (1981) Mass propagation of *Begonia* and *hiemalis* plantlets by shake cultures. Plant and Cell Physiol 22:461–467
13. Wood HN, Braun AC (1961) Studies on the regulation of certain essential biosynthetic systems in normal and crown-gall tumor. Proc Natl Acad Sci USA 47:1907–1913
14. Heble MR, Benjamin BD, Roja PC, Chadha MS (1981) Studies on shoot cultures of *Atropa belladonna* and *Rauwolfia serpentina*. In: Sen SK, Giles KL (ed) Plant cell culture in crop improvement. Plenum Press, New York London, pp 57–63

15. Srivastava AK (1978) *Rauwolfia serpentina* and its cultivation in India. Farm Bulletin No. 10. Central Indian Medicinal Plant organisation, Lucknow, India
16. Roja PC, Benjamin BD, Heble MR, Chadha MS (1984) Indole alkaloids from multiple shoot cultures of *Rauwolfia serpentina* Benth ex. Kurz. Planta Medica (in press)
17. Krueger RJ, Carew DP, Lui JHC, Staba EJ (1982) Initiation, maintenance and alkaloid content of *Catharanthus roseus* leaf organ cultures. Planta Medica 45:56–57
18. Constabel F, Gaudet-La-Prairie P, Kurz WW, Kutney JP (1982) Alkaloid production in *Catharanthus roseus* cell cultures. Plant cell Reports 1:139–142
19. Staba EJ, Chung AC (1981) Quinine and quinidine production by *Cinchona* leaf root and unorganised cultures. Phytochemistry 20:2495–2498
20. Morton JF (1977) Major medicinal plants. Botany, Culture and Uses Thomas Springfield, Illinois
21. Staba EJ, Zito S, Amin M (1982) Alkaloid production from *Papaver* tissue cultures. J Nat Prod 45:256–262
22. Heble MR, Staba EJ (1980) Diosgenin synthesis in shoot cultures of *Dioscorea composita*. Planta Medica Supp, pp 120–123
23. Hirotani M, Furuya T (1977) Restoration of cardenolide synthesis in redifferentiated shoots from callus cultures of *Digitalis purpurea*. Phytochemistry 16:610–611
24. Lui JHC, Staba EJ (1979) Effect of precursors on serially propagated *Digitalis lanata* leaf and root cultures. Phytochemistry 18:1913–1916
25. Hagimori M, Matsumoto T, Obi Y (1982) Studies on the production of *Digitalis* cardenolides by plant tissue culture. Plant Physiol 69:653–656
26. Kuberski C, Scheibner H, Steup C, Diettrich B, Luckner M (1984) Embryogenesis and cardenolide formation in tissue cultures of *Digitalis lanata*. Phytochemistry 23:1407–1412
27. Suffness M, Douros J (1982) Current status of NCI plant and animal product program. J Nat Prod 45:1–4
28. Nittala SS, Lavie D (1981) Chemistry and genetics of withanolides in *Withania somnifera* hybrids. Phytochemistry 20:2741–2748
29. Zeig RG, Zito SW, Staba EJ (1983) Selection of high pyrethrin producing tissue cultures. Planta Medica 84:88–91
30. White PR (1933) Plant tissue cultures: results of preliminary experiments on the culturing of isolated stem tips of *Stellaria media*. Protoplasma 19:97–116
31. Loo SW (1945) Cultivation of *Asparagus* in vitro. Am J Bot 17–32
32. Murashige T (1974) Plant propagation through orgon cultures. Annu Rev Plant Physiol 25: 135–166
33. Kartha KK, Leung NL, Pahl K (1980) Cryopreservation of strawberry meristem and mass propagation of plantlets. J Amer Soc Hort Sci 105:481–484
34. Fowler MW (1984) Time for plant cell culture. Nature 307:504

IV Fermentation and Cryopreservation

Large-Scale Production of Rosmarinic Acid from Plant Cell Cultures of *Coleus blumei* Benth.

B. ULBRICH[1], W. WIESNER[1], and H. ARENS[1]

1 Introduction

From time immemorial, plants have been an inexhaustible source of healing extracts for suffering mankind. Modern drug research has made it possible to isolate and identify the active principles of these extracts and to specify their dose-response relationships. In this way it has become possible to verify the phytotherapeutic action of ancient drugs. Nowadays large screening programs are running in many pharmaceutical companies in order to select new active principles from thousands of plant species. But only few companies have realized that plant cell culture is also a rich source of new active substances [1] and a potent technique to produce natural compounds like shikonin and rosmarinic acid. Its chemical structure was determined in 1958 [2] and its isolation from plant cell lines of *Coleus blumei* Benth. was reported in 1977 [3, 4]. Today we are able to present a procedure of unparalleled high production of rosmarinic acid with our cell line CBL 2B of *Coleus blumei* Benth.

The plant *Coleus blumei* Benth. is a natural hybrid of several *Coleus* species. It attains a maximum height of about 1 m and has dark red blotches on the upper surface of the leaves. The individual plant from which our cell lines originate has a rosmarinic acid content of 1.1% dry wt.

2 Plant Cell Culture

Using literature-based data [3, 4] we initiated in 1979/80 callus cultures from leaves of a typical *Coleus* plant growing on the window sill in our laboratories. After about 1 year of monthly subcultivation the callus had adapted to the constant standard conditions. Maintenance was performed on modified B5-medium [5, 6].

We established the suspension culture in Erlenmeyer shaking flasks under the same conditions as the callus cultures, but rotating at 150 rpm on an orbital shaker. Subcultivation was carried out every week by transferring exactly the same quantity of inoculum to fresh medium. The suspension culture is designated as stable if oscillating in constant (weekly) intervals between fixed limits of biomass, sugar content and pH.

1 A. Nattermann & Cie. GmbH, Biotechnology/Plant Cell Culture, Postfach 350120, 5000 Köln 30, FRG

Primary and Secondary Metabolism of Plant Cell Cultures, ed. by Neumann et al.

These are called "basic cultures" and must be handled in a constant manner during the whole phase of investigation.

3 Production Medium

At present only one running industrial-scale plant cell culture process producing natural compounds exists in the world. The Mitsui Petrochem. Ind. Ltd. [7, 8] produces a mixture of shikonin derivatives from *Lithospermum officinale* var. *erythrorhizon* Sieb. et Zucc. cell culture using a two-stage procedure: mass cultivation of plant cells in growth medium with subsequent transfer of the cells into a special production medium [9–12].

Using a similar concept we started our investigations with a rough screening of ten heterogenous media hoping to find a suitable rosmarinic acid production medium (Fig. 1). Symbols are explained in the Appendix.

The data in Fig. 1 clearly demonstrated that the HI-medium, a modified Heller-medium [13], was a suitable initial medium.

Separating it into five groups of compounds (macro-, microelements, vitamins, phytohormones, and sugars) we studied the effects of simplifying this medium (Fig. 2). Surprisingly, we found that even when we simplified the HI-medium to a mere sucrose solution, productivity was lowered by only about 30%.

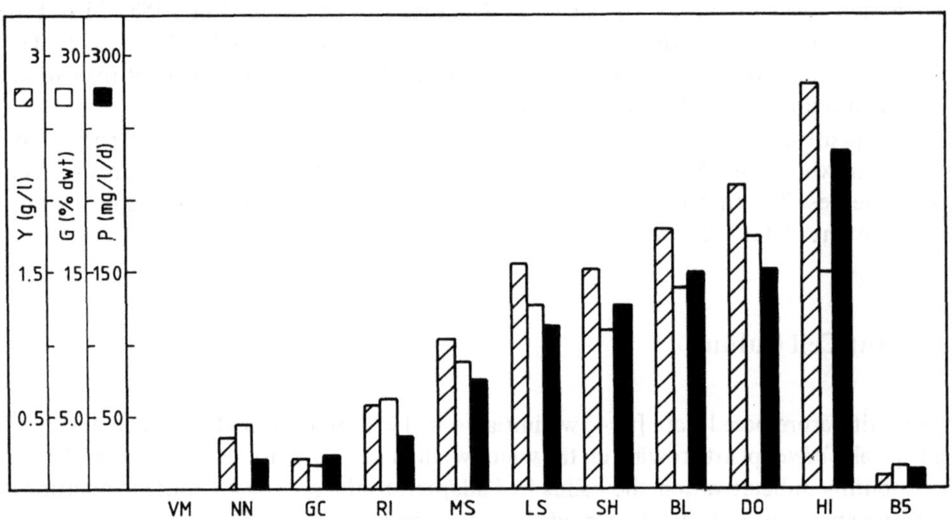

Fig. 1. Rosmarinic acid yields of cell line CBL 1B in 10 different media. Remarks: Shake flask experiments were started with an identical quantity of inoculum and under standard conditions for one passage; parameters were analyzed kinetically, rosmarinic acid data are maximum values, sugar concentration was 50 g l^{-1} in each case. The standard growth medium is B5 with 16 g sucrose l^{-1}. *VM* Velicky 1968; *NN* Nitsch 1969; *GC,B5* Gamborg 1968; *RI* Ringe 1972; *MS* Murashige 1962; *LS* Linsmaier 1965; *SH* Schenk 1972; *BL* Blaydes 1966; *DO* Gresshoff 1972; *HI* Heller 1953; media components in: J. Reinert and Y.P.S. Bajaj, Plant Cell, Tissue, and Organ Culture, Springer Berlin 1977; B.V. Conger, Cloning Agricultural Plants Via In Vitro Techniques, CRS Press Boca Raton 1981

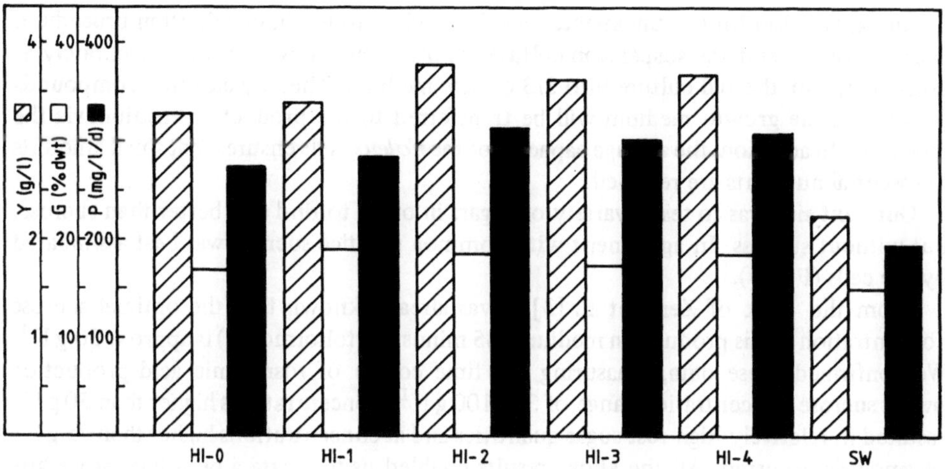

Fig. 2. Rosmarinic acid yields of cell line CBL 1B in various HI-media. Remarks: Simplification of the Heller-medium to a mere sucrose solution (50 g l^{-1}) in shake flask experiments with identical quantities of inocula and under standard conditions for one passage; rosmarinic acid data represent maximum values. *HI-0* HI-medium as control; *HI-1* HI minus phytohormone IAA; *HI-2* HI minus vitamin thiaminium-HCL; *HI-3* HI minus IAA and minus vitamin; *HI-4* HI-3 minus microelements; *SW* HI-4 minus macroelements

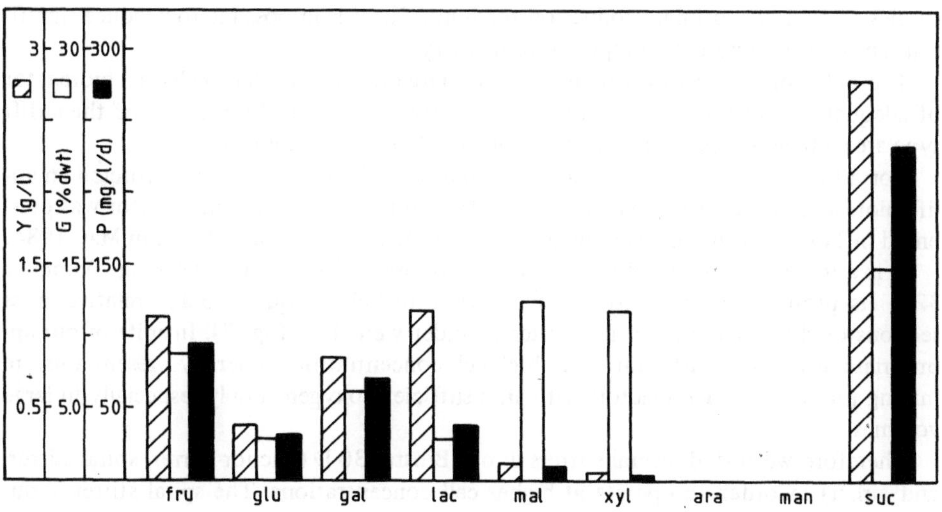

Fig. 3. Rosmarinic acid yields of cell line CBL 2B in HI-medium with different sugars. Remarks: Sugars were all added at a concentration of 50 g l^{-1}. Data were obtained on the day of maximum rosmarinic acid production in sucrose medium under standard conditions for one passage. *fru* fructose; *glu* glucose; *gal* galactose; *lac* lactose; *mal* maltose; *xyl* xylose; *ara* arabinose; *man* mannitol; *suc* sucrose; *ara* and *man* Y < 1 mg l^{-1}

An explanation for this unexpected result could be found in our dilution procedure, because we transfer the suspension cultures to other media by pouring as accurately as possible 1/3 of the old culture into 2/3 of new medium. Thereby essential compounds (anions) of the growth medium will be transferred to the production medium via the inoculum. In addition, the storage capacity of the *Coleus* cells ensures that small amounts of essential nutrients are retained.

Our next aim was to test a variety of sugars in order to find one better than sucrose, but without success. In agreement with common practice sucrose was best assimilated by the cells (Fig. 3).

From the work of Zenk et al. [3] it was already known that the optimal sucrose concentration in his production medium (B5 minus phytohormones) is approx. 50 g l^{-1}. We confirmed these data, measuring the time course of rosmarinic acid production over a sucrose concentration range of 34–100 g l^{-1}. Concentrations higher than 50 g l^{-1} resulted in relatively high rest sugar quantities and at concentrations higher than 75 g l^{-1} plasmolysis occurred. All the above results enabled us to create a two-stage procedure with a simple and low cost production medium for rosmarinic acid yielding a maximum of 270 mg l^{-1} d^{-1} in shake flasks at this step of investigation.

4 Bioreactor Equipment

The experiments described above had to be performed in shake flasks as preliminary studies to find an optimal production medium. But it is impossible to produce significant amounts of a natural compound in this way.

The only suitable instrument to cultivate large masses of plant cells is a bioreactor of adequate dimensions. Up to now it was belived that only bioreactors of the airlift type are suitable for plant cells because of their low shear level [14, 15].

For a long time we were of the same opinion and consequently constructed an airlift bioreactor suitable for our problems. As a utility model [16] the reactor was presented in Tokyo on the occasion of the Deutsche Leistungsschau in Japan in May 1984.

It is a low cost construction with a total volume of 42 l and a working volume of 32 l. Its proportional dimensions of diameter to height equal 1:6 and relative cross sections of flow of the inner and outer cylinder were 1:1 (Fig. 7). In spite of our appropriate airlift we must admit that high cell concentrations generally present a severe mixing problem in such reactors with an insufficient oxygen supply, especially in large volumes.

Therefore we tested several stirrers in our Biostat 30 D (anchor stirrer, spiral stirrer, and airlift) in order to operate at higher cell concentrations. The spiral stirrer is our own construction termed "module spiral stirrer" and is also a utility model [17]. It is built of six modules and a special end module each consisting of one plain metal ring (spiral blade) fixed to the stirrer shaft with two spokes and two rings. The distance between the rings determines the angle of the stirrer blades. In this way we are able to vary the transport speed of the culture broth at constant rotations. The rotation speed of the module spiral stirrer range between 50–100 rpm without significant cell damage. Standard rotation speed was set at 100 rpm (Fig. 4).

System		rosmarinic acid yields		
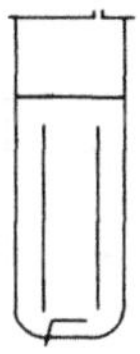	airlift 0.6vvm	2.5g/l	0.31g/l/d	10.3% dwt
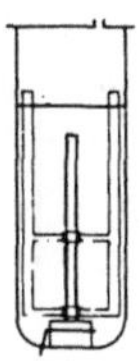	anchor 0.6vvm 40rpm*	2.1g/l	0.21g/l/d	11.7% dwt
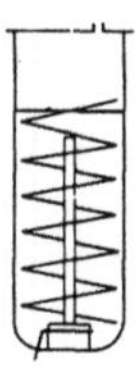	module spiral 0.6vvm 50rpm	3.3g/l	0.41g/l/d	16.6% dwt
	100rpm	3.2g/l	0.45g/l/d	13.6% dwt

Fig. 4. Comparison of different mixing systems

Experimental results were obtained under standard conditions. Aeration was performed by compressed air, sugar concentration was 50 g l^{-1} in each case. * = N > 40 rpm of the anchor stirrer produced cell damage.

5 Fermentation Procedure

Using the module spiral stirrer the productivity of cell line CBL 2B increased to 450 mg l^{-1} d^{-1}. A further increase could only be achieved by increasing the initial cell mass concentrations, thereby shortening the production time.

Thus the inoculum is cultivated in a fed batch process until an appropriate cell concentration is reached. Then about 30%–50% culture broth is transferred from the breeding reactor to the production reactor with sucrose solution (50 g l^{-1}) as production medium (batch).

Using this procedure the yield of rosmarinic acid increases to 5.5 g l^{-1} or 910 mg l^{-1} d^{-1} representing 21% of the dry wt. Maximal rosmarinic acid production appears in the early stationary phase and coincides with a second pH-minimum and the broad pO2-minimum. Afterwards cell mass decreases (cell lysis) and rosmarinic acid vanishes at the same rate as it has been synthesized (Fig. 5).

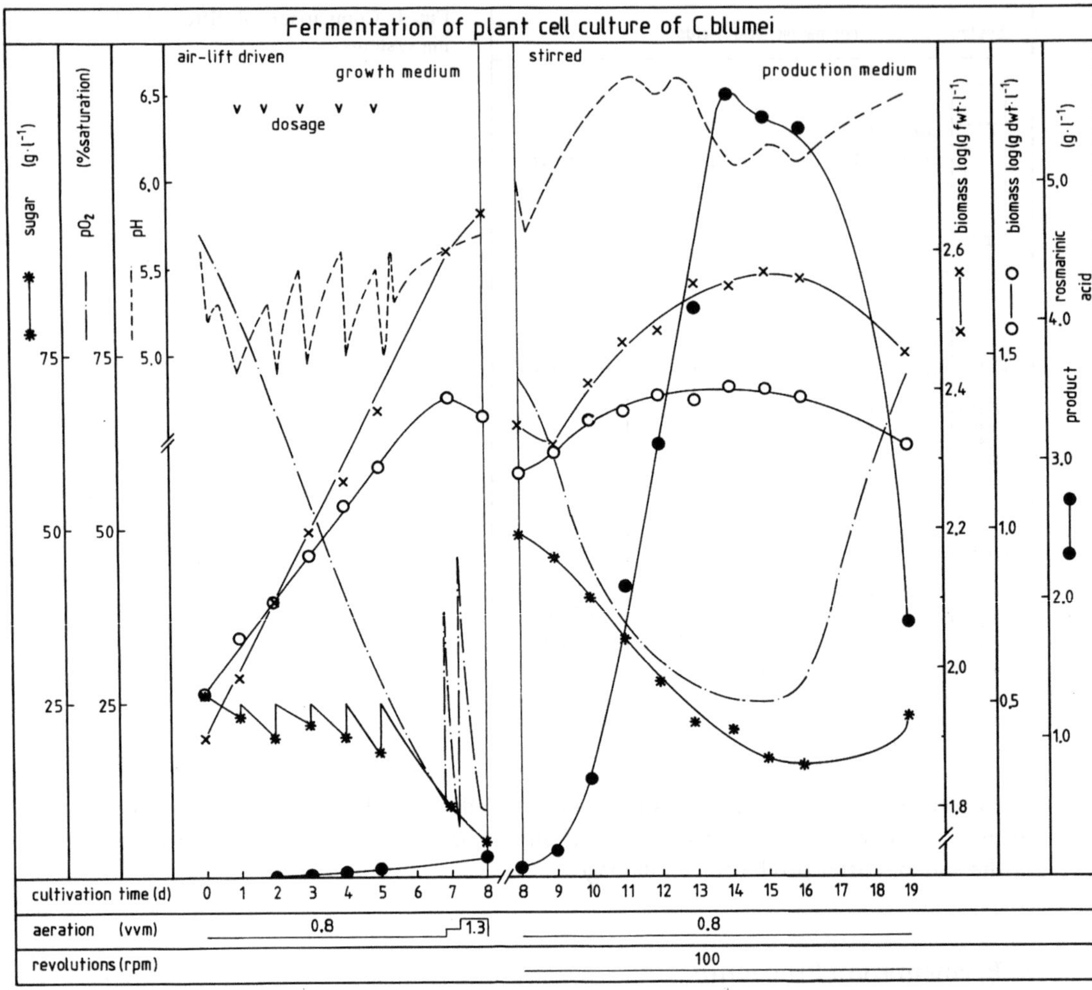

Fig. 5. Two stage process of rosmarinic acid production by cell line CBL 2B. Working volume in both reactors was 32 l

6 Isolation Procedure

We developed a rapid purification procedure for rosmarinic acid.

The plant cell mass is separated from the production medium by centrifugation, extracted in hot water (pH 3, 80 °C) and finally centrifuged once more.

The clear filtrate is poured onto an Amberlite XAD-2 column (diameter 145 mm, heigt 550 mm) in order to adsorb rosmarinic acid. After washing the column with water, elution is performed with aqueous methanol (40%). The eluate is evaporated to dryness and redissolved in highly purified water (about 1 g residue in 20 ml) standing several hours at 4 °C.

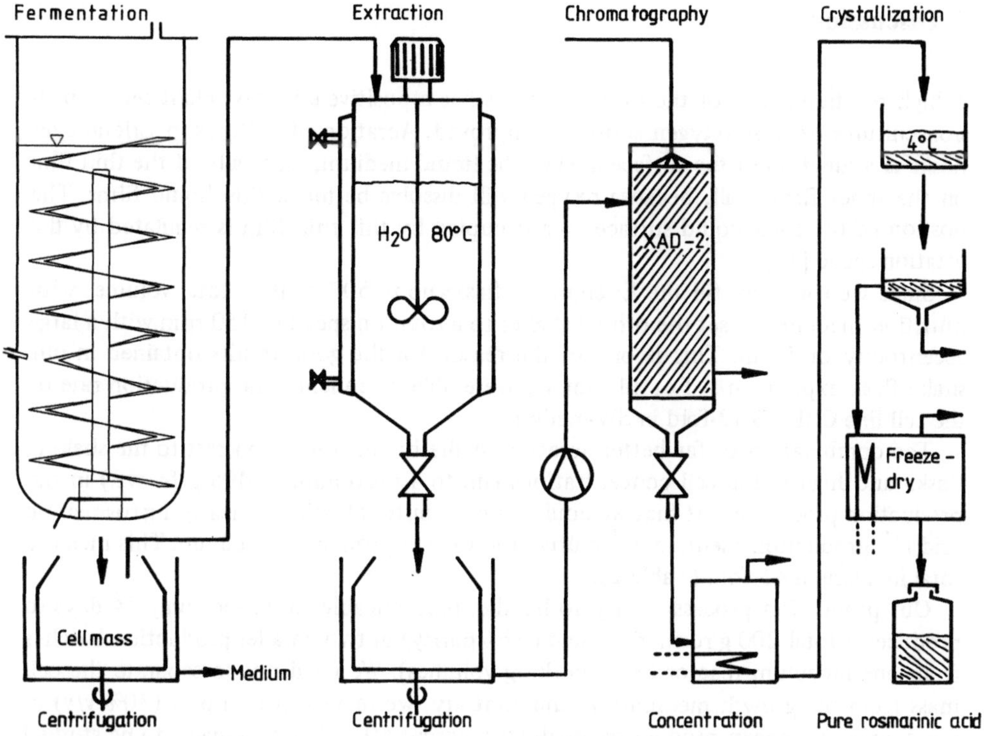

Fig. 6. Diagram projecting the isolation procedure of rosmarinic acid from plant cell line CBL 2B

The precipitated crystals are harvested and washed with ice water and freeze dried (purity 95%). The results of a typical isolation procedure are given in Table 1.

This procedure enables us easily to isolate about 100 g rosmarinic acid from one laboratory scale batch fermentation of a working volume of 32 l production medium.

Table 1. Isolation of pure rosmarinic acid from a typical production fermentation of the cell line CBL 2B

Step	Rosmarinic acid (g)	Yield (%)	Purity (%)
Cell material	152.3	100	20.0
Water extract	146.2	96	30.1
Column eluate	117.8	77	85.5
Crystals	103.7	65	97.0

Remarks: The purity of the crystals is sufficient for requirements in clinical research

7 Discussion

A high rotation speed of the orbital shakers has a positive effect on plant cell suspension cultures because oxygen supply is improved. Aeration of cultures in Erlenmeyer flasks is a function of the surface area of the liquid medium, especially of the thin film on the inner flask wall, because oxygen will dissolve better in thin liquid films. The portion of the total liquid surface area occupied by this thin film is regulated by the rotation speed [18].

Since we routinely fill the Erlenmeyer flasks up to 50% of their total volumes with nutrition medium we set the orbital shaker to a rotation speed of 150 rpm with a large eccentricity of 5 cm. This is one of the reason for the good results obtained in our shake flask experiments. For all that we were able to enhence the production rate of the cell line CBL 2B 13-fold in SW-medium.

The combination of far better aeration in the bioreactor in contrast to the shaking flasks and high initial cell concentrations (up to a maximum of 495 g fr. wt.) in the production process meant that we could raise the output further to nearly 1 g rosmarinic acid l^{-1} production medium d^{-1} cultivation time in production medium. This means a total increase of 45-fold (Table 2).

Our production process is easy to handle, reproducable, and needs only 14 days to produce in total 200 g rosmarinic acid (97% purity) in two parallel production batches from one inoculum reactor (32 l working volumes). We need not to separate the cell mass from the growth medium; on the contrary, we recycle it as a part (30% v/v) of our simple and cheap production medium (sucrose 50 g l^{-1}). It remains to be studied whether sucrose can partially be replaced by molasses.

The stability of our cell lines can be attributed to our strictly controlled constant cultivation conditions. It is absolutely important that no critical parameter is altered.

Table 2. Survey of the single steps of optimization

Step	X (g l^{-1})	Y (mg l^{-1})	G (%)	P (mg l^{-1} d^{-1})
Plant				
Coleus blumei Benth.	–	–	1.1	–
Growth medium				
Basic culture in B5	4.7[a]	150	2.9	20
Production medium				
Sucrose/water	21.3	2,240	15.0	266
Biostat 30 D				
Module spiral stirrer	23.4	3,200	13.6	450
Production process				
Two stage procedure	25.7	5,500	21.4	910
Production medium[b]	23.0	2,900	13.0	220

Remarks: Yields of rosmarinic acid in the various optimization steps with plant cell lines CBL 1B and CBL 2B of *Coleus blumei* Benth.

[a] Extracted dry wt

[b] Calculated from the plot in the paper of Zenk et al. 1977 [3]

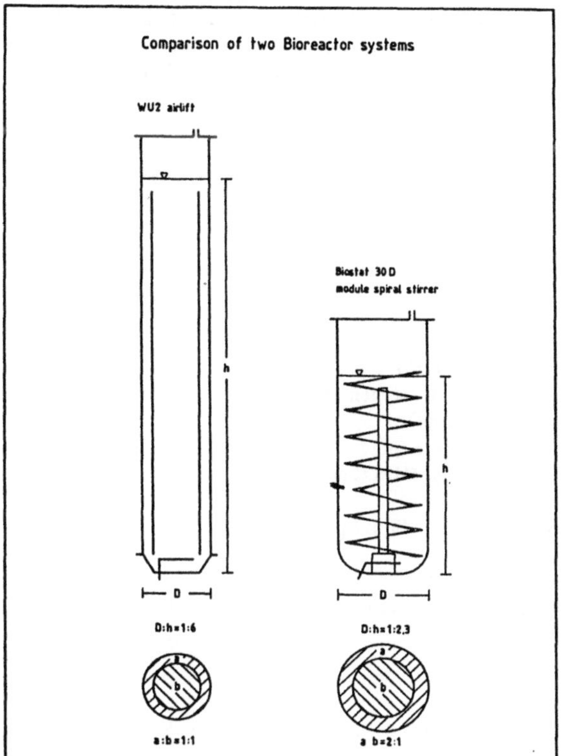

System	rosmarinic acid yields		
WU 2 airlift	5.6 g/l	0.91 g/l/d	21.5 % dwt
Biostat 30D module spiral stirrer	5.5 g/l	0.93 g/l/d	21.4 % dwt

Fig. 7. Two bioreactor systems suitable for mass cultivation of pcc. *D* inner diameter of the outer cyliner; *h* height of the culture broth; *a* outer cross section of flow (down); *b* inner cross section of flow (up)

With regard to the problem of plant cell cultures in bioreactors, we believe that the widespread opinion only airlift-type bioreactors are suitable instruments for mass cultivation of plant cells requires an amendment.

The results under discussion demonstrate the efficiency of our spiral stirrer, which possesses the low shear airlift mixing characteristics as well as providing sufficient transverse mixing of the culture broth. During the production process, volumetric mass transfer coefficients of oxygen ($k_L a$) range between 5–18 h^{-1} as previously reported [19]. Thus we are able to work with compact bioreactors of the Dechema-Norm-type (diameter: height = 1:3), as if it were a tall slim airlift system (Fig. 7). This is an important fact and we hope to solve the scale-up problems of mass cultivation of plant cell cultures of *Coleus blumei* in the next years by using two large bioreactors with 200-l and 450-l capacities.

Our investigations demonstrate the tremendous potential of the plant cell culture technique.

Acknowledgements. The authors gratefully acknowledge the financial support of the Ministry of Research and Technology of the Federal Republic of Germany, Bonn and the cooperation of Prof. Dr. M.H. Zenk, Munich. Dr. M.J. Parnham kindly assisted with the preparation of the manuscript and Dr. K.-H. Gober with computer graphics.

Appendix

Abbreviations:

CBL 1B, 2B	=	cell lines of pcc of C. blumei
G	=	content of rosmarinic acid per dry wt (%)
k_La	=	volumetric mass transfer coefficient of oxygen (h^{-1})
N	=	stirrer speed in rounds per minute (rpm)
P	=	productivity of the cells: yield per time ($mg\,l^{-1}\,d^{-1}$)
p	=	purity of rosmarinic acid fraction in % (HPLC analysis)
pcc	=	plant cell culture
Q	=	gas flow rate in volumes gas per volume liquid per minute (vvm)
SW	=	sucrose/water production medium
X	=	cell concentration as dry weight ($g\,l^{-1}$)
Y	=	yield of natural compound per volume nutrient medium ($mg\,l^{-1}$)
y	=	yield of rosmarinic acid in %

Standard Cultivation Conditions:

Temperature	:	25 °C
Light	:	100 lx of Fluora fluorescent lamp light
Rotations	:	shake flasks 150 rpm, diameter of eccentricity is 5 cm
Rotations	:	N = 100 rpm

Analysis of Rosmarinic Acid:

Method	:	High pressure liquid chromatography (HPLC)
Apparatus	:	Hewlett-Packard 1084 B
Column	:	Lichrosorb RP8 (200 × 4.6 mm i.d.)
Mobile Phase	:	acetonitrile in water (pH 3, phosphoric acid)
Gradient	:	linear
Wave length	:	254 nm detection

References

1. Arens H, Borbe HO, Ulbrich B, Stöckigt J (1982) Detection of pericine, a new CNS-active indole alkaloid from picralima nitida cell suspension culture by opiate receptor binding studies. Planta Med 46:210–214
2. Scarpati LM, Oriente G (1958) Isolamento e costituzione dell'acido rosmarinico (dal rosmarinus off.). Rich Sci 28:2329–2333
3. Zenk MH, El-Shagi HE, Ulbrich B (1977) Production of rosmarinic acid by cell suspension cultures of coleus blumei. Naturwissenschaften 64:585

4. Razzaque A, Ellis BE (1977) Rosmarinic acid production in coleus cell cultures. Planta 137: 287–291
5. Ulbrich B, Wiesner W, Arens H (1982) Verfahren zur Herstellung von Rosmarinsäure aus Pflanzenzellkulturen und von Rosmarinsäure enthaltenden Pflanzenzell-Preßlingen. DOS No 32 47 610.8 (23.12.1982)
6. Gamborg OL, Miller RA, Ojima K (1968) Nutrient requirements of suspension cultures of soybean root cells. Exp Cell Res 50:151–158
7. Mitsui Petrochem Ind Ltd (1982) Jpn Patent J 57039–778
8. Mitsui Petrochem Ind Ltd (1982) Jpn Patent J 57039–779
9. Fujita Y, Hara Y, Ogino T, Suga C (1981) Production of shikonin derivatives by cell suspension cultures of lithospermum erythrorhizon. I. Effects of nitrogen sources on the production of shikonin derivatives. Plant Cell Rep 1:59–60
10. Fujita Y, Hara Y, Suga C, Morimoto T (1981) Production of shikonin derivatives by cell suspension cultures of lithospermum erythrorhizon. II. A new medium for the production of shikonin derivatives. Plant Cell Rep 1:61–63
11. Fujita Y, Maeda Y, Suga C, Morimoto T (1983) Production of shikonin derivatives by cell suspension cultures of lithospermum erythrorhizon. III. Comparison of shikonin derivatives of cultured cells and ko-shikon. Plant Cell Rep 2:192–193
12. Curtin ME (1983) Harvesting profitable products from plant tissue culture. Biotechnology 1: 649–657
13. Heller R (1953) Recherches sur la nutrition minerale des tissus vegetaux cultives in vitro. These, Paris and Ann Sci Nat Bot Biol Veg 14:1–223
14. Wagner F, Vogelmann H (1977) Cultivation of plant tissue cultures in bioreactors and formation of secondary metabolites. In: Barz W, Reinhard E, Zenk MH (eds) Plant tissue culture and its biotechnological application. Springer, Berlin Heidelberg New York, pp 245–252
15. Fowler MW (1982) The large scale cultivation of plant cells. Prog Microbiol 16:207–229
16. A Nattermann & Cie (1983) Bioreaktor mit vertikaler Flüssigkeitsumwälzung. Ger Utility Model G 83 14 233.9 (13.)5.1983)
17. A Nattermann & Cie (1982) Wendelrührer für Bioreaktoren. Ger Utility Model G 82 36 121.5 (23.12.1982)
18. Vogelmann H (1981) Aspects on scale-up and mass cultivation of plant tissue culture. In: Moo-Young M, Robinson CW, Vezia C (eds) Advances in biotechnology, vol I. Pergamon Press, Oxford New York, pp 117–121
19. Smart NJ, Fowler MW (1981) Effect of aeration on large-scale cultures of plant cells. Biotechnol Lett 3:171–176

Immobilised Plant Cell Culture Systems

K. LINDSEY and M.M. YEOMAN [1]

1 Introduction

In the last 5 years in particular there has been a rejuvenated interest in the use of cultured plant cells for the study of secondary metabolism. More specifically, this interest has become apparent as a burgeoning of reports describing the immobilisation of plant cells in a variety of putatively inert matrices, and it is primarily this area with which we will deal in this article. Although, as we will see, immobilisation itself may in some cases act to enhance the biochemical productivity of cell cultures, we believe that it is merely one part of a sequence of "manipulations" which, in concert, will result in the in vitro production of high yields of specific secondary metabolites.

The expression of secondary metabolic activity in cultured plant cells varies *between cells* within a single culture and also *with time* within a single culture - both throughout a growth cycle and over successive culture periods (Lindsey and Yeoman 1984a). This expression can also be modified by almost every conceivable environmental factor - nutrients, growth regulators, pH, temperature, light etc. (see e.g. Mantell and Smith 1983) and as a result much effort has been put into optimising cultural conditions for growth and secondary product formation. The empiricism of this approach is, however, unsatisfactory, and we (Yeoman et al. 1980, 1982; Lindsey and Yeoman 1983a,b) have drawn attention to the relationship, often reported in the literature, between secondary metabolite production and the structural organisation or aggregation of cells in culture. We believe that this phenomenon points to the use of immobilised cells as a means of enhancing yields by artificially producing aggregates, in which metabolic intermediates common to primary and secondary pathways are preferentially used in the latter (Lindsey and Yeoman 1984b). It has been argued (Phillips and Henshaw 1977; Lindsey and Yeoman 1983a) that in rapidly dividing cells some aspects of secondary metabolic activity, such as the accumulation of phenolics or alkaloids, are depressed as a result of the predominant utilization of precursors (e.g. amino acids) in protein synthesis. When the rate of cell division decreases, however, either towards the end of the growth cycle or in aggregates and differentiated structures, the precursors are incorporated primarily in secondary compounds at the expense of primary. It would therefore be expected that the physical entrapment of plant cells in an inert matrix would favour enhanced secondary metabolite production for the following reasons: (1) a high degree of cell-

1 Department of Botany, University of Edinburgh, The King's Buildings, Mayfield Road, Edinburgh EH9 3JH, Scotland, United Kingdom

Primary and Secondary Metabolism of Plant Cell Cultures, ed. by Neumann et al.

cell contact (i.e. partial structural organisation) is encouraged, which would act to limit the rate of cell division (compared with suspended cells) and permit the establishment of gradients of nutrients and other factors, which are important in the regulation of secondary metabolic activity; (2) on a practical level, chemical manipulation is facilitated if cells are immobilised and cultured in a fixed bed bioreactor such as a packed column; such a system readily permits the cyclical and sequential chemical treatment of the cells without recourse to their physical manhandling; and (3) harvest of the product is facilitated in fixed bed systems without destruction of the biomass, if the product is either naturally exported, as is the case for some compounds (e.g. Lindsey et al. 1983; Furuya et al. 1984) or if release can be induced by the permeabilisation of the cell membranes (Felix et al. 1981).

2 Immobilisation Techniques

Since 1979, when Brodelius et al. described the entrapment in calcium alignate gel of viable cells of *Catharanthus roseus, Morinda citrifolia*, and *Digitalis lanata*, a number of immobilisation techniques have been reported. Although microbial and animal cells can be immobilised by adsorptive techniques such as attachment to glass or plastic

Table 1. A summary of techniques for plant cell immobilisation

Species	Immobilisation substratum	Reference
Catharanthus roseus *Morinda citrifolia* *Digitalis lanata*	Calcium alginate	Brodelius et al. (1979)
Digitalis lanata	Calcium alginate	Alfermann et al. (1980)
Catharanthus roseus	Calcium alginate, agarose, agar, carrageenan, gelatin, polyacrylamide, alginate + gelatin, agarose + gelatin	Brodelius and Nilsson (1980)
Solanum aviculare	Polyphenyleneoxide	Jiřku et al. (1981)
Cannabis sativa *Daucus carota* *Ipomoea* sp.	Calcium alginate	Jones and Veliky (1981)
Glycine max	Hollow fibres	Shuler (1981)
Catharanthus roseus	Hypol 3000	Felix and Mosbach (1982)
Catharanthus roseus	Polyacrylamide and alginate	Lambe and Rosevear (1982)
Capsicum frutescens *Datura innoxia* *Solanum nigrum*	Fibrous polypropylene, alginate + nylon, agar + nylon	Lindsey and Yeoman (1983b)
Capsicum frutescens *Daucus carota*	Reticulate polyurethane	Lindsey et al. (1983)
Daucus carota *Petunia hybrida*	Hollow fibres	Prenosil and Pedersen (1983)
Mucuna pruriens	Calcium alginate	Wichers et al. (1983)
Papaver somniferum	Calcium alginate	Furuya et al. (1984)

surfaces (see Lindsey and Yeoman 1983b, 1985) this method is not particularly successful with cellulose-enclosed plant cells. Most techniques have involved the entrapment of cells within polymeric matrices which, apart from the popular calcium alignate, have included gels of agar, polyacrylamide, agarose, gelatin, and carrageenan, as well as synthetic plastics such as nylon, polyurethane, and polyphenyleneoxide (Table 1). Although, in our laboratory, we have used calcium alginate and agar for the entrapment of viable cells of *Datura innoxia, Solanum nigrum*, and *Capsicum frutescens* (Lindsey and Yeoman 1983b), we now favour the use of reticulate polyurethane foam as an immobilisation matrix (Lindsey et al. 1983; Lindsey and Yeoman 1984c). Entrapment occurs as the passive invasion, by suspended cells, of the pores of the foam which is simply immersed in a liquid culture. Not only is the method simple, it requires the involvement of no chemical reagents and, as will be demonstrated below, has no deleterious effects on culture viability. Furthermore, the polyurethane is completely stable in liquid media over prolonged culture periods and since the foam is 97% (v/v) void, there is only a negligible barrier to permeability which is directly attributable to the polyurethane itself.

3 The Manipulation of Cultured Cells

As indicated above, the immobilisation and chemical manipulation of cultured cells are two aspects of the sequence of procedures which, we believe, are necessary to optimise secondary metabolite production. The background against which such treatments should be performed is one of a culture which is capable of responding particularly well to environmental stimuli. In other words, cell lines should be isolated which, when cultured under conditions conducive to secondary metabolite production, accumulate greater than average yields. Such high-yielding lines can, theoretically, be isolated either after treatment of the "parent" culture with a mutagen, to generate variation, or by selection from the heterogeneity that naturally exists in plant cell cultures. Ogino et al. (1978), for example, have demonstrated how the average nicotine yield of tobacco callus cultures can be increased five-fold by continuous selection of "naturally" high-yielding cells. Although much work remains to be done on understanding the origins of natural heterogeneity and the reasons for the stability (or lack of stability) of such cell lines, this technique provides material upon which further physical and chemical manipulations can be performed, to produce a maximum product yield. We will now turn from these general considerations to the more specific, and describe and discuss some of our work on the secondary metabolism of cultured cells of the Chilli pepper, *Capsicum frutescens*.

The secondary products in which we are interested are the capsaicinoids, and capsaicin in particular, which is the major component. These compounds are synthesised from the amino acids phenylalanine and valine, which, respectively, contribute to the aromatic ring and fatty acid chain moieties of the capsaicinoid molecule (Fig. 1). As far as can be detected, capsaicin is only accumulated in the fruit of the intact plant, and gives it its characteristically pungent flavour. When cultured as callus or as liquid-suspended cells, however, tissue derived from the fruit produces no more capsaicin than

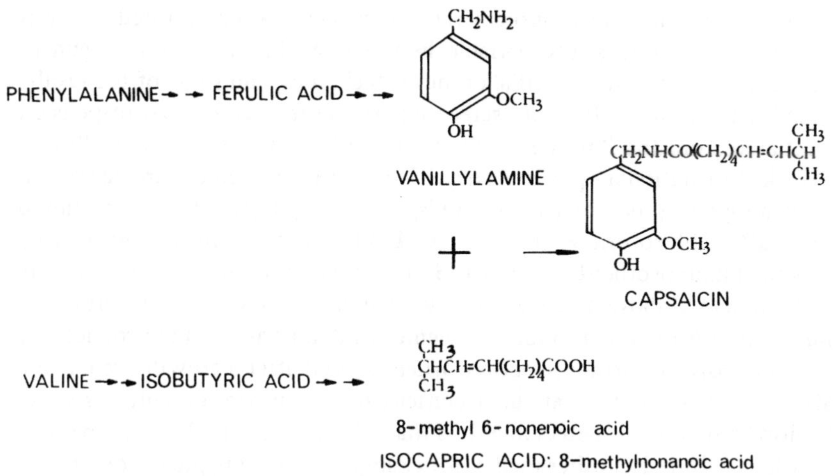

Fig. 1. Aromatic ring and fatty acid chain moieties of the capsaicinoid molecule

does tissue from any other organ (unpublished observations). Furthermore, the cultured cells, grown under a variety of nutritional conditions, accumulate only very low levels of capsaicin (of the order of 10^{-10} M) and release the product into the surrounding medium (Yeoman et al. 1982; Lindsey et al. 1983).

Using a variety of Chilli pepper that accumulates relatively high levels of capsaicin in the fruit, namely *Capsicum frutescens* cv. annuum, we have, in our laboratory, established over 300 cell lines for screening. Each cell line was obtained using one of two methods. The first is a plating technique, in which a fine suspension of cells is diluted and plated out on an agar medium, and colonies derived from single cells are transferred to fresh medium when they have reached a population size of several thousand cells. In the second method, the single-cell origin of cell lines is ensured by the proliferation of cells from protoplasts. Callus is preplasmolysed in a 13% (w/v) mannitol solution, and is subsequently incubated overnight in an enzyme solution comprising 2% (w/v) cellulase, 1% (w/v) macerozyme, and 1% (w/v) rhozyme in 13% mannitol solution. After a series of washes in 13% mannitol and 21% (w/v) sucrose solutions, protoplasts are plated out and maintained in the dark on conditioned SH basal medium (see below) supplemented with 7% (w/v) glucose and solidified with agarose. This medium allows the establishment of colonies of callus from the individual protoplasts.

All 300 or so cell lines are maintained on the same medium, which is primarily a medium favouring the rapid proliferation of the cultures. This is Schenk and Hildebrandt (1972) medium supplemented with 30 g l^{-1} sucrose, 0.5 mg l^{-1} (2.5×10^{-6} M) 2,4-dichlorophenoxyacetic acid (2,4-D), 2.0 mg l^{-1} (10^{-5} M) p-chlorophenoxyacetic acid (CPA), and 0.1 mg l^{-1} (0.5×10^{-6} M) kinetin (SH basal medium). Cells are maintained at 24 ± 1 °C in continuous fluorescent illumination (Warmwhite) of an illuminance of 20 μmol m^{-2} s^{-1}. Even though the callus cultures are grown under homogeneous conditions, much structural and biochemical heterogeneity is apparent. The appearance of the cultures is very variable, ranging from callus of a compact, nodulated form which, in some cases, contains chlorophyll, to a type which is composed primarily of loosely-packed, colourless cells. The relative growth rate of the cells, when cultured either as callus or in liquid suspension, is also variable between cell lines. We have begun to analyse the capsaicin biosynthetic potential of the 300 cell lines, and some of the results are presented in Table 2. The preliminary screening procedure is as follows: suspension cultures are initiated from the "master" calluses grown from single cells, and allowed to grow for 2 weeks to a stationary phase. At day 15 the cells and medium are harvested, and capsaicin is assayed by one of two methods, namely by high-performance liquid chromatography after chloroform extraction (see Lindsey and Yeoman 1984c), or by an enzyme-linked immunosorbent assay (ELISA) (Aitken, Lindsey, and Yeoman, in preparation). As indicated above, this preliminary screening has involved no continuous selection, and the results obtained so far have demonstrated a variation in productivity which provides the basis for further selection. When assayed after a growth cycle maintained by SH basal medium, the capsaicin content of the suspension cultures varies from a level below that which is detectable (less than 10 ng) to about 3 or 4 μg per culture (60 ml). When cultured in other media, for example containing precursors to capsaicin such as iso-capric acid (8-methylnonanoic acid), yields can be increased 100-fold (Lindsey and Yeoman 1984b), although variability in biosynthetic activity between cell lines is still apparent.

Table 2. The appearance, relative growth rate, chlorophyll content and capsaicin content of 20 cell lines of *C. frutescens*. Each measurement was made after suspending approximately 1.5 g fresh wt. of callus of each cell line in 50 ml SH basal medium and culturing on an orbital shaker for a 15-day period. Relative growth rate = Final F.W. – Initital F.W./Initial F.W. Each value presented represents the mean of 3 replicates

Clone No.	Appearance	Relative growth rate	μg Total chlorophyll/ culture	μg Capsaicin/ culture
D3	Green, compact	1.25	4.3	ND
BV8	Green, compact	1.35	5.0	1.08
CA3	Green, compact	1.62	4.9	ND
CF7	Pale, compact and friable	0.25	1.2	2.19
G3	Pale, compact and friable	0.57	1.9	3.56
CB6	Pale, compact and friable	3.85	3.4	0.28
AI1	Green, friable	0.22	5.2	ND
CD5	Pale, friable	0.08	0.8	0.1
CB2	Green, friable	1.28	7.1	ND
BW1	Pale, friable	1.72	1.1	ND
M1	Pale, friable	0.24	ND	0.01
N3	Pale, friable	1.10	ND	0.03
BF4	Pale, compact and friable	1.02	ND	0.37
BQ1	Pale, friable	0.07	ND	0.19
E1	Pale, friable	ND	ND	ND
EA3	Pale, friable	0.40	ND	0.08
BP6	Pale, friable	ND	ND	0.23
BQJ	Pale, friable	ND	ND	0.03
BRJ	Pale, friable	ND	ND	1.05
I1	Pale green, compact	1.03	ND	0.19

ND = Not Detectable

Having briefly described the relevance of cell line selection techniques to a programme of secondary metabolite production, we will now move on to the main subject of this article, namely, systems of plant cell immobilisation. In our laboratory we have developed and characterised two such systems, which share certain important features (Lindsey and Yeoman 1983b,c). The first is a "flatbed" system, in which cells are cultured in a horizontal mode on the surface of fibrous polypropylene, across which liquid nutrient medium is supplied from a reservoir and recirculated. The second type of system comprises a vertical column packed with cells entrapped in a support matrix, and liquid medium is again supplied from a reservoir. Cell entrapment matrices tested have included nylon netting in combination with agar and alginate gels, and, as indicated above, reticulate polyurethane foam. The flatbed and the column systems both allow cells to be cultured with a high degree of physical contact, the cells are physically stationary (two features which encourage partial structural organisation), and the provision of liquid medium from a reservoir allows the facilitated addition and removal of metabolites from the culture. We will now attempt to describe some of the effects of immobilisation on cell metabolism using information derived from our own experimental systems, and make some general comments in the light of other evidence presented in the literature.

4 The Effects of Immobilisation on Primary Metabolism

Very little biochemical work has been performed on the effects of immobilisation on primary metabolic activity. We have found that the rate of protein synthesis (incorporation of [^{14}C]-Phe into soluble protein) in chilli pepper cells immobilised and cultured for 5 days in SH basal medium is approximately half that of freely suspended cells cultured for the same period (log/linear phase), but is almost twice that of day 12 (stationary phase) suspended cells. The initial rate of uptake of [^{14}C]-Phe was no different in immobilised and suspended cells under the culture conditions employed (Lindsey and Yeoman 1984b). Immobilisation in at least some matrices does not appear to affect respiratory activity deleteriously. Brodelius and Nilsson (1980), for example, have investigated the viability of cells of *Catharanthus roseus* entrapped in a variety of gels, namely calcium alginate, agar, polyacrylamide, agarose, gelatin, carrageenan, and copolymers of gelatin with alginate or agarose. Respiratory activity was detectable only in cells immobilised in alginate, agar, agarose, and carrageenan, although intact plasmamembranes were retained by cells immobilised in all gels except polyacrylamide. It is likely that lack of viability in the least successful methods was primarily due to the toxic effects of glutaraldehyde used as a cross-linking and hardening agent. The rate of respiratory activity by cells of *C. frutescens* immobilised in reticulate polyurethane was found by us to be similar to that of freely-suspended cells (Lindsey and Yeoman 1984d). If cells are allowed to invade the foam over a 23-day culture period, the rate of O_2 uptake from air-saturated medium is comparable (0.040 ± 0.011 μmol O_2 g^{-1} fr.wt. min^{-1}) to that of late stationary phase (23 days) suspended cells (0.060 ± 0.013 μmol O_2 g^{-1} fr.wt. min^{-1}). After transfer to fresh SH-basal medium, the rate of O_2 uptake of the immobilised cells increases to 0.062 ± 0.027 μmol O_2 g^{-1} fr.wt. min^{-1} after

3 days incubation, and to 0.102 ± 0.009 μmol O_2 g^{-1} fr.wt. min^{-1} after 7 days, a rate similar to that of linear phase (7 days) suspended cells (0.112 ± 0.002 μmol O_2 g^{-1} fr.wt. min^{-1}). Cells of *Solanum nigrum* retain the ability to respire when cultured in the flatbed system or when entrapped in calcium alginate in glass columns, although respiratory activity was greater in the latter case (Lindsey and Yeoman 1983b).

A third indication of cell viability (though not strictly of "primary metabolic activity") is the fluorescein diacetate (FDA) staining technique of Widholm (1972). According to this technique, a cell is determined as being "viable" if it is capable of cleaving the ester bonds of the FDA molecule, so that the free fluorescein can fluoresce in UV light. Results obtained using this method of analysis indicate that cells entrapped in reticulate polyurethane are not significantly less viable (typically 70%–80% cells staining, throughout a 1 cm × 1 cm × 1 cm foam block) than are freely suspended cells (typically 75%–85% cells staining in a late stationary phase (day 21) culture) (Lindsey and Yeoman 1984d). Cells of *C. frutescens, Datura innoxia*, and *Solanum nigrum* showed 60%–70% fluorescence after immobilisation and culture for 2–10 days in a growth medium on the flatbed or in a 2% calcium alginate gels; this proportion was found to decrease to 50%–60% if cells were entrapped in agar gel, presumably because of the adverse effects of the high temperatures (30°–40 °C) associated with molten agar (Lindsey and Yeoman 1985).

The rate and extent of uptake of essential nutrients from the medium supplying immobilised cells is another aspect of viability we have examined. Cells of *C. frutescens, D. innoxia*, and *S. nigrum* take up orthophosphate, nitrate, ammonium, and sucrose when immobilised on the flatbed, in alginate, in agar or in reticulate polyurethane. When cultured on a Murashige and Skoog growth medium, orthophosphate in particular is rapidly removed from the medium (Lindsey and Yeoman 1983b) so that, typically, after an incubation period of 2 or 3 days 80%–90% of the initial quantity has been taken up. In the case of the alginate-entrapment system, some of the phosphate is sequestered by calcium in the gel matrix, but this accounts for the removal of less than 20% of the total phosphate over 9 days. The rate and extent of nutrient uptake is lower in agar-entrapped compared with flatbed-, alginate- or polyurethane-entrapped cells, presumably reflecting the effects of high-temperature immobilisation. *C. frutescens* cells immobilised in reticulate polyurethane and cultured in SH basal medium typically show a 3- or 4-day lag phase in uptake of nutrients, but in the subsequent 6 or 7 days orthophosphate, nitrate, ammonium, and sucrose are removed at a linear rate, so that by day 10 only 30%, 20%, 35%, and 40% of the initial respective levels remain in the medium.

Perhaps the most obvious general index of primary metabolic activity is an increase in fresh and dry weight and cell number. Alfermann et al. (1980) have described how, by 61 days of culture, the division of cells of *Digitalis lanata* had caused a breakdown in the structure of the alginate gel in which they were immobilised. Similarly, Morris and Fowler (1981) have described the breaking away of viable cells from spheres of alginate, due to cell division. If cultured in SH basal medium, cells of *Daucus carota* and *C. frutescens*, immobilised in reticulate polyurethane, divide and expand to fill all the available free space in the foam, and ultimately grow out of the surface to form compact aggregates in which the foam matrix is completely obscured. The rate of increase in the fresh weight and of protein synthesis of the immobilised cells can be finely controlled by manipulation of the culture conditions (Lindsey, unpublished observations).

5 The Effects of Immobilisation on Secondary Metabolism

There are, in general, two aspects of secondary metabolism which are of interest to tissue culturists and have been examined in immobilised cells. The first is one- or two-step *biotransformations* such as hydroxylations, glycosylations, acetylations, and demethylations. The greatest volume of published data so far has been on the hydroxylation or glycosylation of cardiac glycosides in either *Digitalis lanata* (Brodelius et al. 1979; Alfermann et al. 1980) or in *Daucus carota* (Jones and Veliky 1981). Brodelius et al. (1979) found that the 12,β-hydroxylation of digitoxin to digoxin occurred in cells immobilised in a calcium alginate gel with the same efficiency of conversion as freely suspended cells, and activity was maintained for 33 days. Alfermann et al. (1980) have reported how similarly immobilised *D. lanata* cells glycosylate digitoxin to purpurea glycoside A and hydroxylate β-methyldigitoxin to β-methyldigoxin in a way which is qualitatively identical to freely-suspended cells. The hydroxylation activity was only 50% that of suspended cells, but was maintained for at least 61 days. Jones and Veliky (1981) have similarly found that freely suspended cells of *Daucus carota* hydroxylate digitoxigenin to periplogenin at a faster rate than cells immobilised in calcium alginate, although the rate of hydroxylation of gitoxigenin to 5,β-hydroxygitoxigenin could be increased by permeabilising the entrapped cells with chloroform; activity was retained for more than 30 days. Recently Furuya et al. (1984) have shown that alginate-entrapped opium poppy cells *(Papaver somniferum)*, when cultured in shake flasks, transform (–) codeinone to (–) codeine more efficiently (70.4% conversion in 3 days) than do suspended cells (60.8% conversion).

These results therefore indicate that cells immobilised in calcium alginate, at least, retain the potential for performing certain biotransformations, and in some cases perform them as well as freely-suspended cells. In those cases in which the activity of a particular reaction is somewhat reduced as a result of entrapment, there is some evidence that the activity can, however, be maintained for extended periods of time.

The second aspect of the secondary metabolism of immobilised plant cells to be discussed is the *complex (multi-enzyme) synthesis* of specific products; this is the area in which we, at Edinburgh, are particularly interested. To date there have been relatively few reports on the efficiency of multistep reactions in immobilised cells, and a number of basic characteristics must be established; for example, does immobilisation allow the retention of multienzyme activity, and under what conditions? If activity is retained, then for how long? Does synthesis proceed de novo, or must precursors to the product be supplied, and how distant from the product can they be? A problem which is immediately apparent is the lack of knowledge of the structure and regulation of plant secondary metabolism, and until more basic biochemistry is carried out, progress in this field will be greatly handicapped. Nevertheless, some headway is being made in answering some of the questions posed.

As would be predicted from the hypothesis concerning the relationship between cellular organisation and secondary product formation, there is evidence that, in some cases, immobilised cells are more efficient than suspended in performing complex syntheses. Two examples of this phenomenon are given by Brodelius et al. (1979). Cells of *Catharanthus roseus* were more efficient at synthesising ajmalicine from the precursors tryptamine and secologanin when immobilised in calcium alginate than when freely suspended, and immobilised cells of *Morinda citrifolia* accumulated higher yields of anthraquinone pigments than did suspended.

We (Lindsey and Yeoman 1983b) have found that *Solanum nigrum* cells cultured on the flatbed accumulate, in the absence of exogenously supplied precursors, higher yields of steroidal glycoalkaloids than do the rapidlydividing suspension cultures from which the flatbed cultures are established. Cells of *Datura innoxia*, immobilised in nylon + calcium alginate, accumulate a greater number and higher total amounts of tropane alkaloids than do suspended cells. The biosynthetic potential of cells of *C. frutescens* immobilised in reticulate polyurethane has been studied in more detail (Lindsey and Yeoman 1984b,d). Immobilised cells typically produce, in SH basal medium, between two and three orders of magnitude higher yields of capsaicin than freely suspended cells over a 5- or 10-day culture period, as determined by HPLC analysis of the nutrient medium. In one experiment, by 5 days after subculture, suspended cells had produced only 0.002 ± 0.000 mg capsaicin g^{-1} dry wt. l^{-1} medium, compared with 1.589 ± 0.256 mg g^{-1} dry wt. l^{-1} for immobilised cell cultures. By day 10, the cumulative yield of the suspended cell cultures (now in the stationary phase) had increased tenfold to 0.020 ± 0.006 mg g^{-1} dry wt. l^{-1}, although the yield of the immobilised cells had also increased to 3.184 ± 0.297 mg g^{-1} dry wt. l^{-1}. This phenomenon is reflected by a greater (sevenfold) increase in the incorporation of L-[U-^{14}C]-phenylalanine into capsaicin in day 5-immobilised than in day 5-suspended cell cultures over a 24-h incubation period, although the rate and extent of uptake of the label is not different. If precursors to capsaicin are used to supplement SH basal medium, then the yield of capsaicin can be increased, in both immobilised and freely suspended cells cultured on that medium. For example, the presence of 5 mM isocapric acid in the medium resulted, after a 10-day culture period, in a yield of 2.38 ± 0.72 mg capsaicin g^{-1} dry wt. l^{-1} in suspension cultures and 124.71 ± 6.5 mg g^{-1} dry wt. l^{-1} in immobilised cell cultures, compared with 18.68 ± 2.65 μg g^{-1} dry wt. l^{-1} and 2.82 ± 0.38 mg g^{-1} dry wt. l^{-1} respectively, in the absence of the precursor. One mM phenylalanine also causes an increase in capsaicin production (Lindsey and Yeoman 1984b). These results demonstrate clearly that, when cultured under identical, growth promoting conditions (i.e. on SH basal medium), immobilised pepper cells produce significantly higher yields of capsaicin than do suspended cells, and they also appear to be more efficient at utilising precursors for the synthesis of capsaicin. In view of the fact that, even under growth conditions, the growth index of the immobilised cells is very much lower than that of suspended cells, the results presented can be interpreted in terms of the "organisation and growth rate" hypothesis outlined above. Indeed, the results described in Lindsey and Yeoman (1984b) demonstrate an inverse correlation between the incorporation of [^{14}C]-Phe into protein and into capsaicin in both immobilised and suspended cells.

The capsaicin-synthetic activity of immobilised pepper cells is stable over prolonged culture periods (Lindsey and Yeoman 1984d). Batch cultures of entrapped cells, maintained on SH basal medium, retain esterase activity and the ability to synthesise capsaicin if given exogenous isocapric acid after a 12-week culture period. During this period cells were transferred to fresh medium at 14-day intervals. Column-cultured immobilised cells have further been found to retain their synthetic capacity after a 6-month culture period. Over a shorter time period, it was found that immobilised pepper cells could synthesise up to about 11 mg capsaicin g^{-1} dry wt. l^{-1} from endogenous precursors over a series of at least 2-day batch cultures. After 7 or 8 transfers to fresh medium, the levels of capsaicin accumulated in the medium began to decline, until the 13th batch, when capsaicin was no longer detectable. Synthesis could be restored at day 36 (the 18th batch) by the addition of 5 mM isocapric acid to the medium.

The data described above have been obtained using cells cultured in SH basal medium. This medium is, as indicated above, a growth medium, and as might be expected, is not optimum for secondary metabolite production, even in relatively slow-growing immobilised cell cultures. Indeed, by omitting constituents of the medium which are essential for maintaining cell division, it is possible to enhance the yields of capsaicin further still. Yeoman et al. (1980) have described how nitrogen and sucrose stress can increase production in callus cultures of *C. frutescens*, and, in a series of papers, Knobloch and co-workers have indicated the relationship between a high intracellular phosphate concentration, high culture growth rate and low secondary metabolite production in cell cultures of *Nicotiana tabacum* and *Catharanthus roseus* (e.g. Knobloch et al. 1981). Nutrient stress (see Lindsey, unpublished observations) and manipulation of the levels of growth regulators (Lindsey and Yeoman 1984b) can enhance capsaicin production by immobilised pepper cells, either in batch culture, or as a sequence of treatments in packed columns (Lindsey, unpublished observations).

6 Conclusions

The work discussed in this article demonstrates that, if the right entrapment technique is used, the potential for primary and secondary metabolic activity of cells needed not be lost on immobilisation. Rather, there is evidence that the production of specific secondary metabolites may actually be enhanced as a direct result of immobilisation. When combined with procedures for the selection of high-yielding cell lines and for the chemical manipulation of the cells, immobilisation would appear to have much potential as an integral part of a large-scale production process. Before such a concept is realised, however, much basic research remains to be carried out. Perhaps the most obvious potential difficulty is in inducing the release of products which, unlike capsaicin, are retained intracellularly; this is essential if the biomass is to be used catalytically in a continuous or semi-continuous process, as we would advocate. Some work has been reported in this area (e.g. Felix et al. 1981), but much more remains to be done. We envisage that the controlled permeabilisation of cells would be incorporated into a sequence of growth and metabolite production phases, each effected by a one- or two-step manipulation of the chemical composition of the nutrient medium which would be continuously circulated over the physically stationary biomass. In this way, immobilisation techniques could provide the basis for a successful commercial process, as well as fundamental information on the mechanisms by which secondary metabolism is regulated in plants.

References

Alfermann AW, Schuller I, Reinhard E (1980) Biotransformation of cardiac glycosides by immobilised cells of *Digitalis lanata*. Planta Med 40:218–223

Brodelius P, Nilsson K (1980) Entrapment of plant cells in different matrices. FEBS Lett 122: 312–316

Brodelius P, Deus B, Mosbach K, Zenk MH (1979) Immobilised plant cells for the production and transformation of natural products. FEBS Lett 103:93–97

Felix HR, Mosbach K (1982) Enhanced stability of enzymes in permeabilised and immobilised cells. Biotechnol Lett 4:181–186

Felix HR, Brodelius P, Mosbach K (1981) Enzyme activities of the primary and secondary metabolism of simultaneously permeabilised and immobilised plant cells. Anal Biochem 116:462–470

Furuya T, Yoshikawa T, Taira M (1984) Biotransformation of codeinone to codeine by immobilised cells of *Papaver somniferum*. Phytochemistry 23:999–1002

Jirku V, Macek T, Vanek T, Krumphanzl V, Kubanek V (1981) Continuous production of steroid glycoalkaloids by immobilised plant cells. Biotechnol Lett 3:447–450

Jones A, Veliky IA (1981) Examination of parameters affecting the 5-β-hydroxylation of digitoxigenin by immobilised cells of *Daucus carota*. Eur J Appl Microbiol Biotechnol 13:84–89

Knobloch KH, Beutnagel G, Berlin J (1981) Influence of accumulated phosphate on culture growth and formation of cinnamoyl putrescines in medium-induced cell suspension cultures of *Nicotiana tabacum*. Planta 153:582–585

Lambe CA, Rosevear A (1982) Production of chemical compounds from viable cells. UK Patent No. GB2 096 169A

Lindsey K, Yeoman MM (1983a) The relationship between growth rate, differentiation and alkaloid accumulation in cell cultures. J Exp Bot 34:1055–1065

Lindsey K, Yeoman MM (1983b) Novel experimental systems for studying the production of secondary plant metabolites by plant tissue cultures. In: Mantell SH, Smith H (eds) Plant biotechnology. Soc Exp Biol Semin Ser, vol 18. Cambridge Univ Press, Cambridge, pp 39–66

Lindsey K, Yeoman MM (1984a) Dynamics of plant cell cultures. In: Vasil IK (ed) Cell culture and somatic cell genetics of plants. II. Academic Press, London New York, in press

Lindsey K, Yeoman MM (1984b) The synthetic potential of immobilised cells of *Capsicum frutescens* Mill. cv. annuum. Planta 162:495–501

Lindsey K, Yeoman MM (1984c) Techniques for the immobilisation of plant cells. In: Mosbach K (ed) Immobilised enzymes and cells. Methods in enzymology. Academic Press, London New York, in press

Lindsey K, Yeoman MM (1984d) The viability and biosynthetic activity of cells of *Capsicum frutescens* Mill. cv. annuum immobilised in reticulate polyurethane. J Exp Bot 35:1684–1696

Lindsey K, Yeoman MM (1985) Immobilised plant cells. In: Yeoman MM (ed) Plant cell culture technology, botanical monograph series. Blackwell, Oxford, in press

Lindsey K, Yeoman MM, Black GM, Mavituna F (1983) A novel method for the immobilisation and culture of plant cells. FEBS Lett 155:143–149

Mantell SH, Smith H (1983) Cultural factors that influence secondary metabolite accumulation in plant cell and tissue cultures. In: Mantell SH, Smith H (eds) Plant Biotechnology. Soc Exp Biol Semin Ser, vol 18. Cambridge Univ Press, Cambridge, pp 75–108

Morris P, Fowler MW (1981) A new method for the production of fine plant cell suspension cultures. Plant Cell Tissue Org Cult 1:15–24

Ogino T, Hiraoka N, Tabata M (1978) Selection of high nicotine-producing cell lines of tobacco callus by single cell cloning. Phytochemistry 17:1907–1910

Phillips R, Henshaw GG (1977) The regulation of synthesis of phenolics in stationary phase cell cultures of *Acer pseudoplatanus* L. J Exp Bot 28:785–794

Prenosil JE, Pedersen H (1983) Immobilised plant cell reactors. Enzyme Microb Technol 5:323–331

Schenk RU, Hildebrandt AC (1972) Medium and techniques for induction and growth of monocotyledonous and dicotyledonous plant cell cultures. Can J Bot 50:199–204

Shuler ML (1981) Production of secondary metabolites from plant tissue culture – problems and prospects. Ann NY Acad Sci 369:65–79

Wichers HJ, Malingre TM, Huizing HJ (1983) The effect of some environmental factors on the production of L-DOPA by alginate-entrapped cells of *Mucuna pruriens*. Planta 158:482–486

Widholm JM (1972) The use of fluorescein diacetate and phenosafranine for determining viability of cultured plant cells. Stain Technol 47:189–194

Yeoman MM, Miedzybrodzka MB, Lindsey K, McLauchlan WR (1980) The synthetic potential of cultured plant cells. In: Sala F, Parisi B, Cella R, Cifferi O (eds) Plant cell cultures: results and perspectives. Elsevier/North Holland, Amsterdam, pp 327–343

Yeoman MM, Lindsey K, Miedzybrodzka MB, McLauchlan WR (1982) Accumulation of secondary products as a facet of differentiation in plant cell and tissue cultures. In: Yeoman MM, Truman DES (eds) Differentiation in vitro. Br Soc Cell Biol Symp, vol IV. Cambridge Univ Press, Cambridge, pp 65–82

Biotransformation of Cardiac Glycosides by *Digitalis* Cell Cultures in Airlift Reactors [1]

A.W. ALFERMANN, H. SPIELER, and E. REINHARD [2]

1 Introduction

The production of valuable plant products by biotransformation from inexpensive precursors which cannot be transformed effectively by chemical or microbial means is an interesting field for practical application of plant cell cultures (Reinhard and Alfermann 1980). Among the reactions observed biotransformation by cardiac glycosides is of special pharmaceutical importance because these are widely used in medicine for treatment of heart diseases. Digitoxin and digoxin and their derivatives which can be extracted from leaves of *Digitalis lanata*, are the most interesting ones. Nowadays, digitoxin is used in therapy to a lesser extent than digoxin due to special pharmacological characteristics. *D. lanata* plants, however, always contain substantial amounts of digitoxin, up to about one quarter as compared to the amount of digoxin. Digoxin differs from digitoxin only by an additional hydroxyl function at C-12 (Fig. 1). Undifferenti-

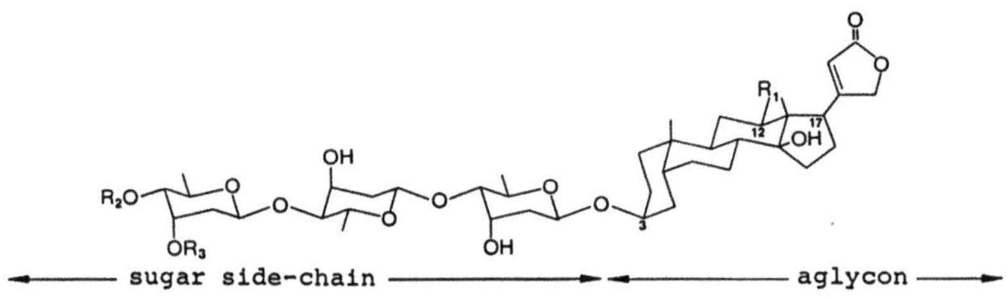

	R_1	R_2	R_3
Digitoxin	H	H	H
Purpureaglycoside A	H	glucose	H
β-Methyldigitoxin	H	CH_3	H
Digoxin	OH	H	H
β-Methyldigoxin	OH	CH_3	H

Fig. 1. Structure of some natural and semisynthetic cardenolides

1 Dedicated to Professor Dr. O.H. Volk (Würzburg) on the occasion of his 80th birthday

2 Pharmazeutisches Institut (Pharmazeutische Biologie), Universität Tübingen, Auf der Morgenstelle 8, 7400 Tübingen, FRG

Primary and Secondary Metabolism of Plant Cell Cultures, ed. by Neumann et al.

ated cell cultures of *Digitalis* do not produce cardiac glycosides, however they are able to perform special biotransformation reactions on substrates added into the medium (Reinhard et al. 1975).

2 Cultural Conditions and Analytical Procedures

Medium, cell strain, cultural conditions, and analytical procedures were as published earlier (Helmbold 1979; Alfermann et al. 1983) with the exception that the medium was prepared with tap water, casamino acids were omitted and KH_2PO_4 concentration was increased to 340 $mg\,l^{-1}$.

3 Biotransformation of Digitoxin

Digitoxin would be an ideal substrate for such a biotransformation process. Figure 2, however, shows that digitoxin is transformed to a series of compounds, of which purpureaglycoside A is the main product. Those compounds hydroxylated at C-12 like digoxin, deacetyllanatoside C or lanatoside C, are produced only in minor amounts. After testing a series of digitoxin derivatives we found β-methyldigitoxin (Fig. 3) to be a usable substrate (Reinhard et al. 1975).

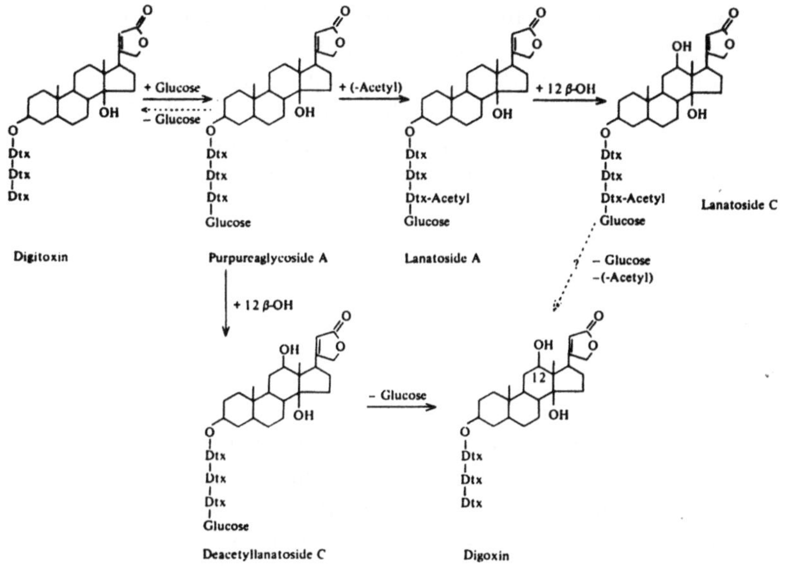

Fig. 2. Biotransformation of digitoxin by different cell lines of *Digitalis lanata*. Purpureaglycoside A is the main product, whereas those hydroxylated at C-12 like digoxin, deacetyllanatoside C and lanatoside C are only produced in minor amounts. Dtx – digitoxose molecule

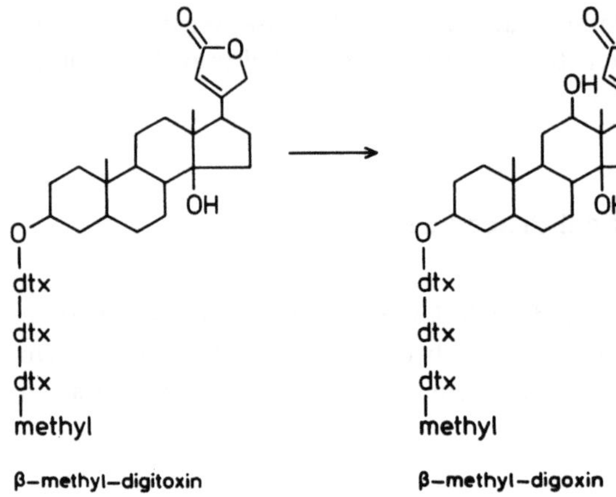

Fig. 3. β-Methyldigitoxin is transformed by selected cell lines only to β-methyloxin. *dtx* digitoxose molecule

4 Biotransformation of β-Methyldigitoxin

β-Methyldigitoxin can be prepared from digitoxin by chemical methylation (Heins et al. 1978). Until now β-methyldigoxin, which is used in medicine, is prepared by chemical methylation of digoxin. During intensive screening work we have tested a large number of cell lines for their hydroxylating capacity. All cell lines tested had been initiated from *Digitalis* plants containing high amounts of digoxin derivatives. We assumed that there was a correlation between the digoxin content of the plant and the hydroxylating capacity of the cell culture derived from it. However, we found that different cell lines performed various reactions to a very different extent. Some lines could not perform any biotransformation reaction at all; others demethylated β-methyldigitoxin and glucosylated digitoxin produced as an intermediate to purpureaglycoside A, eventually accompanied by additional acetylation to lanatoside A. Besides this, a third group also hydroxylated at C-12. Only very few lines were able to hydroxylate β-methyldigitoxin to β-methyldigoxin without any byproduct (Heins 1978; Friedrichson 1979; Baumann 1980; Hu 1984).

5 Development of a Bioreactor Process

5.1 Batch Culture

Those cell lines with high hydroxylating capacity have been used for studies in bioreactors. As compared with microorganisms, plant cell cultures are very sensitive to shear stress (Wagner and Vogelmann 1977). Therefore, most of our experiments have been done in airlift reactors known to have a low shear rate (Wahl 1978; Helmbold 1979; Barthlen 1983). However, normal stirred tank reactors can also be used for plant cell cultures. This may even be of advantage if high yields of biomass are achieved (i.e. Fujita et al. 1982; Ulbrich et al., this volume; Spieler 1985). As compared with microbial fermentation the stirrer velocity used is quite low. In most cases the energy input by the stirring should not exceed 0.2 kW m^{-3} (J. Wahl, personal communication).

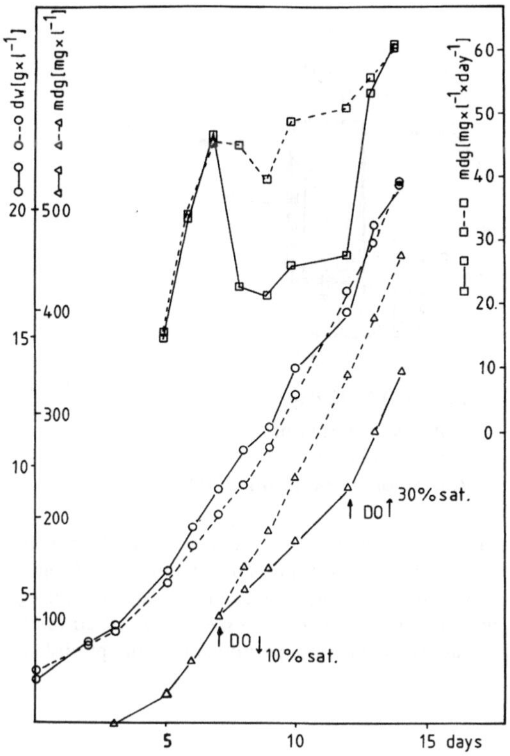

Fig. 4. Influence of dissolved oxygen (DO) on hydroxylation of β-methyldigitoxin by cell cultures of *Digitalis lanata* in a 20-l airlift reactor. In one set of experiments the oxygen tension (% of saturation, using air) was maintained at 30%, in the other the oxygen tension was reduced to 10% on day 7 and raised again to 30% on day 12

In order to increase product yields we tried manipulating various parameters, i.e. adding the substrate during different growth phases in one or more batches and changing the substrate, glucose, and phosphate levels. As a result we decided to use an elevated phosphate level (340 mgl^{-1} KH_2PO_4), methanol as solvent for methyldigitoxin and a glucose level kept at 5 gl^{-1} or above. Furthermore, as shown in Fig. 4 the hydroxylation rate is increased by a high tension of dissolved oxygen. The experiment was started using a constant oxygen tension of 30% saturation (using air for aeration). Biotransformation was started on day 3, on day 7 the oxygen tension was reduced to 10%. This was followed by a substantial reduction in biotransformation. On day 12 the oxygen tension was increased again to 30%, and the biotransformation rate increased again to the control value. Consequently, the following experiments have been done using an oxygen tension of 30% of saturation. Higher oxygen tensions can only be achieved by such high aeration rates that the foaming produced leads to problems.

If one wants to use such a process for technical application, it is necessary to test whether a scale-up to larger volumes is possible. Recently, we have have installed a normal, commercial bioreactor of a working volume of 200 l modified as an airlift reactor. Figure 5 demonstrates the scaling up, Fig. 6 shows a typical biotransformation in this reactor. On day 5 of cultivation, the cells were incubated with the substrate. Product formation parallels growth, in this case reaching a maximum after a further 13 days. Four hundred and thirty mg of β-methyldigoxin, which is about 70% as compared to the substrate added, can be extracted from 1 l of medium. About 10% of the substrate remains untransformed, the rest of product and substrate is stored in the cells.

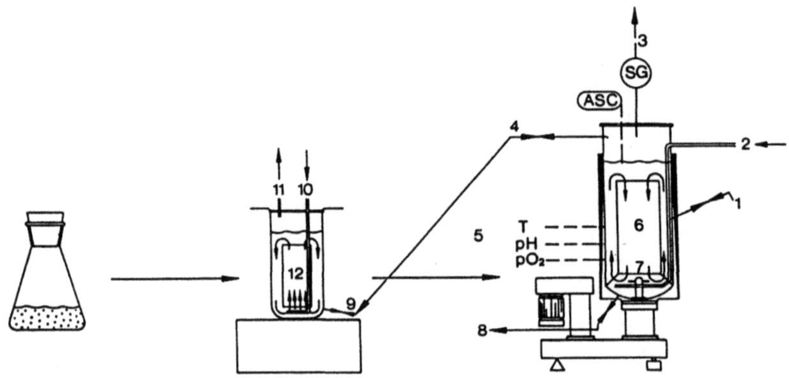

Stage 1: 7 days	Stage 2: 10 days	Stage 3: 15–20 days
Multiplication of cells	Multiplication of cells	Biotransformation of
Shake culture,	Bioreactor,	β-methyldigitoxin
300 ml medium	working vol 30 l	Bioreactor, working vol. 200 l

Fig. 5. Scaling up of *Digitalis* cell cultures to a working volume of 200 l. For stages 2 and 3 conventional bioreactors converted into airlifts were used. *1* sampling device; *2* air inlet; *3* air outlet; *4* transfer valve and transfer line; *5* control for temp., pH, diss. oxygen; *6* draft tube; *7* stirrer (during sterilisation); *8* harvesting line; *9* sampling device and valve for transfer, respectively; *10* air inlet; *11* air outlet; *12* draft tube (Alfermann 1983, changed, with kind permission of the publisher, Metzler Verlag, Stuttgart)

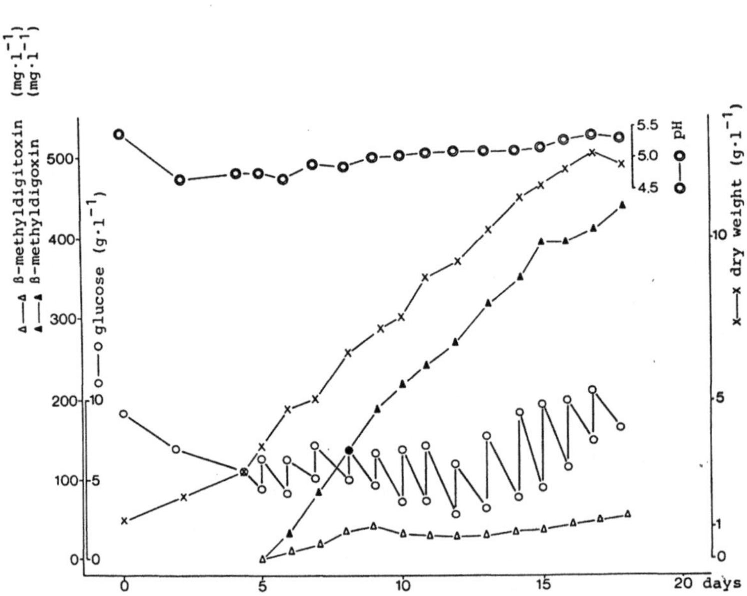

Fig. 6. Biotransformation of β-methyldigitoxin to β-methyldigoxin by cell cultures of *Digitalis lanata* in a 200-l-airlift reactor. (Alfermann 1983, changed, with kind permission of the publisher, Metzler Verlag, Stuttgart)

5.2 Semicontinuous Culture (Repeated Batch)

There are two main reasons why plant cell cultures are not widely used in industry. First, in most cases the interesting compounds are not produced in sufficiently high amounts that practical application could be considered. Second, these plant cell cultures multiply very slowly. From Fig. 5 one can calculate that scaling up of our plant cell culture process from the Erlenmeyer flask to a commercial working volume of 50 m^3 would take about 40 days. This would be very expensive.

Therefore, to reduce the costs for this long time of scaling-up by using the cells during a prolonged production phase, we have tested a semicontinuous (repeated batch) cultivation of *Digitalis* cells for biotransformation of β-methyldigitoxin.

The scale-up was as already shown in Fig. 5. Five days after the transfer into the 200 l reactor the cells were incubated with β-methyldigitoxin. On day 14 about 500 mg l^{-1} of β-methyldigoxin were produced. About 85% of the culture was harvested for product isolation. Then the reactor was refilled with fresh medium and the biotransformation process started again. Table 1 shows that six such runs were possible without reduction in productivity. During these three months of cultivation about 0.5 kg of β-methyldigoxin was produced. This would be sufficient for preparing 5 million cardiovascular tablets to treat 1,000 patients for more than 7 years.

Table 1. Semicontinuous cultivation of *Digitalis lanata* cells in a 200-l airliftreactor and biotransformation of β-methyldigitoxin to β-methyldigoxin

Run	I	II	III	IV	V	VI
Cultivation time (days)	14	13	14	15	16	17
β-Methyldigoxin produced g l^{-1}	0.4	0.45	0.49	0.48	0.54	0.5
g harvested	80	82	94	96	108	98

In total during these 3 months of biotransformation 505 g β-methyldigoxin were produced

6 Outlook

Recent success of other groups (Fujita et al. 1982) has shown that plant cell cultures can be used for technical production of secondary plant products. We are still convinced that we can to improve the product yields of our biotransformation process that practical application is possible.

Acknowledgements. We thank Mrs. C. Figur and Mr. D. Schwantag for skilfull technical assistance and the Federal Minister for Research and Technology, Bonn, for sustaining this work by a research grant.

References

Alfermann AW (1983) Gewinnung von Arzneistoffen durch pflanzliche Zellkulturen. In: Dohmen K (ed) Biotechnologie. Metzlersche Verlagsbuchhandlung, Stuttgart, pp 47–62

Alfermann AW, Bergmann W, Figur C, Helmbold U, Schwantag D, Schuller I, Reinhard E (1983) Biotransformation of β-methyldigitoxin to β-methyldigoxin by cell cultures of *Digitalis lanata*. In: Mantell SH, Smith H (eds) Plant biotechnology. Cambridge Univ Press, Cambridge, pp 67–74

Barthlen U (1983) 12β-Hydroxylierung von β-Methyldigitoxin in Fermenter- und Suspensionskulturen von *Digitalis lanata*. Untersuchungen zur Nährmediumsoptimierung. Doct Thesis, Univ Tübingen

Baumann L (1980) Biotransformationsversuche mit Digitalis lanata Gewebekulturen. Selektion von 12β-hydroxylierenden Zellstämmen. Dipl Thesis, Univ Tübingen

Friedrichson JG (1979) Versuche zur Hydroxylierungskapazität von Blättern, Suspensions- und Fermentationskulturen von *Digitalis lanata*. Doct Thesis, Univ Tübingen

Fujita Y, Tabata M, Nishi A, Yamada Y (1982) New medium and production of secondary compounds with the two-staged culture method. In: Fujiwara A (ed) Plant tissue culture 1982. Jpn Assoc Plant Tissue Cult, Tokyo, pp 399–400

Heins M (1978) Screening of *Digitalis lanata* plants and cell cultures for hydroxylating capacity. In: Alfermann AW, Reinhard E (eds) Production of natural compounds by cell culture methods. Ges Strahlen- und Umweltforsch, München, pp 39–47

Heins M, Wahl J, Lerch A, Kaiser F, Reinhard E (1978) Preparation of β-methyldigoxin by hydroxylation of β-methyldigitoxin in fermenter cultures of *Digitalis lanata*. Planta Med 33:57–62

Helmbold U (1979) Versuche zur Steigerung der 12β-Hydroxylierung von β-Methyldigitoxin durch Fermenterkulturen von Digitalis lanata Zellstämmen. Doct Thesis, Univ Tübingen

Hu Z (1984) Untersuchungen zur Biotransformation von Cardenoliden durch spezielle Zellstämme von *Digitalis lanata*. Doct Thesis, Univ Tübingen

Reinhard E, Alfermann AW (1980) Biotransformation by plant cell cultures. In: Fiechter A (ed) Advances in biochemical engineering, vol 16. Springer, Berlin Heidelberg New York, pp 49–83

Reinhard E, Boy M, Kaiser F (1975) Umwandlung von *Digitalis*-Glykosiden durch Zellsuspensionskulturen. Planta Med Suppl, pp 163–168

Spieler H (1985) Untersuchungen zur Sauerstoffversorgung von *Digitalis lanata* Fermenterkulturen. Auswirkungen auf Wachstum und Hydroxylierung. Doct Thesis, Univ Tübingen

Wagner F, Vogelmann H (1977) Cultivation of plant tissue cultures in bioreactors and formation of secondary metabolites. In: Barz W, Reinhard E, Zenk MH (eds) Plant tissue culture and its biotechnological application. Springer, Berlin Heidelberg New York, pp 245–252

Wahl J (1978) Biotransformation of β-methyldigitoxin by *Digitalis lanata* fermenter cultures. In: Alfermann AW, Reinhard E (eds) Production of natural compounds by cell culture methods. Ges Strahlen- und Umweltforsch, München, pp 48–59

Cryopreservation of Plant Cell Cultures

U. SEITZ, I. REUFF, and E. REINHARD[1]

1 Introduction

Plant cell cultures are potential sources for the production of specific secondary compounds. These cultures are commonly maintained by continuous subculture, which includes the disadvantages of high costs, requirement of shaking capacity, risk of contamination and, above all, the possible loss of biosynthetic capacity in long-term culture (Zenk 1978). Consequently, a culture with particular yield characteristics may not be readily available if serial passage is the only means of preservation. Among the attempts to eliminate these problems freeze preservation seems to be the most successful method in the case of cell suspensions. Until now little work has been carried out with callus cultures. The progress in this area has been reviewed recently by Withers (1980, 1982) and Henshaw et al. (1980).

In the majority of cases where cryogenic storage has been achieved, viability of the cells and regrowth capacity were the only tests for the success of the procedure (see Table 3 and Withers 1982). There are a few reports related to the maintainance of biosynthetic (Dougall and Whitten 1980, Watanabe et al. 1983; Diettrich et al. 1982; Seitz et al. 1983) and/or morphogenetic capacities (see review of Withers 1982) during cryogenic storage.

In the following section we will describe the behaviour of different cell cultures during the freeze-thaw cycle: *Daucus carota, Digitalis lanata, Berberis wilsonae, Panax ginseng*. These cultures show different growth characteristics, degrees of vacuolation and accumulation of secondary products. In one case *(B. wilsonae)* it is at the moment impossible to achieve viable cells after freezing to −196 °C. The failure of this culture to survive freezing poses the question of whether certain species may be unsuitable for preservation at ultra-low temperatures.

In this paper we will especially concentrate on the preservation of biosynthetic capacities in frozen *D. carota* and *D. lanata* cells and on the susceptibility of cells which contain secondary products in various concentrations. All experiments were carried out with cell suspensions. In an Appendix we give a list of plant species successfully freeze-preserved as cell suspensions (Table 3).

1 Pharmazeutisches Institut (Pharmazeutische Biologie), Universität Tübingen, Auf der Morgenstelle 8, 7400 Tübingen, FRG

Primary and Secondary Metabolism of Plant
Cell Cultures, ed. by Neumann et al.

2 Plant Material and Growth Conditions

All species used during this work were cultivated as shake cultures in liquid media at 25 °C in the dark. Media and growth conditions in detail were as follows: An Afghan variety of *Daucus carota* was cultivated as described by Noé et al. (1980), growth conditions of *Digitalis lanata* (strains 287 and 10/56-S50-Se) were as described elsewhere (Alfermann et al. 1977). *Berberis wilsonae* cells were propagated in Murashige and Skoog's medium with 0.1 mg l^{-1} 2,4D, 0.2 mg l^{-1} IAA, 2.0 mg l^{-1} kinetin, 40 g l^{-1} sucrose. For *Panax ginseng* Murashige and Skoog's medium was used with the following additives: 2.0 mg l^{-1} 2,4D, 1.0 mg l^{-1} 2-isopentenyladenin, 1.0 mg l^{-1} kinetin, and 44 g l^{-1} glucose. Subculture regime of the different cells strains: *D. carota* and *D. lanata* 7 days, *B. wilsonae* 10.5 days, *P. ginseng* 10–11 days.

2.1 Cryopreservation Protocol

The standard procedure (Fig. 1) was performed according to Withers and King (1980). In order to prepare cells for freezing they were taken from the beginning of the growth phase and suspended in nutrient medium containing 6% mannitol. The inoculum was about twice the size compared with routine subculture. The preculture period was 3–4 days.

The cryoprotectant mixture was prepared in culture medium at twice the final concentration and sterilized by filtration: 1 M DMSO, 1 M glycerol, 2 M sucrose (DGS), pH 5.6–5.8. Cryoprotectant solution (10 ml) and cell suspension (10 ml) were cooled on ice and mixed, thus halving the molarities given above. Because of the high viscosity of the cryoprotectant solution, it is important to mix thoroughly. If possible, the cell suspension was further concentrated before the cryoprotectant treatment. After an incubation period of 1 h the suspensions were distributed in sterile polypropylene

Cryopreservation protocol (Cell suspensions)

Pregrowth

Transfer of exponentially growing cells to medium supplemented with 3%–6% mannitol. Pregrowth period: 3–4 days.

Cryoprotection

0.5 M DMSO plus 0.5 M glycerol plus 1.0 M sucrose prepared in culture medium. Treatment: 1 h, 0°–4 °C.

Freezing conditions

Cell suspension (1 ml samples) in freezing ampoules. Slow freezing at 1 °C min^{-1} to –35 °C, 40 min at –35 °C. Immersion in liquid nitrogen.

Storage

In liquid nitrogen.

Thawing

Rapid thawing in a water bath (+40 °C).

Post-thaw treatments

No washing. Transfer of cells and suspending medium to culture over semi-solid medium.

Fig. 1. Standard protocol of the freezing-thawing procedure ampoules (1 ml per 2 ml ampoule) and kept on ice. The cultures were frozen in a cooling apparatus (Type BF-R 200/101, Messer Griesheim, Düsseldorf, FRG) at a rate of 1 °C per min and then kept for 40 min at − 35 °C. In some experiments the specimens were cooled down to − 100 °C and were then immersed immediately in liquid nitrogen.

After the chosen storage period, cell suspensions were thawed rapidly by agitation of the ampoules in warm (40 °C) water. Immediately after thawing, cell suspensions were spread on ca. 8 ml culture medium solidified with agar (0.7%) in a 5 cm diameter Petri dish. Once growth had been re-established, the cells were transferred into liquid medium and cultivated under standard conditions.

3 Results and Discussion

3.1 Viability of Frozen Cells

As a routine test in each freezing experiment the percentage of viable cells before and after freezing was determined with fluorescein diacetate or phenosafranine staining (Widholm 1972). Table 1 shows a list of post-thaw viabilities. An anthocyanin-accumulating strain of carrot, a fast growing culture with small aggregates and a high ratio of cytoplasm/vacuole could be frozen without difficulties. The post-thaw viabilities in all experiments were high. Viability of *Digitalis lanata* cells dependend upon the culture strain used. Strain 10/56-S50-Se displayed high freezing tolerance in connexion with post-thaw viability values of about 70%, whereas strain 287 showed mean values of about 50%. Diettrich et al. (1982), who reported cryopreservation of *Digitalis lanata* cell cultures for the first time, used conditions similar to those employed here, apart from a lower mannitol concentration and a modified cryoprotection mixture (without DMSO). These authors reported cell survival of 40%–50%. It is generally accepted that

Table 1. Viability of cell cultures immediately after thawing

Species	Viability
Daucus carota	60%–70%
Digitalis lanata	
Strain 287	50%
Strain 10/56-S50-Se	70%–75%
Panax ginseng	
Experiment 1	8%–10%
Experiment 2	24%
Experiment 3	31%

Cell suspensions were treated as described in Fig. 1, except *P. ginseng*, experiments 2 and 3, where a long preculture in highly concentrated sucrose at low temperature was applied. As cryoprotectant only sucrose was used in these experiments (Butenko et al. 1984). An aliquot of the thawed cell suspension was stained with fluorescein diacetate or phenosafranine *(P. ginseng)* and the percentage of viable cells was determined small cell aggregates survive better than large ones (Dougall and Wetherell 1974; Withers 1982). Contrary to this rule, the result with strain 10/56-S50-Se is striking, because this highly tolerant strain contains much larger aggregates than strain 287. The cell suspension of *Panax ginseng*, which is a more slowly growing culture with a subculture regime of 10–11 days, showed low post-thaw viabilities and resumed growth only after a lag period of several weeks. Therefore, the preculture treatment published by Butenko et al. (1984) was applied. This method includes an increase of sucrose concentration up to 20% in connexion with a decrease of temperature from +25 °C to +4 °C over a period of 18 days. In consequence of this kind of "cold hardening" the post-thaw viabilities became significantly higher (see Table 1), but the regrowth capacity still remained very low. Correlation between growth regime, stage of culture growth and regrowth after freeze-preservation has been reported by many workers. Therefore, one could suggest that it would be of great advantage to speed up the growth of *P. ginseng* cell culture, thus having a subculture regime of 6–7 days before introducing the cells to the preculture treatment.

A viability test is in fact a valuable aid for determining rapidly the success of a cryopreservation experiment. Yet cells in the early recovery period seem to be particularly vulnerable (Cella et al. 1978; Withers 1982), and part of the cell population, which is more severely damaged, will not survive. Consequently, a decrease of cell viability is observed during the early recovery phase (Withers and Street 1977). This is at the moment a severe problem in the case of *Berberis wilsonae* cultures (see below).

3.2 Viability of Frozen Berberis wilsonae Cells

Cells of *Berberis wilsonae* (strain WSG_3) cultivated in suspension accumulated alkaloids of the protoberberine type with jatrorrhizine as a main component. The total alkaloid content at the end of the growth period was 4%–5% yield of cell dry weight (Rothenberger 1982). It would be of interest to preserve this high-producing culture strain by cryostorage. Unfortunately, after freezing the cells according to the protocol described in Fig. 1, no viable cells were obtained. Therefore, a great number of variations of this method were tested to increase the freezing resistance. Different preculture treatments, cryoprotectant additives, cooling rates, and regrowth conditions were applied. As an example, the influence of preculture conditions on cell survival is listed in Table 2. Even in those cases when satisfying viability values were obtained immediately after thawing, the cultures were not able to grow; most of them became dark coloured and died a few hours later. In some specimens it was observed during the early regrowth period that the surrounding medium turned yellow, thus indicating that a portion of the cell population had become leaky, and the alkaloid-containing cell sap had poured into the medium. On the other hand, addition of berberine (up to 1 mg/ml) to the cell suspension grown under standard conditions had no damaging effect. Alkaloid-producing periwinkle cells in freezing experiments benefit from a short preculture treatment with sorbitol and a combination of 1 M sorbitol plus 5% DMSO as cryoprotectants (Chen et al. 1984). *Berberis* cells frozen under similar conditions showed very low viability, and they died soon after the transfer on agar medium. At the moment, *B. wilsonae* seems to be a nonresponsive culture. The difference to *Catharanthus* may be a consequence of the very high alkaloid content of *Berberis* cells, which might influence

Table 2. Influence of preculture treatments on the viability of *Berberis wilsonae* cell suspensions (strain WSG_3)

Preculture	Cryoprotectant additives	Viability after cryoprotectant treatment	Post-thaw viability
1 % Mannitol	DGS	90%	5%
2 % Mannitol	DGS	10%	1%
3 % Mannitol	DGS		15%
6 % Mannitol	DGS	15%	0%
2 % Sorbitol	DGS	20%	0%
3 % Sorbitol	DGS	30%	0%
6 % Sorbitol	DGS	n.d.	15%
6 % Sorbitol 16–24 h	Sucrose	90%	10%
6 % Sorbitol 24 h	1 M Sorbitol plus 2.5% DMSO	80%	0%
1 M Sorbitol 20 h	1 M Sorbitol plus 5 % DMSO	40%	0%
1 M Sorbitol 20 h	1 M Sorbitol	40%	20%
6.5% Sucrose	DGS	98%	25%
7 % Sucrose	DGS	98%	25%
8 % Sucrose	DGS	95%	20%
7 % Sucrose	dextran 5%–30%	85%	0%
7 % Sucrose	70% Sucrose	97%	10%
2 M Methanol	0.5 M DMSO plus 1.0 M Sucrose	70%	0%
6.6% Proline	DGS	90%	15%
10 % Proline	DGS	70%	30%

If not otherwise mentiond, the preculture period was 3–4 days. The viability of the cells was determined after fluorescein diacetate staining. The following additives were used as cryoprotectants: sucrose (70%), sorbitol (1 M), dextran 40,000 (5%–30%), DMSO (2.5%–5.0%), DGS (0.5 M DMSO, 0.5 M glycerol, 1 M sucrose)

(in a so far unknown manner) its freezing resistance. High yield alkaloid production in *Catharanthus* starts only after the cells have been transferred into a production medium. Cryopreservation and alkaloid accumulation can therefore be separated temporally.

3.3 Cryostorage and Stability of Biosynthetic Capacities

Cells of an Afghan variety of *Daucus carota* accumulate anthocyanin in large amounts (Noé et al. 1980). This culture was used as a model system to test the biosynthetic capacity of frozen cells. It was treated as described in the scheme of Fig. 1, thawed after a storage period of 5–7 days, cultivated for a short period on semi-solid medium and then as usual in liquid medium. The anthocyanin content of a frozen culture and of a non-frozen control were compared. The result is depicted in Fig. 2. The maximum value and the accumulation kinetic are both identical. This is in complete agreement with the results published by Dougall and Whitten (1980), who investigated anthocyanin production in different strains of *Daucus carota* after freezing.

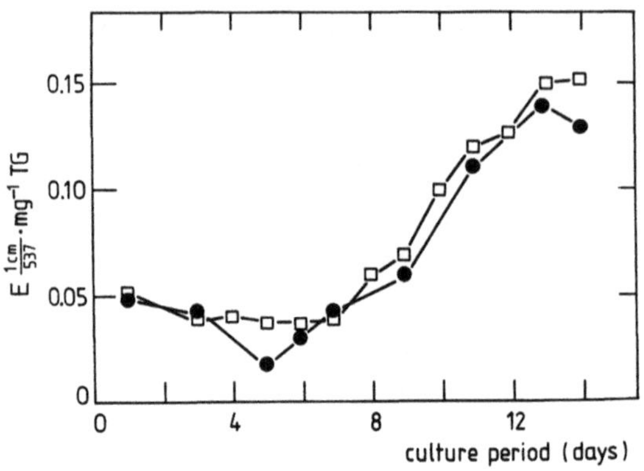

Fig. 2. Time course of anthocyanin accumulation in *Daucus carota* cell cultures. Freeze-dried cell material was extracted with methanol-HCl and the absorbance at 537 nm was determined. ● frozen culture, □ control

Digitalis cell cultures hydroxylate β-methyldigitoxin to β-methyldigoxin, a component which is widely used in medicine.

The cell strain 287, which is the product of intensive screening work, was shown to be biochemically unstable. Although the decline in biochemical potential is not more than 20% during 50 passages, repeated selection is necessary.

Digitalis cell suspensions were frozen according to the scheme of Fig. 1. After a regrowth phase of 4–6 weeks cells were transferred from agar to liquid medium and cultivated for 3–5 passages. After this period the suspension was re-established and the material sufficient for biotransformation experiments. β-Methyldigitoxin was used as substrate for the biotransformation reaction. The predominant portion of the product (β-methyldigoxin) could be obtained from the culture medium. The biotransformation capacity was tested in 300 ml shake cultures. The *Digitalis* strain from which the frozen cells had been taken was used as a control. Results are given in Fig. 3. The time curves of β-methyldigoxin accumulation from cultures treated in two separate freezing experiments are identical. A comparison with the control curves revealed that no reduction of the biotransformation capacity occurred during the freeze-thaw procedure.

In addition, the behaviour of cryopreserved cells was tested in 20-l airlift reactors. Since the growth pattern of the culture was unchanged, the scale up was attained without difficulties. Both tolerance against β-methyldigitoxin and the maximum yields of β-methyldigoxin were equal in recultivated and control cells. This result was published in a previous paper (Seitz et al. 1983). Moreover, Diettrich et al. (1982) have demonstrated that the glucosylation capacity of *D. lanata* cells could be preserved during cryostorage. Recently, these authors published long-term experiments with *D. lanata* (Diettrich et al. 1985). After a 3-year storage period, the recultivated cells showed the same growth rate, DNA content, and rate of glucosylation of added digitoxin and embryogenic capacity as did the original cell population.

Gathering the reports concerning the biosynthetic (biotransformation) capacity of cryopreserved cell cultures (Diettrich et al. 1982, 1985; Dougall and Whitten 1980; Seitz et al. 1983; Watanabe et al. 1983), we conclude that cryogenic storage is a useful method for preserving various biochemical characteristics in otherwise unstable plant cell cultures.

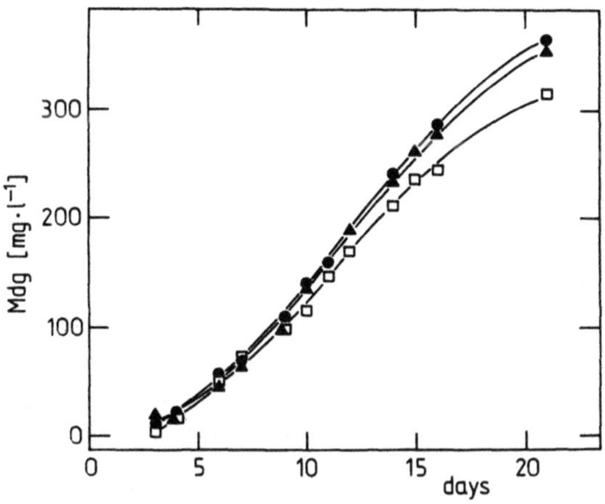

Fig. 3. Biotransformation of β-methyldigitoxin to β-methyldigoxin (*Mdg*) in a 300 ml shake culture of *Digitalis lanata* (strain 287). Glucose was fed from day 8 to maintain a glucose level of between 2 and 5 g l^{-1}. An aliquot of culture medium was diluted with methanol and analyzed by HPLC using a Hewlett Packard 1084 B Liquid Chromatograph equipped with a stainless steel column (25 cm long, 4.6 mm internal diameter), and LiChrosorb RP 8, 7 μm (Merck, Darmstadt, FRG) as stationary phase. The liquid phase was acetonitrile/water (a gradient from 40% to 50% acetonitrile until 6.0 min, thereafter held constant at 50%. Measuring wavelength was 220 nm, reference wavelength 360 nm. ●, ▲ frozen cells, □ control

3.4 Freezing Resistance of Cardenolide-Containing Digitalis Cells

Digitalis lanata cells are able to transform digitoxin to glucosylated products, mainly purpureaglucoside A (PA) and lanatoside A (LA) (Reinhard et al. 1975). These products are accumulated in the vacuoles (Kreis and Reinhard, unpublished results). Considering the observation that several secondary products reduce the freezing tolerance, we have asked if cardenolides accumulated in the vacuole might influence the survival of frozen-thawed cells.

Digitalis lanata (strain 10/56-S50-Se) was incubated with different digitoxin concentrations during the preculture period. An aliquot of the cell suspension was used for extraction and determination of the cardenolide content. Freezing was performed according to Fig. 1. The result is shown in Fig. 4. At the end of the preculture period (series A), a reduction of cell viability occurred from 94% (control) to 61% (samples with high cardenolide content). After incubation with cryoprotectants (DMSO/glycerol/sucrose) similar viability values were observed (B). Post-thaw viabilities (C) are equal in controls and in the samples with relatively low cardenolide content. A reduction (61% compared with ca. 80%) occurred only in sample 3 (total cardenolide content more than 1,200 μg g^{-1}). On the other hand, all specimens of series C are capable of resuming growth, namely, control, 1 and 2 with almost equal growth rates, sample 3 with slight delay (data not shown). It is noteworthy that fluorescein diacetate staining revealed unexpected high viability values in series B and C compared with samples A. We suggest that the cryoprotectants, presumably DMSO, enhances to some extent the uptake of fluorescein diacetate, thus simulating a higher viability. Obviously, a direct comparison of

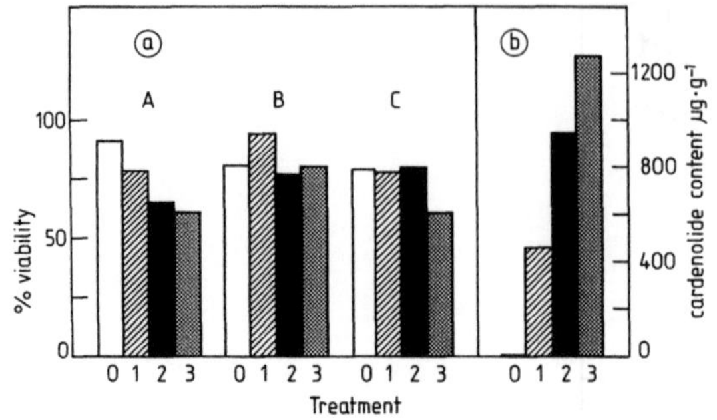

Fig. 4a,b. Freezing resistance and cardenolide content of *Digitalis lanata* cells (strain 10/56-S50-Se). The cells were fed with various amounts of digitoxin during the preculture period and were then frozen according to the scheme of Fig. 1. An aliquot of the cell suspension (0.4 g fresh weight) was homogenized in the presence of methanol (0.8 ml). After centrifugation (20 min, 13,000 g) the supernatant was analyzed by HPLC (see Fig. 3). Viability (**a**) was determined after the preculture period (*A*), after the cryoprotectant treatment (*B*), and immediately after thawing (*C*). *Abscissa* Digitoxin treatment: *0* control; *1* 40 $\mu g\,ml^{-1}$; *2* 80 $\mu g\,ml^{-1}$; *3* 120 $\mu g\,ml^{-1}$. The overall cardenolide content (purpureaglucoside A plus lanatoside A plus digitoxin) was calculated on the basis of fresh weight (**b**)

viability values from samples with different histories is misleading, and it takes us to the limits of this method. As was mentioned above, only a slight decrease in cell survival occurred in cardenolide-containing cells after freezing. On the other hand, compared with concentrations of secondary products for example of alkaloids in *B. wilsonae* (approx. 100 mg g^{-1} dry wt.), the glucosylation of digitoxin yielded only relatively low amounts of cardenolide (ca. 12. mg g^{-1} fresh wt). Therefore, we conclude that the nature of secondary products, their concentrations in the cells and, probably, some unknown cytological characteristics which are in connection with the accumulation of secondary products, may all interfere with freezing resistance of plant cell cultures.

4 Appendix

Table 3 gives a list of plant cell cultures regrown successfully after cryopreservation. Possible secondary product formation and/or regeneration of plantlets are noted.

Table 3. Summary of reports of successful cryopreservation (–196°) of cell suspensions. Only those species are listed from which recovery growth was communicated. Regeneration data and biochemical capacities are given

Species	Biochemical capacity	References
Acer pseudoplatanus		Sugawara and Sakai 1974; Withers and King 1980; Withers and Street 1977
Atropa belladonna		Nag and Street 1975

Table 3 (continued)

Species	Biochemical capacity	References
Brassica napus		Weber et al. 1983
Catharanthus roseus		Kartha et al. 1982
Catharanthus roseus	Alkaloid production	Chen et al. 1984
Datura innoxia		Weber et al. 1983
Daucus carota		Dougall and Wetherell 1974; Hauptmann and Widholm 1982; Weber et al. 1983; Withers and Street 1977
Daucus carota (+)		Nag and Street 1973
Daucus carota	Anthocyanin accumulation	Dougall and Whitten 1980
Dianthus caryophyllus		Anderson 1979
Digitalis lanata	Biotransformation	Dietrich et al. 1982, 1985; Seitz et al. 1983
Dioscorea deltoidea	Steroid biosynthesis	Butenko et al. 1984
Glycine max		Weber et al. 1983
Hyoscyamus muticus (+)		Withers and King 1980
Lavandula vera	Biotin biosynthesis	Watanabe et al. 1983
Nicotiana plumbaginifolia		Maddox et al. 1982/83
Nicotiana sylvestris		Maddox et al. 1982/83
Nicotiana tabacum		Hauptmann and Widholm 1982; Maddox et al. 1982/83
Nicotiana tabacum (+)		Bajaj 1976
Oryza sativa		Cella et al. 1978, 1982; Sala et al. 1979
Panax ginseng		Butenko et al. 1984
Rosa Pauls' Scarlet		Withers and King 1980
Saccharum sp.		Finkle and Ulrich 1979
Zea mays		Withers 1980; Withers and King 1979, 1980

(+) Regeneration of plantlets was achieved. For literature on freezing of callus, shoot-tips, embryos etc. see review of Withers (1982)

Acknowledgements. This work was supported by the Deutsche Forschungsgemeinschaft.

References

Alfermann AW, Boy HM, Döller PC, Hagedorn W, Heins M, Wahl J, Reinhard E (1977) Biotransformation of cardiac glycosides by plant cell cultures. In: Barz W, Reinhard E, Zenk MH (eds) Plant tissue culture and its bio-technological application. Springer, Berlin Heidelberg New York, pp 125–141

Anderson JO (1979) Cryopreservation of apical meristems and cells of carnation *(Dianthus caryophyllus)*. Cryobiology 16:583 (Abstr)

Bajaj YPS (1976) Regeneration of plants from cell suspensions frozen at –20, –70 and –196 °C. Physiol Plant 37:263–268

Butenko RG, Popov AS, Volkova LA, Chernyak ND, Novov AM (1984) Recovery of cell cultures and their biosynthetic capacity after storage of *Dioscorea deltoidea* and *Panax ginseng* cells in liquid nitrogen. Plant Sci Lett 33:285–292

Cella R, Sala F, Nielsen E, Rollo F, Parisi B (1978) Cellular events during the regrowth phase after thawing of freeze-preserved rice cells, Abstract. Meet Fed Eur Soc Plant Physiol, Edinburgh 1978, pp 127–128
Cella R, Colombo R, Galli MG, Nielsen E, Rollo F, Sala F (1982) Freeze-preservation of rice cells: A physiological study of freeze-thawed cells. Physiol Plant 55:279–284
Chen THH, Kartha KK, Leung NL, Kurz WGW, Chatson KB, Constabel F (1984) Cryopreservation of alkaloid-producing cell cultures of periwinkle *(Catharanthus roseus)*. Plant Physiol 75:726–731
Diettrich B, Popov AS, Pfeiffer B, Neumann D, Butenko R, Luckner M (1982) Cryopreservation of *Digitalis lanata* cell cultures. Planta Med 46:82–87
Diettrich B, Haack U, Popov AS, Butenko RG, Luckner M (1985) Cryopreservation of *Digitalis lanata* cell cultures. Long-term storage in liquid nitrogen of an embryogenic cell strain of *Digitalis lanata*. Biochem Physiol Pflanz 180:33–43
Dougall DK, Wetherell DF (1974) Storage of wild carrot cultures in the frozen state. Cryobiology 11:410–415
Dougall DK, Whitten GH (1980) The ability of wild carrot cell cultures to retain their capacity for anthocyanin synthesis after storage at –140 °C. Planta Med Suppl 129–135
Finkle BJ, Ulrich JM (1979) Effects of cryoprotectants in combination on the survival of frozen sugarcane cells. Plant Physiol 63:598–604
Hauptmann RM, Widholm JM (1982) Cryostorage of cloned amino acid analog-resistant carrot and tobacco suspension cultures. Plant Physiol 70:30–34
Henshaw GG, Stamp JA, Westcott RJ (1980) Tissue cultures and germplasm storage. In: Sala F, Parisi B, Cella R, Cifferi O (eds) Plant cell cultures: results and perspectives. Elsevier/North Holland Biomed Press, Amsterdam New York Oxford, pp 277–282
Kartha KK, Leung NL, Gaudet-LaPrairie P, Constabel F (1982) Cryopreservation of periwinkle, *Catharanthus roseus* cells cultures in vitro. Plant Cell Rep 1:135–138
Maddox AD, Consalves F, Shields R (1982/83) Successful preservation of suspension cultures of three *Nicotiana* species at the temperature of liquid nitrogen. Plant Sci Lett 28:157–162
Nag KK, Street HE (1973) Carrot embryogenesis from frozen cultured cells. Nature (London) 245: 270–272
Nag KK, Street HE (1975) Freeze preservation of cultured plant cells. II: The freezing and thawing phases. Physiol Plant 34:261–265
Noé W, Langebartels C, Seitz HU (1980) Anthocyanin accumulation and PAL activity in a suspension culture of *Daucus carota* L. Inhibition by L-α-aminooxy-β-phenylpropionic acid and t-cinnamic acid. Planta 149:283–287
Reinhard E, Boy M, Kaiser F (1975) Umwandlung von *Digitalis*-Glycosiden durch Zellsuspensionskulturen. Planta Med Suppl: 163–168
Rothenberger S (1982) Untersuchungen zur Bildung von Protoberberin-Alkaloiden in Zellkulturen von *Berberis wilsoniae* Hemsl. & Wils. Dissertation, Univ Tübingen
Sala F, Cella R, Rollo F (1979) Freeze-preservation of rice cells. Physiol Plant 45:170–176
Seitz U, Alfermann AW, Reinhard E (1983) Stability of biotransformation capacity in *Digitalis lanata* cell cultures after cryogenic storage. Plant Cell Rep 2:273–276
Sugawara Y, Sakai A (1974) Survival of suspension-cultured sycamore cells cooled to the temperature of liquid nitrogen. Plant Physiol 54:722–724
Watanabe K, Mitsuda H, Yamada Y (1983) Retention of metabolic and differentiation potentials of green *Lavandula vera* callus after freeze-preservation. Plant Cell Physiol 24:119–122
Weber G, Roth EJ, Schweiger H-G (1983) Storage of cell suspensions and protoplasts of *Glycine max* (L.) Merr., *Brassica napus* (L.), *Datura innoxia* (Mill.), and *Daucus carota* (L.) by freezing. Z Pflanzenphysiol 109:29–39
Widholm JM (1972) The use of fluorescein diacetate and phenosafranine for determining viability of cultured plant cells. Stain Technol 47:189–194
Withers LA (1980) The cryopreservation of higher plant tissue and cell cultures – an overview with some current observations and future thoughts. Cryo-Lett 1:239–250
Withers LA (1982) Storage of plant tissue cultures. In: Withers LA, Williams JT (eds) Crop genetic resources – the conservation of difficult material. Int Union Biol Sci/Int Board Plant Genet Res, Rome, pp 49–82

Withers LA, King PJ (1979) Proline: a novel cryoprotectant for the freeze preservation of cultured cells of *Zea mays* L. Plant Physiol 64:675–678

Withers LA, King PJ (1980) A simple freezing-unit and routine cryopreservation method for plant cell cultures. Cryo-Lett 1:213–220

Withers LA, Street HE (1977) Freeze preservation of cultured plant cells. III: The pregrowth phase. Physiol Plant 39:171–178

Zenk MH (1978) The impact of plant tissue culture on industry and agriculture. In: Thorpe TA (ed) Frontiers of plant tissue culture 1978, Calgary. The Bookstore, Univ Calgary, Canada, pp1–13

V Herbicides

Altered Amino Acid Biosynthesis in Amino Acid Analog and Herbicide-Resistant Cells

R.A. GONZALES[1] and J.M. WIDHOLM[2]

1 Introduction

This laboratory (University of Illinois) has been working for a number of years on the selection of plant cell strains which are resistant to various compounds. Such resistance should be useful as markers in genetic manipulation experiments. In these studies cell strains with alterations in amino acid biosynthetic enzymes have been recovered by selecting for resistance to several amino acid analogs and to the herbicide glyphosate [N-(phosphonomethyl)-glycine]. Several of our most thoroughly studied resistant strains will now be discussed.

2 5-Methyltryptophan Resistance

The first successful selection work done in this laboratory was with the tryptophan analog, 5-methyltryptophan (5MT). This analog is not incorporated into plant cell protein (Sasse et al. 1983), but does false feedback inhibit the tryptophan control enzyme anthranilate synthase (Widholm 1972a) (Fig. 1). Unmutagenized tobacco and carrot suspension cultured cells were inoculated into liquid medium containing completely inhibitory levels of 5MT and were incubated for two months (Widholm 1972b,c). Within this time growth was noted in some flasks indicating a resistance frequency in the range of 10^{-6} to 10^{-7}. Studies with mutagens have shown that this frequency can be increased by tenfold or more with carrot (Widholm 1977a) and *D. innoxia* (Ranch et al. 1983). Generally the selected resistance is stable and is about 100-fold that of the wild type.

Amino acid analysis has shown that most 5MT-resistant strains contain 30- to 40-fold higher levels of free tryptophan than the wild type. Studies of anthranilate synthase, the first enzyme in the branch leading to tryptophan in the aromatic amino acid biosynthetic pathway (Fig. 1), show that in most cases the enzyme in the 5MT-resistant cells is less sensitive than the wild type to feedback inhibition by tryptophan and 5MT. This altered control at the feedback control step in the tryptophan branch would explain the 5MT resistance and the resulting overproduction of free tryptophan.

1 Samuel Roberts Noble Foundation, P.O. Box 2180, 2510 Hwy 70 E., Ardmore, OK 73402, USA
2 University of Illinois, Agronomy Department, 1102 S. Goodwin Ave., Urbana, IL 61801, USA

Primary and Secondary Metabolism of Plant Cell Cultures, ed. by Neumann et al.

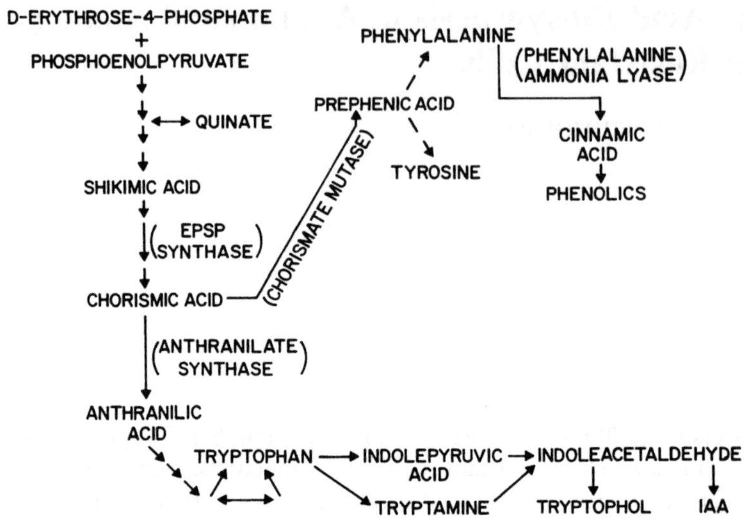

Fig. 1. Aromatic amino acid biosynthetic pathway. *EPSP* 5-enolpyruvylshikimate-3-phosphate synthase; *IAA* indoleacetic acid

Further studies using preparative polyacrylamide gel electrophoresis with potato cell extracts (Carlson and Widholm 1978) and HPLC with tobacco cell extracts (J.E. Brotherton and J.M. Widholm, unpublished) show that these species normally have two forms of anthranilate synthase. One form is tryptophansensitive while the other is tryptophan resistant. The wild-type cells contain predominantly the sensitive form, while the 5MT-resistant cells contain predominantly the resistant form. Plants regenerated from the 5MT-resistant tobacco cells were not resistant to 5MT as determined by the ability to root in its presence, and contained only the sensitive form of anthranilate synthase. The resistance to 5MT and the altered form of anthranilate synthase reappeared in callus formed from leaves of the regenerated plants. This phenomenon has been followed through several plant regeneration and callus formation cycles.

In contrast to the potato and tobacco cases, carrot and *D. innoxia* have different characteristics. First of all only one form of anthranilate synthase can be detected from wild-type carrot cells using preparative polyacrylamide gel electrophoresis and HPLC (gel permeation). Secondly, in the case of *D. innoxia*, plants have been regenerated which express the trait (Ranch et al. 1983) and pass the resistance on to their progeny (J.M. Widholm and J.P. Ranch, unpublished). Thus, in the case of the diploids, carrot and *D. innoxia*, only one enzyme form appears to be normally present while the tetraploids, tobacco, and potato have two forms.

We have also observed that in some species, like carrot and potato, cell strains which are 5MT-resistant and overproduce tryptophan are often auxin autotrophic, i.e. require no auxin for growth (Widholm 1977b). In the case of carrot, five of ten 5MT-resistant strains tested required no auxin for growth. However, none of nine carrot lines selected as auxin autotrophic were 5MT-resistant. None of the 5MT-resistant tobacco (Widholm 1977b) or *D. innoxia* strains (Ranch et al. 1983) were auxin-autotrophic, however. At least in the case of carrot and potato the correlation of high free tryptophan with auxin autotrophy indicates that the increased levels of the indoleacetic acid (IAA) precursor, tryptophan, somehow causes more flow through the pathway (Fig. 1) to synthesize IAA. Sung (1979) has reported that a 5MT-resistant carrot line with increased free tryptophan was auxin autotrophic and contained higher than normal levels of free IAA.

3 p-Fluorophenylalanine Resistance

The phenylalanine analog, p-fluorophenylalanine (PFP), is toxic to plants apparently because it can be utilized by plant aminoacyl-tRNA synthetases to charge plant phenylalanyl tRNA in place of phenylalanine in vitro (Smith and Fowden 1968) and be incorporated into protein in vivo (Palmer and Widholm 1975). Proteins containing an amino acid analog are likely to be structurally altered and be inactive biologically. We have selected PFP-resistant carrot and tobacco strains from suspension cultures (Palmer and Widholm 1975). The carrot strain showed increases in free phenylalanine while the tobacco did not but did contain increased levels of phenolics due apparently to a 10- to 20-fold increase in phenylalanine ammonia lyase (Berlin and Widholm 1977) (Fig. 1). Increased phenylalanine ammonia lyase may cause the cell strain to be resistant to PFP since the enzyme can detoxify PFP by conversion to P-fluorocinnamic acid (Berlin et al. 1982). This illustrates a positive selection method for selecting cell strains which overproduce a specific class of secondary compounds.

4 Hydroxyproline Resistance

A suspension-cultured carrot cell strain was selected as resistant to hydroxyproline (Widholm 1976). Hydroxyproline is probably toxic to plants due to its incorporation into protein as demonstrated by Holleman (1967) with cultured sycamore cells. The selected carrot strain was also resistant to another proline analog, azetidine-2-carboxylate (A2C), and contained from 15 to 30 times the normal free proline levels. The enzymatic alteration in the resistant cells has not been studied since an assay for the probable feedback control enzyme, glutamate kinase, has not been developed yet in plants.

5 Ethionine Resistance

Ethionine is a structural analog of the amino acid methionine. The difference being the presence of an ethyl-thioester instead of the methyl-thioester found in methionine. Although ethionine is readily incorporated into protein, toxicity is mainly attributed to replacement of methionine in other cellular functions, including incorporation into the S-adenosylmethionine (SAM) pool as S-adenosylethionine (SAE) (Alix 1982). The ubiquitous nature of SAM and its importance in mjaor metabolic functions make it especially vulnerable.

We have selected a *Nicotiana tabacum* cell line that is resistant to the inhibitory effects of ethionine (Gonzales et al. 1984). These variants were obtained by inoculating cells from suspension culture into growth medium supplemented with twice the minimum lethal concentration of ethionine (0.2 mM). The selected lines were able to tolerate 27-fold higher levels of ethionine than the wild type. A profile of the free amino acid pools showed a dramatic increase in the levels of the aspartate-derived amino acids methionine (110 X), threonine (18 X), lysine (5 X), and isoleucine (5 X). Glycine and serine levels were also elevated. This may be due in part to their involvement in the C_1 pathway, which contributes the S-methyl group in the last step of methionine synthesis.

In ethionine-resistant bacterial and fungal mutants, loss of activity for methionine adenosyltransferase resulted in overproduction of methionine (Hobson and Smith 1973; Jacobson et al. 1977).

This enzyme converts methionine into SAM. However, in our ethionine-resistant tobacco cells there was no difference in activity for this enzyme. There was also no difference in activity for homoserine dehydrogenase. This feedback regulated enzyme is located at the step in the methionine pathway where lysine biosynthesis branches off.

On the other hand, activity for aspartate kinase (AK), the first enzyme common to all aspartate amino acids, was elevated ninefold in the resistant cells. AK is usually present in two isoforms, one feedback regulated by lysine and one by threonine. From inhibition profiles, activity for each isozyme can be calculated. As shown in Fig. 2, the

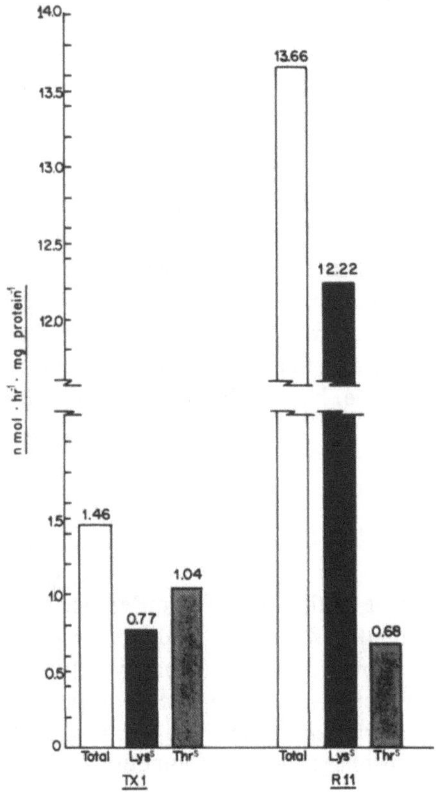

Fig. 2. *Aspartate Kinase* activity in (*TX1*) ethionine-sensitive and (*R11*) resistant cultured tobacco cells. (Data compiled from Gonzales et al. 1984)

increased activity in the resistant cells can be attributed solely to a 16-fold increase in the activity of the lysine-sensitive isozyme. Preliminary enzyme kinetics and co-elution in DEAE-cellulose anion exchange chromatography, suggest lysine-sensitive AK activities in ethionine-sensitive and -resistant cells represent the same enyzme.

We are continuing efforts in our laboratories to identify the specific cause of the elevated lysine-sensitive AK activity in the ethionine-resistant tobacco cells. Simple immunochemical experiments are planned to distinguish between activation or increased expression of the enzyme. Preliminary evidence from protoplast fusion experiments suggest the latter.

Our protoplast fusion protocol uses a double inhibitor selection scheme. Each of the parental lines contain different resistance markers. Only the fusion hybrids should have a combination of the two. Therefore, hybrids can be differentially selected in media containing both inhibitors. Using the ethionine-resistant tobacco cells, we selected for resistance to a second amino acid analog 5MT. This resistance marker was used in the hybrid selection step to allow a blind test of ethionine resistance. A number of putative fusion hybrids showing double resistance were obtained. In every case the hybrids showed a loss of the ethionine-resistance marker. However, still to be proven is that the hybrids are true heterokaryons and that the ethionine-resistance marker is in fact a heritable genetic trait.

6 Glyphosate Resistance

We have recently selected a carrot suspension culture resistant to the herbicide glyphosate [N-(phosphonomethyl)glycine]. The growth inhibition caused by this nonselective herbicide is known to be reversed by the addition of the aromatic amino acids, phenylalanine, tyrosine, and tryptophan to the growth medium of *Lemna* (Jaworski 1972) and carrot and tobacco cells (Haderlie et al. 1977). Only recently, however, has the target enzyme in the shikimate pathway, 5-enolpyruvylshikimic acid-3-phosphate synthase (EPSPS) (Fig. 1) been identified (Steinrücken and Amrhein 1980).

The resistant carrot strain was selected by inoculating suspension cultured cells sequentially into increasing levels of glyphosate until a 50-fold increased resistance was attained (the 50% inhibition concentration was increased from 0.12 mM for the wild type to 6.31 mM for the selected strain) (Nafziger et al. 1984). The resistance was not lost when the cells were cultured away from glyphosate for 2 years.

The glyphosate resistance was not due to a decrease in uptake but was apparently caused by a 12-fold increase in the activity of EPSPS. The increased EPSPS activity appears to be due to increased levels of the normal type enzyme as determined by glyphosate inhibition characteristics. The activities of two other enzymes of the pathway, shikimate dehydrogenase and anthranilate synthase, were not changed. Even though EPSPS should not be a control enzyme, since it is situated internally in the pathway (Fig. 1), the cells with the higher activity also have two or more fold the normal levels of free tryptophan, phenylalanine, and tyrosine. The reasons for this alteration in the pathway control has not been clarified as yet.

7 Conclusion

Most of the resistant strains described here overproduce amino acids as a result of some alteration in the enzyme forms or activity level. Only a few strains have been recovered which are resistant due to decreased uptake. This might be due to the possibility that most of the enzyme alterations selected for cause resistance in a dominant fashion while a deficiency in uptake might be recessive. One can reason that a feedback-altered enzyme would lead to resistance being dominantly expressed in the heterozygous condition since only some of the control enzyme activity needs to be active to allow amino acid overproduction. As mentioned below, such alterations are indeed usually expressed in protoplast fusion hybrids which would be a genetically heterozygous condition. However, the one case where an uptake resistant strain (aminoethylcysteine, Matthews et al. 1980) was adequately tested in fusion experiments, the resistance was also expressed in the hybrids (Harms et al. 1981).

Protoplast fusion studies have shown that the 5MT, A2C, and glyphosate resistances are expressed in the fusion hybrids and thus can be used to select the hybrids from the parental cells carrying single resistances (Harms et al. 1981; R.M. Hauptmann and J.W. Widholm, unpublished). In contrast, the ETH resistance of the tobacco strains appears to not be expressed in fusion hybrids (R.A. Gonzales and J.M. Widholm, unpublished). This might indicate that the overproduction of aspartokinase in the ETH-resistant cells is due to the lack of an active repressor molecule which is supplied by the other protoplast in a fusion hybrid.

The observation that most of the amino acid analog resistant cell lines overproduce the corresponding natural amino acid is consistent with the general action of the analogs. Analogs are compounds similar enough in structure to the normal amino acid to mimic some functions but different enough to interfere with the proper action of the amino acid or product made from it. If a compound acts truly as an analog, i.e. the toxic effect can be prevented by the addition of the normal amino acid, then it is possible to select cells which overproduce the normal amino acid by selecting for analog resistance since the mutant overproducers will be resistant to the analog. Wild-type cells will not be able to grow in the presence of the analog. Such selection systems can be very effective, since millions of cells can be screened quite readily.

The selection and biochemical characterization studies described here show that it is possible to select for alterations in amino acid biosynthetic pathways and to learn something about the controls of the pathway by studying the enzyme alterations. It is also possible to select for alterations in the production of secondary compounds (phenolics) and hormones (IAA) with these techniques.

Acknowledgments. This research has been supported by funds from the Illinois Agricultural Experiment Station and the National Science Foundation.

References

Alix JH (1982) Molecular aspects of the in vivo and in vitro effects of ethionine, an analog of methionine. Microbiol Rev 46:281–295

Berlin J, Widholm JM (1977) Correlation between phenylalanine ammonia lyase activity and phenolic biosynthesis in p-fluorophenylalanine-sensitive and -resistant tobacco and carrot tissue cultures. Plant Physiol 59:550–553

Berlin J, Witte L, Hammer J, Kukoschke KG, Zimmer A, Pape D (1982) Metabolism of p-fluorophenylalanine in p-fluorophenylalanine sensitive and resistant tobacco cell cultures. Planta 155: 244–250

Carlson JE, Widholm JM (1978) Separation of two forms of anthranilate synthetase from 5-methyltryptophan-susceptible and -resistant cultured *Solanum tuberosum* cells. Physiol Plant 44:251–255

Gonzales RA, Das PK, Widholm JM (1984) Characterization of cultured tobacco cell lines selected for resistance to a methionine analog, ethionine. Plant Physiol 74:640–644

Haderlie LC, Widholm JM, Slife FW (1977) Effect of glyphosate on carrot and tobacco cells. Plant Physiol 60:40–43

Harms CT, Potrykus I, Widholm JM (1981) Complementation and dominant expression of amino acid analogue resistance markers in somatic hybrid clones from *Daucus carota* after protoplast fusion. Z Pflanzenphysiol 101:377–390

Hobson AC, Smith DA (1973) S-Adenosylmethionine synthetase in methionine regulatory mutants of *Salmonella typhimurium*. Mol Gen Genet 126:7–18

Holleman J (1967) Direct incorporation of hydroxyproline into protein of sycamore cells incubated at growth-inhibitory levels of hydroxyproline. Proc Natl Acad Sci USA 57:50–54

Jacobson ES, Chen GS, Metzenberg RL (1977) Unstable S-adenosylmethionine synthetase in an ethionine-resistant strain of *Neurospora crassa*. J Bacteriol 132:747–748

Jaworski EG (1972) Mode of action of N-phosphonomethylglycine: inhibition of aromatic amino acid biosynthesis. J Agric Food Chem 20:1195–1198

Matthews BF, Shye SH, Widholm JM (1980) Mechanism of resistance of a selected carrot cell suspension culture to S(2-aminoethyl)-L-cysteine. Z Pflanzenphysiol 96:453–463

Nafziger ED, Widholm JM, Steinrucken HC, Killmer JC (1984) The selection and characterization of a carrot cell line tolerant to glyphosate. Plant Physiol 76:571–574

Palmer JE, Widholm JM (1975) Characterization of carrot and tobacco cell cultures resistant to p-fluorophenylalanine. Plant Physiol 56:233–238

Ranch JE, Rick S, Brotherton JE, Widholm JM (1983) The expression of 5-methyltryptophan-resistance in plants regenerated from resistant cell lines of *Datura innoxia*. Plant Physiol 71: 136–140

Sasse F, Buchholz M, Berlin J (1983) Site of actin of growth inhibitory tryptophan analogues in *Catharanthus roseus* cell suspension cultures. Z Naturforsch 38c:910–915

Smith IK, Fowden L (1968) Studies on the specificities of the phenylalanyl- and tyrosyl-sRNA synthetases from plants. Phytochemistry 7:1065–1075

Steinrücken HC, Amrhein N (1980) The herbicide glyphosate is a potent inhibitor of 5-enolpyruvylshikimic acid-3-phosphate synthase. Biochem Biophys Res Common 94:1207–1212

Sung ZR (1979) Relationship of indole-3-acetic acid and tryptophan concentration in normal and 5-methyltryptophan-resistant cell lines of wild carrots. Planta 145:339–345

Widholm JM (1972a) Tryptophan biosynthesis in *Nicotiana tabacum* and *Daucus carota* cell cultures: site of actin of inhibitory tryptophan analogs. Biochim Biophys Acta 261:44–51.

Widholm JM (1972b) Cultured *Nicotiana tabacum* cells with an altered anthranilate synthetase which is less sensitive to feedback inhibition. Biochim Biophys Acta 261:52–58

Widholm JM (1972c) Anthranilate synthetase from 5-methyltryptophan-susceptible and -resistant cultured *Daucus carota* cells. Biochim Biophys Acta 279:48–57

Widholm JM (1976) Selection and characterization of cultured carrot and tobacco cells resistant to lysine, methionine, and proline analogs. Can J Bot 54:1523–1529

Widholm JM (1977a) Isolation of biochemical mutants of cultured plant cells. In: Rubenstein I, Phillips RL, Green CE, Desnick R (eds) Molecular genetic modification of eucaryotes. Academic Press, London New York, pp 57–65

Widholm JM (1977b) Relation between auxin autotrophy and tryptophan accumulation in cultured plant cells. Planta 134:103–108

Acetohydroxyacid Synthase Inhibitors as Herbicides

D. SCHEEL[1] and J.E. CASIDA[2]

1 Introduction

1.1 Diverse Actions of Herbicides

Herbicides are notable for their diverse chemical types and modes of action. Major classes of herbicides act as hormones or antihormones, block photosynthetic electron flow, interfere with microtubular systems, disrupt membranes, or inhibit biosynthetic pathways (Fedtke 1982). They also vary greatly in potency and in toxicity to nontarget species, including crops and mammals. Selectivity usually depends on species-dependent differences in rates of detoxification or activation or on intrinsic differences in the target site. Plant cell suspension cultures are playing an increasing role in unravelling the intricacies of herbicide metabolism and mode of action (Gressel et al. 1978).

1.2 Inhibition and Induction of Amino Acid Synthesis

Amino acid biosynthesis is an ideal target for selective herbicides. Plants synthesize each of the 20 amino acids whereas animals are unable to synthesize, de novo, about half of them, including the branched chain amino acids. Many herbicides and herbicide antidotes alter amino acid synthesis in plants. Glyphosate blocks the synthesis of aromatic amino acids by inhibiting 5-enolpyruvyl-shikimate-3-phosphate synthase (Berlin and Witte 1981; Haderlie et al. 1977; Steinrücken and Amrhein 1980). Amitrole interferes with histidine biosynthesis (Fedtke 1982). Many herbicides induce phenylalanine ammonia-lyase which leads to alterations in levels of aromatic amino acids and several secondary metabolites (Fedtke 1982). Dichloroacetamides acting as herbicide antidotes enhance the synthesis of cysteine (Adams et al. 1983). The sulfonylurea and imidazolinone herbicides inhibit acetohydroxyacid synthase (AHAS) (EC 4.1.3.18) and biosynthesis of branched chain amino acids as detailed in this review. Chlorsulfuron (CS) and sulfometuron methyl (SM) are the most potent commercial herbicides (Levitt

Abbreviations. AHAS acetohydroxyacid synthase; CS chlorsulfuron herbicide; dRN deoxyribonucleoside; RDP ribonucleoside diphosphate; SM sulfometuron methyl herbicide; TPP thiamine pyrophosphate

1 Max-Planck-Institut für Züchtungsforschung, Abt. Biochemie, Egelspfad, 5000 Köln 30, FRG

2 Pesticide Chemistry and Toxicology Laboratory, Department of Entomological Sciences, University of California, Berkeley, CA 94720, USA

Primary and Secondary Metabolism of Plant Cell Cultures, ed. by Neumann et al.

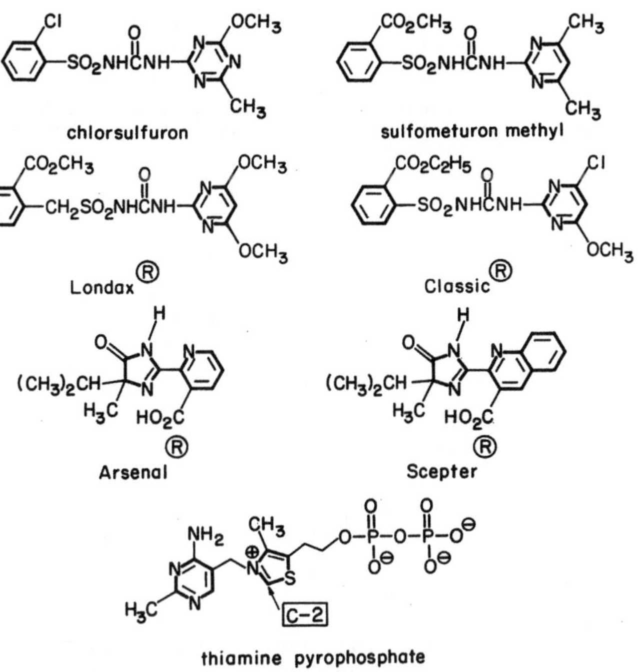

Fig. 1. Structures of four sulfonylurea and two imidazolinone herbicides which are inhibitors of acetohydroxyacid synthase (AHAS) (EC 4.1.3.18) and of the coenzyme

1983; Levitt et al. 1981). The sulfonylureas vary greatly in their crop tolerance, i.e., SM is essentially non-selective, whereas CS, Londax and Classic are used or proposed for use to control weeds in wheat, rice and soybean, respectively (Beyer 1984). Arsenal is a broad-spectrum herbicide (Reider and Shaner 1984) and Scepter is selective for weed control in soybean (American Cyanamid Co. 1983) (Fig. 1).

2 Acetohydroxyacid Synthase

2.1 Functions, Properties, and Assay

The biosynthetic pathways for valine, leucine, and isoleucine and the relevant enzymes given in Fig. 2 are applicable to nearly all bacteria, fungi, and plants that have been studied. The first common enzyme of these pathways is AHAS which catalyses the condensation of two molecules of pyruvate for the ultimate synthesis of valine and leucine or of pyruvate and 2-ketobutyrate for isoleucine biosynthesis. Thiamine pyrophosphate (TPP, Fig. 1) is the coenzyme in these condensation reactions. An adduct formed between the TPP C-2 carbanion and pyruvate undergoes decarboxylation to the stabilized anion of hydroxyethyl-TPP, which in turn acts as a nucleophile on the 2-keto group of a second molecule of pyruvate or 2-ketobutyrate, releasing TPP and

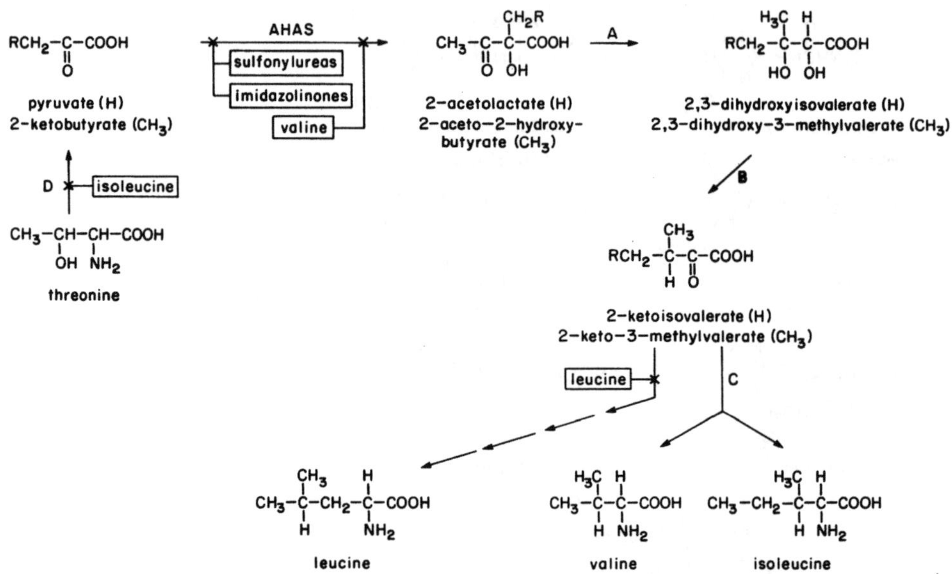

Fig. 2. Pathways for biosynthesis of branched chain amino acids showing sites of inhibition by the endproduct amino acids and by sulfonylurea and imidazolinone herbicides. The first step is catalyzed by AHAS which is also referred to as acetolactate synthase when the product is 2-acetolactate. Other enzymes are: *A* acetohydroxyacid isomeroreductase (EC 1.1.1.86), *B* dihydroxy acid dehydratase (EC 4.2.1.9), *C* branched-chain amino acid amino-transferase (EC 2.6.1.42), and *D* threonine deaminase (EC 4.2.1.16) for 2-ketobutyrate synthesis

2-acetolactate or 2-aceto-2-hydroxybutyrate (Walsh 1979). AHAS usually appears in the soluble fraction but it is also localized in the mitochondria with *Euglena gracilis* (Oda et al. 1982), *Neurospora crassa* (Kuwana and Date 1975; Cassady et al. 1972), and yeast (Ryan and Kohlhaw 1974).

AHAS is normally assayed with pyruvate, TPP, and divalent cations (Mg^{2+} or Mn^{2+}) at pH 6 to 8 (Bauerle et al. 1964; Davies 1964; Miflin 1969, 1971; Satyanarayana and Radhakrishnan 1963) sometimes with addition of ATP (Oda et al. 1982). FAD is an additional requirement for AHAS from some microorganisms (Semeraro and Wixom 1977; Stormer and Umbarger 1964). Product formation is conveniently measured spectrophotometrically after converting acetolactate to acetoin (Westerfeld 1945).

2.2 Inhibitors

Biosynthesis of branched chain amino acids is regulated by end product inhibition of the first specific enzyme of the respective pathway (Fig. 2) (Umbarger 1983). The synthesis of AHAS is repressed by excess valine and leucine in bacteria (Umbarger 1983) but apparently not in higher plants (Miflin and Cave 1972). Relatively weak AHAS inhibitors, requiring mM levels in vitro, include 2-aminobutyrate (Bauerle et al. 1964; Oda et al. 1982), L-threonine, L-methionine, L-tryptophan, L-2-chloroisocaproate, 4-methylvalerate, trifluoromethylcrotonate (Bauerle et al. 1964), 2-thiazolealanine

(Tsuchida and Momose 1975), and sulfhydryl reagents (Miflin 1971). The herbicides Scepter and Arsenal, derived from 2-methylvaline, inhibit AHAS at μM levels (Shaner et al. 1984; Anderson and Hibberd 1984) with much higher potency for (R)- than (S)-Scepter (Shaner et al. 1984). The most potent known AHAS inhibitors are the sulfonylurea herbicides with 50% inhibition at nM levels for enzymes from many different plant species (Ray 1984) and *Salmonella typhimurium* (LaRossa and Schloss 1984).

2.3 Isozymes

Ripening pea seeds appear to contain two AHAS isozymes with maximum activities at pH 8 to 8.5 and 6 to 7; only the former isozyme is inhibited by valine and isoleucine (Davies 1964). *S. typhimurium* (O'Neill and Freundlich 1972) and *N. crassa* (Takenaka and Kuwana 1972) have one AHAS isozyme which is sensitive to inhibition by valine and another which is insensitive. *Escherichia coli* contains three AHAS isozymes (I, II, and III) which vary in their substrate affinities and responses to repression (De Felice et al. 1979). AHAS I is most effective in condensation of two molecules of pyruvate, whereas AHAS III favors the condensation of pyruvate and 2-ketobutyrate, indicating that AHAS I is linked to the valine/leucine pathway and AHAS III to isoleucine. Both isozymes are sensitive to inhibition by valine. Various enteric bacteria possess a third valine-insensitive isozyme, AHAS II. *E. coli* K-12 carries the structural genes for these three isozymes, but AHAS II is not expressed. There is some indication of differential sensitivity of AHAS isozymes to sulfonylureas in both plants (Scheel and Casida 1985) and bacteria (Yadav and Benard 1984; LaRossa and Schloss 1984).

3 Consequences of Acetohydroxyacid Synthase Inhibition

3.1 Depletion of Branched Chain Amino Acids

1. Amino Acid Analysis. L-valine applied to duckweed cultures *(Spirodela polyrhiza)* lowers the isoleucine level and elevates leucine and 2-aminobutyrate; leucine probably accumulates due to its formation from valine (Borstlap 1981). CS at 14–28 nM in soybean cell suspension cultures drastically lowers free valine and leucine with a small decline in isoleucine, without affecting any other free amino acid (Table 1). Even 30 min after CS application the amount of valine and leucine is lower than in control soybean cells (Scheel and Casida 1985). Arsenal at 10 μM specifically lowers the levels of free valine, leucine, and isoleucine in maize cell suspension cultures 35 h after application (Anderson and Hibberd 1984). Glyphosate inhibition of the growth of carrot cell suspension cultures is reversible with aromatic amino acids yet glyphosate does not lower the cellular levels of these amino acids (Haderlie et al. 1977). In contrast, glyphosate completely inhibits incorporation of the label from [2,3,4,5n-^{14}C]shikimic acid into phenylalanine and tyrosine in tobacco cell suspension cultures (Berlin and Witte 1981). These results indicate that in plants there appear to be different cellular pools of aromatic amino acids but not of branched chain amino acids.

Table 1. Reduced levels of free branched chain amino acids in soybean cells 5 h after treatment with chlorsulfuron. (After Scheel and Casida 1985)

Free amino acids	Amino acid levels, % of control	
	14 nM CS	28 nM CS
Branched chain		
Isoleucine	60	49
Leucine	28	29
Valine	28	4
Others[a]	88–127	91–132

[a] Alanine, 4-aminobutyrate, arginine, asparagine, aspartate, glutamine, glutamate, glycine, histidine, lysine, phenylalanine, proline, serine, threonine, tryptophan, and tyrosine

2. Reversal of Growth Inhibition. Growth inhibition due to suboptimal AHAS activity is in principle reversible by branched chain amino acids (Borstlap 1981; De Felice et al. 1979; Umbarger 1983). Thus isoleucine specifically overcomes the end product inhibition by valine in plants (Borstlap 1981) and bacteria (De Felice et al. 1979). The reversal pattern differs for plants and bacteria after growth inhibition by sulfonylurea herbicides (Tables 2 and 3). In soybean cells leucine, valine, 2-ketoisovalerate, and mixtures of branched chain amino acids alleviate the inhibition, whereas isoleucine and pyruvate are inactive (Scheel and Casida 1985). A similar relationship exists in pea root cultures

Table 2. Reversal of chlorsulfuron-induced growth inhibition in plants and bacteria by branched chain amino acids and their precursors

Additive[a]	Reversal rating[b]		
	Soybean[a]	Pea[a]	Bacteria[c]
Branched chain amino acids			
Isoleucine	0	0	+
Leucine	+	0	0
Valine	+	+	–
Isoleucine, leucine, valine	+	+	
Other amino acids	0	0	0[d]
Precursors of branched chain amino acids			
2-Ketoisovalerate	+		
2-Ketoisovalerate, 2-keto-3-methylvalerate		+	
Pyruvate	0		
Pyruvate, threonine		0	

[a] Data for soybean cell suspension cultures from Scheel and Casida (1985) and for excised pea root cultures from Ray (1984). See Table 3 for test conditions

[b] + = Partial or complete reversal, 0 = no effect, and – = enhanced growth inhibition

[c] Data for *E. coli* from Scheel and Casida (1985). Similar reversal with isoleucine and enhancement with valine is obtained for SM-induced growth inhibition in *S. typhimurium* and *Citrobacter freundii* (La Rossa and Schloss 1984)

[d] Aromatic amino acids alleviate CS- and SM-induced growth inhibition in *E. coli*, whereas cysteine enhances the bacteriostatic activity of CS (Scheel and Casida 1985)

Table 3. Reversal of chlorsulfuron-induced growth inhibition in soybean cell suspension cultures and excised pea roots by branched chain amino acids and their precursors

Additive[a]	Growth relative to control, %[b]			
	Additive alone		Additive + CS	
	Soybean	Pea	Soybean	Pea
Control				
None	100	100	30	20
Branched chain amino acids				
Isoleucine	95	66	34	19
Leucine	98	49	71	23
Valine	77	90	81	40
Isoleucine, leucine	98	47	86	21
Isoleucine, valine	97	85	110	80
Leucine, valine	81	96	102	36
Isoleucine, leucine, valine	109	81	113	90
Casein amino acids	97		107	
Precursor of branched chain amino acids				
2-Ketoisovalerate	108		82	
2-Ketoisovalerate, 2-keto-3-methylvalerate		60		42
Pyruvate	105		44	
Pyruvate, threonine		103		13

[a] Additive concentration in soybean cell suspension cultures (*Glycine max* L. cv. Merrill var. Mandarin) was 0.05 mg ml^{-1} for each individual amino acid or 1 mg ml^{-1} for casein amino acids. CS level was 5 ppm relative to initial cell dry weight and 14 nM in culture medium, respectively (Scheel and Casida 1985). In excised pea root cultures (*Pisum sativum* L. var. Alaska) amino acids were at 100 μM, pyruvate, threonine, 2-ketoisovalerate, and 2-keto-3-methylvalerate at 1 mM, and CS at 28 nM (Ray 1984)

[b] Increase in dry weight after one growth cycle of 7 days for soybean and increase in root length after 96 h for pea

except leucine alone and with isoleucine does not protect (Ray 1984), a difference possibly attributable to interference from the ~50% direct growth inhibition from leucine. In *E. coli* K-12 (Scheel and Casida 1985) and *S. typhimurium* (LaRossa and Schloss 1984) sulfonylurea-induced growth inhibition is specifically reversed by isoleucine (Table 2).

3.2 Indirect or Secondary Effects

1. Inhibition of Growth. Inhibition of AHAS in both plants (Borstlap 1981) and bacteria (De Felice et al. 1979) results in growth retardation, apparently caused by deficiencies of one or more branched chain amino acids. The most potent AHAS inhibitors, the sulfonylurea herbicides, inhibit plant growth but not by affecting cell elongation (Ray 1982a, 1984; Scheel and Casida 1985).

2. Inhibition of DNA Synthesis. CS is a potent inhibitor of DNA synthesis in plants (Ray 1982a) but not in *E. coli* K-12 (Scheel and Casida 1985). The potency and rapidity

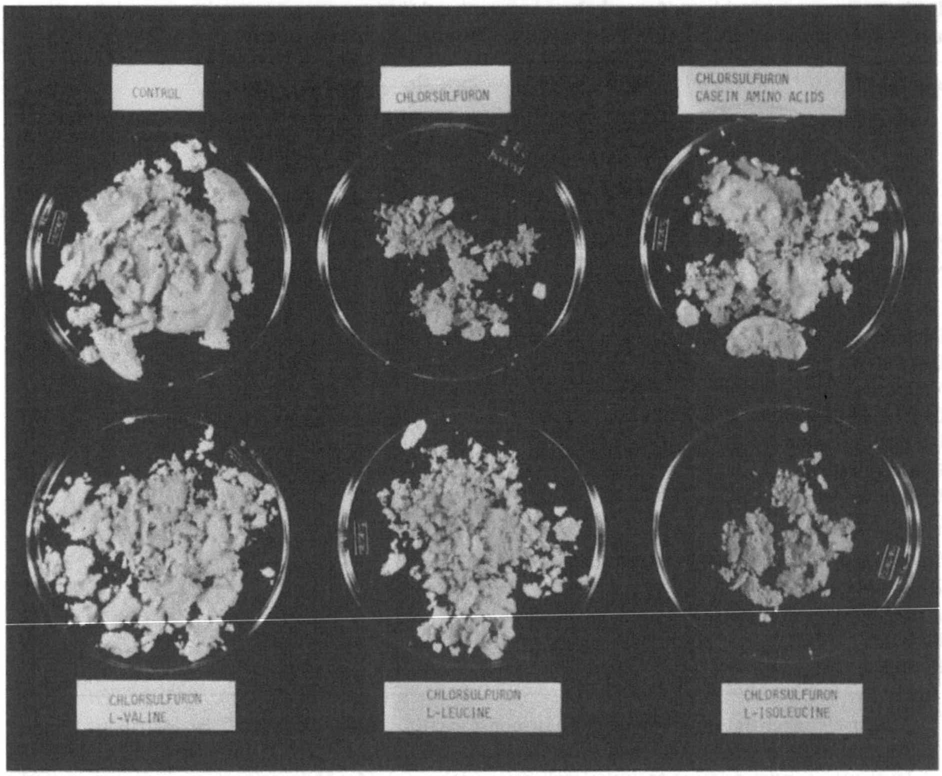

Fig. 3. Reversal of sulfonylurea action by branched chain amino acids. Soybean cell suspension cultures (*Glycine max* L. cv. Merrill var. Mandarin) were treated with chlorsulfuron (14 nM) and individual amino acids (0.05 mg ml^{-1}) or casein amino acids (1 mg ml^{-1}) with cell harvest after one growth cycle (7 days)

of CS action in peas and corn are indicated by inhibition of [^{3}H]thymidine incorporation into DNA without effect on precursor uptake as follows: significant inhibition 1 h after 2.8 μM CS with no further incorporation after 4 h (Rost 1984); 40% reduction at 1 h and 80% at 6 h after 28 μM CS and 40% inhibition at 6 h at 28 nM CS (Ray 1982a). CS does not act directly on plant DNA synthesis, i.e., there is no inhibition of thymidine kinase, DNA polymerase or of incorporation of [^{3}H]thymidine into DNA of isolated plant nuclei. CS-induced inhibition of DNA synthesis in corn seedlings is not alleviated by addition of purine or pyrimidine dRNs to the treatment solution (Ray 1982b). However, the reversal of CS growth inhibition may require application of all four dRNs near the time of onset of DNA replication since on earlier application they may be metabolized (Odmark 1969) before they are utilized for DNA synthesis.

RDP reductase is suggested as a possible indirect target for sulfonylureas in blocking DNA synthesis (Scheel and Casida 1985). RDP reductase reaches maximum activity just before the onset of DNA replication (Hovemann and Follmann 1979). dRNs applied externally at this time partially alleviate CS action in soybean cells, yet RDP reductase extracted from these cells is not inhibited by CS or SM in vitro (Scheel and Casida 1985).

CS may therefore act indirectly to reduce DNA precursor synthesis by lowering branched chain amino acid levels. An analogous effect is observed in regenerating rat liver where branched chain amino acids are apparently involved in the regulation of RDP reductase activity (Chiba et al. 1983).

3. Cell Cycle Block. The cell cycle consists of sequential stages G_1, S, G_2, and M with DNA synthesis and mitosis occurring during the S and M phases, respectively. The mitotic index (percent cells in M phase) is drastically reduced by CS in a dose dependent manner in the roots of both pea and broad bean *(Vicia faba)* (Ray 1982a; Rost 1984). The frequency distribution of the various mitotic stages is not changed, and no aberrant mitotic figures occur, suggesting that the mitotic apparatus itself is unaffected and that the block is not during M phase. The entry of pea root cells labeled with $[^3H]$thymidine from G_1 into S is inhibited by CS (2.8 to 280 μM) and no cells previously in S are able to pass through G_2 into M. An additional 43% inhibition of $[^3H]$uridine incorporation into RNA after 24 h leads to the hypothesis that CS blocks cell cycle progression at the G_2 and G_1 transition points by inhibition of cell cycle specific RNA synthesis (Rost 1984). The cell cycle inhibition by CS is completely reversed by a mixture of valine and isoleucine (Rost and Reynolds 1985).

4. Diversity of Other Actions of Unknown Origin. Sulfonylureas at nM levels directly inhibit AHAS (Ray 1984), deplete cellular branched chain amino acids (Scheel and Casida 1985) and indirectly inhibit DNA synthesis (Ray 1982a,b; Rost 1984). At μM levels they inhibit RNA, protein and lipid synthesis and photosynthesis (De Villiers et al. 1980; Hatzios and Howe 1982; Ray 1982a; Rost 1984). Delayed actions of CS include increases in phenylalanine ammonia-lyase activity, anthocyanin content, ethylene production, and cellulase activity followed by accelerated leaf abscission (Hageman and Behrens 1984a; Suttle and Schreiner 1982). RNA and protein synthesis are reduced in *E. coli* K-12 at CS levels (280 to 840 μM) and treatment times (15–60 min) that do not affect DNA synthesis (Scheel and Casida 1985).

4 Tolerance to Acetohydroxyacid Synthase Inhibitors

4.1 Selectivity Among Crops and Weeds

The sulfonylureas are pre- and postemergence herbicides effective at rates as low as 5 g active ingredient h^{-1}. CS is used to control most broadleaved weeds and some grasses in wheat and SM for johnsongrass in bermudagrass (Levitt 1983). Scepter is proposed for control of broadleaved weeds, sedges, and grasses in soybean and tobacco (American Cyanamid Co. 1983) and Arsenal for total vegetation control (Reider and Shaner 1984).

4.2 Target Site

1. Direct – AHAS Isozymes. Different AHAS isozymes confer susceptibility or tolerance to sulfonylureas and other inhibitors. Valine-resistant tobacco mutants, regenerated from UV-mutagenized mesophyll protoplasts of haploid tobacco, contain an AHAS less sensitive to valine and leucine (Wallsgrove et al. 1984). AHAS from CS-tolerant and -resistant plants is equally sensitive to this herbicide in vitro (Ray 1984). In contrast, CS- and SM-resistant tobacco mutants isolated from cell cultures (Chaleff and Ray 1984) produce an AHAS that is insensitive to inhibition by the two herbicides (Chaleff and Mauvais 1984). Therefore selection of resistant mutants from cell cultures or protoplasts or transfer of genes coding for a sulfonylurea insensitive AHAS into susceptible plants seem to be promising methods for producing tolerant crops.

E. coli K-12 mutants that express the valine-insensitive AHAS II are resistant to growth inhibition by valine (De Felice et al. 1979). On the other hand in *E. coli* (Yadav and Benard 1984) and *S. typhimurium* (LaRossa and Schloss 1984) AHAS II appears to be the isozyme most sensitive to SM and resistant mutants produce an isozyme less sensitive to SM inhibition. The valine-sensitive AHAS III appears to be the site of action for CS and SM in *E. coli* K-12 in which AHAS II is not expressed (Scheel and Casida 1985).

2. Indirect – DNA Synthesis. Germinating wheat, which is tolerant to CS, does not follow the normal sequence of biosynthesis of DNA precursors and DNA itself, i.e., ribonucleotide reduction lags well behind DNA formation, indicating that the onset of DNA replication may be subject to different controls in different plant species (Hovemann and Follmann 1979). This possible difference in the significance of RDP reductase, combined with herbicide detoxification (Sweetser et al. 1982), may contribute to the CS resistance of wheat.

4.3 Metabolism

Metabolic detoxification is a major factor contributing to the selective action of CS, i.e., tolerant plants like wheat, barley, wild oats, and eastern black nightshade metabolize CS much more rapidly than sensitive plants like cotton, soybean, mustard, sugar beet, and velvetleaf (Hageman and Behrens 1984b; Sweetser et al. 1982). Metabolism of CS in wheat involves initial hydroxylation at the 5-position of the phenyl ring to yield 5-hydroxy-CS and then conjugation at this hydroxyl group with glucose. Both 5-hydroxy-CS and its glucoside conjugate lack significant herbicidal activity (Sweetser et al. 1982). In nightshade the triazine-methyl of CS is hydroxylated followed by conjugation with glucose (Erbes 1984). While both glucosides are only weak AHAS inhibitors, the hydroxylated metabolites show the same potency as the parent compound (Erbes 1984). The detoxification step for CS in wheat and nightshade is therefore the final conjugate formation. Crop tolerance is conferred for Londax in rice by O-demethylation and glucoside formation and for Classic in soybean by replacement of the chloro substituent with homoglutathione (Beyer 1984).

The selectivity of Scepter is also linked to differences in metabolism rate, i.e., three- to tenfold faster metabolism in the tolerant soybean than the sensitive cocklebur and velvetleaf (American Cyanamid Co. 1983).

4.4 Herbicide Antidotes

The herbicide antidote 1,8-naphthalic anhydride increases the tolerance of corn, sorghum and rice to CS whereas N,N-diallyl-2,2-dichloroacetamide and two nitrile derivatives give only marginal effects in corn (Hatzios 1984; Parker et al. 1980).

5 Conclusions

AHAS is the first common enzyme of the biosynthetic pathways leading to the branched chain amino acids, valine, leucine, and isoleucine. AHAS isozymes of plants and bacteria differ in their substrate affinities and sensitivity to inhibition by branched chain amino acids. Repression of this enzyme activity results in selective depletion of branched chain amino acids and growth inhibition which is alleviated by external application of branched chain amino acids or the valine precursor 2-ketoisovalerate.

The most potent AHAS inhibitors are sulfonylureas such as CS and SM which are commercial herbicides used at extraordinarily low rates of application. Sulfonylurea-induced growth inhibition is reversed by valine or leucine in soybean cells, by valine alone or mixed with isoleucine in pea roots, and by isoleucine in bacteria. Tobacco mutants selected for resistance to sulfonylurea herbicides produce an AHAS which is insensitive to inhibition by CS and SM. Sulfonylurea herbicides block cell cycle progression and DNA synthesis in plants but not in *E. coli*. It is suggested but not established that the indirect inhibition of plant DNA synthesis by CS is mediated via a lack of induction of RDP reductase caused by depletion of cellular branched chain amino acids. A variety of plant processes are disrupted after long periods or at high sulfonylurea levels. Detoxification is a major basis for the natural tolerance of plants to sulfonylureas but different control mechanisms for the onset of DNA synthesis in resistant species may also participate. Marginal improvement in crop tolerance is achieved with herbicide antidotes. A more permanent solution to crop safety involves changing the herbicide structure for greater selectivity or engineering the crop for more rapid detoxification or reduced AHAS sensitivity to the sulfonylureas.

The herbicidal imidazolinone Scepter is also an AHAS inhibitor with crop tolerance attributed to selective detoxification.

Acknowledgements. We thank Baruch Rubin and John Huppatz for helpful comments. The studies at Berkeley were supported by Grant PO1 ES00049 from the National Institutes of Health.

References

Adams CA, Blee E, Casida JE (1983) Dichloroacetamide herbicide antidotes enhance sulfate metabolism in corn roots. Pestic Biochem Physiol 19:350–360

American Cyanamid Co (1983) Scepter herbicide. Tech Inf Rep, Princeton, NewJersey

Anderson PC, Hibberd KA (1984) Evidence for the interaction of the imidazolinone herbicides with leucine, valine, and isoleucine metabolism. Poster pres 60th Annu Meet Am Soc Plant Physiol, Davis, Calif, August 1984

Bauerle RH, Freundlich M, Stormer FC, Umbarger HE (1964) Control of isoleucine, valine and leucine biosynthesis. II. Endproduct inhibition by valine of acetohydroxy acid synthetase in *Salmonella typhimurium*. Biochim Biophys Acta 92:142–149

Berlin J, Witte L (1981) Effects of glyphosate on shikimic acid accumulation in tobacco cell cultures with low and high yields of cinnamoyl putrescines. Z Naturforsch 36c:210–214

Beyer EM (1984) Du Pont contributions to agricultural science. Agricultural Sciences Symposium, Du Pont Life Sci, Wilmington, Delaware, September 1984

Borstlap AC (1981) Interactions between the branched-chain amino acids in the growth of *Spirodela polyrhiza*. Planta 151:314–319

Cassady WE, Leiter EH, Bergquist A, Wagner RP (1972) Separation of mitochondrial membranes of *Neurospora crassa*. II. Submitochondrial localization of the isoleucine-valine biosynthetic pathway. J Cell Biol 53:66–72

Chaleff RS, Mauvais CJ (1984) Acetolactate synthase is the site of action of two sulfonylurea herbicides in higher plants. Science 224:1443–1445

Chaleff RS, Ray TB (1984) Herbicide-resistant mutants from tobacco cell cultures. Science 223: 1148–1151

Chiba T, Kawai K, Ikenaka K, Nakamura H, Okamoto M, Harada N, Tanaka H, Shirasaka T, Fujii S (1983) Effect of an exogenous energy source and amino acids on DNA synthesis in regenerating rat liver. Biochim Biophys Acta 755:420–427

Davies ME (1964) Acetolactate and acetoin synthesis in ripening peas. Plant Physiol 39:53–59

De Felice M, Levinthal M, Iaccarino M, Guardiola J (1979) Growth inhibition as a consequence of antagonism between related amino acids: effect of valine in *Escherichia coli* K–12. Microbiol Rev 43:42–58

De Villiers OT, Vandenplas ML, Koch HM (1980) The effect of DPX-4189 on biochemical processes in isolated leaf cells and chloroplasts. Proc 1980 Br Crop Prot Conf Weeds 1:237–242

Erbes DL (1984) The effect of some sulfonylurea metabolites on acetolactate synthase. Plant Physiol Suppl 75:49

Fedtke C (1982) Biochemistry and physiology of herbicide action. Springer, Berlin Heidelberg New York

Gressel J, Zilkah S, Ezra G (1978) Herbicide action, resistance and screening in cultures vs plants. In: Thorpe TA (ed) Frontiers of plant tissue culture. Int Assoc Plant Tissue Cult, Calgary, Canada, pp 427–436

Haderlie LC, Widholm JM, Slife FW (1977) Effect of glyphosate on carrot and tobacco cells. Plant Physiol 60:40–43

Hageman LH, Behrens R (1984a) Chlorsulfuron induction of leaf abscission in velvetleaf *(Abutilon theophrasti)*. Weed Sci 32:132–137

Hageman LH, Behrens R (1984b) Basis for response differences of two broad-leaf weeds to chlorsulfuron. Weed Sci 32:162–167

Hatzios KK (1984) Interactions between selected herbicides and protectants on corn *(Zea mays)*. Weed Sci 32:51–58

Hatzios KK, Howe CM (1982) Influence of the herbicides hexazinone and chlorsulfuron on the metabolism of isolated soybean leaf cells. Pestic Biochem Physiol 17:207–214

Hovemann B, Follmann H (1979) Deoxyribonucleotide synthesis and DNA polymerase activity in plant cells (*Vicia faba* and *Glycine max*). Biochim Biophys Acta 561:42–52

Kuwana H, Date M (1975) Solubilization of valine-sensitive acetohydroxy acid synthetase from *Neurospora* mitochondria. J Biochem (Tokyo) 77:257–259

LaRossa RA, Schloss JV (1984) The sulfonylurea herbicide sulfometuron methyl is an extremely potent and selective inhibitor of acetolactate synthase in *Salmonella typhimurium*. J Biol Chem 259:8753–8757

Levitt G (1983) Sulfonylureas: new high potency herbicides. In: Miyamoto J, Kearney PC (eds) Pesticide chemistry: human welfare and the environment, vol I. Pergamon Press, Oxford New York, pp 243–250

Levitt G, Ploeg HL, Weigel RC Jr, Fitzgerald DJ (1981) 2-Chloro-*N*-[(4-methoxy-6-methyl-1,3,5-triazin-2-yl)aminocarbonyl]benzene-sulfonamide, a new herbicide. J Agric Food Chem 29: 416–418

Miflin BJ (1969) Acetolactate synthetase from barley seedlings. Phytochemistry 8:2271–2276

Miflin BJ (1971) Cooperative feedback control of barley acetohydroxyacid synthetase by leucine, isoleucine, and valine. Arch Biochem Biophys 146:542–550

Miflin BJ, Cave PR (1972) The control of leucine, isoleucine, and valine biosynthesis in a range of higher plants. J Exp Bot 23:511–516

Oda Y, Nakano Y, Kitaoka S (1982) Properties and regulation of valine-sensitive acetolactate synthase from mitochondria of *Euglena gracilis*. J Gen Microbiol 128:1211–1216

Odmark G (1969) Studies on deoxyribonucleotide synthesis in *Vicia faba*. Physiol Plant 22:161–170

O'Neill JP, Freundlich M (1972) Two forms of biosynthetic acetohydroxy acid synthetase in *Salmonella typhimurium*. Biochem Biophys Res Commun 48:437–443

Parker C, Richardson WG, West TM (1980) Potential for extending the selectivity of DPX 4189 by use of herbicide safeners. Proc 1980 Br Crop Prot Conf Weeds 1:15–22

Ray TB (1982a) The mode of action of chlorsulfuron: a new herbicide for cereals. Pestic Biochem Physiol 17:10–17

Ray TB (1982b) The mode of action of chlorsulfuron: the lack of direct inhibition of plant DNA synthesis. Pestic Biochem Physiol 18:262–266

Ray TB (1984) Site of action of chlorsulfuron: inhibition of valine and isoleucine biosynthesis in plants. Plant Physiol 75:827–831

Reider ML, Shaner DL (1984) Uptake of Scepter and Arsenal into soybean leaf discs. Plant Physiol Suppl 75:50

Rost TL (1984) The comparative cell cycle and metabolic effects of chemical treatments on root tip meristems. III. Chlorsulfuron. J Plant Growth Regul 3:51–63

Rost TL, Reynolds T (1985) Reversal of chlorsulfuron-induced inhibition of mitotic entry by isoleucine and valine. Plant Physiol 77:481–482

Ryan ED, Kohlhaw GB (1974) Subcellular localization of isoleucine-valine biosynthetic enzymes in yeast. J Bacteriol 120:631–637

Satyanarayana T, Radhakrishnan AN (1963) Biosynthesis of valine and isoleucine in plants. I. Formation of α-acetolactate in *Phaseolus radiatus*. Biochim Biophys Acta 77:121–132

Scheel D, Casida JE (1985) Sulfonylurea herbicides: growth inhibition in soybean cell suspension cultures and in bacteria correlated with block in biosynthesis of valine, leucine or isoleucine. Pestic Biochem Physiol 23:398–412

Semeraro RJ, Wixom RL (1977) Studies in valine biosynthesis. X. The acetolactate synthase from *Rhodopseudomonas spheroides*. Microbios 20:7–14

Shaner DL, Anderson PC, Stidham MA (1984) Imidazolinone herbicides are inhibitors of acetohydroxy acid synthase. Plant Physiol Suppl 75:50

Steinrücken HC, Amrhein N (1980) The herbicide glyphosate is a potent inhibitor of 5-enolpyruvylshikimic acid-3-phosphate synthase. Biochem Biophys Res Commun 94:1207–1212

Stormer FC, Umbarger HE (1964) The requirement for flavine adenine dinucleotide in the formation of acetolactate by *Salmonella typhimurium* extracts. Biochem Biophys Res Commun 17:587–592

Suttle JC, Schreiner DR (1982) Effects of DPX-4189 (2-chloro-*N*-[(4-methoxy-6-methyl-1,3,5-triazin-2-yl)aminocarbonyl]benzenesulfonamide) on anthocyanin synthesis, phenylalanine ammonia lyase activity, and ethylene production in soybean hypocotyls. Can J Bot 60:741–745

Sweetser PB, Schow GS, Hutchison JM (1982) Metabolism of chlorsulfuron by plants: biological basis for selectivity of a new herbicide for cereals. Pestic Biochem Physiol 17:18–23

Takenaka S, Kuwana H (1972) Control of acetohydroxy acid synthetase in *Neurospora crassa*. J Biochem (Tokyo) 72:1139–1145

Tsuchida T, Momose H (1975) Genetic changes of regulatory mechanisms occurred in leucine and valine producing mutants derived from *Brevibacterium lactofermentum* 2256. Agric Biol Chem 39:2193–2198

Umbarger HE (1983) The biosynthesis of isoleucine and valine and its regulation. In: Herrmann KM, Somerville RL (eds) Amino acids: biosynthesis and genetic regulation. Addison-Wesley, London, pp 245–266

Wallsgrove RM, Bright SWJ, Bourgin J-P (1984) Acetohydroxyacid synthase in a valine-resistant mutant of *Nicotiana tabacum*. Plant Physiol Suppl 75:35

Walsh C (1979) Enzymatic reaction mechanisms. Freeman, San Francisco

Westerfeld WW (1945) A colorimetric determination of blood acetoin. J Biol Chem 161:495–502

Yadav NS, Benard SA (1984) Mutations in cloned *E. coli* gene for acetolactate synthase II confer resistance to sulfometuron methyl herbicide by enzyme alteration. Abstr 11th Aharon Katzir-Katchalsky Conf Plant Mol Biol, Jerusalem, Israel, Jan 1984

A Glyphosate-Tolerant Plant Tissue Culture

N. AMRHEIN, D. JOHÄNNING, and C.C. SMART [1]

1 Introduction

1.1 Herbicide Resistance in Plant Cell Cultures

The use of cell cultures in the screening for variant cell types and for the isolation of potentially useful phenotypes in crop plants has been emphasized (e.g. [5]), and selection for drug and, in particular, herbicide resistance at the cell level and subsequent regeneration of resistant plants have met with certain success (see [14, 17]). The biochemical characterization of resistant cell lines, and the genetic characterization of the regenerated mutant plants may allow the identification of the site of action of a herbicide, a striking example being the recent demonstration that the target of sulfonylurea herbicides in higher plants is the first enzyme specific for the biosynthesis of the branched chain amino acids, acetolactate synthase [6, 7]. Resistance to a herbicide may, however, be acquired by a number of mechanisms, not all of which need be related with the primary site of action of the particular agent. For example, reduced uptake and translocation, or increased detoxification may determine the differential intraspecific responses to herbicides [16]. Target-related mechanisms of resistance to drugs and other metabolic inhibitors can involve either a reduced affinity of the target site for binding the inhibitor, or an increased production of target sites. The latter phenomenon has been intensely studied in bacteria and, in particular, in cultured animal cells, in which specific genes can undergo amplification when the cells are placed under selective conditions (see [22, 26]). Only in a few cases has resistance to metabolic inhibitors in plant cells been associated with increased activities of their respective target enzymes: aminopterin-resistant soybean and maize cell cultures were shown to have an increased dihydrofolate reductase activity [19, 24], and acetohydroxamate resistance in tobacco cells was correlated with high levels of urease [30]. In the course of our studies on the mode of action of the herbicide glyphosate (see below) we have isolated cell lines of the higher plant *Corydalis sempervirens* capable of growth in the presence of high herbicide concentrations. We present evidence here that these cells overproduce the enzyme 5-enolpyruvylshikimate-3-phosphate synthase which is specifically inhibited by glyphosate.

---

1 Lehrstuhl für Pflanzenphysiologie, Ruhr-Universität, Postfach 102148, 4630 Bochum, FRG

Primary and Secondary Metabolism of Plant
Cell Cultures, ed. by Neumann et al.

1.2 The Site of Action of Glyphosate

Glyphosate (N-[phosphonomethyl]glycine, Fig. 1) is the herbicidal component of Roundup which is widely used as a broad spectrum, non-selective postemergence weed killer [11]. Interference of glyphosate with the biosynthesis of the aromatic amino acids had been deduced by Jaworski [15] from the observation that aromatic amino acids, in particular phenylalanine, alleviate the inhibition of growth by glyphosate in the bacterium *Rhizobium japonicum* and the aquatic higher plant *Lemna gibba*. In our work, we found that glyphosate inhibited the formation of chorismate-derived aromatic amino acids and secondary plant metabolites [1, 2, 12]. The simultaneous massive accumulation of shikimic acid in glyphosate-treated plant tissues implicated a biosynthetic step between shikimate and chorismate as the target of glyphosate. This was confirmed by the demonstration that the 5-enolpyruvylshikimate-3-phosphate (EPSP) synthase (3-phophoshikimate 1-carboxyvinyl transferase; EC 2.5.1.19) of *Aerobacter aerogenes* (= *Klebsiella pneumoniae*) was subject to potent inhibition by the herbicide [27]. The reaction catalyzed by EPSP-synthase is shown in Fig. 2. The potent and specific inhibition of EPSP-synthase by glyphosate, which is competitive with phosphoenolpyruvate has been confirmed for the enzymes of *Escherichia coli, Neurospora crassa*, pea seedlings, tissue cultures of *Corydalis sempervirens* and *Nicotiana silvestris* (see [21, 28]). While bacteria possessing a glyphosate-sensitive EPSP-synthase excrete predominantly shikimate-3-phosphate (Fig. 2) into the medium when exposed to glyphosate [23], plant cells dephosphorylate this compound and store the released shikimate in the vacuole, as was shown for cultured buckwheat cells [13].

Resistance to glyphosate in bacteria has been found to correlate either with the overproduction of a glyphosate-sensitive EPSP-synthase [3, 9, 20], or with the possession of a glyphosate-insensitive EPSP-synthase [8, 23, 25]. While a glyphosate-insensitive EPSP-synthase has not yet been identified in a higher plant, it has been suggested that the introduction and expression in plant cells of a dominant (EPSP-synthase) gene conferring glyphosate resistance might result in herbicide resistance [8]. Alternatively, this goal might directly be achieved by the selection of glyphosate-resistant plant cell lines.

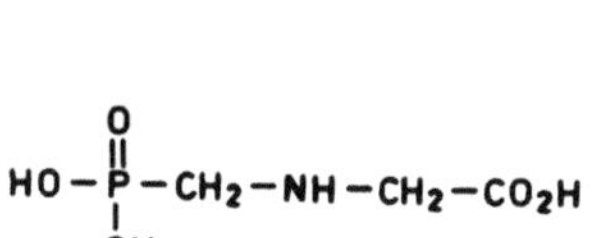

Fig. 1. Structure of glyphosate

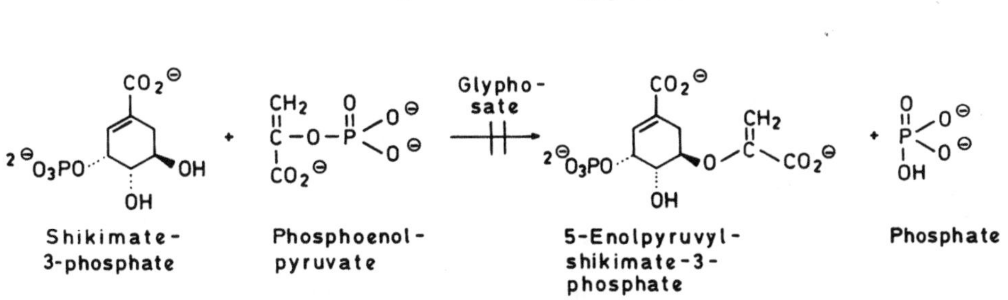

Fig. 2. Reaction catalyzed by 5-enolpyruvylshikimate-3-phosphate synthase

2 Glyphosate-Tolerance in Cultured Cells of Corydalis sempervirens Pers.

2.1 Selection for Glyphosate-Tolerance

In the initial stage of our work on EPSP-synthase, a tissue culture of *Corydalis sempervirens* Pers. (Papaveraceae family) was chosen as a source of the higher plant enzyme because it possessed a relatively high specific activity and did not require exogenous amino acids for growth. The latter property was essential for the selection of glyphosate-tolerant cells, because the aromatic amino acids were known to reverse growth inhibition by glyphosate in plant cell cultures (e.g. [10]). Glyphosate inhibited the growth of the cells 50% at a concentration of ca. 0.2 mM, and little, if any, growth occurred at concentrations > 0.5 mM (Fig. 3). In 0.5 mM glyphosate, growth was severely inhibited for nearly two weeks, and the cells accumulated shikimic acid, but growth was then resumed at a slow rate. By a stepwise increase of the glyphosate concentration at each transfer during a period of several months, we were able to obtain cells growing in liquid suspension culture in the presence of up to 10 mM glyphosate [3]. At the present time, cell lines capable of growth in the presence of 20 mM glyphosate (liquid medium) or 30 mM glyphosate (solid medium) are cultivated in our laboratory, but we routinely use cells growing in the presence of 5 mM glyphosate for our biochemical studies, because these cells are completely adapted to this concentration of the herbicide.

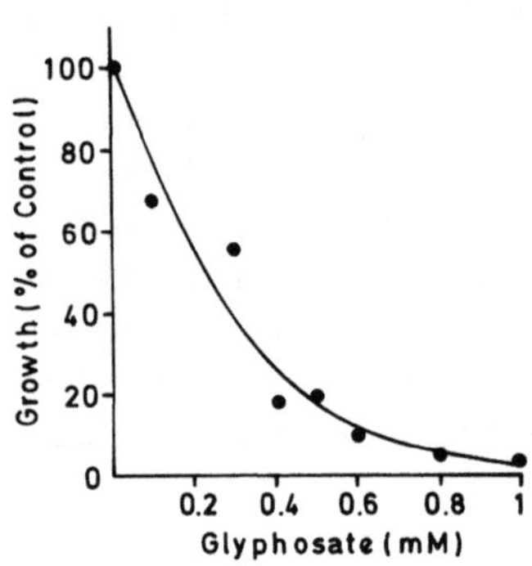

Fig. 3. Effect of glyphosate on the growth of cultured cells of *Corydalis sempervirens*

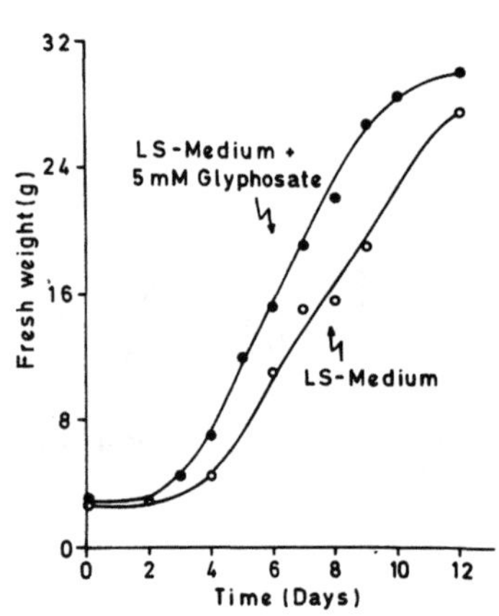

Fig. 4. Growth kinetics of *Corydalis sempervirens* cells. *Open symbols* glyphosate-sensitive cells growing in the absence of glyphosate; *closed symbols* glyphosate-tolerant cells growing in the presence of 5 mM glyphosate. *LS-medium* medium of Linsmaier and Skoog (see [3])

In some experiments, the growth of these cells in the presence of 5 mM glyphosate was found to be even more vigorous than the growth of non-adapted cells in the absence of glyphosate (Fig. 4). We prefer the term *tolerance*, as compared to *resistance*, to describe the growth of the cells in the presence of glyphosate, because any further increase of the glyphosate concentration will again reduce their growth until further "adaptation" has occurred.

2.2 Biochemical Basis for Glyphosate-Tolerance: Overproduction of 5-Enolpyruvylshikimate-3-Phosphate Synthase

C. sempervirens cells continuously cultured in the presence of 5 mM glyphosate have a specific EPSP-synthase activity of ca. 15 nkat mg^{-1} protein as compared to a specific activity of 0.5 nkat mg^{-1} protein in cells which had never been exposed to glyphosate. This 30-fold increase in the extractable activity of EPSP-synthase could either be due to an overproduction of the enzyme existing in the cells prior to their exposure to glyphosate, or to the formation of a new molecular species of the enzyme. The second alternative was excluded by a number of criteria: EPSP-synthases from cultures growing either in the absence or presence of glyphosate were purified to homogeneity and shown to produce bands with identical molecular weights (M_r = 45,500) when subjected to SDS-gel-electrophoresis. In a double-immunodiffusion assay with antiserum raised against the highly purified enzyme from cells growing in 5 mM glyphosate, no differences in the antigenic determinants of the EPSP-synthases from the two sources were detected. Furthermore, no significant differences were found in the following properties of the enzymes: pH and temperature optima, isoelectric point, saturation curves for both substrates, and sensitivity to glyphosate. Thus, it can be safely assumed that *C. sempervirens* cells growing in the absence or presence of glyphosate contain indentical molecular species of EPSP-synthase. When the soluble proteins of the two strains were subjected to high-resolution two-dimensional electrophoresis [18], drastically increased amounts of a polypeptide were detected in extracts from the cells growing in the presence of 5 mM glyphosate, which had the same charge and molecular weight as purified EPSP-synthase. Furthermore, when the soluble proteins were separated by SDS-gel-electrophoresis, blotted onto nitrocellulose [4] and EPSP-synthase protein visualized by reaction first with an antiserum against EPSP-synthase and then with peroxidase-coupled protein A [29], overproduction of the EPSP-synthase protein could be unequivocally established for the cells growing in 5 mM glyphosate (Fig. 5).

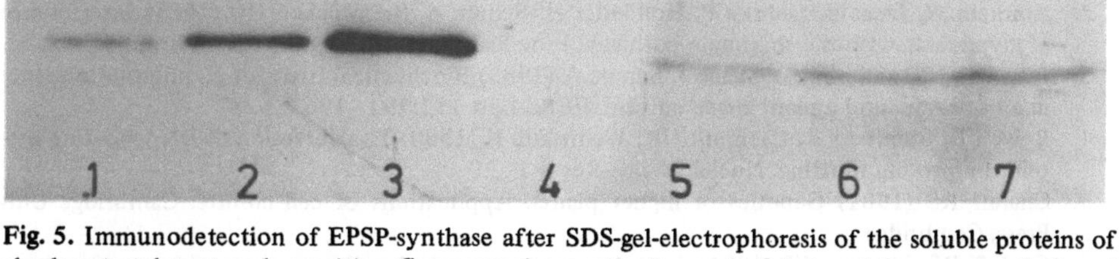

Fig. 5. Immunodetection of EPSP-synthase after SDS-gel-electrophoresis of the soluble proteins of glyphosate-tolerant and -sensitive *C. sempervirens* cells. Lanes *1–3* 1,2 and 5 μg, respectively, of protein from glyphosate-tolerant cells; lanes *4–7* 10, 50, 100, and 200 μg, respectively, of protein from glyphosate-sensitive cells

A cautions estimate indicates a ca. 40-fold increased content of EPSP-synthase protein during early logarithmic growth of the glyphosate-tolerant cells as compared to the sensitive cells. EPSP-synthase protein in the glyphosate-tolerant cells constitutes ⩾ 2.5% of their total soluble protein which is in agreement with the observation that a ca. 40-fold purification is required to obtain a near-homogenous EPSP-synthase preparation from these cells. It can be assumed that the amount of EPSP-synthase protein, as measured by quantitative immunodetection, represents enzymatically active protein since the time-courses of EPSP-synthase activity and EPSP-synthase protein during growth of the cell cultures parallel one another.

3 Conclusions

We have shown that glyphosate-tolerant plant cell lines can be obtained by a stepwise selection procedure and that the tolerant cells overproduce the enzyme EPSP-synthase, which is inhibited by glyphosate and which we consider, in the absence of contrary evidence, to be the site of action of the herbicide. The correlation of herbicide tolerance with the overproduction of EPSP-synthase provides further evidence, though no definitive proof, for our assumption that EPSP-synthase is, indeed, the target of glyphosate.

In glyphosate-tolerant cells, none of the other enzymes of the shikimate pathway which we have investigated so far – DAHP-synthase, dehydroquinate hydrolyase, shikimate dehydrogenase, shikimate kinase, and chorismate mutase – are overproduced (Müller and Amrhein, unpublished) which indicates the specificity of the cells' response to glyphosate. Our present efforts are now directed towards the elucidation of the mechanism underlying the overproduction of EPSP-synthase: gene amplification (see Introduction) or specific enhancement of transcription and/or translation.

Acknowledgements. Support by the Deutsche Forschungsgemeinschaft, Bonn, and Monsanto Agricultural Products Co., St. Louis, is gratefully acknowledged.

References

1. Amrhein N, Deus B, Gehrke P, Steinrücken HC (1980) The site of the inhibition of the shikimate pathway by glyphosate. II. Interference of glyphosate with chorismic acid formation in vivo and in vitro. Plant Physiol 66:830–834
2. Amrhein N, Deus B, Gehrke P, Holländer H, Schulz A, Steinrücken HC (1981) Interference of glyphosate with the shikimate pathway. Proc Plant Growth Regul Soc Am 8:99–106
3. Amrhein N, Johänning D, Schab J, Schulz A (1983) Biochemical basis for glyphosate-tolerance in a bacterium and a plant tissue culture. FEBS Lett 157:191–196
4. Bowen B, Steinberg J, Laemmli UK, Weintraub H (1980) The detection of DNA-binding proteins by protein blotting. Nucleic Acids Res 8:1–20
5. Chaleff RS (1981) Genetics of higher plants. Applications of cell culture. Cambridge Univ Press, Cambridge
6. Chaleff RS, Mauvais CJ (1984) Acetolactate synthase is the site of action of two sulfonylurea herbicides in higher plants. Science 224:1443–1445

7. Chaleff RS, Ray TB (1984) Herbicide-resistant mutants from tobacco cell cultures. Science 223:1148–1151
8. Comai L, Sen LC, Stalker DM (1983) An altered aroA gene product confers resistance to the herbicide glyphosate. Science 221:370–371
9. Duncan K, Lewendon A, Coggins JR (1984) The purification of 5-enolpyruvylshikimate-3-phosphate synthase from an overproducing strain of *Escherichia coli*. FEBS Lett 165:121–127
10. Haderlie LC, Widholm JM, Slife FW (1977) Effect of glyphosate on carrot and tobacco cells. Plant Physiol 60:40–43
11. Hoagland RE, Duke SO (1982) Biochemical effects of glyphosate [N-(phosphonomethyl)glycine]. In: Moreland DE, St John JB, Hess FD (eds) Biochemical responses induced by herbicides. ACS Symp Ser 181:175–205
12. Holländer H, Amrhein N (1980) The site of the inhibition of the shikimate pathway by glyphosate. I. Inhibition by glyphosate of phenylpropanoid synthesis in buckwheat (*Fagopyrum esculentum* Moench). Plant Physiol 66:823–829
13. Holländer-Czytko H, Amrhein N (1983) Subcellular compartmentation of shikimic acid and phenylalanine in buckwheat cell suspension cultures grown in the presence of shikimate pathway inhibitors. Plant Sci Lett 29:89–96
14. Hughes K (1983) Selection for herbicide resistance. In: Evans DA, Sharp WR, Ammirato PV, Yamada Y (eds) Techniques for propagation and breeding. Handbook of plant cell culture, vol I. MacMillan, New York, pp 442–460
15. Jaworski EG (1972) Mode of action of N-phosphonomethylglycine: inhibition of aromatic amino acid biosynthesis. J Agric Food Chem 20:1195–1198
16. Jensen KIN (1982) The roles of uptake, translocation, and metabolism in the differential intraspecific responses to herbicides. In: Le Baron HM, Gressel J (eds) Herbicide resistance in plants. Wiley, New York, pp 133–162
17. Meredith CP, Carlson PS (1982) Herbicide resistance in plant cell cultures. In: Le Baron HM, Gressel J (eds) Herbicide resistance in plants. Wiley, New York, pp 275–291
18. O'Farrell (1975) High resolution two-dimensional electrophoresis of proteins. J Biol Chem 250:4007–4021
19. Ohyama K (1976) A basis for bromodeoxyuridine resistance in plant cells. Environ Exp Bot 16:209–216
20. Rogers SG, Brand LA, Holder SB, Sharps ES, Brackin MJ (1983) Amplification of the aroA gene from *Escherichia coli* results in tolerance to the herbicide glyphosate. Appl Environ Microbiol 46:37–43
21. Rubin JL, Gaines CG, Jensen RA (1984) Glyphosate inhibition of 5-enolpyruvylshikimate-3-phosphate synthase from suspension-cultured cells of *Nicotiana silvestris*. Plant Physiol 75: 8390–8465
22. Schimke RT (ed) (1982) Gene amplification. Cold Spring Harbor Lab, Cold Spring Harbor
23. Schulz A, Sost D, Amrhein N (1984) Insensitivity of 5-enolpyruvylshikimic acid-3-phosphate synthase to glyphosate confers resistance to this herbicide in a strain of *Aerobacter aerogenes*. Arch Microbiol 137:121–123
24. Shimamoto K, Nelson OE (1981) Isolation and characterization of aminopterin-resistant cell lines in maize. Planta 153:436–442
25. Sost D, Schulz A, Amrhein N (1984) Characterization of a glyphosate-insensitive 5-enolpyruvylshikimic acid-3-phosphate synthase. FEBS Lett 173:238–242
26. Stark GR, Wahl GM (1984) Gene amplification. Annu Rev Biochem 53:357–388
27. Steinrücken HC, Amrhein N (1980) The herbicide glyphosate is a potent inhibitor of 5-enolpyruvylshikimic acid-3-phosphate synthase. Biochem Biophys Res Commun 94:1207–1212
28. Steinrücken HC, Amrhein N (1984) 5-Enolpyruvylshikimate-3-phosphate synthase of *Klebsiella pneumoniae*. 2. Inhibition by glyphosate [N-(phosphonomethyl]glycine). Eur J Biochem 43:351–357
29. Towbin H, Staehelin T, Gordon J (1979) Electrophoretic transfer of proteins from polyacrylamide gels to nitrocellulose sheets. Procedure and some applications. Proc Natl Acad Sci USA 76:4350–4354
30. Yamaya T, Filner P (1981) Resistance to acetohydroxamate acquired by slow adaptive increases in urease in cultured tobacco cells. Plant Physiol 67:1133–1140

Plant Cell Culture – Future Perspectives

M.W. FOWLER[1,2]

With the development and operation of industrial-scale plant cell culture processes, cell culture technology may be said to have "come of age." One or two processes operating commercially do not, however, "make an industry." Much yet still remains to be done if plant cell culture technology is to have a major impact upon process technology in the chemical industry. With this in mind, and with the wide range of papers presented at the Giessen Symposium, it was felt appropriate to review developments in plant cell culture and to provide a perspective of the challenges to be faced in the years ahead. To this end a small group of people were asked to put on paper their ideas, concerns and visions relating to the future of plant cell culture. The following text is an edited version of their thoughts.

The review begins by considering the relationship between primary and secondary metabolism. Secondly, the mutually interactive areas of biosynthesis and biotransformation are discussed. The third section is specifically concerned with the complex area of cell selection, variability, and stability. The direction then changes towards photoautotrophism and cell immobilisation. This finally leads into production strategies and the biotechnology of plant cell culture.

Fundamental to all work on natural product synthesis is an understanding of the underlying biochemistry. A difficulty frequently encountered in the production of useful substances by plant cell culture is that secondary metabolite synthesis is often suppressed or reduced. Unfortunately our knowledge of this area is much restricted and a great deal of effort will be required to elucidate mechanisms of expression or repression. Undoubtedly attention will be focussed on this area in the years ahead with implications not only for plant cell culture, but also for whole plant biochemistry and the regulation of secondary metabolite synthesis.

Two aspects of the regulation of secondary metabolite synthesis need to be taken into account; first the interaction between primary and secondary metabolism, and secondly the relationship between secondary metabolism and biochemical differentiation.

Secondary metabolism could be regulated by switching on or off metabolite flux from primary metabolism. Study of the primary/secondary metabolite interface is therefore of major importance, as is also the regulation of the fate of exogenously supplied carbon sources, typically sucrose or glucose. In the latter case it is also not

1 University of Sheffield, Wolfson Institute of Biotechnology, Sheffield, S10 2T, United Kingdom

2 Edited from Contributions by D.K. Dougall, M.W. Fowler, W. Hüseman, A. Komamine, W. Kurz, M.M. Yolman

Primary and Secondary Metabolism of Plant Cell Cultures, ed. by Neumann et al.

completely clear whether or not sucrose and glucose have different effects on biomass production and secondary metabolism. The whole of this area is in need of fundamental study. The interaction of primary metabolism with secondary metabolism through the regulation of interconnecting pathways, e.g., the shikimate pathway, is also key. Regulatory mechanisms of enzymes at branching points, e.g., anthranilate synthetase have been investigated in cultured plant cell metabolically as well as genetically. However, the regulatory mechanisms of branching from such important pathways as prephenate to the phenylalanine pathway and to the tyrosine pathway are still unclear. The former pathway leads to phenolics production and the latter is more important in alkaloid production. 3,4-Dihydroxyphenylalanine (DOPA) is a precursor of many alkaloids and betalaines. Little attention has however been focused on the biosynthesis of DOPA from tyrosine. The enzyme catalyzing this conversion is tyrosine hydroxylase which has not been purified and characterized. However, it is possible that this enzyme might act as a key catalyst in alkaloid production, like PAL or TAL in phenolics production. Furthermore, PAL, TAL, and tyrosine hydroxylase are enzymes at branch points from phenylalanine and tyrosine to protein synthesis and to phenolics and alkaloids respectively. The regulatory mechanisms of metabolite flow from primary to secondary metabolism at these branch steps require further investigation.

The regulation of secondary metabolism linked with induction of morphological differentiation is potentially also of great significance. It is well known that many secondary metabolic pathways are expressed in specifically differentiated cells or tissues, suggesting that the expression of secondary metabolism may be induced by the same trigger as that for induction of morphological differentiation; for instance, the induction of anthocyanin synthesis and somatic embryogenesis in a carrot suspension culture, and also in the formation of cardiac glycosides and adventive embryogenesis in *Digitalis* cell cultures. In some cases therefore, the expression of secondary metabolism can be regarded as metabolic differentiation and also as a biochemical marker of morphological differentiation. It is possible that the gene expression mechanisms operating during morphological differentiation could be approached through investigation of expression mechanisms of secondary metabolism in cultured cells. To allow this, experimental systems in which induction or reduction of secondary metabolism is regulated by definite trigger(s) are required. Such systems have been established in carrot suspension cultures for example where anthocyanin synthesis is clearly regulated by exogeneous auxin and also in parsley suspension cultures where flavonoid synthesis is induced by UV. There is evidence to suggest that the key enzyme of the induction of flavonoid synthesis is chalcone synthase. Regulation mechanisms at the transcriptional level have been investigated using antisera and cDNA prepared from the purified enzyme. However, it is still not clear how the trigger, auxin in the carrot system or UV light in the parsley system, regulates the transcription of mRNA for chalcone synthase. This is a most important problem which needs early elucidation.

Investigations into the molecular mechanisms of regulation need to be extended to other secondary metabolite areas including alkaloid, terpenoid, and steroid synthesis. An understanding of these mechanisms should allow the controlled expression of secondary metabolism under defined conditions leading to production of useful substances by plant cell cultures. Whether or not it will allow profitability in a process is more difficult to establish.

That an understanding of biosynthetic mechanisms is a key element in the development of productive plant cell cultures is axiomatic. Such an understanding is also crucial for the identification and amplification of biotransformation systems. It is probably not going too far to say that the elucidation and control of biosynthetic pathways will, in many cases, take precedence over the selection of cell lines with high capabilities for production. No improvement of cell lines by any means, including genetic engineering, will be of value if we are unable to express the genetic potential of cells to ensure maximum efficiency of production. The biosynthetic instability of cell lines is only to a lesser extend based on genetic lability but is more a question of repression/derepression of metabolic blocks. Only by elucidating and regulating these mechanisms will we gain control over the production of commercially valuable metabolites and make employment of cell cultures in industry more attractive.

As the industrial use of cell cultures will be a question of economics, additional aspects to those mentioned above will have to be considered and addressed. Specific areas include the shortening of long generation and production periods especially when compared to microbial fermentations. Such remarkable increases in the rate of synthesis and accumulation will significantly promote plant cell systems, making them likely candidates for performing reactions, which, could not be easily carried out organic-chemically, would not be stereospecific, or would otherwise be too costly to produce. Most likely candidates for such biotransformations will be synthetic precursors and analogous to high value compounds such as pharmaceuticals, flavours, and fragrances.

An increasingly important aspect of work with plant cell cultures concerns variability, selection, and stability. Variability in the behaviour of plant cell cultures and in the individual cells of a culture is widespread and can be classified in three categories. At one end of the range is physiological variability or responsiveness, in which changed conditions in the cultures lead to a change in the behaviour of the culture. These changes occur within one to three passages, i.e. at extremely high frequency, and are reversed when the condtitions of the culture are changed back to the original conditions. At the other end of the range are classical, spontaneous mutations which arise randomly and at frequencies of the order of 10^{-6} per cell generation in cultures. These are not reversible in a strict sense, but may be phenotypically reversed by a second mutation at the same general frequency. In the middle of the range is the category which has been referred to as "heritable variation" or "epigenetic changes." Changes in this class are characterized by being non-randon, directed changes in phenotype of cells occurring at frequencies of 10^{-3} per cell generation which can be maintained through mitosis, and which can be reversed by specific treatments of the cells. This category is not the result of irreversible changes in the genome but is a probabilistic phenomenon.

Physiological variability in plant cell cultures is a major cause of concern but is the easiest to address because of the test of rapid reversibility. There are numerous reports in the plant cell culture literature of changed properties of cultures, for example, changes in yield of specific chemicals when a culture is grown in a different vessel. Such a change can be readily diagnosed as physiological if on returning the culture to its original conditions the change in characteristics is rapidly reversed. Such a demonstration immediately implicates an additional significant variable in the performance of the cultures which must then be identified. Situations with these characteristics will appear time and time again as our experience with plant cell cultures increases and the criteria by which we judge culture performance increase in specificity and sensitivity. Each one, with the necessary dissection and analysis, will add to our knowledge of variables affecting cell culture performance and increase our control over the behaviour of cells.

The role of mutations in the variable behaviour of cell cultures depends on the growth advantage the mutation confers. The growth advantage can be changed by changing the culture medium composition. For example, resistant mutants can be isolated by growing cultures in the presence of an inhibitor of growth. Clearly, auxotrophs whose growth is dependent on an essential nutrient and mutants in secondary product formation where the chemical of interest is not required for growth, are difficult to isolate. The proof that changed properties of a culture are due to a mutation requires demonstration that the change is the property of single cells, i.e. cloning, and demonstration that the changed phenotype is inherited in a rational pattern in sexual crosses. This latter requirement can only be achieved when regeneration of plants from the culture is achievable, a requirement that is elusive in a number of plant cell cultures. There is no doubt that mutants will continue to arise in plant cell cultures and remain a source of variation in plant cell culture behaviour. This calls for accelerated development of methods such as storage in liquid nitrogen to preserve special cultures. Mutants can be exploited to give important information about primary metabolism, and they may give information about secondary metabolism if ways can be found to actively select for these types of mutation rather than having to find them by "brute force."

The role of heritable variation or epigenetic changes in plant cell cultures is the most difficult to evaluate because of lack of information. However, this type of variation in plant cell culture is potentially one of major importance in understanding and utilizing cultures. The best example is cytokinin habituation of tobacco cell cultures studied by Meins and co-workers. If the idea of cells existing in alternative states of cytokinin production and cytokinin non-production can be generalized to other metabolites a cell produce, it has major implications for the productivity of cell cultures and their exploitation.

Whether the cytokinin model can be generalized to other cellular metabolites has not been established, though circumstantial data suggests that it can. The extent of the generalization is not known. Even with the best documented case of cytokinin habituation there is no data on the mechanism of the change between the two states. By analogy with methotrexate resistance in animal cell cultures, some workers hypothesize that gene amplification is a mechanism for such changes. There are some difficulties with this hypothesis for all cases, because the selection of methotrexate-resistant lines requires a series of increasing concentrations of methotrexate to achieve gene amplification. Some workers have advanced the idea that heritable or epigenetic changes have parallels with developmental phenomena. While this may provide insights into both phenomena in the future, the current parlous state of knowledge of either provides little comfort.

The heritable changes in plant cell cultures, if sufficiently general will in the short-term provide the greatest difficulty for people wanting to use or exploit plant cell cultures. However, the effort to describe the phenomena and its generality, and to understand the mechanisms available to plant cells by which they can alter their phenotypes may in the longer term be the best investment. It may open opportunities to control these changes and exploit them to achieve increased control over the behaviour of plant cells in culture.

Over the years a great deal of effort has been applied to the development of photosynthetic or photoautotrophic cultures. Photoautotrophic plant cell cultures "reduce" a complex photoautotrophic organism to the single-cell level in so far as growth requirements and photosynthetic capacities are concerned. Photoautotrophic cell cultures, whether batch-propagated or grown as steady-state continuous cultures, facilitate measurements of chloroplast-associated metabolic pathways and the relationship of cell growth and chloroplast differentiation.

It has been found possible to grow cell cultures of various dicotyledonous species under photoautotrophic conditions. However no successful photoautotrophic cell cultures of the gymnosperms, the monocotyledons, C_4-plants, or CAM plants exhibiting their specific pattern of photosynthesis, have so far been reported. To achieve the latter requires a detailed knowledge of the prerequisites necessary to obtain a well-balanced ratio of cell multiplication and cell differentiation which in turn will allow chloroplast development to proceed. More attention needs to be focussed on the composition of the gas phase (P_{CO_2}/P_{O_2} ratio) supplied to the cells and on the benefits of sugar starvation in triggering photosynthetic development of the cells.

Light plays a crucial role in chloroplast development. In contrast to the intact plant, within cultured cells chloroplast differentiation is strictly blue light-dependent. The availability of photoautotrophic cell cultures now offers new perspectives in analysing the blue light syndrome in chloroplast differentiation, excluding any unspecific changes in the pattern of chloroplast development, possibly caused by the sugars of the culture medium under photoheterotrophic growth conditions. Preliminary results so far indicate that it should prove possible to relate the blue light effect in chloroplast differentiation with transcription of specific genes and the formation of specific m-RNA and chloroplast proteins.

Green callus cultures of various C_4-plants grown with an exogenous sugar in the light often lack typical C_4-photosynthesis. It is also well known that exogenous sugars will alter the photosynthetic characteristics of cultured plant cells, as well as the pathway for the formation of phosphoenolpyruvate (PEP), by CO_2 fixation catalysed through PEP-carboxylase. Therefore, only studies with C_4-plant cells cultured photoautotrophically will provide answers to the question of whether C_4-photosynthesis may occur on a single cell level.

In general, photoautotrophic cell cultures have only been grown at CO_2-concentrations (0.5%–2% v/v), which are markedly above ambient levels. There is an urgent need to select cell variants that will grow photoautotrophically under normal air. However, now some first reports are available on callus cultures plated on sugar free agar-medium and growing photoautotrophically at CO_2-concentrations of normal air (*Arachis* callus cultures).

High PEP-carboxylase activity and correspondingly high proportions of CO_2 incorporation into C_4-acids are supposed to contribute to a C_4-pathway of CO_2 fixation in photosynthesizing cell cultures of C_3-plants. As previously mentioned, exogenously supplied sugars may alter the pattern of photosynthetic carbon metabolism in cultured cells. In consequence, only photoautotrophic cell cultures are appropriate for investigations of the proposed role of PEP-carboxylase in cells of C_3-plants where through anaplerotic carboxylation reactions the role of the enzyme system is to provide carbon skeletons for the citric acid cycle and so maintain its synthetic function. Photoautotrophic cell culture systems will also be useful for measurements of carbon flow of photosynthetic intermediates into dark respiratory metabolic pathways, depending of course on the physiological state of the cells.

These approaches will advance our knowledge in basic problems of cell metabolism which have not been fully understood till now.

The high photosynthetic potential of photoautotrophically cultured plant cells may now be used as an effective marker to select for protoplast fusion products. It still needs to be ascertained however, whether the transferred trait "photoautotrophic growth in vitro" will be expressed in early cell divisions of the fusion product.

Another question concerns whether or not the morphogenetic potential of the cells to regenerate intact plants is maintained during the process of selection for photoautotrophic growth.

The usefulness of photosynthesizing plant cell cultures in analyzing chloroplast-associated pathways of secondary metabolism is also becoming more obvious. Light-enhanced secondary product formation may be due to increased substrate availability as well as to enzyme activation by reduced thioredoxin. For example, the enhanced formation of quinolizidine alkaloids in green cell cultures of *Lupinus polyphyllus* of vitamin B_6 synthesis in green cell cultures of *Cytisus scoprius*, of volatile oils production in chlorophyllous cell cultures of *Ruta graveolens* and finally the switch from vitamin K to anthraquinone synthesis in cell cultures of *Morinda lucida* all by changing from a photoautotrophic to heterotrophic mode of nutrition, testify the involvement of chloroplasts in secondary product formation. These promising approaches should be extended to other secondary compounds, such as steroids, alkaloids, cardenolides, and isoprenoids. If these compounds are mainly produced in the green parts of the plants, they should, if at all, preferentially be formed in cell cultures possessing functioning chloroplasts.

Mutants of C_3-plants possessing reduced photorespiration have long been the goal of plant scientists. Deficiencies in the oxygenase activity of the RuPB-carboxylase/oxygenase enzyme, are likely sites of mutation needed for a stable reduction in photorespiration without injurious effects on the carbon flow through the Calvin cycle.

Photoautotrophic cell cultures may also be of use in selecting for and creating new cell types with lowered photorespiration using conventional screening methods (high P_{O_2} ; P_{CO_2} below the compensation point). This is provided that the cells can be grown at normal ambient CO_2 concentrations and not at elevated levels; until now generally necessary for photoautotrophic growth of cultured plant cells, and the active site of the RuPB-carboxylase/oxygenase is saturated and unvailable for O_2. Trying to reach an understanding of the reason(s) why cultured higher plant cells need such high CO_2 concentrations to grow photoautotrophically is therefore a major task for future research.

Two questions need to be answered:

1. Will it be generally possible to select cell variants with an improved CO_2 uptake and concentration system, so that photoautotrophic growth of cultured cells may occur under normal air?

2. Will it be possible by advanced knowledge of molecular biology and genetic engineering techniques to clone those genes that are responsible for CO_2 capture and concentration in aquatic plants, and to transfer and express them in photosynthesizing cultured cells of higher plants?

If so, it should be possible to grow the cells on a broad scale photoautotrophically under normal air, the increase in the cellular CO_2 concentration in turn inhibiting photorespiration.

The carboxylase/oxygenase ratio of the Rubisco, encoded by the chloroplast genome, is obviously not immutable, but variations have occurred during the natural evolution of photosynthesis. At least in theory, these findings involve, eventually, selecting chloroplast mutants with deficiencies in the oxygenase activity provided an efficient screening system is available. This would need to allow selection for the desired mutation on a single cell level under the appropriate selection pressure.

Finally, in vitro mutagenesis of cloned RuBP-carboxylase/oxygenase genes may be another approach to reduce oxygenase activity of this enzyme. Again, photoautotrophic cell cultures would be the best system for such selection and transformation studies.

Cell cultures also offer possibilities as a screening system for pesticides, and especially herbicides. Also green cell cultures (for obvious reasons) reveal the effect of toxins on plant seedlings better than chlorophyll-free cell cultures. For example, simazine as well as metribuzine discriminate strongly between photosynthesizing and non-photosynthesizing cells. Presumably, if secondary modes of action of herbicides not affecting photosynthesis are involved, they can now be assayed with photoautotrophic and heterotrophic cell lines of the same plant species. Even without plant regeneration, herbicide-resistant photoautotrophic cell lines will provide a clue as to the basics for herbicide toxicity, herbicide metabolism and detoxifying strategies of the cells as well as for the mode and site of herbicide action at the level of the single cell. This will enable us to screen for herbicide action under quite different metabolic states of the cells. Thus, photoautotrophic cell cultures will support greenhouse and field studies with intact plants for herbicide action.

A potential approach in transferring herbicide resistance factors is through protoplast fusion. If the traits controlling herbicide resistance are encoded in the chloroplasts, for example triazine resistance, organelle transfer may be a further possible way of transfer of resistance factors. This might be realized in two ways.

1. The direct transfer of chloroplasts carrying the herbicide resistant trait to protoplasts of sensitive species using the micro-injection method.
2. The fusion of enucleated microplasts carrying the herbicide resistance with protoplasts of sensitive plant species.

Whether the introduced genes will be expressed and replicated during the first cell divisions of the fusion products, is of course an open question.

So far the discussion has been principally concerned with the fundamental base of plant cell culture. Alongside this have occurred other developments in plant cell culture, particularly aimed towards process development. A particular area is concerned with immobilised cell technology.

The advantages of immobilized cell systems have been recognized and exploited for a number of years in microbiological processes, and it is only relatively recently that the application of similar techniques to plant cell cultures has been considered seriously. The rationale for the immobilization of yeasts, bacteria and fungi is, in certain respects, different from that for plant cell entrapment, and is associated primarily with the advantages related to process engineering. Such advantages include high biomass concentrations, ease of biomass recovery and improved cell handling, resistance to washout effects and high productivity at high throughput rates. Although many of these characteristics are also applicable to plant cell systems, there would appear to be other advantages of immobilization which are peculiar to higher plant cells and are of a physiological rather than an engineering nature. Essentially, the intercellular communication and the reduced growth rate characteristic of immobilized cells seems to be important for the expression of many secondary metabolic pathways, and, as we have seen, can result in the significant enhancement of product yields in some systems – primarily those in which the product is synthesised in a complex multi-step process from relatively distant precursors.

For the large-scale production of such types of compounds in which the relationship between primary and secondary metabolic activity appears to be a major factor determining product yield, immobilization presents itself as a valuable technique. Before one can say with certainty whether such multicellular organization of cells is essential for the observed benefits, much more metabolic and enzymological work must be performed to determine the nature of the metabolic blocks which operate in freely-suspended cells. If immobilization is characterized, metabolically, by an alteration in a single environmental factor such as reduced O_2 tension, for example, then it is possible that the experimental treatment of freely suspended cells in the same way would result in a similar enhancement in yield. However, it is believed likely that no single controlling factor operates but that the situation is much more complex – it may be very difficult, if not impossible, to mimic in a suspension culture the microenvironmental effects of multicellular organisation which seem to be so important in controlling the expression of metabolism.

From an academic viewpoint, therefore, immobilization techniques provide us with a means of studying not only the effects of sequential chemical treatments on product yield, as facilitated by a continuous or semi-continuous flow-through system, but also provides us with an experimental system to study how organisation (i.e. cell aggregation and communication), or the lack of it, may effect the expression of a specific metabolic pathway, which can be termed "biochemical differentiation."

In a similar vein, immobilization techniques can be used to entrap not just cells but also intact organs for developmental studies. For example, methods have been developed to study cellular interactions during graft formation in internodes entrapped in agar gel, and similar techniques can be used to study environmental and nutritional factors affecting a variety of biological phenomena. These might include, for example, interactions between symbiotic partners – algae and fungi in lichens; fungi and roots; *Rhizobium* bacteria in roots. Furthermore, continuous culture systems lend themselves well to studies on the production of extracellular metabolites, including proteins and polysaccharides, as well as secondary products.

A miscellany of other potential applications exists. The immobilization of protoplasts on agarose via lectin molecules, for example, has been proposed as a tool for the recognition and isolation of specific antigen-bearing cell lines. Also, immobilized cells may in some cases be resistant to low temperature damage, and therefore be of value in cell line storage. A further application in cell line storage is related to the possibility that the biosynthetic and morphogenetic instability of cultured immobilized cells may be attenuated because of their lower rates of cell division, compared with freely suspended cells.

The most widely recognised application of immobilized plant cells is in the industrial-scale production of specific secondary metabolites, in which there can be recognised several potential practical advantages over the use of freely-suspended cells, besides those of a physiological nature. Firstly, relatively small amounts of potentially reuseable biomass can be processed, and can be used in either fermenter-type reactors (air-lift, stirred tank) where immobilized cells would be expected to be more resistant to shear force effects than suspended, or in fixed bed reactors (packed-bed, fluidised bed). It is also possible to use together immobilized particles possessing different biochemical characteristics (e.g. O_2-producing cells, precursor-producing cells) which can be maintained as stable mixed cultures – because of the limited growth rate of the cells there would not occur the preferential selection of cell lines which is observed in suspension culture. The low growth rate also favours the establishment of continuous or semi-continuous processes, although much more work remains to be performed on inducing the release of product into the medium; the obvious separation of cells and medium in immobilized systems lends itself well to this type of process. Furthermore, the continuous removal of product will reduce negative feedback inhibition effects which may regulate the synthesis of a product. Moreover, as with microbial systems, immobilized plant cells would be expected to be more resistant to washout than are freely-suspended cells, and biosynthetic activity would be correspondingly prolonged.

Central to scale-up of plant cell culture processes is of course the bioreactor configuration used to grow up plant cell biomass in the first instance. As stated above, plant cells, in contrast to microorganisms, function within a highly organised and in parts, specialized structure – the plant. Because of their size and nature, they are sensitive to shear stress, they require in culture a minimal cell density to initiate cell division and growth, and interaction between neighbouring cells is believed to be necessary to perform certain metabolic functions. It is therefore necessary to investigate growth and metabolitic production in plant cell cultures as a function of bioreactor design and culture environment. The effects of the gas phase, the mode of agitation, the scale and type of culture system or the development of cell support matrices for immobilized culture systems will all be important factors in enhancing biosynthesis and biotransformations in plant cell cultures. Development of new techniques to recover metabolites stored in vacuoles without damaging or destroying the cells will have to be part of product optimization.

Given all the above it is nonetheless encouraging that a number of groups around the World have managed to grow plant cells in a wide variety of bioreactor configurations up to a volume of some 15,000 l and in continuous culture for 2–3 months. Such progress bodes well for the future.

Out of all the different aspects of cell culture technology reviewed above, one stands out as central to the whole development of plant cell culture towards industrial application – that is product yield and productivity. Without a significant enhancement of these two, plant cell culture will remain an academic discipline. Major inputs are required into this area if the ultimate goal of significant commercial application for plant cell culture is to be achieved.

Subject Index

MIX
Papier aus verantwortungsvollen Quellen
Paper from responsible sources
FSC® C105338

If you have any concerns about our products,
you can contact us on
ProductSafety@springernature.com

In case Publisher is established outside the EU,
the EU authorized representative is:
Springer Nature Customer Service Center GmbH
Europaplatz 3, 69115 Heidelberg, Germany

Printed by Libri Plureos GmbH
in Hamburg, Germany